Engineering for Sustainable Human Development

Other Titles of Interest

Climate Change Modeling, Mitigation, and Adaptation, edited by Rao Y. Surampalli, Tian C. Zhang, C.S.P. Ojha, B.R. Gurjar, R.D. Tyagi, and C.M. Kao (ASCE Technical Report, 2013). Presents the most current thinking on the environmental mechanisms contributing to global climate change and explores scientifically grounded steps to reduce the buildup of greenhouse gases in the atmosphere.

Field Guide to Environmental Engineering for Development Workers: Water, Wastewater, and Indoor Air, by James R. Mihelcic, Lauren M. Frye, Elizabeth A. Myre, Linda D. Phillips, and Brian D. Barkdoll (ASCE Press, 2009). Explains sustainable engineering techniques for application in preparing for and executing international engineering service projects.

Sustainability Guidelines for the Structural Engineer, edited by Dirk M. Kestner, P.E.; Jennifer Goupil, P.E.; and Emily Lorenz, P.E. (ASCE Technical Report, 2010). Offers guidelines to advance the understanding of sustainability in the structural community and to incorporate concepts of sustainability into structural engineering standards and practices.

Sustainable Wastewater Management in Developing Countries: New Paradigms and Case Studies from the Field, by Carsten Laugesen, Ole Fryd, Hans Brix, and Thammarat Koottatep (ASCE Press, 2010). Draws upon the authors' experiences in Malaysia, Thailand, and other countries to examine the failures of traditional planning, design, and implementation, and to offer localized solutions that will yield effective sustainable management systems.

Sustainable Engineering Practice: An Introduction, by the Committee on Sustainability. (ASCE Technical Report, 2004). Provides a broad, fundamental understanding of sustainability principles and their application to engineering work.

Toward a Sustainable Water Future: Vision for 2050, edited by Walter M. Grayman, Ph.D., P.E., D.WRE; Daniel P. Loucks, Ph.D.; and Laurel Saito, Ph.D., P.E (ASCE Technical Report, 2012). Contains essays by more than 50 experts in environmental and water resource issues who describe their visions of the field in 2050 and the steps necessary to make those visions a reality.

Engineering for Sustainable Human Development

A Guide to Successful Small-Scale Community Projects

Bernard Amadei, Ph.D., NAE

ASCE PRESS

Library of Congress Cataloging-in-Publication Data

Amadei, Bernard, 1954-
  Engineering for sustainable human development : a guide to successful small-scale community projects / Bernard Amadei, Ph.D., NAE.
     pages cm
  Includes bibliographical references and index.
  ISBN 978-0-7844-1353-1 (paper : alk. paper)—ISBN 978-0-7844-7840-0 (pdf)—ISBN 978-0-7844-7841-7 (epub)   1. Sustainable construction.   I. Title.
  TH880.A44 2014
  624.068'4—dc23
          2014011726

Published by American Society of Civil Engineers
1801 Alexander Bell Drive
Reston, Virginia, 20191-4382
www.asce.org/bookstore | ascelibrary.org

Cover photo credit: Joe Thiel, from the Engineers Without Borders USA Montana State University Chapter, working with residents of the Khwisero District in Kenya. Photo courtesy of Dolan Personke; reproduced with permission.

Contents

Preface   xi

1. Introduction   **1**
  1.1   Context   1
  1.2   Scope   8
  1.3   Proposed Framework Goal and Objectives   16
  1.4   Framework Characteristics and Caveats   18
  1.5   Book Content   21
  References   25

2. International Development   **29**
  2.1   Toward a Sustainable World   29
  2.2   The Spectrum of Human Needs   34
  2.3   Poverty   37
  2.4   Development and Human Development   52
  2.5   Sustainability and Sustainable Development   69
  2.6   Frameworks for Sustainability   80
  2.7   Progress in Human Development   84
  2.8   Concluding Remarks   87
  References   91

3. Engineers and Development   **101**
  3.1   Context   101
  3.2   Engineers Indispensable to Development   103
  3.3   Engineering and Society   104
  3.4   Engineering for Sustainable Human Development   109
  3.5   Design for Sustainable Human Development   112
  3.6   Engineering Education for Sustainable Human Development   114
  3.7   The Making of the Global Engineer   119
  3.8   Chapter Summary   126
  References   127

4. Development Project Frameworks   **131**
  4.1   Guiding Principles   131
  4.2   Project Life-Cycle Management   137
  4.3   Project Design   139

4.4	Project Life-Cycle Frameworks	140
4.5	Review of Major Development Frameworks	143
4.6	Proposed Framework	147
4.7	Rights-Based Approach	155
4.8	Uncertainty in Development Projects	156
4.9	Project Delivery in Complex Systems	160
4.10	Chapter Summary	165
	References	166

5.   Defining and Appraising the Community   169

5.1	About Appraisal	169
5.2	Appraisal Outcome	171
5.3	Community Diagnostic Tools	179
5.4	A Reality Check: Challenges and Biases	185
5.5	Building a Support Team	188
5.6	Data Collection	189
5.7	Designing and Carrying Out the Appraisal	192
5.8	Analysis and Presentation of Data	195
5.9	Problem Identification and Ranking	208
5.10	Social Network Analysis	210
5.11	Chapter Summary	212
	References	212

**6.   A System Dynamics Approach to Community
Development   217**

6.1	Communities as Systems	217
6.2	Systems and Systems Thinking Basics	221
6.3	What Systems Thinking Is Not About	227
6.4	Systems Components and Archetypes	228
6.5	Modeling Systems Dynamics Using *iThink* and STELLA	232
6.6	Systems and Community Development	234
6.7	Illustrative Example Using *iThink* and STELLA	239
6.8	Chapter Summary	239
	References	241

7.   From Appraisal to Project Hypothesis   244

7.1	Preliminary Design	244
7.2	Causal Analysis: Problem and Solution Trees	245
7.3	Preliminary Solutions	249
7.4	Chapter Summary	257
	References	257

8.   Focused Strategy and Planning   258
    8.1   Comprehensive Planning   258
    8.2   Strategy   259
    8.3   Operation—Logistics and Tactics   271
    8.4   Planning of Management Activities   279
    8.5   Project Quality Planning   279
    8.6   Refining the Work Plan   281
    8.7   Behavior Change Communication   281
    8.8   Chapter Summary   286
    References   289

9.   Capacity Analysis and Capacity Development   291
    9.1   From Development Aid to Capacity   291
    9.2   Capacity Assessment   296
    9.3   Capacity Development Response   306
    9.4   Chapter Summary   316
    References   317

10.   Risk Analysis and Management   319
    10.1   Capacity, Vulnerability, and Risk   319
    10.2   About Risks   321
    10.3   Risk Management in Sustainable Community
            Development   323
    10.4   Project Impact Assessment   333
    10.5   Chapter Summary   335
    References   335

11.   Community Resilience Analysis   337
    11.1   About Resilience   337
    11.2   Resilience to Major Hazards and Disasters   340
    11.3   Resilience as Acquired Capacity   343
    11.4   Measuring Community Resilience   347
    11.5   U.S. Frameworks for Community Resilience   349
    11.6   International Resilience Frameworks   350
    11.7   A Systems Framework for Community Resilience   353
    11.8   Chapter Summary   356
    Note   357
    References   357

12.   Project Execution, Assessment, and Sustainability   360
    12.1   From Work Plan to Project Execution   360
    12.2   Project Assessment   361

12.3   Project Monitoring and Evaluation   364
12.4   From Assessment to Corrective Action   374
12.5   Exit Strategy, Ensuring Long-Term Benefits,
        and Scaling Up   375
12.6   Chapter Summary   378
References   379

13.   Service Delivery in Development Projects   381
13.1   Delivering Services Rather than Technology   381
13.2   Service and Service Capacity   384
13.3   Appropriate and Sustainable Technology   386
13.4   From Crisis to Development   394
13.5   Chapter Summary   400
References   401

14.   Energy Services for Development   405
14.1   Climbing the Energy Ladder   405
14.2   Using Biomass More Efficiently   413
14.3   Using Renewable Sources of Energy   417
14.4   Grid Extensions   426
14.5   Chapter Summary   427
References   428

**15.   Water, Sanitation, and Hygiene Services for
        Development   431**
15.1   The WASH Health Nexus   431
15.2   Climbing the WASH Ladder   436
15.3   Sustainability of WASH Services   440
15.4   Basic Water and Sanitation Requirements   451
15.5   The Water of Ayolé   461
15.6   Two Paths of WASH Interventions   462
15.7   Community-Based WASH Interventions   463
15.8   Chapter Summary   468
References   469

16.   Conclusions   476
16.1   Development Engineering   476
16.2   Poverty Is Not Normal   478
16.3   From Household Livelihood Crunch to Release   482
16.4   Project Success in Complex and Uncertain
        Environments   484

16.5   Global Engineering for a Small Planet   487
16.6   Sustainability and Development for All   489
References   492

Index   495
About the Author   505

Preface

In spring 2000, I had the opportunity to visit several Mayan villages in Belize. One of them, by the name of San Pablo, caught my attention. Little did I know at the time that the people in San Pablo would change my life and professional career forever. During my visit, I was introduced to some young girls who, I was told, spent a lot of time doing basic house chores, including fetching water from the river located 30 m below the village. As a result, they could not go to school. Because the villagers heard that I was a civil engineer, I was asked if I could do something about bringing water to the village using an alternative method. The problem was simple from a technical point of view but not from a social aspect. The community was poor and could not afford the fuel to drive a pump. The government of Belize had not provided electricity to the village at that time. Furthermore, the village was somewhat illegal, in the sense that it was created by migrant workers from Central America who decided to set up camp on someone else's land. San Pablo was my first introduction to engineering for sustainable human development, or development engineering.

The trip to San Pablo changed my life and led to the creation of Engineers Without Borders in the United States (EWB–USA) and the development of a program on engineering for developing communities (EDC) at the University of Colorado at Boulder (CU–Boulder). It turns out that I was not the only one interested in helping the people of San Pablo; many of the engineering students at CU–Boulder wanted to be of help to the community as well. My office became a gathering point for the project, which quickly took shape. We decided, with the technical assistance of a local engineer, Dennis Walsh, to build a ram pump that would convert the potential head from a waterfall located upstream of San Pablo into enough pressure head to reach the village. The students raised the funding necessary to go to Belize in May 2001 to install the pump and build a small water distribution system. Upon our return from Belize, the students were even more excited about doing real voluntary projects as part of their education. They conveyed to me their frustration at doing the same virtual engineering work in the classroom; they wanted more of what they called at the time "meaningful engineering." Interestingly enough, I was also trying to integrate more experiential learning in my teaching. The students and I had similar interests.

The trip to Belize made me realize that I would better serve humanity by working on projects that improve people's lives rather than writing a steady stream of academic publications that few people read. I quickly realized that projects like the one in San Pablo could be integrated into the engineering classroom and could be a powerful way of training our students to address real problems and come up with solutions that build on the fundamentals they learn in the classroom.

The San Pablo project and other projects over the past decade have made me aware that delivering engineering solutions to problems in developing communities in less than predictable conditions and in different cultural contexts is not easy and requires special skills that are not taught in engineering schools. The problems are often not well defined, and they involve technical and nontechnical issues. After all, the issue of water in San Pablo was never presented to me in a technical form; it all came about from wanting to help and do something so the girls in the village could stop fetching water and go to school. After the trip to Belize, I quickly realized that a need existed to train engineering students to address not only the needs of the richest segments of the world but also those living in poverty; this notion is addressed extensively in this book.

New ideas are challenging, and the first years of growing EWB–USA were difficult. The concept of providing experiential learning opportunities to students in an international setting was not always supported by my academic colleagues. I was told by my direct supervisors that doing work in the developing world was not part of the mission of the university and that I should focus on publishing in well established fields of engineering. Thanks to my students and friends, I was able to overcome many of the barriers deliberately created by some colleagues in my home institution. Since then, EWB–USA has grown into a fully operational 501(c)(3) not-for-profit organization with more than 14,000 student and professional members involved in 350 projects in 45 countries. Its success over the past 10 years is a true reflection of the vision and hard work of the staff, led by Executive Director Cathy Leslie; its various board members, sponsors, and donors; and all of its dedicated members and volunteers.

While moving EWB–USA forward, I came to the realization that a need existed to train young engineers to work on projects in developing countries. Even though the fundamentals of engineering are the same in projects in the developed and the developing world, the implementation of solutions to local problems can be quite different. Furthermore, in the context of development projects, engineers are required to possess hard technical skills, and other skills, such as dealing with people, culture, governance, and policy. Projects and their solutions need to be contextual, integrated, and multidisciplinary and need to account for uncertainty and complexity. These critical nontechnical skills are rarely provided or required in traditional engineering curricula, which I fear occurs for the simple reason that the teaching faculty members do not have the resources and/or experience to do so or are just not interested in them. The Engineering for Developing Communities (EDC) program at CU–Boulder was born out of the necessity to train engineers to address the needs of all, not just the richest part of the world's population. In 2010, EDC became the Mortenson

Center in Engineering for Developing Communities (MCEDC) at CU–Boulder, thanks to a generous endowment by the Mortenson family.

As a faculty member in the MCEDC, my responsibility over the past five years has been the teaching of core graduate courses, such as Sustainable Community Development I and II, and the coordination of a practicum where our students work as interns in various nongovernmental organizations around the world. This book has been a way for me to assemble the material taught in those courses. It introduces a framework and guidelines for conducting small-scale development projects in communities that are vulnerable to a wide range of adverse events and have low capacity to handle the stress associated with those events. The projects take place in medium- to high-risk and low-resilience environments. The framework presented in this book is called ADIME-E and represents an extension of the framework used by CARE International, combined with components from other frameworks used by different development agencies.

The framework presented in this book is unique, in the sense that it combines concepts and tools that have been traditionally used by development agencies and other tools more specifically used in engineering project management. It also emphasizes the importance of integrating systems thinking, risk analysis, capacity analysis, and resilience analysis in decision making. When combined, these tools and concepts from seemingly independent fields have the potential to better handle and model the complexity and uncertainty inherent in community development projects and the issues faced by households in these communities.

This book emphasizes the idea that human development calls for a new generation of global engineers who can operate in unpredictable and complex environments that are different from those encountered in the developed world. Engineering for sustainable human development is about the delivery of projects that are *done right* from a performance (technical) point of view and are also the *right projects* from a social, environmental, and economic (nontechnical) point of view. The book emphasizes that engineering for sustainable human development is not just about technology; it is also about people, values, ethics, culture, commitment, engagement, passion, and other issues that are not traditionally associated with engineering education and practice.

This book presents a framework and guidelines for small-scale engineering projects in developing communities. The methodology presented in the framework is meant to be robust, rigorous, and at the same time flexible, allowing for change and updates to be made based on experience and input. The reader should see the guidelines as a dynamic living document or a work in progress, but not as standards. Furthermore, the guidelines are presented as practical and comprehensive. Several illustrative case studies conducted by my graduate students or by me have been incorporated in the text.

This book is intended *primarily* for engineers, students, and professionals interested in small-scale human development projects, whether they reside in developed or developing countries. Development workers and practitioners may also find parts of this book useful, especially if they are interested in the technical aspect of small-scale development projects.

I want to thank Jenny Starkey, Tamara Stone, and Lauren Szenina for editing this book and Shawna Epps for drafting the various figures and tables. Special thanks go to the reviewers, who probably spent countless hours reviewing the first manuscript. Their feedback and suggestions are greatly appreciated.

I want to especially thank friends and colleagues who have supported the writing of this book, have encouraged me in pursuing my vision, and have taught me valuable tools in the various aspects of sustainability and international development. They include Bud Ahearn, Barry Bialek, George Bugliarello, Paul Chinowski, Robert Davis, Steve Forbes, Anne Heinz, Keyvan Izadi, Matt Jelacic, Rita Klees, Cathy Leslie, Hunter Lovins, Andrew Reynolds, Don Roberts, Robyn Sandekian, Mark Schueneman, David Silver, Mark Talesnick, Bill Wallace, Andy Yager, and Alex Zahnd. Special thanks go to Mort and Alice Mortenson and their family members for supporting the vision of the EDC program at CU–Boulder and believing in the future of engineering education and the betterment of humanity.

To all of the people who continue to work day in and day out to keep the mission and vision of EWB–USA alive, thank you for your commitment, vision, hard work, and more importantly, for your compassion and friendship. I also want to recognize the current and past staff of EWB–USA and its many board members over the past 13 years. I am glad to report that the future of the engineering profession is indeed in good hands.

I am also grateful to the graduate students with whom I have interacted through EWB–USA and the EDC program at CU–Boulder over the past 15 years. Their enthusiasm, energy, and dedication to the great work of making the world a better place are admirable. They have a bright future and will most likely become instruments of change in institutions that need creative disruption.

Finally, I especially thank my wife Robin and our children, Elizabeth Ann and Alex, for their support, patience, and love.

1

Introduction

1.1 Context

During my lifetime, I have had the privilege of witnessing the progress of the space program since its inception. If there is one thing that has always impressed me at the start of each rocket launch, it is the sheer amount of energy it takes to lift a rocket and its payload to overcome the gravitational pull of the Earth. Just taking a visit to Cape Canaveral, Florida, and walking under the Saturn V rocket is enough to impress anyone about the technology and the engineering that is necessary to meet the challenge of sending people into space well beyond low Earth orbit. Think about it: 7.6 million lbf (35 million Newtons) of thrust at launch are needed to send several astronauts into space; it is not very efficient, but it is effective. It also demonstrates that gravity is not easy to overcome.

Back on Earth, poverty is also not easy to overcome. In many ways, we need a large amount of energy to lift hundreds of millions of people out of poverty and to make sure that they do not regress back into it. We also need the same shared vision, will, commitment, and leadership that were necessary to successfully send people into space and bring them back. Not only do we need the necessary thrust for lifting the poor off the ground, but we also need the fuel necessary to generate a continuous thrust to keep the mission going. Once in place, the next challenge is to learn to live sustainably and peacefully with limited resources.

Like gravity in the space program, a counteracting pull from our planet prevents us from living this way. It can be argued that there are many *additive* (and *addictive*) *forces* that have multiple names such as selfishness, apathy, intolerance, dogmatism, greed, competitiveness, lack of compassion, fear, violence, power, ego, consumption, corruption, ignorance, denial, hatred, exclusion, competition, and politics, among many others that keep us from rising above this challenge. These forces, which have been responsible for the greatest evils in human history (and poverty is certainly one of them), have infiltrated all of our systems (e.g., environmental, economic, financial, and political) and institutions. When combined with unique human characteristics such as (1) "bounded rationality, limited certainty, limited predictability, indeterminate causality, and evolutionary change" (Hjorth and Bagheri 2006); (2) a simplified heuristic approach to complexity and change (Simon 1979); and (3) pathological forms of behavior (Korten 2006), it is nearly impossible to envision a sustainable, equitable, and peaceful world any time soon, short of espousing a drastically different mindset and redefining what it is to be truly human on this planet.

Despite global prosperity over the past 50 years, pervasive poverty still affects 3–4 billion people on our planet (Prahalad 2006). Most of that prosperity, supported by advances in technology and economic models, has benefited only the richest segments of the world's population (1–2 billion people). It has been mostly driven by the Western world in a never-ending quest for "better, bigger, and more complex" everything. As a result, the solutions for the 1–2 billion have been obtained with a production–consumption model that consumes too many resources, lacks flexibility, and is elitist and insular (Radjou et al. 2012). As remarked by Polak (2008), a need existed to consider the design (of solutions) for (or with) the other 5–6 billion people. A question arises as to whether the models of human development and economic development used for the richest segments of the world's population should also be used for the poorest. And if not, what models of development are more appropriate? A third question is how should science, technology, and engineering (STE) adapt to address the sustainable and human development issues for the vast majority of the world's population? These three questions are very relevant as the economic development and technical tools that have helped the 1–2 billion and have provided them with longer life expectancy and unlimited material wealth have not always been free of unintended social, psychological, and environmental consequences. These consequences include environmental degradation and pollution, waste generation, reduced technical and cultural diversity, loss in biodiversity, less resource equity between rich and poor, increased inequities between genders, and centralization of power. These externalities have not been accounted for in the balance sheets of the world's economy and have a profound effect on society. To these externalities, we can also add the unhappiness that is often associated with affluence and want, as described in the "World Happiness Report" (Helliwell et al. 2012).

Other consequences of economic development for the 1–2 billion have been shared with the developing world as well, such as top-down decision making in labor and resource exploitation, pollution, the destruction of native cultures and their social fabric, loss of autonomy, and a general lack of concern for human development. We are living in a globalized world today, where we are slowing learning that the actions of one group can have substantial effect on others very rapidly (Singer 2004). A case in point is how the effect of CO_2 emission on climate change does not stop along the geographical borders where the emission takes place. As reported in the 2011 Human Development Report of the United

Nations Development Programme (UNDP/HDR 2011), the CO_2 emission of a person in developed countries is 30 times that of a person in developing countries. It affects the climate over the entire planet and disproportionately affects the poor segments of the world's population, whereas the rich people have resources to adapt better to a changing climate. Simply put, promoting the use of such practices in countries with emerging markets does not work for the rest of the world. They have devastating consequences, despite short terms gains that only benefit a few, mostly in the developed world, or an elite class in the developing world. The concept of sustainability, with its emphasis on the interaction between society, the environment, and the economy, has to become an integral part of the discussion in human development (UNDP/HDR 2010).

Unfortunately, the 1–2 billion people who benefit the most from the current system do not have much interest in changing the status quo for the others, despite their continuous rhetoric about the need for change. As remarked by Schmidheiny 20 years ago regarding the challenge in adopting the concept of sustainable development in practice,

> *The painful truth is that the present is a relatively comfortable place for those who have reached positions of mainstream political or business leadership. This is the crux of the problem of sustainable development, and perhaps the main reason why there has been great acceptance of it in principle, but less concrete actions to put it into practice: many of those with the power to effect the necessary changes have the least motivation to alter the status quo that gave them the power.* (Schmidheiny 1992)

The same could be said of sustainable development and human development today. As noted by Prahalad (2006), it has to do with a "dominant logic" that permeates aid agencies, politicians, public policy establishments, the private sector, nongovernmental organizations (NGOs), and civil society organizations with regard to empowering the poor. Such logic is based on a combination of vested economic and political interests, complacency, a lack of vision and compassion, ignorance, and a range of outdated concepts that are out of touch with today's needs and reality. There is indeed a need for a new mindset for the way humanity is operating on the planet.

This new mindset needs to recognize that human and economic development are closely linked to sustainable development. It also implies that building a sustainable, equitable, and peaceful planet requires a different world view and a more mature level of consciousness in the day-to-day management and operation of our institutions and in our occupations (Korten 2006; Huesemann and Huesemann 2011; Jaworski 2012).

Development can mean different things to different people. In general, it represents a transformation of society through its betterment, creating a better life for all (Peet and Hartwick 2009). More often than not, economic growth, a key indicator of *economic development*, is seen as *the* measure of human well-being. Over the past 20 years, there has been a realization in the international community that *human development* must be included as well; economic development alone does not guarantee human development, and both must be addressed *simultaneously* (UNDP/HDR 1990). Human development is about providing people access to services to address their basic human needs (food, health, safe shelter, and access to services and resources) as preconditions for subsequent economic growth in the form of poverty reduction, aid delivery, debt reduction, individual empowerment, community participation, small-scale technology, and local capacity building. As summarized in the 2010 UNDP/HDR report, human development has the potential for expanding people's choices when combined with equity, empowerment, and sustainability.

Development is about such things as freedom (Sen 1999); transformation; well-being (material, physical, and social); hope; voice; empowerment; giving people a chance; knowledge; resources; assets; security; and treating people with dignity and respect. Development also means women's empowerment and gender equality (WDR 2012). Development is about improving people's quality of life in a safe and secure environment where rule of law and stable governance are the norm.

Development is about "creating an environment in which people can develop their full potential and lead productive, creative lives in accord with their needs and interests" as defined in the 1990 UNDP/HDR report. Development is also about "tapping into the ingenuity and creativity of the poor, and enabling them to express their own hopes for their families and their communities," according to Narayan (1993). Development has also been presented as a human right since the UN "Declaration on the Right to Development" was proclaimed in 1986 (UN 1986). However, as remarked by Kirchmeier (2006), the right to development is still perceived by the world community as "soft law," "but not legally binding."

If properly conceptualized and implemented, development can actually be the fuel (recall the thrust of the rocket discussed at the beginning of this chapter) needed to counteract the additive and addictive forces mentioned earlier in this chapter and allow a few billion people to regain hope and live a dignified life. Finally, development must recognize that "people are the real wealth of a nation," as concluded in the UNDP/HDR (1990, 2006) reports.

Furthermore, if one were to look (with the current mindset) at development from *just* a market point of view, "4 billion people in search of an improved quality of life will create one of the most vibrant growth markets we have ever seen" (Prahalad 2006) and unique opportunities for innovation and technology transfer. Some numbers reveal the scale of the potential market. As summarized in an article entitled "A New Alliance for Global Change" by Drayton and Budinich (2010), the market for joint ventures between the private and public sectors in the low-income world has been estimated as being worth $202 billion in health care, $424 billion in low-cost housing, $553 billion in energy, and $36 trillion in agricultural products and food. Likewise, de Soto (2003) estimated that the unregistered capital in urban villages and city slums was worth about $9 trillion. These figures show that many opportunities exist for the private and public sectors in the developed and developing world to *do well by doing good*, a statement that is sometimes attributed to Benjamin Franklin based on his book, *The Way to Wealth*, initially published in 1758 (Franklin 2011). Today, that statement is best manifested by the social entrepreneurship movement, a global phenomenon mostly driven by a third sector called the *citizen sector* by Bornstein (2007). This movement is gaining speed, especially in the emerging markets of India, China, and Brazil, and is likely to be very disruptive to the traditional way of looking at development as only economic growth (Kotler et al. 2010). However, without including sustainability principles, a danger exists that those market opportunities may negatively affect the planet.

Development is about addressing poverty straight on and recognizing that poverty is *unnecessary pain*: physical, psychological, and mental pain. As remarked by Kenny (2011), "poorer people die younger, their children die more frequently, they lack access to education, and they face higher rates of crime and violence." Luckily, we have today the knowledge and resources to provide dignity to everyone on our planet. For instance, in 2008 the Food and Agricultural Organization of the UN estimated that it would cost about $30 billion a year to end the world's hunger (FAO 2008). In 2011, the International Energy Agency (OECD/IEA 2011) concluded that universal access to energy services could be achieved by 2030 with an annual investment of $48 million per year. Likewise, the cost of meeting the UN Millennium Development Goals (see Chapter 2) has been estimated to vary between $40 and $60 billion (Devarajan et al. 2002) and $150 billion per year (Sachs 2006), simply a drop in the bucket compared with the well accepted expenses that we humans have legitimized spending year after year. Compared with the worldwide budget for military expenditures ($1.4 trillion in 2012), the $40–150 billion range is equivalent to about 10 to 38 days of military expenditures per year. Absurd priorities permeate our richest societies, whether it is the annual spending for alcohol in Europe ($105 billion), perfumes in Europe and America ($12 billion), or the $52 billion annual price tag of the pet industry in the United States. Pets in America today are better fed and treated than millions of children in the developing world! Statistics abound that confirm the pathological forms of behavior and misguided priorities of the Western world.

Development also implies a much-needed long-term commitment from rich nations to promote cooperation in solving global economic and social problems so that those at the bottom of the pyramid have a chance to live dignified lives. It has been demonstrated that universal access to basic social services can easily be achieved if all rich countries agree to have their official development assistance (ODA) level at 0.7% of gross national income (GNI), a measure of aid introduced by the Development Committee of the Organisation for Economic Co-operation and Development in 1970 (OECD 2012). As of 2009, only five countries in the European Union (Sweden, Norway, Luxemburg, Denmark, and the Netherlands) have met the 0.7% of GNI target, and 16 countries have pledged to meet that target by 2015. Finally, as remarked by Korten (2006), "had the benefits of the sixfold increase in global economic output achieved since 1950 been equitably shared among the world's people, poverty would now be history, democracy would be secure, and war would be but a distant memory."

So the bottom line is that achieving an equitable, peaceful, and sustainable planet has nothing to do with the availability of technology and resources, because both are plentiful; it has everything to do with changing the existing dysfunctional economic, educational, and political systems and the moral bankruptcy and inaction that drive those systems. Their immature consciousness is unnecessarily forcing a large segment of humanity behind in their development.

The immature consciousness of our institutions creates an inner poverty, which according to Meister Eckhart, a theologian of the 14th century, is reflected in their outer work, and vice versa. This state applies at all scales, ranging from individuals to institutions. Thus, addressing material (external) poverty cannot be done without addressing the inner dimension of poverty that is dominant at all scales in our Western institutions today. The former is simply an outcome of the latter. Both forms of poverty are equally destructive to each one of us and to the planet in general.

Because we cannot expect the consciousness of our institutions to change overnight, my take is that, in the meantime, it is an obligation for each one of us to consider what it is to be human on our planet. Alternatives to business as usual in our institutions, how we see the world (as an interconnected system of systems rather than isolated parts), and how we interact with each other in that whole, are much needed to end extreme poverty in our lifetime on one side and reduce or even eliminate nonextreme poverty on the other. Development requires humanity to adopt and promote a *new normal* of what represents dignified life for all. As noted by Kenny, we are aware of what is and is not normal:

> It should not be normal for children to have diarrhea. It should not be normal for girls to be kept at home as their brothers go to school. It should not be normal for teachers to be absent, for clinics to be empty of supplies and staff, and for police to demand bribes. (Kenny 2011)

To this list, we could add a series of other issues that are not acceptable in a so-called civilized world with regard to justice, human rights, and equalities.

Such statements are not meant to be political but to bring a sense of reality to the nature of the systems at play on our planet and the attitude of denial (and denial of denial) of mankind in general. We see poverty daily in the media, along with its consequences, such as terrorism, violence, abuse, and enslavement. The bottom line is that we can do something about it. As my colleague Andrew Yager (previously from the UN) would summarize, we have the capacity today to eliminate extreme poverty that affects 1.3 billion people who make less than $1.25 a day (Chen and Ravallion 2012) (at 2005 purchasing power parity [PPP], which is "the conversion rate for a given currency into a reference currency (invariably the $US) with the aim of assuring parity in terms of purchasing power over commodities, both internationally traded and non-traded" [Chen and Ravallion 2008]). Being able to increase their capital from $1.25 to $2 per day (or to multiply it manyfold) and to provide opportunities at that economic model would change the face of our planet instantly. Simple achievable goals such as better child nutrition at birth and an improved diet for women have the potential to create benefits that last a lifetime for the poor. For instance, it has been estimated that within a year,

> Every dollar spent promoting breastfeeding in hospitals yields returns of between $5–67. And every dollar spent giving pregnant women extra iron generates between $6–14. Nothing else in development policy has such high returns on investment.
> (Economist 2012)

Finally, development and peace are interrelated. As stated in the 1992 UN report by the Secretary General Boutros Boutros-Ghali titled "An Agenda for Peace: Preventive Diplomacy, Peacemaking and Peace-Keeping," peace is essential for development, and development is essential for long-lasting peace. Weak states with limited or no governance and rule of law and impoverished communities in a globalized world are more vulnerable to terrorist networks around the world and are more prone to dysfunctional behavior and extremism. There is enough evidence that "in terms of development, conflict can undo years of investment costing millions of dollars, limit sustainability, and restrict the humanitarian space needed for successful [development] programs" (USAID 2011). In their book, *Breaking the Conflict Trap*, Collier et al. describe the relationship between war, violent conflicts, and development as follows:

> War retards development, but conversely, development retards war.... Where development succeeds, countries become progressively safer from violent conflict, making subsequent development easier. Where development fails, countries are at high risk of becoming caught in a conflict trap in which war wrecks the economy and increases the risk of future war. Collier et al. (2003)

Simply put, addressing poverty is a win-win for everyone. However, despite that evidence, our desire to build a more sustainable and equitable planet seems contrary and disconnected to our behavior and actions (or lack thereof) and to our funding priorities. It is time to realize that a sustainable planet is a peaceful, equitable, and compassionate planet for *all*—or the planet will not be! As summarized by the National Academy of Engineering (2008), "A world divided by wealth and poverty, health and sickness, food and hunger, cannot long remain a stable place for civilization to thrive."

1.2 Scope

Engineering and Development

This book is intended primarily for engineers (students and professionals) interested in small-scale human development projects, whether they reside in developed or developing countries. Development workers and practitioners may also find parts of this book useful, especially if they are interested in the technical aspects of small-scale development projects.

In writing this book, I had three broad objectives in mind. First, I wanted to emphasize the connection between human development and sustainable development, two concepts that are closely linked in the overall discussion on poverty reduction. Combined under the theme of *sustainable human development* in the 2011 UNDP/HDR report, both concepts contribute to securing healthy, productive, and meaningful lives for all. I deliberately avoided the field of charity work. I have never believed much in charity except in critical situations when it is necessary to provide basic human needs for people to get back on their feet; don't expect someone to lift a shovel if they don't have food in their stomachs and are not healthy. Simply put, charity has a role to play in poverty reduction, especially when in-country institutional support is lacking, but it is not an end to a mean. Charity, in my opinion, creates dependency, and dependency is the antithesis of empowerment.

Second, I wanted to build awareness of the *role of engineering in human development.* This awareness cannot be built by writing another scholarly treatise on poverty and development. It is my opinion that poverty is not an academic exercise and that addressing poverty should not be an academic exercise either. Instead, I wanted to offer a more practical approach for the engineering profession to not only stand against poverty but also to do something about it through the application of sound engineering skills, knowledge, and practice. Engineering is presented here as an entry point into the overall discussion on poverty reduction and development in general, but not as an end in itself.

The third objective in writing this book is to convince my engineering and nonengineering colleagues that when it comes to lifting billions of people out of poverty and contributing to reducing their daily pain, the engineering profession (in partnership with other professions) has a major role to play in developing solutions for (or with) the poor. Addressing the needs of the 4–5 billion people whose job it is to stay alive by the end of the day is no longer an option for the engineering profession; it is a professional and personal *obligation*. Engineering has contributed to sending people into space and has helped improve the living standards and economic development in the developed world over the past 200 years. One of its next great challenges is to contribute to the relief of the endemic problems afflicting developing communities worldwide by providing knowledge, resources, and appropriate and sustainable solutions. Engineering, like economics, health, and other disciplines, needs to be seen as an integral part of development activities rather than as a means to provide just technical solutions. Yet, it is rarely perceived that way. If engineering is so critical to sustainable human development, questions arise as to (1) how to educate engineers to address global issues, (2) the role of engineering practice in developing tools that can be used to address those issues, and (3) how to conduct sound and meaningful projects in developing communities.

It is my belief that sustainable human development calls for a *demystification* of engineering practice as usual, which traditionally provides value-neutral technical solutions to well defined problems irrespective of any social context. Instead, engineers are also called to be change makers, peacemakers, social entrepreneurs, and facilitators of sustainable development. Sustainable human development also calls for a *new epistemology of engineering practice and education*, based on the idea of reflective and adaptive practice, systems thinking, engagement, and fieldwork. It requires looking at problems in a more holistic and positive (strength-based) way and being able to interact with a wide range of technical and nontechnical stakeholders from various disciplines and walks of life rather than remaining in traditional silos of technical expertise. Finally, sustainable human development requires a humanization of the engineering profession and the realization that engineering in this context is above all—and has always been—about people.

The writing of this book originated from my own experience with the Engineers Without Borders (EWB) movement in the United States and worldwide over the past 15 years and my experience in spearheading a program of engineering for developing communities (EDC) at the University of Colorado at Boulder over the past 10 years. In both instances, my experience has shown me that today's youth are ready to walk the walk in addressing the world's problems. They may not have a lot of professional skills and experience, but they have enthusiasm. They may be perceived as naive and entitled, but I know from experience that they are courageous. They demand more meaningful education and seek engagement. Most of them have witnessed global events unfolding in the media. They have been bombarded with major issues throughout their short lives; no wonder they want to effect change and not stay on the sidelines. The same level of enthusiasm cannot be said, however, of the engineering profession and the academic world in general. I hope that this book will create some awakening and promote positive change in engineering practice and education.

Small-Scale Development Projects

More specifically, this book is about the field of sustainable community development in *low-income* and *lower-middle-income* countries (and economies), as defined by the World Bank (Figure 1-1), and, more generally, countries (and economies) classified as *developing*. This regrouping was done for convenience because the book is not about theories of development. The reader should be aware that there is a lot of discussion in the literature on how countries should be classified as "developed" or "developing" and how income levels relate to levels of development. Another group of countries called "least developed" countries also appears in the development literature and refers to countries in extreme poverty, facing ongoing conflict, and/or lacking political and social stability. These countries have not yet reached the status of developing countries.

This book provides practical guidelines for conducting engineering projects in developing countries through a combination of community development practices, technology, engineering management, and system dynamics tools. Though emphasizing more of the technical and engineering side of community development projects, the guidelines highlight the multidisciplinary, participatory, integrated, contextual, and systemic nature of those projects and the need for incorporating nontechnical issues into such projects.

The guidelines focus on *small-scale community projects* only and mostly in communities that can be described as being of *high-medium risk and low capacity* (or ability) to respond to various adverse events. The projects used as illustrative case studies in this book are mostly borrowed from the fields of civil and environmental engineering, which are my own fields of interest. The proposed guidelines do *not* address large-sector programs consisting of multiple interventions or projects (including policy development). Finally, this book is *not* about development at the country level and development policy. It limits itself to development projects benefiting communities, households, and their participants. The rationale in deciding on the scope of the book and its content was based on my own observation that small-scale projects in the developing world have never been of much interest to large engineering companies and to engineering professionals in general. Yet, a huge demand exists for such projects because they have the capacity to do well and lift many out of poverty using engineering as an entry point.

The challenge for the engineering profession to address the needs of the developing world is not that engineers are incapable of coming up with

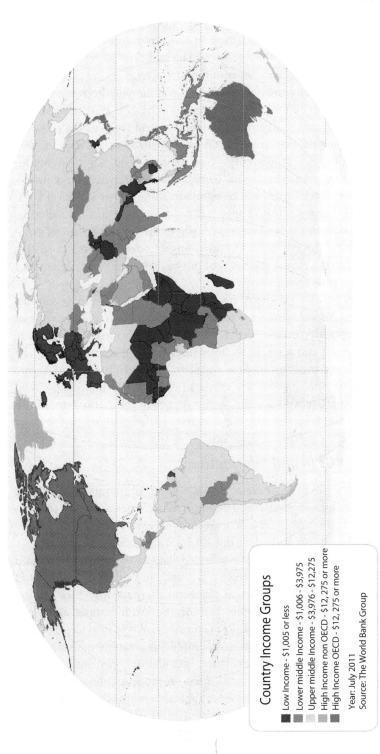

Figure 1-1. The World by Country Income Groups (2011 data)

Note: The most recent 2012 data (World Bank 2013) divide economies based on their gross national income per capita as low income (<$1,035); lower middle income ($1,036–$4,085); upper middle income ($4,086–$12,615); and high income (>$12,616).

Source: ChartsBin statistics collector team (2011) <http://chartsbin.com/view/2438> (Feb 5. 2012); with permission from ChartsBin.

Country Income Groups

- Low Income - $1,005 or less
- Lower middle Income - $1,006 - $3,975
- Upper middle Income - $3,976 - $12,275
- High Income non OECD - $12, 275 or more
- High Income OECD - $12, 275 or more

Year: July 2011
Source: The World Bank Group

appropriate technical solutions for basic humanitarian problems. That could certainly be done with some restructuring in the way traditional engineering firms plan, manage, and deliver their projects. The main reason for their lack of interest is more about reconciling a for-profit model of engineering practice with the manner in which development projects have traditionally been handled by development agencies. As best explained by Radjou et al. (2012), it relates to the misconceived view that companies in the developed world see "creating products and services for segments that are typically marginalized as a social mission rather than a core business opportunity." That mindset is quickly changing for the better in emerging markets in India, China, and Brazil, where large investments in technology development are being made at the global and local levels (Kotler et al. 2010). However, this change has not taken root in the traditional engineering profession in the Western world, where the for-profit part of engineering practice is still directly linked to providing engineering services (with a good profit return) to the richest segments of the world's population, and mostly in the form of large projects.

Furthermore, small-scale engineering in communities in the developing world is challenging because project planning, design, and execution have to be made in partnership with communities that have limited resources (e.g., human, financial, and material) and assets at the local, regional, and country levels. The low-income and lower-medium-income countries (and economies) in Figure 1-1, have a gross national income per capita in 2011 of less than $1,005 and $3,975 per year, respectively. As a result, communities in these countries have a limited ability to invest in development projects.

A Spiky World of Data and Information

The design and execution of engineering projects at the community level require collecting data at the local, regional, and national scales. When analyzed, the data provide valuable information necessary to create a preliminary community baseline. But information at one scale does not always transfer to another scale.

Many development agencies have collected data at the country level and published country-specific statistical reports that tend to be difficult to read and analyze. The reader should be aware that graphical alternatives to such reports are available on the Internet. Examples include the Worldmapper and Gapminder sites. The electronic platform called Gapminder World provides dynamic representation of statistics using innovative graphs on a variety of issues (e.g., health, education, economy, infrastructure, population, environment, and energy) for which indicators are available. Custom-made graphs can easily be created to show trends by country and geographical region and over time.

It must be noted that the statistics presented in platforms such as *Gapminder World* or in traditional reports are at the country level and only give an average (and often a false) picture of well-being at the local (community) level within a country. As remarked by Barton et al. (1997), "economic growth measured by increases in gross national product has little to do with the distribution of wealth within a country." Regrettably, these statistics are often used by policy makers to rank countries and decide whether a country needs assistance. The problem with that approach is that it does not truly reflect what is happening at the community grassroots level and the daily challenges faced by the poor, which are often much worse than the average. A case in point is the large disparities in living conditions that still exist in post-apartheid South Africa (Aliber 2001) despite the fact that the country is rated as a medium-rich country.

This book acknowledges that at the level of a community and its households, the world is not *flat* (Friedman 2005) but rather *spiky* (Florida 2005). It is at that scale that issues important to the well-being of community members must be addressed by the engineering profession in partnership with other disciplines. And, it is the scale that this book addresses.

Capacity, Vulnerability, and Risk

Communities in developing countries have their own share of problems. In general, when faced with adverse events and hazards, their households are more at risk when subject to various forms of stress and shock associated with those events. This risk is illustrated in Figure 1-2 as part of a Household Livelihood Crunch Model, which was adapted from the Disaster Crunch Model proposed by the Tearfund relief and development agency in the United Kingdom (Tearfund 2011).

The situation can range from ensuring that every day's basic human needs are met (e.g., access to water, sanitation, energy, shelter, education, and health) to facing infrequent extreme events, such as those associated with natural hazards

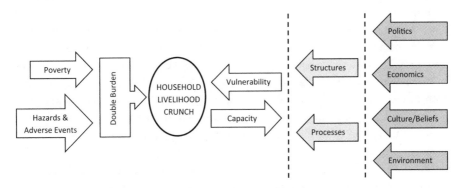

Figure 1-2. Household Livelihood Crunch Model

Source: Adapted from Tearfund (2011); with permission from Tearfund.

(e.g., earthquakes, droughts, and floods) or nonnatural hazards (e.g., war and conflict, accidental or human-induced technical issues, and collapse of social welfare systems). Such hazards have the potential to affect regions, countries, and cross geopolitical boundaries. In between those two extremes, communities may face small-scale and medium-scale hazard events, such as changes in weather patterns, variability in local or global market economics, living in vulnerable locations, or environmental threats (e.g., deforestation or extraction of natural resources) related to the unintended consequences associated with resources used by the communities and others. Exposure to those events can vary from low to high or local to global and can be periodic from one season to the next or occur on an annual basis. As remarked in the UNDP/HDR (2011) report, the combination of everyday events and other uncontrollable events with the inherent isolation, powerlessness, physical and psychological weakness, and precarious livelihood associated with poverty creates a *double burden* for the world's most disadvantaged people. In turn, this increased vulnerability (lack of security) creates ill health, less access to natural resources as sources of income, and endangered livelihoods.

However, as illustrated in Figure 1-2, at the same time that communities and their households are exposed to various forms of vulnerability, they also have some form of capacity (wealth and strength) that can be built upon once it has been identified and mapped. That enabling environment has many expressions, such as skills, understandings, attitudes, values, relationships, behaviors, motivations, and resources, among many others. Capacity enables "individuals, organizations, networks/sectors and broader social systems to carry out functions and achieve their development objectives over time" (Bolger 2000). Capacity is multidimensional and has many forms, including institutional, human resource, technical, economic and financial, energy, environmental, and sociocultural. Capacity is indeed a great place to start in development work because it builds on the existing strengths of the community, which always exist in one form or another.

As we see in this book, both the *enabling* and *constraining* environments need to be identified when appraising communities and defining community baselines. The difference between vulnerability and capacity in Figure 1-2 defines in many ways what each day looks like for the members of each household in a community, the risks they are likely to face when coping with change, and their ability to adapt to change. As shown in Figure 1-2, the enabling and constraining environments also depend on social *structures* (e.g., government, businesses, or important individuals) and their influence (*processes*) at different scales (local, national, and international), which themselves are influenced by politics, economics, culture, the environment, and other factors.

It is important to note that in the constrained environment portrayed in Figure 1-2, a slight change in internal and external conditions within or without the control of the household is enough to tip the balance against itself when its assets (e.g., people, property, facilities, or the environment) are at stake because of their exposure to various events, big or small as they might be. For instance, the illness or death of a breadwinner in a household may throw it into "a cycle of poverty from which it cannot escape" (WHO 2001). In other cases, a slight change in internal and external conditions can turn out to be for the better, more often for a short period of time until a new crisis arises. Using a definition of resilience adopted by the National Research Council (NRC 2012), the households in poor communities have limited acquired capacity to "prepare and plan for, absorb, recover from or more successfully adapt to actual or potential adverse events." In conducting projects in developing communities, the low level of community resilience can have drastic effects on the sustainability of engineering solutions being implemented.

Throughout this book, the dynamic that exists among hazard events (or adverse events), exposure, vulnerability, and capacity in developing communities is captured using the following equation:

$$\text{Risk} = \text{Hazard} \times \text{Exposure} \times (\text{Vulnerability} - \text{Capacity})$$

Even though this equation is conceptual and cannot be used in a mathematical fashion, it captures the components that enter into the definition of risk and their dynamic interactions. As a result, the community projects discussed in this book take place in *high-medium risk* and *low capacity* situations. Furthermore, the aforementioned equation needs to be looked at for different physical scales (individuals, households, community, region, and country) and temporal scales, especially when extreme events are involved. This scale is particularly important when understanding the dynamic of risk within the context of rapid response, recovery, development, and sustainable development after extreme events. Over that continuum, the risks are different. Therefore, the strategies, logistics, and tactics to address those risks vary with the scale being considered. No guarantee exists that what works at one scale works in a different context.

It should be noted that understanding the dynamic among risk, hazard, capacity, and vulnerability in communities in the developing world is as critical as understanding it in poor communities in the *developed world*. Whether they are in developed or developing countries, poor communities seemed to be doomed to carry a larger burden of risk than rich ones (Cuny 1983). In New Orleans, for instance, at the time of the 2005 Hurricane Katrina disaster, more than 120,000 people (27% of New Orleans households in 2005) had no mobility and were not able to evacuate because they were poor and had no access to transportation means (Brodie et al. 2006). Today, one can still visit poor neighborhoods in New Orleans that are lagging behind in their reconstruction, despite an average recovery level at the larger metropolitan scale (NRC 2012).

The project guidelines and methodology presented herein contribute to a broad strategy for sustainable community development designed to reach an overarching goal, which is to build (or strengthen) communities that have the potential to become over time more stable, equitable, secure, and above all more *resilient, peaceful, prosperous,* and *healthy.* Throughout this book, health is defined as "a state of complete physical, mental and social well-being, and not merely the absence of disease or infirmity" (WHO 1948). In general, such communities must have the resources and knowledge to be able to acquire over time the *capacity* (or ability) to (1) address their own problems, (2) be self-motivated and self-sustaining, (3) cope with and adapt to various forms of stress and shocks, (4) satisfy their own basic needs, and (5) demonstrate livelihood security for current and forthcoming generations.

As a result of an increased capacity, these communities are capable of transitioning over time from an uncertain, high-risk, and low-capacity enabling environment to a more stable, lower risk, and higher capacity environment, as per the aforementioned equation. Another way of looking at the proposed guidelines is to replace the *crunch model* of Figure 1-2 by a *release model* (Figure 10-1). Using the terminology proposed by the Tearfund (2011), that release model is about increasing capacity, decreasing vulnerability, and addressing and reducing the effect of structures and processes. Another important outcome of sustainable community development is that of contributing to peace building at the community level, because development and peace are related (Norwegian Ministry of Foreign Affairs 2004; Ricigliano 2012).

Finally, the guidelines presented in this book emphasize that strategies for community development must ensure livelihood security at multiple physical scales ranging from the entire community to its basic units (i.e., the households) and ultimately to the individuals who comprise those units. The households can be single or multifamily units. The guidelines incorporate, at their core, the household livelihood security (HLS) model proposed by CARE (Caldwell 2002). Within that model, household security implies (1) the possession of human capabilities, (2) access to tangible and intangible assets, (3) the existence of economic activities, and (4) the interconnectedness of seemingly diverse units (households) in the community. To be long-lasting, household security must incorporate the concept of sustainability and its three pillars: people (equity), planet (environment), and profit (economics).

1.3 Proposed Framework Goal and Objectives

The goal of this book is to present an *integrative* and *participatory* framework that can be used as a guide for engineers involved in small-scale development projects in communities described as being of high-medium risk and low capacity. The framework is intended to be systemic, adaptive, reflective, and contextual. Main emphasis is placed on civil and environmental infrastructure projects, i.e., projects that provide the utilities, facilities, and systems upon which society depends for its normal functions and for addressing basic human needs. In meeting the goal of this book, the framework is expected to have a broader impact by contributing to more meaningful, robust, structured, higher quality, effective, and therefore more successful community development projects in which engineering can play a more active role. Using a metaphor suggested by the Institute for Sustainable Infrastructure (ISI 2012) and Bill Wallace (personal communication, 2012) for sustainable infrastructure projects, another way at looking at the goal of this book is to demonstrate how engineers can deliver sustainable community development projects that are both *done right* from a performance (technical) point of view and are also the *right projects* from a social, environmental, and economic (nontechnical) point of view. Another way of looking at it is the delivery of projects that are both technically correct and correctly done from a nontechnical point of view.

There are five main objectives in developing the framework. The guidelines are designed to

- Emphasize the role that engineering plays in small-scale community projects in developing countries and human development in general;
- Help define the components of a new epistemology of engineering education and practice;
- Encourage users to become more knowledgeable in the steps involved in planning and managing small-scale engineering projects in developing communities;
- Serve as a common platform for collaboration on small-scale community projects; and
- Be a dynamic living document (open source approach): a work in progress to be used, updated, and improved over time through users' contributions.

The following activities are carried out in this book to reach the aforementioned goal and objectives:

- The guidelines are framed within the context of human development and sustainable development.
- The role of engineering in sustainable community development is defined.
- Existing development frameworks used by development agencies are reviewed, and their advantages, disadvantages, and ranges of application are assessed.
- System dynamics tools and other tools not traditionally used in development projects are introduced. It is shown how they could help better understand the complex interactions and inherent uncertainty

in communities and the interconnections of their basic units, i.e., households.
- The integration of basic engineering project management tools with appropriate technology and solutions into community development practices is demonstrated.
- A common road map for small-scale community projects in developing countries is outlined.
- Case studies illustrating the range of applications (and limitations) of the framework in civil and environmental projects are presented.

1.4 Framework Characteristics and Caveats

Listed below are the main characteristics of the framework and guidelines proposed in this book and their limitations.

1. This book is about *suggested guidelines* and is not about proposing standards. The methodology presented in the guidelines is meant to be robust, rigorous, but at the same time flexible, thus allowing for change and update to be made based on experience and input. The reader should see the guidelines as a dynamic living document or a work in progress. Furthermore, the guidelines are presented as practical and comprehensive tools.

2. This book covers many topics; one could have an entire book devoted to each topic. Therefore, no attempt was made to provide a comprehensive literature review on each topic because so much is available in the existing development literature. Likewise, the book cuts across many disciplines, each one with its own share of data, information, knowledge, and range of experts. As appropriate, the reader is referred to additional references that were published before the middle of calendar year 2013.

3. The proposed guidelines emphasize the necessary *integrated* (not piecemeal) and *participatory* approach to small-scale community projects (at all levels) in developing countries, thus encouraging active community involvement, participation, and influence, what Prahalad (2006) calls "co-creation." One of the many assumptions in the guidelines is that ingenuity, talent, and creativity already exist at the local level (indigenous knowledge and capacity). The challenge is to build on native capability (Hart and London 2005) and real wealth (Korten 2006). This positive approach also acknowledges positive deviance (Pascale et al. 2010), which is "based on the observation that in every community there are certain individuals or groups [changemakers] whose uncommon behaviors and strategies enable them to find better solutions to problems than their peers, while having access to the same

resources and facing similar or worse challenges." Furthermore, para-phrasing Schumacher (1973), it is about finding out what people do and teaching them to do it better. It is *not* about telling them what to do.

4. The framework presented herein combines (and extends) several frameworks and concepts already available in the existing literature dealing with the design and execution of development and engineering projects. It is not about reinventing the wheel but rather reviewing, strengthening, integrating, and complementing what is already available in the literature.

5. The proposed framework recognizes the multidisciplinary and cross-disciplinary nature of the problems and correspondingly the solutions at stake. Of particular interest are the integration of sustainable prac-tices into human development issues and the recognition that sustain-able community development happens when seemingly independent issues are being addressed in a systemic way. Engineering in the devel-oping world is about technology but also such things as health, gover-nance, rule of law, entrepreneurship, and security.

6. The framework recognizes that the guidelines for small-scale projects in developing communities are *contextual* in nature and variable in physical and time. Simply put, no two development projects are alike. Some of the underlying issues may be the same, but addressing those issues requires using tools that are appropriate to the context in which the issues arise (Caldwell 2002). For instance, this is the case when designing and planning water, energy, shelter, and other solutions within the context of rapid response, recovery, and development after some hazard events. The issues are the same, but the time and nature of the execution are different. The same could be said of the various steps in project management. They are essentially the same in projects in developed and developing countries, at least in principle. But in developing communities, the implementation of those steps is quite different, because of the increased uncertainty and complexity inher-ent in small-scale development projects and the high-risk, high-vulnerability, and low-capacity context in which the projects unfold. Development project managers have to be able to manage challenging and sometimes seemingly competing tasks that are not often found in projects in developed countries.

7. The framework guidelines recognize the complex and uncertain nature and the changing environment of small-scale development projects. Such projects cannot be addressed and executed with blueprint rational and predicable methods dictated from the top down. Because of the inherent *uncertainty* associated with human systems and other

systems, the guidelines need to be adaptive to the context in which they are developed. There is no such thing as "a *unique* best possible solution" in uncertain and complex systems (Callebaut 2007). Complex adaptive systems, such as communities, need to be addressed with tools (assessment, design, planning, monitoring, and evaluation) that are appropriate to the dynamic of change, thus emphasizing adaptation instead of specification, *satisficing* (i.e., reaching good enough and adequate solutions) instead of *optimizing* (i.e., reaching optimal solutions) (Simon 1972), and reflection-in-action practices (Schön 1983). Special emphasis is placed on using interactive (rather than directive) project planning methods and a design-as-you-go (rather than rigid) approach in project management and execution. For this reason, the guidelines are not meant to be read as "best practices" in the sense that "best" would imply something that is of highest quality and superior. Simply, best practices make no sense in community development projects. But good, effective, and satisficing practices are recommended and need to be appropriate and contingent to the situation at stake. However, bad and inappropriate practices should be avoided.

8. The guidelines proposed in this book do *not* enter into the details of engineering design and project execution or the nuts and bolts of specific technologies. They emphasize, however, the importance of the *appropriate* and *sustainable* nature of technical solutions in development projects. More specifically, they look at the execution of projects as providing *services* to address specific needs (e.g., water, sanitation, hygiene, energy, shelter, or health) rather than providing *technology* as an end in itself. The guidelines also emphasize the interdependence of those various needs and how multiple solutions may have to be combined to address those needs.

9. Finally, the guidelines recognize the bottom-up approach to development projects that is more attuned to local conditions, needs, innovations, and local knowledge and more adaptable, meaningful, and relevant. The top-down approach should not be ignored, as long as it does not become the *dominant* mindset. Among other things, it has a role to play because it may ensure some level of quality control in data collection and analysis. It may also establish relevant policies and encourage duplication and diffusion of good practices. Excluding the top-down approach is likely to result in failed projects. The guidelines call for a middle adaptive ground solution that reconciles and synergistically combines top-down and bottom-up approaches (Patton 2011). The two are not exclusive and can be complementary, if used properly.

1.5 Book Content

The guidelines developed herein borrow concepts and tools from the existing fields of development (as done by development agencies), sustainable development, technology, engineering project management, and system dynamics. They also leverage what works well in those approaches. For instance, engineering project management is well suited to provide much needed logic and structure to development projects. However, social and health science ethnographic tools developed within the context of participatory action research and used for community appraisal and the monitoring and evaluation of projects are better defined in the field of development because they incorporate nontechnical (e.g., social, economic, or political) issues into project decision making. Such tools are rarely used in traditional engineering practice. System dynamics modeling tools fit better within the complexity and uncertainty of communities rather than traditional linear cause-and-effect tools. They help in combining the multiple issues that enter into development projects, provide whole-system design principles, and can be used to explore "what if" questions. In general, combining practices from seemingly independent fields provides a unique opportunity to build stronger development projects.

Chapter 2 presents an overview of various models of development, human development, and sustainable development. It emphasizes the link between human development and sustainable development. The spectrum of human needs at the individual and household levels is discussed. Poverty, along with its definitions and characteristics as perceived by those who have been defined as poor, are reviewed. Over the past 50 years, the approach used in development projects has evolved from development aid to development capacity. It has also revealed the complex nature and multidimensional nature of development. The chapter makes an attempt at defining the characteristics of sustainable community development within that context.

Chapter 3 explores the role played by engineering in society and more specifically in human development. Interestingly, engineers have had a limited role in small-scale projects in developing communities in the past. A need exists for training a new generation of engineers who could better meet the challenges and address the needs of the developing world. The challenge is the education of *global engineers* who (1) have the skills and tools appropriate to address the issues that our planet is facing today and is likely to face within the next 20 years in a holistic way; (2) are aware of the needs of the developing world; and (3) can contribute to the relief of the endemic problems afflicting developing communities worldwide by providing knowledge, resources, and appropriate and sustainable solutions. A need also exists for the engineering profession to develop analysis and design tools that address, more specifically, issues faced by communities in the developing world. The tools can be adapted from those used in the developed world combined with the talent already existing at the community level. There is a huge demand and market at the bottom of the pyramid for innovative and frugal solutions that also include sustainability principles.

In Chapter 4, various development frameworks are reviewed and compared. The proposed framework is presented as a combination of development and project management tools. Components of the proposed framework, its contextual nature, and its range of application and limitations are discussed. Special emphasis is placed on the terminology used in the management of development projects compared with that used in traditional engineering management projects. The chapter also discusses the need to integrate rights-based tools into project design and management. It also discusses the use of more flexible tools, such as interactive planning, reflective practice, development evaluation, and a design-as-you-go approach to better account for the complex, uncertain, and adaptive issues encountered in development project planning and execution.

Chapter 5 covers community appraisal, the collection of community data (qualitative and quantitative), and the analysis necessary for project design. Tools of participation action research (PAR) and participatory research used in social, health, and agricultural sciences are an integral part of the community data collection, analysis, and synthesis process. The chapter also explores the various dimensions of "community" and how to establish a community baseline, i.e., how to define the community environment (enabling and constraining) in which to build capacity, reduce vulnerability, and create resilience. The baseline is established and defined within geographical boundaries (area of influence) and time boundaries.

Chapter 6 deals with systems thinking and dynamic modeling and gives a brief overview of systems basics. The application of systems theory to community development is addressed. Systems theory is better suited than reductionist predictable models to account for the uncertainty, nonlinearity, emerging nature, complexity, and chaotic behavior of communities. Systems thinking is particularly relevant to the field of engineering for sustainable human development where development workers work within interdisciplinary teams, interact with various decision makers, and integrate and process issues from various disciplines in an uncertain environment. They also deal with human and environmental systems that constantly change, evolve, and are being reconfigured. The reader is introduced to the *iThink* and STELLA systems dynamic software from ISEE systems.

Chapter 7 presents a methodology to analyze the community appraisal data and identify and prioritize main issues and core problems in the community. This chapter consists of three parts: (1) the systematic identification and prioritization of problems through *causal analysis* (cause–effect relationships), (2) the planning of preliminary solutions, and (3) the development of a preliminary action plan (feasibility stage) in consultation with the stakeholders. These three parts contribute to the preliminary project design phase and the layout of a *project hypothesis* (or project statement). It summarizes for each problem (1) the problem being addressed, (2) the anticipated outcomes in solving the problem, (3) the critical causes of the problem, (4) the relationships between the problem's causes and effects, (5) the effects of possible interventions, (6) a rating of the various interventions, and (7) the assumptions and preconditions necessary to support the project hypothesis.

Chapter 8 presents a logical framework (i.e., logic model) and project hierarchy for development projects: a focused strategy that helps in project planning (strategy, logistics, and tactics) and execution. The chapter reviews the steps and issues involved in the management and operation of development projects toward the selection and recommendation of sound steps of project intervention (e.g., timeline, resources, and materials) and management activities that are ready for execution. This chapter also discusses the importance of integrating project quality planning in project management. Finally, it concludes with the need to include behavior change communication in the project plan to promote positive behaviors and create a supportive environment so that the behaviors are sustainable for the long term and eventually become habits.

Chapter 9 describes capacity analysis tools that can be used to understand the dynamic that exists between the current capacity of a community, its ability to handle future solutions and action plans, and the necessary steps in capacity building and development. Capacity building is seen as a key strategy to sustainable community development. It first requires a clear definition of the enabling environment of the community through capacity appraisal. This definition is followed by assessing whether the community has the strength, resources, and capability to (1) accept the proposed solutions and recommendations outlined in the focused strategy and planning stages of the project, (2) implement those solutions, and (3) carry out the corresponding action plan in a sustainable way. An outcome of that analysis is the identification of the community weak links and/or potential challenges that could prevent implementation of the recommended solutions. This outcome leads to the formulation of a capacity development program necessary to overcome the limitations and the development of a strategy on how the community will progress in its development over time and become more resilient.

Chapter 10 is about risks associated with various adverse events faced by developing communities before projects are executed and those that may arise in projects conducted in those communities. In the medium-high risk and complex environment of development projects, risk is a given and can take multiple forms. It can be internal or external to a project. The chapter presents the different steps in risk management: (1) risk identification, (2) risk analysis and prioritization (mapping), (3) development of risk management strategies,

(4) the implementation of risk strategies, and (5) monitoring and evaluation of strategies. The chapter concludes with a discussion of project impact assessment.

Chapter 11 explores the resilience of developing communities to risk associated with adverse events. Building community resilience is presented as a development activity. In general, developing communities are extremely vulnerable to adverse events, whether they are small or cross geopolitical boundaries. Unlike communities in the developed world, developing communities are high-risk and low-capacity environments for which building resilience requires a tremendous amount of effort and time. Resilience is presented as an acquired capacity for a community to cope with and adapt to unusual and unscheduled conditions and transient dysfunctions and to return to a level of functional balance or a new normal. The acquired capacity of a community needs to be compared with its vulnerability. Various indices proposed in the literature to measure resilience are reviewed.

Chapter 12 is about project assessment and, more specifically, the role of monitoring and evaluation strategies in keeping track of how project execution is taking place. It also discusses the steps that are necessary to (1) close a project, (2) ensure its long-term benefits (sustainability), and (3) decide whether to scale up.

Chapter 13 approaches the execution of development projects as providing multiple interconnected services to address community needs rather than providing "just" technology. The outcome-oriented service approach emphasizes the importance of appropriate and sustainable technologies in development projects and more specifically those that enter into three categories of services: (1) energy; (2) water, sanitation, and hygiene (WASH); and (3) health. Other services that are necessary in the day-to-day activities of a community, such as providing shelter, food and nutrition, transportation, or clothing, are only briefly discussed. This chapter also demonstrates that appropriate technology is a perfect entry point into social innovation and social entrepreneurship. Finally, this chapter concludes with a discussion of service delivery for communities facing an emergency situation or recovering from a major crisis associated with natural or nonnatural disasters.

Chapters 14 and 15 look respectively at various forms of energy and water and sanitation and hygiene services in development projects. These chapters do not go into the details of various technologies but rather address issues such as the availability, accessibility, affordability, sustainability, and scalability of energy and WASH services.

Chapter 16 draws key conclusions on major themes addressed in this book. They include (1) the abnormality of poverty on our planet today, (2) understanding household livelihood and real wealth, (3) defining project success in uncertain and complex conditions, (4) improving the future practice and education of global engineers, and (5) defining sustainability and sustainable development within the context of addressing the needs of developing communities.

It is noteworthy that this book was written for a wide range of readers but primarily for engineers who are interested in becoming acquainted with development tools and want to learn how to incorporate these tools into small-scale community projects. In reading this book, development workers will learn how to integrate engineering practice in their development work. Chapters 4–12 were designed to be read sequentially, whereas the other chapters can be explored on an individual basis based on the readers' individual needs, interests, and backgrounds.

References

Aliber, M. (2001). "Study of the incidence and nature of chronic poverty in South Africa: An overview." Chronic Poverty Research Center, University of Western Cape, South Africa. http://www.chronicpoverty.org/uploads/publication_files/WP03_Aliber.pdf (Oct. 10, 2012).

Barton, T., et al. (1997). *Our people, our resources: Supporting rural communities in participatory action research on population dynamics and the local environment*, International Union for Conservation of Nature (IUCN) Publications Services, Gland, Switzerland, and Cambridge, U.K.

Bolger, J. (2000). "Capacity development: why, what and how." Capacity development occasional papers 1(1). Canadian International Development Agency (CIDA) policy branch. <http://www.impactalliance.org/ev_en.php?id=4048_201&id2=do_topic> (Sept. 15, 2013).

Bornstein, D. (2007). *How to change the world*, Oxford University Press, New York.

Brodie, M., et al. (2006). "Experiences of Hurricane Katrina evacuees in Houston shelters: Implications for future planning." *American Journal of Public Health*, 96(8), 1402–1408.

Caldwell, R. (2002). *Project design handbook*, Cooperative for Assistance and Relief Everywhere (CARE), Atlanta.

Callebaut, W. (2007). "Herbert Simon's silent revolution." *Biological Theory*, 2(1), 76–86.

ChartsBin statistics collector team. (2011). "Country income groups (World Bank classification)." *ChartsBin.com*, <http://chartsbin.com/view/2438> (Oct. 16, 2013).

Chen, S., and Ravallion, M. (2008). "The developing world is poorer than we thought, but no less successful in the fight against poverty." *Quarterly J. Economics*, 125(4), 1577–1625.

Chen, S., and Ravallion, M. (2012). "An update to the World Bank's estimates of consumption poverty in the developing world." <http://siteresources.worldbank.org/intpovcalnet/resources/global_poverty_update_2012_02-29-12.pdf> (Feb. 12, 2013).

Collier, P., et al. (2003). *Breaking the conflict trap: Civil war and development policy*, Oxford University Press, New York.

Cuny, F. (1983). *Disasters and development*, Oxford University Press, New York.

de Soto, H. (2003). *The mystery of capital: Why capitalism triumphs in the West and fails everywhere else*, Basic Books, New York.

Devarajan, S., Miller, M. J., and Swanson, E. V. (2002). "Goals for development: History, prospects and costs." <elibrary.worldbank.org/doi/pdf/10.1596/1813-9450-2819> (Mar. 28, 2014).

Drayton, W., and Budinich, V. (2010). "A new alliance for global change." *Harvard Business Review*, <http://hbr.org/2010/09/a-new-alliance-for-global-change/ar/1> (Oct 9, 2012).

Economist. (2012). "The nutrition puzzle." <http://www.economist.com/node/21547771> (Feb. 18, 2012).

Florida, F. (2005). "The world is spiky." *The Atlantic Monthly*, 48–51.

Food and Agriculture Organization of the United Nations (FAO). (2008). "The world only needs 30 billion dollars a year to eradicate the scourge of hunger." <http://www.fao.org/newsroom/en/news/2008/1000853/index.html> (Oct. 1, 2012).

Franklin, B. (2011). *The way to wealth*, Best Success Book Publ. First published in 1758.

Friedman, T. (2005). *The world is flat: A brief history of the 21st century*, Farrar, Straus and Giroux, New York.

Hart, S., and London, T. (2005). "Developing native capability: What multinational corporations can learn from the base of the pyramid." *Stanford Social Innovation Review*, <http://www.ssireview.org/articles/entry/developing_native_capability/> (March 10, 2012).

Helliwell, J., Layard, R., and Sachs, J., eds. (2012). "The world happiness report." <http://www.earthinstitute.columbia.edu/sitefiles/file/Sachs%20Writing/2012/World%20Happiness%20Report.pdf> (March 13, 2013).

Hjorth, P., and Bagheri, A. (2006). "Navigating towards sustainable development: A system dynamics approach." *Futures*, 38, 74–92.

Huesemann, M., and Huesemann, J. (2011). *Techno-fix: Why technology won't save us or the environment*, New Society Publishers, Gabriola Island, BC, Canada.

Institute for Sustainable Infrastructure (ISI). (2012). *Envision: A rating system for sustainable infrastructure* (Version 2.0), Institute for Sustainable Infrastructure, Washington, DC. <http://www.sustainableinfrastructure.org/portal/workbook/GuidanceManual.pdf> (Jan. 28, 2013).

International Energy Agency (OECD/IEA). (2011). *Energy for all: Financing access for the poor*, International Energy Agency, Paris. <http://www.worldenergyoutlook.org/media/weowebsite/energydevelopment/weo2011_energy_for_all.pdf> (Sept. 8, 2012).

Jaworski, J. (2012). *Source: The inner path of knowledge creation*, Berrett-Koehler Publishers, San Francisco.

Kenny, C. (2011). *Getting better: Why global development is succeeding and how we can improve the world even more*, Basic Books, New York.

Kirchmeier, F. (2006). *The right to development: Where do we stand?* Friedrich Ebert Stiftung, Geneva, Switzerland. <http://library.fes.de/pdf-files/iez/global/50288.pdf> (Dec. 10, 2012).

Korten, D. (2006). *The great turning: From empire to Earth community*, Berrett-Koehler Publishers, San Francisco.

Kotler, P., Kartajaya, H., and Setiawan, I. (2010). *Marketing 3.0: From products to customers to the human spirit*, John Wiley & Sons, Hoboken, NJ.

Narayan, D. (1993). *Participatory evaluation: Tools for managing change in water and sanitation*, The World Bank, Washington, DC. <http://www.chs.ubc.ca/archives/files/Participatory%20Evaluation%20Tools%20for%20Managing%20Change%20in%20Water%20and%20Sanitation.pdf> (May 10, 2012).

National Academy of Engineering (NAE). (2008). *Grand challenges for engineering*, National Academies Press, Washington, DC.

National Research Council (NRC). (2012). *Disaster resilience: A national imperative*, National Academies Press, Washington, DC.

Norwegian Ministry of Foreign Affairs. (2004). "Peacebuilding: A development perspective." Norwegian Ministry of Foreign Affairs, Oslo, Norway. <http://www.regjeringen.no/upload/UD/Vedlegg/Utvikling/peace-engelsk.pdf> (May 25, 2012).

Organisation for Economic Co-operation and Development (OECD). (2012). "The 0.7% ODA/GNI target—A history." <http://www.oecd.org/dac/stats/the07odagnitarget-ahistory.htm> (Feb. 1, 2013).

Pascale, R., Sternin, J., and Sternin, M. (2010). *The power of positive deviance: How unlikely innovators solve the world's toughest problems*, Harvard University Review Press, Cambridge, MA.

Patton, M. Q. (2011). *Developmental evaluation: Applying complexity concepts to enhance innovation and use*, Guilford Press, New York.

Peet, R., and Hartwick, E. (2009). *Theories of development: Contentions, arguments, alternatives*, 2nd ed., Guilford Press, New York.

Polak, P. (2008). *Out of poverty: What works when traditional approaches fail?* Berrett-Koehler Publishers, San Francisco, CA.

Prahalad, C. K. (2006). *The fortune at the bottom of the pyramid: Eradicating poverty through profits*, Wharton School Publishing, Upper Saddle River, NJ.

Radjou, N., Prabhu, J., and Ahuja, S. (2012). *Jugaad innovation*, Jossey-Bass Publishers, San Francisco.

Ricigliano, R. (2012). *Making peace last: A toolbox for sustainable peacebuilding*, Paradigm Publishers, Boulder, CO.

Sachs, J. (2006). *The end of poverty: Economics possibilities for our time*, Penguin Press, New York.

Schmidheiny, S. (1992). *Changing course: A global business perspective on development and the environment*, MIT Press, Boston.

Schön, D. A. (1983). *The reflective practitioner: How professionals think in action*, Basic Books, New York.

Schumacher, E. F. (1973). *Small is beautiful*, Harper Perennial, New York.

Sen, A. (1999). *Development as freedom*, Anchor Books, New York.

Simon, H. A. (1972). "Theories of bounded rationality." *Decision and organization*, C. B. McGuire and R. Radner, eds., North-Holland Pub., Amsterdam, Netherlands, 161–176.

Simon, H. A. (1979). *Models of thought*, Yale University Press, New Haven, CT.

Singer, P. (2004). *One world—The ethics of globalization*, Yale University Press, New Haven, CT.

Tearfund. (2011). *Reducing risks of disasters in our community*, Tearfund, Teddington, U.K. <http://tilz.tearfund.org> (Oct. 22, 2012).

United Nations (UN). (1986). "Declaration on the right to development: A/RES/41/128." <http://www.un.org/documents/ga/res/41/a41r128.htm> (Oct. 2, 2012).

United Nations (UN). (1992). "An agenda for peace: Preventive diplomacy, peacemaking and peacekeeping." Report of the Secretary-General pursuant to the statement adopted by the Summit Meeting of the Security Council on 31 January 1992. <http://unrol.org/files/A_47_277.pdf> (March 16, 2014).

United Nations Development Programme Human Development Report (UNDP/HDR). (1990). *Concepts and measurement of human development*, United Nations Development Programme, New York.

United Nations Development Programme Human Development Report (UNDP/HDR). (2006). *Beyond scarcity: Power, poverty, and the global water crisis*, United Nations Development Programme, New York.

United Nations Development Programme Human Development Report (UNDP/HDR). (2010). *Real wealth of nations—Pathways to human development*, United Nations Development Programme, New York.

United Nations Development Programme Human Development Report (UNDP/HDR). (2011). *Sustainability and equity, a better future for all*, United Nations Development Programme, New York. <http://hdr.undp.org/en/media/HDR_2011_EN_Complete.pdf> (April 22, 2012).

U.S. Agency for International Development (USAID). (2011). *Systems thinking in conflict assessment: Concepts and application.* USAID, Washington, DC. <http://www.usaid.gov/what-we-do/working-crises-and-conflict/technical-publications> (Jan, 5, 2013).

World Bank. (2013). "How we classify countries." <http://data.worldbank.org/about/country-classifications> (Oct. 1, 2013).

World Development Report (WDR). (2012). *World Development Report—Gender equality and development*, The World Bank, Washington, DC.

World Health Organization (WHO). (1948). "Preamble to the Constitution of the World Health Organization as adopted by the International Health Conference." World Health Organization, New York, June 19–July 22, 1946; signed on July 22, 1946, by the representatives of 61 states (Official Records of the World Health Organization, No. 2, p 100) and entered into force on April 7, 1948.

World Health Organization (WHO). (2001). *Non-communicable diseases and mental health*, World Health Organization, Geneva. <http://www.who.int/mip2001/files/2008/NCDDisease Burden.pdf> (July 23, 2012).

2

International Development

2.1 Toward a Sustainable World

Some time at the end of October 2011, the population of the world crossed the 7 billion mark. Looking ahead, the United Nations (UN) expects the population to reach 8 billion by 2025, 9.6 billion by 2050, and 10 billion sometime between the years of 2085 and 2100 (UNFPA 2011). Another opinion in the literature is that the world's population will stabilize around 9 billion or even less by mid-century because of marked decreases in fertility rates in both developed and developing countries (Altaf 2011; Goldman 2011). Today, the annual world population growth is about 1.1%; it takes about 13 years to add 1 billion extra humans to the planet with an increase of about 80 million people per year (or 211,000 per day). Furthermore, the world population is about equally distributed between urban and rural areas, and it is expected that about two-thirds of humanity will live in towns and cities by 2050 (UN-HABITAT 2006).

Most of the population growth expected over the next 40–50 years is likely to occur in the developing world and in cities, which will result in various challenges for every continent. In Africa alone, population is expected to increase from about 1 billion today to 2 billion in 2050. Urban population in Asia and Africa is expected to double from 1.7 billion in 2000 to 3.4 billion in 2030 (PEP 2008). This population growth across our planet will create unprecedented demands for energy, food, land, water, transportation, materials, waste disposal, earth moving, public health care, environmental cleanup, telecommunications, and infrastructure. The role of engineers will be critical in fulfilling those demands at various scales, ranging from small remote communities to large urban areas (megacities), mostly in the developing world. In the video *Lagos: Wide & Close* (van der Haak 2006), these demands and challenges, along with the human condition, are captured while detailing the complexity of the problems faced by large cities in the developing world that are well beyond capacity. With a population of 15 million today, Lagos is likely to become the third largest city in the world by 2020 and must immediately deal with the increasing demands of an expanding population, as must many other megacities.

It must be emphasized that the demands associated with population growth in the near future supplement what we already have today; a planetary *baseline* where 11% of the world's population lacks access to improved sources of clean water, 36% lacks access to improved sanitation, 43% either doesn't have enough to eat or has too much, 25% of adults are illiterate, 20% lacks access to clean energy, and 20% lacks adequate housing. To keep up with the demand of the world's population, it is estimated that food production will have to increase by 50% by 2030 and 75% by 2050, which will double the demand for water (UNDP/HDR 2011). In regard to water, the OECD (2008) has estimated that by 2030, almost half of the population will be living in areas that are water stressed, with water demand outstripping *current* supplies by 40%.

Population growth will also require investment in human capital among the planet's young people (today 43% of world population is below the age of 25 according to the UNFPA [2011], and 1.3 billion of them live in developing countries) by improving the quality of education and health services they receive and giving them capabilities to invest in their future and start a productive and healthy working life (WDR 2007).

As the population continues to grow, so too does the amount of waste produced. Currently, 80% of sewage in developing countries is untreated and disposed into the environment, resulting in the pollution of surface water and groundwater (WWDR 2009). The disposal of waste is expected to double over the next 20 years in lower-income countries (World Bank 2012c).

Because of population growth, stress on the environment and on our natural resources is expected to increase in the near future and will be more damaging to the Earth's human population and its life-support systems than ever before. The resulting environmental degradation will not only be a problem for industrially wealthy nations, it will become a major issue for developing countries as they become trapped in a downward spiral of ecological and economic decline and become more vulnerable to natural and nonnatural hazards. As remarked 20 years ago at the 1992 Rio Summit and reaffirmed at the Rio+20 meeting in 2012, we are living in a world in which human populations are more densely populated, consume more, are more connected, and in many places are more diverse than at any time in history. Many of our living systems and cultural systems are in jeopardy as the increasing population results in a reduction in biodiversity, increasing ecological stress, and general impoverishment of life on Earth.

Dramatic variations in climate change leading to extreme drought or floods, environmental degradation associated with rogue economic development, and the recent international economic crisis are exacerbating the situation, especially for those at the bottom of the economic pyramid (Global Adaptation Institute 2011; Tearfund 2011; UNDP/HDR 2011; World Bank 2012b; World Bank 2013a). As noted in the *2005 World Resources* report by the World Resource

Institute (WRI 2005), "ecosystems are—or can be—the wealth of the poor," and damaging ecosystems has huge consequences for their livelihood. In a recent report entitled "Inclusive Green Growth," the World Bank (2012a) remarked that "the damage done by environmental degradation is costly for an economy: equivalent to 8% of GDP across a sample of countries representing 40% of the developing world's population."

Over the past 20 years, much discussion has taken place about sustainable development and how people can live within existing resources and life-support systems. According to the Global Footprint Network (Borucke et al. 2012), on average we are using the equivalent of 1.5 planets for the resources we need and the disposal and absorption of our waste. Another way of looking at the global footprint is that it takes our planet one year and six months to regenerate what we use (and want) in one year. Projecting this over the next 20 years, it is estimated that we will need two planets by 2030. This threshold has already been exceeded in rich countries such as the United States, which is operating as if it has five planets available to support its lifestyle (ISI 2012). This problem poses a very serious question: How do *all* humans, under such constraints, have fulfilling lives, meet their basic needs, and live with dignity without degrading the ecosystems and their services in the years to come?

Another global issue is the nonequitable distribution of income, wealth, health, and social power between rich and poor countries and even among people living within a given country (rural vs. urban), developed or developing. Those who have too much need to consume less and in a more sustainable way, and those who do not have enough need to be given a chance to acquire more to be able to live dignified lives. The consumption of commercial energy per capita is a case in point. As remarked by Barnes et al. (1997), at the end of the 20th century, U.S. energy consumption was "80 times higher than in Africa, 40 times higher than in South Asia, 15 times higher than in East Asia, and 8 times higher than in Latin America." Today, the average residential electrical system in the United States ranges in size between 3,000 and 5,000 watts. In comparison, "average home systems in the developing world are 50–75 Watts, and 'large' systems may be 120 W" according to Ratterman (2007). This disparity in having access to energy services is best illustrated by the picture of the world at night (Figure 2-1). This combined photograph not only shows where the *energy poor* live (e.g., Africa) but also the places of lost opportunity and unrealized socioeconomic and human potential associated with the lack of access to energy in terms of human and economic development.

Similar inequalities have been reported by the World Health Organization (WHO) in the field of public health (Dodd and Munck 2002) for life expectancy at birth and infant mortality rates. "In the least developed countries, life expectancy is just 49 years, and one in ten children does not reach their first birthday.

Figure 2-1. Compounded View of the World at Night (Nov. 27, 2000)

Source: C. Mayhew and R. Simmon (NASA/GSFC) <http://apod.nasa.gov/apod/ap001127.html> (March 2012).

In high-income countries, by contrast, the average life span is 77 years and the infant mortality is six per 1000 live births."

In the world today, the health conditions that contribute to the burden of disease differ with the level of economic development (WHO 2001; Birley 2011; WHO 2013b). In low- and middle-income economies, the health conditions related to communicable diseases (e.g., malaria, tuberculosis, cholera, pneumonia, and diarrhea); maternal and perinatal conditions (related to childbirth); and nutritional disorders are dominant. In contrast, in high-income economies, most of the health conditions are related to noncommunicable diseases (e.g., cancer, anxiety, and heart problems). The third group of health conditions corresponding to injuries (intentional and nonintentional) is found across the range of economies. Finally, large health inequalities also exist within countries, whether rich or poor.

Even though 89% of the world's population has access to improved water supplies as of 2010, disparities still exist among regions of the world. They also exist within a given country, e.g., between rural and urban areas and poor and rich neighborhoods, according to the Third World Academy of Sciences (TWAS 2002). As remarked in the WHO/UNICEF (2012) report, access to improved water supply is only 61% in sub-Saharan Africa and 63% in the least developed countries. In general, "the poorest people not only get access to less water, and to less clean water, but they also pay some of the world's highest prices" (UNDP/HDR 2006). In the field of water consumption in developed and developing countries, the inequalities are huge. The 2006 UNDP/HDR report mentions people in Europe and the United States, respectively, using about 200 and 400 L daily, compared with 5 L per day for those categorized as lacking access to clean water.

In addition, the quality of housing and the livable space vary substantially with household incomes. The inequity in shelter conditions in the world is best illustrated on the Gapminder website in their Dollar Street application, which shows photos of households at different income levels around the world. About 50% of the population in developing countries, the majority of rural populations, and at least 20% of urban and suburban populations still live in earth homes (Houben and Guillaud 2005). In Sudan alone, soil construction methods are used in 80% of urban dwellings and more than 90% in rural areas (Adam and Agib 2001). Even though earthen homes can be better than modern homes if properly designed, they are still perceived as carrying a social stigma in affluent communities.

Gender inequalities have also been observed around the world and within given countries. Today, women represent 70% of the world's poor (UN-WOMEN 2013). Even though progress has been made over the past 20 years to close some gender gaps in terms of education, life expectancy, and labor force participation (WDR 2012), major disparities still remain. According to WDR (2012), the gaps manifest in terms of unequal access to health care, more deaths of girls and women, disparities in access to education, unequal access to economic opportunities, and an absence of voice and opinion in household decision making and in society in general. Women and girls bear the vast majority of chores at the household level and resource collection, which translates into lost time for personal growth, large demands of personal energy, and poor health. As remarked in the 2011 UNDP/HDR report, women are more likely to bear the effects associated with environmental degradation, and girls are less likely to receive basic education. Out of 793 million illiterate adults in the world, 67% are women. According to the 2013 WHO/UNICEF report, it has been found that in 25 poor countries, "women spend a combined total of at least 16 million hours *each day* collecting drinking water; men spend 6 million hours; and children 4 million hours." Fetching water is time consuming and tiring, can lead to injury, and severely affects women's use of time.

Large disparities also exist in the well-being of children around the world. In the State of the World's Children report by UNICEF (2005), it was estimated that there are 2.2 billion children in the world; half of them live in extreme deprivation associated with poverty, war, and HIV/AIDS. Statistics released by the World Health Organization (WHO 2013a) indicated that "6.6 million children under the age of five died in 2012" (i.e., 18,000 per day) and that "more than half of these early child deaths are due to conditions that could be prevented or treated with access to simple, affordable, interventions." Furthermore, "children in low-income countries are 18 times more likely to die before the age of five than children in high-income countries." Among the major causes of death is malnutrition, which is responsible for 45% of all deaths. Next are pneumonia (17%), prematurity (17%), birth asphyxia (11%), diarrhea (9%), malaria (7%), and others.

In 2005, the UNDP/HDR report deplored the inequality between rich and poor in the world. "The richest 50 individuals in the world have a combined income greater than that of the poorest 416 million." Out of 73 countries analyzed, the 2005 UNDP/HDR report found that 53 countries recorded an increase in inequality of wealth distribution. The 2005 UNDP/HDR report also called for a fundamental reform of global international trade rules, aid structures, and security to reduce international inequities. When it comes to poverty, human justice and empowerment, and opportunities to rise on the economic ladder, the world is *spiky* (Florida 2005) and not flat (Friedman 2005) because of the concentration of activities and resources (physical, human, financial, and intellectual) in the hands of a few, a condition that breeds more harm than good. As noted by Oxfam (2013), inequality, along with extreme wealth, has a cost because it is "economically inefficient … politically corrosive … socially divisive … environmentally destructive, and unethical."

Over the past 20 years, sustainable development has been presented as an alternative to development as usual and a potential road map for a better planet for all. This chapter looks at human development and sustainable development, two issues that are closely intertwined. One cannot talk about creating a sustainable planet without including human development, and vice versa. Both concepts have been regrouped under sustainable human development, as suggested in the 2011 UNDP/HDR report. Of particular interest is the integration of sustainable practices and human rights into human development: "Sustainable human development is the expansion of the substantive freedoms of people today while making reasonable efforts to avoid seriously compromising those of future generations."

Even though the concept of sustainable human development may sound like a utopian ideal at first, we need to realize that no other alternatives have been proposed to replace the current dysfunctional global situation, which benefits the richest segments of the world and hurts the rest of humanity, mainly the poor.

2.2 The Spectrum of Human Needs

Human Needs

Development is about addressing basic human needs, which can take multiple forms. In 1943, Maslow characterized these needs in his paper entitled "A Theory of Human Motivation." Needs were divided into five categories in a hierarchical manner and were selected toward the formulation of a general theory of human motivation. They include in order of priority:

- *Physiological needs* are those associated with survival and maintenance of the human body; they are the most potent of all needs. They include basic life needs such as air, food, water, sanitation, hygiene, health, sex, and sleep.

- *Safety needs* are about possessing shelter, staying warm, being removed from danger, and the inclination toward familiar and known things rather than unfamiliar and unknown ones. These needs are about protection, security, rule of law, and stability.
- *Love and belonging needs* are about belonging to a group or community, being accepted and caring for others, receiving and giving affection, and having relationships.
- *Esteem needs* are about self-esteem, self-respect, confidence, esteem from others, and a sense of being useful and necessary in the world through achievements.
- *Self-actualization needs* are about achievements, self-fulfillment, potential, and personal growth.

This five-stage model of needs is often displayed in the form of a pyramid (Figure 2-2), even though that representation was never proposed by Maslow himself. In Maslow's approach, the hierarchical nature of these needs implies that lower needs in the hierarchy must be met before securing higher needs. Furthermore, it is assumed that the more fundamental the needs are, the more a person is likely to abandon higher needs to secure lower ones. These two assumptions have been questioned in the human psychology literature. They also assume a Western mindset, where higher forms of needs are not a priority.

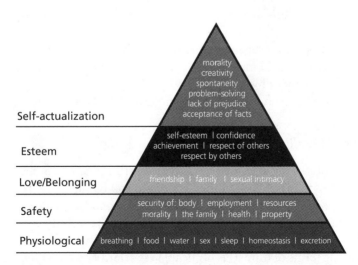

Figure 2-2. Pyramid Representation of Maslow's Basic Human Needs

Source: Wikimedia Commons <http://en.wikipedia.org/wiki/File:Maslow%27s_Hierarchy_of_Needs.svg> (Feb 5. 2012).

In non-Western cultures, there is no conflict in fulfilling love, esteem, and self-actualization needs while trying to fulfill basic needs. For instance, the Sarvodaya Shramadana Movement (SSM) in Sri Lanka, led by Dr. A. T. Ariyaratne (and originally proposed by M. Gandhi), has a unique way of looking at poverty reduction through human empowerment. Since 1972, the essence of the SSM approach has been that a *whole person* approach to development starts with the concepts of equality and spiritual development. In many ways, it taps into the wealth of human beings by addressing and strengthening the higher needs of love, self-esteem, and self-actualization while *concurrently* addressing education and basic needs. A similar example relates to how the country of Bhutan sees the primordial importance of happiness in its 2020 vision for human development (Bhutan 1999), a higher need indeed.

Since Maslow presented the components of his theory of human motivation in the 1940s, several variants of that theory have been proposed. A seven-stage model of needs was proposed in the 1970s by adding *cognitive needs* (search for meaning and knowledge) and *aesthetic needs* (search for beauty, form, and balance) to Maslow's original hierarchy of needs. An eight-stage model was further proposed in the 1990s by adding *transcendence needs* (an excellent review of Maslow's work and modifications of his theory of needs can be found at http://www.businessballs.com/maslow.htm). More recently, Max-Neef (1991) revisited Maslow's concept of human needs and divided them into two main categories: (1) *existential needs,* such as being, having, doing, and interacting; and (2) *axiological needs,* such as subsistence, protection, affection, understanding, participation, idleness, creation, identity, and freedom. Furthermore, he concluded that human needs enter into three postulates of human scale development:

1. Development is about people and not about objects.
2. Fundamental human needs are the same in all cultures and have been so throughout history.
3. Fundamental human needs are finite, few, and classifiable.

Household Livelihood Security

Another way of looking at human needs is to consider their role in the *livelihood security* (lack of vulnerability) of the basic social and economic units of a community, i.e., the *households*. In these units, resources and knowledge are distributed to meet the needs of their members. Preserving the security of those units by meeting their basic needs and keeping their assets (natural, human, social, and economic forms of capital) is of the highest priority because functional and healthy households are critical to the livelihood of poor people. They often represent the only safety net people have. Different forms of security must be considered at the household level, such as food, health, and habitat (shelter). As suggested by Chambers, livelihood

> comprises the capabilities, assets (stores, resources, claims and access) and activities required for a means of living; a livelihood is sustainable which can cope with and recover from stress and shocks, maintain or enhance its capabilities and assets, and provide sustainable livelihood opportunities for the next generation. (Chambers 1983)

The household livelihood security (HLS) concept was proposed by CARE in 1994 as a basic framework to guide its programming of international projects. It evolved from a household food and nutritional security approach introduced in the late 1980s (Frankenberger et al. 2000). It is about adequate and sustainable access to assets, forms of capital, knowledge, and resources necessary to satisfy basic human needs where needs represent gaps and discrepancies between "what is" and "what should be" in the community (Caldwell 2002). Household livelihood security is also about resilience to shock and stress associated with adverse events.

The three main attributes of household livelihood security can be summarized as (1) possession of human capabilities (e.g., education, skills, and health); (2) access to tangible and intangible assets (capital); and (3) the existence of economic activities. The HLS model is people centered and requires having an enabling environment already in place to work. Conditions for that enabling environment include such elements as good governance, human rights recognition, participation, risk management, rule of law, and gender equity. It also implies that the enabling environment is sustainable at different physical scales and time scales.

Furthermore, household livelihood security implies that the community consists of a system of interconnected (and not isolated) social and economic units called households. The strength of those units, combined with their relationship through participation and common vision, defines the overall community's security level, the well-being of its members, its resilience, and its overall level of peace. The importance of recognizing the systems nature of communities in the management of development projects is further discussed in Chapter 6, which deals with a systems approach to development.

2.3 Poverty

Defining Poverty

The terms *poverty, poverty reduction, poverty eradication, absolute poverty,* and *relative poverty* permeate the entire development literature. Countless articles, reports, and books have been written on the subject of poverty and what poverty is and its sublevels: extreme poverty, severe poverty, and moderate poverty. Statistics about the various expressions of poverty abound (Global Issues 2013). We have reached a point where more time and energy have been spent on writing about poverty than on addressing it head on. Questions abound in the academic literature about this issue: What is poverty? What are its causes? How do we measure it and at what level? What are the indicators of poverty? It is understandable that a need exists for academic discussion in addressing and measuring poverty. After all, if we cannot measure poverty, we cannot manage it. However, where does the academic rhetoric stop and tangible action begin?

Another question in the overall conversation on poverty is determining whose voices need to be heard when developing poverty reduction strategies, besides the usual voices from politicians, economists, and individuals in power:

> We must understand poverty from the perspective of the poor and explore the interlocked barriers poor women and men have to overcome, many of which have to do with social norms, values and institutional roles and rules beyond their individual control. Yet to take local action, the details and contours of the patterns have to be understood in each location, for each social group, for each region, for each country in a particular institutional context at a particular time in history. (Narayan et al. 1999)

This statement echoes and reinforces Chambers's (1983) call for "putting the last first" in the grand discussion of human development and the recommendation of Schumacher (1973) that "You cannot help a person if you yourself don't understand how that person manages to exist at all."

As remarked by Forbes (2009), poverty is not new and "has been the dominant social class throughout history. All cultures have been plagued with poor health, disease, and wars from time immemorial." In general, history has shown that there are no universal benefits of poverty. Poverty still remains a grand challenge today, as it has always been in the past. The difference is that in our media-driven culture, poverty is thrown in our face on a daily basis, a not-so-gentle reminder that "we can't not be involved" in addressing it, as remarked by Nolan (2011). But can we truly understand poverty by not walking in the shoes of someone whose job it is to just stay alive until the end of the day?

Poverty is a concept that is hard to define, especially by development professionals and politicians in both poor and rich countries and by those who are not poor. It is a concept "loaded with meaning and historical baggage" (Prahalad 2006) and is often used to describe people in a demeaning, pejorative, and negative way. For those who are indeed poor, poverty is pain in its multiple forms: physical, emotional, and moral (Narayan et al. 1999). Poverty is rarely about the lack of one thing, but it starts with a lack of accessibility to basic and essential needs. It is more than a lack of wealth, though. As summarized by Narayan et al., poverty is about *ill-being* and its many characteristics, including

> *material lack and want (of food, housing and shelter, livelihood, assets and money); hunger, pain and discomfort; exhaustion and poverty of time; exclusion, rejection, isolation and loneliness; bad relations with others, including bad relations within the family; insecurity, vulnerability, worry, fear and low self-confidence; and powerlessness, helplessness, frustration and anger.* (Narayan et al. 1999)

Several definitions of poverty have been proposed in the literature, and it is not our intent here to review all of them. Among the many, three are highlighted below.

- According to Haughton and Khandker (2009), poverty is defined as "pronounced deprivation in well-being." One limited way of looking at well-being is in terms of being able to meet financial needs. In that perspective, the poor are those who do not have enough income or consumption to put them above an adequate minimum threshold. Another way is to look at well-being in terms of "capability to function in society," which encompasses more than income, as suggested by Sen (1999). On its website, the World Bank has extended the definition of poverty to include a broader definition of well-being beyond just economic security, where

> *poverty is pronounced deprivation in well-being, and comprises many dimensions. It includes low incomes and the inability to acquire the basic goods and services necessary for survival with dignity. Poverty also encompasses low levels of health and education, poor access to clean water and sanitation, inadequate physical security, lack of voice, and insufficient capacity and opportunity to better one's life.* (World Bank 2011)

- According to the United Nations, poverty is much more than a lack of financial security. It is

> *a denial of choices and opportunities, a violation of human dignity. It means lack of basic capacity to participate effectively in society. It means not having enough to feed and clothe a family, not having a school or clinic to go to; not having the land on which to grow one's food or a job to earn one's living, not having access to credit. It means insecurity, powerlessness and exclusion of individuals, households and communities. It means susceptibility to violence, and it often implies living on marginal or fragile environments, without access to clean water or sanitation.* (UN 1998)

This statement was signed in 1998 by the heads of all UN agencies.

- UNESCO defines poverty as it

relates primarily to the limited access of poor people to the knowledge *and* resources *with which to address their* basic human needs, *and* promote sustainable development *in such areas as water supply and sanitation, food production and processing, housing and construction, energy, transportation and communication, income generation, and employment creation.* (UNESCO 2003)

This definition is interesting because it links sustainable development with human development and capacity building.

Poverty is a relative concept because it implies that some are poor and others are not. Despite the eloquent definitions above, poverty is mostly seen in the literature and the media in terms of economic security. For instance, the well known *economic pyramid* initially proposed by Prahalad and Hart (2002) has been used many times to show the distribution of humans on our planet based on how much they have available to live on in a given day. A histogram presentation of that distribution is shown in Figure 2-3.

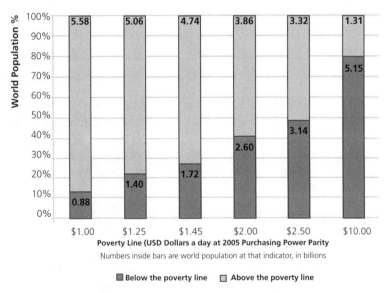

Figure 2-3. Percent of People in the World at Different Poverty Levels

Note: Chart created based on 2008 World Development Indicators Poverty Data (World Bank 2008).

Source: Shah (2013) <http://www.globalissues.org/article/26/poverty-facts-and-stats> (Jan. 10, 2013), with permission from globalissues.org.

Since 2000, statistics on human economic security have been changing for the better very rapidly, especially for those at the bottom of the economic pyramid living in *extreme poverty* on less than $1.25 per day. A report published by Chen and Ravallion (2012) from the Development Research Group at the World Bank concluded that "between 2005 and 2008, the percentage living below $1.25 a day and the number of people in extreme poverty fell in all six regions. This is the first time this has happened over the three-yearly intervals since 1981." The report concluded that the percentage living below the extreme poverty line was 22% in 2008, compared with 43% in 1990 and 52% in 1981. In terms of numbers, this translates into 1.29 billion people in 2008, compared with 1.91 billion in 1990 and 1.94 billion in 1981. It is noteworthy that a large part of the reduction in extreme poverty can be attributed to China's rapid economic rise since 2000 where 662 million fewer people live below the extreme poverty line. Excluding China, the reduction in extreme poverty is not that dramatic but is still encouraging.

The second category in the income pyramid consists of people who live on less than $2 per day and are classified as *moderately poor*. The $2 per day poverty line defines the Bottom of the Pyramid (or BoP) group, according to Prahalad and Hart (2002). The report by Chen and Ravallion (2012) showed that 2.47 billion lived on less than $2 per day in 2008, down from 2.59 billion in 1981 and 2.86 billion in 1990. Percentagewise, this set of numbers translates into 43% of the world's population living on less than $2 per day in 2008, compared with 70% in 1981 and 65% in 1990. Finally, about 5 billion people (70% of the world's population) live on less than $10 per day.

In general, the $2 poverty line breaks humanity into two major categories (Table 2-1). As noted by Forbes (2009), below that line are those for whom survival is a matter of life and death. Above that line are those for whom survival is a matter of maintaining sufficient income. It must be recognized that the $2 poverty line is a purely arbitrary threshold and is meaningless to those who are poor. It is hard to accept that those who live on slightly more than $2 a day are better off than those who live on $2 or slightly less per day. They are "still poor by the standards of middle-income developing countries, and certainly by the standards of what poverty means in rich countries" (Chen and Ravallion 2012).

In richer countries, the poverty thresholds are higher and depend on the country's economy. For instance, according to the U.S. Census Bureau (DeNavas-Walt et al. 2012), the poverty threshold in 2011 was $23,021 for a family of four people. The U.S. poverty rate in 2011 was 15%; a total number of 46.2 million people were living in poverty (the highest level since 1993), with a large inequity according to race. It was found to vary between 22.2% in New Mexico (followed by Louisiana; Washington, DC; South Carolina; Arkansas; and Georgia) and 7.6% in New Hampshire (U.S. Census Bureau 2011).

Table 2-1. Two Different Worlds Below and Above the $2 Poverty Line

Above the $2 poverty line
- Adults wake up to seek ways to improve financial security, social status, convenience, comfort, entertainment, and discretionary time.
- The children go to school to prepare for the same pursuit.
- Success means to attain financial freedom and reach personal goals (typically material wealth) over time.
Survival is a matter of maintaining sufficient income.

Below the $2 poverty line
- Adults and children wake up and go in search of water, firewood, food, and work.
- Children go to school when they can—if at all.
- Success means making it through another day, owning a cow, and having excess food. Reaching the national life expectancy is a major accomplishment and commands great respect.
Survival is a matter of life and death.

Source: Adapted from Forbes (2009), with permission from Stephen Forbes.

In theory, the economic pyramid and other statistical representations (e.g., Figure 2-3) help visualize economic disparities on our planet. In practice, though, it is of limited use because the $1.00, $1.25, $2.00, and $2.50 a day poverty lines do not reflect the daily challenges faced by those who have to survive in precarious conditions. Unfortunately, the income pyramid is often used as a starting point when developing policy on poverty reduction. It fails to recognize that poverty is rarely about a lack of one thing (i.e., financial security) because of its multidimensionality. As remarked by Narayan et al. (1999), even poor people acknowledge that poverty is not uniform and compartmentalized. It can take multiple forms, such as dependent poor, resource poor, temporary poor, and working poor. A broader discussion of what represents poverty is therefore needed.

Poverty is, above all, not being able to meet basic human needs or ensure basic livelihood and not having access to knowledge and resources, as implied in the World Bank's, United Nations', and UNESCO's definitions of poverty. Maslow's and Max-Neef's models and the livelihood security model cover most of those basic needs. But poverty reduction is even more than meeting individual needs. As remarked by Narayan et al. (1999), "poor people rarely speak of income but focus instead on managing assets—physical, human, social, and environmental—as a way to cope with their vulnerability, which in many areas takes on gendered dimensions." Poverty reduction is about ensuring the well-being of those in need,

which is often expressed as having enough; bodily wellbeing, which includes being strong, well and looking good; social wellbeing, including caring for and settling children; having self-respect, peace and good relations in the family and community; having security, including civil peace, a safe and secure environment, personal physical security and confidence in the future; and having freedom of choice and action, including being able to help other people in the community. Wealth and wellbeing are seen as different, and even contradictory. (Narayan et al. 1999)

Poverty reduction is also about addressing the context in which people's needs arise and the multiple constraints and reasons (e.g., social, economic, political, geographic, and climatic) that prevent those needs from being met or even being addressed. It is also about dealing with the unplanned (and sometimes deliberately planned) consequences that arise when unmet needs start interacting with each other. Poverty can be seen as an *undesirable property* emerging from the interaction of various systems (human, economic and financial, political, and environmental), some more dysfunctional than others, with all of them characterized by uncertainty and complexity.

Voices of the Poor

Three reports titled "Voices of the Poor" published by the World Bank (Narayan et al. 1999, 2000; Narayan and Petesch 2002) analyzed poverty from the perspective of the poor based on the results of participatory research interviews (participatory poverty assessment) conducted with tens of thousands of people living in a wide range of countries around the world. The interviews revealed that poor people throughout the world are *trapped* because of various barriers that exist irrespective of gender, culture, and community size and location. Their chances of escaping those traps are slim. The results of those interviews are as relevant today as they were 10 years ago. Those barriers have been regrouped below into several categories using the information from the three aforementioned World Bank reports combined with that from other references (Chambers 1983; Prahalad 2006; WHO 2001):

- *Precarious livelihoods* result from limited assets and limited or no access to basic and affordable services related to such things as clean water supply, sanitation, energy, health, housing, and supporting land. This precariousness is compounded by the vulnerability and limited capacity of households or families to respond to unforeseen conditions ranging between smaller events (e.g., illness or injury) and larger ones associated with natural and nonnatural hazards. In poor communities, it does not take much for people to fall into extreme conditions and deprivation and continue falling down the poverty ladder. According to Huesemann and Huesemann (2011), "a global increase in food prices of only 10%

would translate into 160 million more hungry people," which is roughly half the population of the United States.

- *Isolation* results from poor households located in remote rural areas with little or no contact with the outside world and social mobility. Households may also be located in undesirable land areas around large cities (e.g., unstable hillsides or floodplains). Isolation translates into vulnerability and no or limited access to information and representation, education, work, health care, trade, and services provided by governments, states, and administrations.

- *Physical weakness* creates physical limitations of what people can do and how they can react to various situations. This weakness can be the result of poor health, malnutrition, and no access to health services. It translates into high child mortality, weakened or disabled adults, and low life expectancy.

- *Gender relationships* exist in many places where women and children (especially girls) are the most at-risk group at home or in the community. Poverty affects women more harshly than men. Many reasons lead to this observation: women's traditional child-rearing responsibilities, cultural norms and restrictions on women's rights and access to resources, no or limited decision-making power in the household and in community issues, the burden of domestic chores and resource collection, dependence on men, and women remaining subject to men's will and temperament.

- *Psychological weakness* is related to "powerlessness, voicelessness, dependency, shame, and humiliation." It arises when people are faced with abuse, predation, exclusion, denial of identity, distress, insecurity, and exploitation from elites or corruption. It forces poor communities to resignation, acceptance of the status quo, and reluctance to question authority for fear of reprisal. It means no representation in decision making, no opportunity to negotiate power, and limited opportunity to decide on one's future. It also means less self-esteem and hope of positive change in the future; denial of basic human dignity; lack of freedom (political, economic, or personal); and absence of being valued as a human being. For women, it is also associated with the fear of being objects of abuse and crime. In some extreme cases, there is exposure to crime, addiction, violence, and abuse (e.g., slavery and human trafficking).

- *Weak state institutions* (national, state, and local governments; district and regional administrations; police; schools; and health clinics) fail to provide services to the poor and protect their citizens. They tend to be ineffective, inaccessible, and disempowering, thus making them irrelevant in improving poor people's livelihood.

- *High vulnerability and limited assets* exist in unpredictable and insecure environments, whether the environment is political, economic, social, or geographical.
- *Weak community* (civil society) institutions may keep poor people together and provide a collective support platform, lifeline, and social solidarity. At the local level, they may be expressed in terms of community-based organizations or neighborhoods. But local institutions are weak safety nets and do not have much negotiating power.

These barriers are not isolated from each other because a fair amount of interaction exists among them. Using an interaction matrix representation, Table 2-2 shows the linkage and double causality (feedback) that exist among five components of poverty: poverty itself, physical weakness, isolation, vulnerability, and powerlessness. These five components are located on the diagonal part of the matrix. The off-diagonal boxes in the matrix are nonsymmetric. For instance, the effect of poverty on physical weakness is different from the effect of physical weakness on poverty. In some instances, the two interacting components create a reinforcing loop (or vicious circle), as discussed in Chapter 6 on system dynamics.

Poverty and poor health are related as well; poverty is a determinant of health in developed and developing countries. In a report titled *Dying for Change*, which followed the *Voices of the Poor* reports, Dodd and Munck (2002) looked more specifically at the relationship between poverty and poor health from the perspective of the poor. The report concluded that "poverty creates ill-health and ill-health leads to poverty" in many different ways, specifically the ways listed here:

- Poverty creates ill-health because it forces people to live in environments that make them sick, without decent shelter, clean water, or adequate sanitation.
- Poverty creates hunger, which in turn leaves people vulnerable to disease.
- Poverty denies people access to reliable health services and affordable medicines and causes children to miss out on routine vaccinations.
- Poverty creates illiteracy, leaving people poorly informed about health risks, and forces them into dangerous jobs that may harm their health.
- Ill health is an impediment to reaching one's potential and limits people's capabilities.
- Ill health represents an obstacle for people to step out of poverty and perpetuates poverty.

As summarized by Huesemann and Huesemann,

there is significant evidence that lower socio-economic status and poverty are associated with higher incidence of disease and mortality. The reason for this phenomenon is that the

Table 2-2. Double Causality Among Five Components of Poverty

	Poverty	Physical Weakness	Isolation	Vulnerability	Powerlessness
Poverty		Lack of available food, leading to malnutrition and hunger, resulting in physical limitations and low immune systems. People cannot pay for health services. Low life expectancy, high child mortality.	Limited or no access to health, jobs, education, opportunities, and representation.	No or limited assets are available. No or limited contingencies and capacity to absorb shocks and stresses and unforeseen conditions.	Low status in society, no voice, no representation, no access to information. Put down. No hope.
Physical Weakness	Low productivity because of weak labor, unable to work over long hours and maintain consistent jobs.		Cannot attend meetings and contribute to decision process, cannot travel, limited mobility.	Increases vulnerability because of limitation to undertake hard work, start new activities.	No time and energy to fight for representation and organize themselves.

	Poverty	Physical weakness	Isolation	Vulnerability	Powerlessness
Isolation	No services and information make it to the community.	Increased weakness at the community level as able bodies leave for big cities. No access to information that could improve well-being.		More vulnerability because community cannot be informed and has no access to needed services and outside assistance.	Not informed about decisions and opportunities. No representation from political leaders and administrative agents.
Vulnerability	Selling of limited assets to survive in response to crisis or unusual conditions.	Handling contingencies takes time and energy away from work and may reinforce physical weaknesses.	More isolation arises after crises.		Need to request help and loans to survive.
Powerlessness	Exploitation by powerful. No interest from state for assistance.	Queuing to receive services. Low priority in receiving services such as health care.	Do not attract government aid and services. Voices don't count.	More vulnerable to abuse and effects of corruption.	

Note: This table was created based on the discussion by Chambers (1983, p 111) on the links among poverty, physical weakness, isolation, vulnerability, and powerlessness (diagonal terms). Off-diagonal terms show the feedback between one component and the other four.

poor have less nutritious food, more hazardous jobs, higher stress and lower levels of education and literacy, the latter preventing them from learning about or achieving a healthy lifestyle. (Huesemann and Huesemann 2011)

As remarked by UNICEF (2013), malnourished children are "less able to fight off illness, less likely to get the most out of schooling, and often become physically and mentally stunted. Malnutrition keeps children trapped in the cycle of poverty" throughout their lifetimes.

Other forms of linkages and double causalities have been recognized, such as a decrease in national mean IQ with an increase in infectious and parasitic diseases (Hassall and Sherratt 2011); poverty and inequality; poverty and environmental degradation (UNDP/HDR 2011); poverty and oppression and entrapment because of belief systems such as caste systems in India and Nepal (Bista 2001); poverty and conflict (UN 1992; WDR 2003; Norwegian Ministry of Foreign Affairs 2004; Ricigliano 2012); and poverty and vulnerability to natural and non-natural disasters (Cuny 1983; Sphere Project 2011). Finally, higher forms of linkages arise when consequences of poverty become root causes to more problems and create cascading effects that become deeper traps.

Changing the Discussion

Much discussion in the past 50 years has been about the development of social programs and strategies for poverty reduction. As remarked by Narayan et al. (2000) and Prahalad (2006), using specific terminology such as "poverty reduction" projects a stigma onto those who are officially defined as members of "the poor." Words have implications, and in this case, "poverty reduction" is not necessarily the most optimistic paradigm to use when trying to show respect for and empowering others. The term "poverty reduction" fails to recognize the inherent wealth that people living in poverty have access to, i.e., their capacity (or strengths). Wealth has nothing to do with money and material items, as many Western societies have wrongly defined it. However, it has everything to do with the human dimension of development and community. As remarked in the 1990 UNDP/HDR report, "people are the real wealth of a nation."

So, as suggested by my colleague Hunter Lovins (personal communication, 2011), it is time to change the discussion on how to strengthen communities through *wealth enhancement* rather than *poverty reduction*—a phrase that invokes a more positive attitude toward improving humanity. To start with, a strategy of wealth enhancement needs to make a clear distinction between *financial* wealth and *real* wealth. The two are not exclusive but are rarely found together. Korten proposed an excellent definition of real wealth as follows:

Real wealth consists of those things that have actual utilitarian or artistic value: food, land, energy, knowledge, technology, forests, beauty, and much else. The natural systems of the planet are the foundations of all real wealth, for we depend on them for our very lives.
(Korten 2006)

Real wealth can take multiple forms and deliver *assets* (capital) that bring *meaning* to people and provide them with a dignified quality of life. As discussed by Narayan et al. (1999), wealth seen in a broader context crosses all forms of capital, besides just economic capital, such as

- Human capital in the form of labor power, literacy, and know-how;
- Environmental capital in the form of land, trees, forests, water, and other natural resources;
- Social capital expressed through relationships between community members as part of social networks, ties, and safety nets and through having time for social interaction, for family gathering and support, and for cultural identity; and
- Physical capital in the form of access to land (size and quality), as the ability to self-provision; household property; and valuables.

In his book, *The Fortune at the Bottom of the Pyramid*, Prahalad (2006) advocates real wealth at the bottom of the pyramid and emphasizes that the capacity is there and is latent. All it takes is to use a different perspective or mindset: "If we stop thinking of the poor as victims or as a burden and start recognizing them as resilient and creative entrepreneurs and value-conscious consumers, a whole new world of opportunity will open up."

In other words, a huge potential exists to lift millions of people out of poverty through what Prahalad calls a *cocreation* process that emphasizes the strength and collective intelligence of communities and their members. By partnering and leveraging indigenous talent, skills, and knowledge; the private sector, and multiple stakeholders through market-based solutions, this in turn creates social transformation. The cocreation process requires a complete revision of what drives bottom of the pyramid (BoP) market consumers in terms of product *affordability, accessibility, availability, reliability,* and *sustainability* that guarantee a dignified quality of life. It also redefines how the private sector needs to respond to the consumer needs by

- Delivering products and services in response to market demand;
- Accounting for the characteristics of those services and products (scalability, hybrid nature, price performance, resource conservation, functionality, adaptation to skill level and education, rapidity, and characteristics of the consumer in different settings); and
- Collaborating with nontraditional partners such as nongovernmental organizations (NGOs), microenterprises, local firms, and government agencies.

Another way of looking at the cocreation process is a transformation of asset-rich communities from capital-poor to capital-rich, as suggested by de Soto (2003). As we discuss in Chapter 13 in relation to social entrepreneurship and innovation, a huge potential exists to lift millions out of poverty and help them regain hope and a sense of dignity through social innovation (disruptive and frugal), grassroots empowerment, and micro-financing (*Economist* 2005; Bornstein 2007). Furthermore, that opportunity is likely to positively affect both the developed world and the developing world and force markets to be more grassroots conscious.

It is clear from the previous discussion that wealth enhancement and cocreation begin "with the presumption that, for the most part, people are competent and resourceful and strive for connection and meaning" (Bornstein 2005), a very empowering and promising concept indeed. But having said that, it must also be acknowledged that, unlike in rich countries, these forms of wealth are in constant flux and are highly vulnerable to adverse events (climatic and seasonal) and to decisions (political) that are well beyond the capacity of poor communities and individuals to control and handle because of their powerlessness, voicelessness, and isolation. The events can occur locally, in country, or halfway around the world. As a case in point, poor people are vulnerable to climate variation and its associated consequences (such as hunger and environmental changes). The 2010 State of Environmental Migration report (IDDRI) remarked that migrations related to environmental and climatic causes are already occurring and exceed those related to conflict. In 2008, about 4.6 million people were displaced in their own countries because of internal violent conflict. In that same year, about 20 million people had to relocate to other areas because of catastrophic natural events. The problem is that most internally displaced persons (IDPs) and refugees are not able to define where they are supposed to relocate.

Too often, the aforementioned events exacerbate challenges related to existing precarious living conditions (e.g., arid and high-altitude conditions, landslide areas on steep hillsides near big cities, and floodplain areas) and the limited social, environmental, and human assets associated with the powerlessness, voicelessness, and weak bargaining power poor people have, a "double burden" of risks (UNDP/HDR 2011). Even though communities have wealth, they need to be resilient and have enough capacity to handle risks (real and perceived) associated with multiple adverse events, as further discussed in Chapters 10 and 11.

A Bigger Picture of Poverty

Figure 2-4 shows an attempt at conducting a cause-and-effect analysis (or causal analysis) of poverty. It uses a method that is discussed in detail in Chapter 7, the

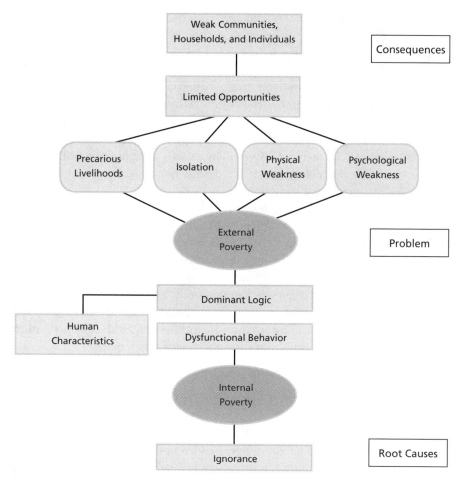

Figure 2-4. Possible Cause-and-Effect Analysis of Poverty

problem tree. A problem is attached to the trunk of a tree; in our case it is (external) poverty. The branches of the tree represent the consequences (effects) of the problem, and its roots represent the root causes.

The problem tree shows some of the barriers (i.e., precarious livelihood, isolation, physical weakness, and psychological weakness) proposed by Narayan et al. (1999, 2000, 2002) that trap the poor as consequences of external poverty. As discussed previously, these barriers are interconnected (Table 2-2) and, when combined, lead to limited opportunities (socioeconomic or self-empowerment) and weak communities, households, and individuals.

Figure 2-4 shows that deeper root causes of poverty could be attributed to ignorance and *internal poverty,* mostly of decision-making institutions (and those who make decisions) at various levels in society (local, regional, national, and international). Internal poverty expresses itself in the form of dysfunctional behavior (greed, fear, selfishness, and apathy). This behavior in turns imposes a dominant logic with priorities in society that create and perpetuate external poverty. These priorities include among others (1) exploitation of nature; (2) possession; (3) unfair competition; and (4) a view of humanity as consisting of objects of economic development and an afterthought to the production of goods, rather than as interconnected subjects (Schumacher 1973). Human characteristics (such as bounded rationality, lack of will and commitment, and lack of global vision) do not help in questioning these priorities and often reinforce them. As discussed further in Chapter 6, by using a systems approach, poverty can be seen as an *emerging property* of a dysfunctional society, its components, and the links among those components.

A new mindset is therefore needed to break the cycle mentioned, a paradigm that looks at a *whole person* and a fulfilling approach to development as suggested for instance by the Sarvodaya Shramadana Movement in Sri Lanka. This movement focuses on equality and spiritual development rather than promoting the traditional concepts of inequality, ignorance, possession, and unfair competition. This new paradigm confronts head on the inner poverty of our institutions and taps into the *real wealth* of human beings by addressing and strengthening higher needs of compassion, love, self-esteem, and self-actualization while concurrently addressing education and basic human needs.

2.4 Development and Human Development

A Brief History of Development

Development can mean different things to different people. In general, development represents a transformation of society through its betterment, making a better life for everyone (Peet and Hartwick 2009). Other terms that define development include meeting basic needs, participation, improvement of people's lives, poverty eradication, and empowerment (Nolan 2002). The traditional view of development implies some form of interaction involving recipients of development (insiders) and those providing development (outsiders). Forbes (2009) introduces the term "extensionalist" to describe someone who is a facilitator, trainer, and coordinator, compared with the "intervener," who is someone who intervenes or intrudes.

Different forms of interaction between insiders and outsiders groups have developed over time, leading to different models of development. It should be noted that this chapter does not discuss the pros and cons of the various models of development and human development over the past 60 years. Several authors have contributed to this genre of academic discussion, and the reader is referred to several publications on this subject (Stiglitz 1998; Nolan 2002; Sachs 2006; Collier 2007; Easterly 2007; Forbes 2009; Chang 2010; Kenny 2011). Even more discussion can be found in the various World Bank World Development Reports and UNDP Human Development Reports that have been published annually since 1990.

Development is mostly a concept that emerged in Western countries after World War II. Before that, there was not much of a sense of moral or ethical obligation for one nation to help another. The general view was that the economy of Western nations depended on less developed countries in the form of resource exploitation and cheap labor, a process that had been in place since the 15th century. In return, Western nations were under the impression that they were destined to bring civilization to natives deemed inferior and uncivilized. That colonial mindset lasted well into the 20th century, with the mandate system established by the victorious nations after World War I and unsuccessfully enforced by the League of Nations (1918–1943) shortly thereafter.

The history of development over the past 50 years is well documented (Nolan 2002; Collier 2007; Chang 2010; Mansouri and Rao 2012). In summary, after World War II, the world emerged as consisting of three groups of countries. Western industrialized countries under reconstruction represented the *First World*. The *Third World* consisted of countries that were agriculturally focused with low per capita incomes, high mortality, low life expectancy, and high population growth. In between, centrally planned economies of the Soviet bloc comprised the *Second World*.

As Chang (2010) notes, the *traditional* consensus about development after the end of World War II was that of "transformation in productive structure and the development of social and technological capacities that are both the causes and consequences of such transformation" to be achieved through rapid *industrialization* and *urbanization*. This purely top-down economic strategy of development, combined with technology, was designed to yield higher incomes and growth and at the same time guarantee better living conditions and increased life expectancy for all, as quickly as possible.

Development as economic growth after the success of the Marshall Plan (1948–1952) in Europe and the Bretton Woods agreements in July 1944 (which led to the establishment of the International Monetary Fund [IMF] and the early version of the World Bank) was seen as a preferred model (or template) to be promoted around the world. Rapid disbursing of foreign aid combined with an ethnocentric approach to foreign cultures by Western powers and the belief that Western concepts of development, technology, economy, and management were superior became the only recipe and "magic pill" for rapid growth and development worldwide. It also helped "legitimize the 'development' of 'underdeveloped' people against their will" (Huesemann and Huesemann 2011). That global vision of a better and peaceful world, where economic growth would guarantee an increase in living standards (e.g., health, nutrition, and education) and reduce poverty, was further reinforced with the establishment in 1945 of the United

Nations mandate to "maintain international peace and promote cooperation in solving international economic, social and humanitarian problems."

Over the past 60 years, four major groups of players have focused on development issues (Nolan 2002): multilateral agencies (e.g., the UN, the World Bank, regional development banks, and the IMF); bilateral in-country agencies (e.g., the U.S. Agency for International Development [USAID], EuropeAid, and other in-country aid organizations); nongovernmental organizations (NGOs and international NGOs [INGOs]); and the private sector. As remarked by Valadez and Bamberger (1994), development players can also be regrouped as "international agencies (donors, NGOs, and research foundations); national and sectorial agencies (central government ministries, financial agencies, line ministries, local NGOs, and national consulting and research groups); project implementation agencies; and intended beneficiaries." Because of the various players' interests and orientations, development as an enterprise has become quickly enmeshed in complex geopolitical and economic agendas, often empowering special groups or individuals, rather than those needing assistance.

From an ideological viewpoint, the business of economic development has often been presented (packaged or sold) by development players as a combination of *aid* and *transformation* (economic, social, and political) toward democratic ideals guaranteeing human progress and personal, economic, and political freedom. Still attached to that model, however, was the precondition that it had to be dictated by outsiders, mostly from rich Western countries, and had to follow the Western model of development. The inaugural address of President Harry Truman on January 19, 1949, presented development as a heroic task of bringing democracy to everyone and improving the standards of living for all people, with the noble principle of international order and justice. To be more exact, it was also about counteracting the "false philosophy of communism" and the expansion of the Soviet Union's influence through development and other means (Rostow 1960). During the Cold War, both Western powers and the USSR struggled for world dominance using development and ideology as a convenient facade. That approach to development often resulted in the support of controversial dictatorships friendly to Western powers or the USSR but with myopic views of human rights, labor rights, and democracy, in general.

Jumping forward in time, the World Bank introduced in 1980 the structural adjustment program (SAP) as a way to provide loans to developing countries to address their economic imbalances (which themselves originated from cumulative debts associated with previous aid loans) but not necessarily poverty reduction. These loans came with conditions: they required in-country internal changes (such as privatization, deregulation, social spending cuts, and cash crops) and external changes, especially in the form of reducing trade barriers and deregulating markets. Penalties were imposed if countries deviated from any agreement. According to Easterly (2005), such structural adjustment loans did not prove "to be effective in achieving widespread policy improvements and in raising growth potential."

Another benchmark was the Washington Consensus, proposed in 1989, originally designed to aid Latin American countries (and later sub-Saharan Africa) in their economic development (Williamson 2004). Consisting of 10 principles deemed necessary as policy reform to incite economic growth, the consensus (supported by the U.S. government and international financial institutions) could be summarized as three words of advice to developing countries: "stabilize, privatize, and liberalize" (Rodrik 2006). In the 1990s, the Washington Consensus became a source of major controversy and was seen as synonymous with neoliberal and globalization strategies enforced by Western countries (mostly the United States) onto the developing world in general. It was also seen as benefiting the wealthy elite in developing countries that did not show much interest in promoting change for their people.

Failure of the SAP and the Washington Consensus to improve the living conditions of the world's poor led to a post–Washington Consensus area in the late 1990s, a more poverty-focused strategy to support the poor through social community-based development programs, which emphasized more social spending on education and health but with no well defined blueprints and modest reforms (Rodrik 2006). The rise of emerging markets in India and China in the 1990s and 2000s, based on policies antithetic to those of the Washington Consensus, further confirmed the inadequacy of the traditional Western market economic policies of the 1980s and 1990s. Since 1981, China has been able to move, using its own model of development, about 660 million people out of poverty and recently sustain a very good annual economic growth.

As noted by Nolan (2002), a secondary effect of the development mindset of the 1980s and 1990s was that the development players focused mostly on the policy of development, rather than on addressing grassroots issues, thus becoming less effective in addressing the world's poverty and relying more on outside consultants. They often lost direct experience with their projects and programs. Another compounding effect was that donors were becoming less likely to contribute to development that did not show tangible results. Last, but not least, globalization was becoming a new game in town that confused rather than clarified the situation. According to Collier (2007), globalization benefited the economy of some developing countries such as China, Brazil, and India but kept the bottom billion (those in extreme poverty) out of the discussion. This situation begs the question, how can the poor with no voice, power, or capital participate in the discussion regarding globalization?

More recently, in parallel with the model of economic development, the concept of *human development* emerged. In 1990, the UNDP/HDR report defined human development as a process "of enlarging people's choices," with two sides to it: "The formation of human capabilities such as improved health, knowledge and skills—and the use people make of their acquired capabilities—for leisure, productive purposes or being active in cultural, social and political affairs."

Over the past 20 years, it has become clear that economic development alone is not a sufficient condition to improve human well-being. Human development is about providing people access to services to address their basic human needs as preconditions for subsequent economic growth in the form of wealth enhancement, aid delivery, debt reduction, individual empowerment, community participation, small-scale technology, and local capacity building. As a new paradigm, human development encompasses the major components of the "social development programs" of the 1990s, which according to Valadez and Bamberger refers to

> *An array of programs designed to improve the quality of life by improving the capacity of citizens to participate fully in social, economic, and political activities at the local and national levels. On the one hand, these programs may focus on improving physical well-being and access to services; protecting vulnerable groups from the adverse consequences of economic reform and structural adjustment; or providing education, literacy and employment and income-generating opportunities. On the other hand, they may focus directly on local empowerment and equity issues by strengthening community organizations, encouraging women to participate in development, or alleviating poverty.*
> (Valadez and Bamberger 1994)

Human development is about people, the communities to which they belong (e.g., community development), and the other institutions with which those communities interact. In the 2011 UNDP/HDR report, the definition of human development was updated to include the concepts of sustainability and equity as follows:

> *Human development is the expansion of people's freedoms to live long, healthy and creative lives; to advance other goals they have reason to value; and to engage actively in shaping development* equitably and sustainably on a shared planet. *People are both the beneficiaries and the drivers of human development, as individuals and in groups.*
> (2011 UNDP/HDR)

The Millennium Development Initiative discussed in the following is a good example of a major human development framework that has been endorsed by all development players throughout the world. Recent figures by the World Bank (Chen and Ravallion 2012) seem to show promising trends in poverty reduction. However, there seems to be some ongoing confusion in the development world as to how the two development tracks (economic growth and human development) are supposed to work together.

The Millennium Development Goals

In September of 2000, the world's governments met at the 65th session of the United Nations and made a general commitment (or global common responsibility) to fight poverty, hunger, and disease through the Millennium Declaration with a pledge:

> To spare no effort to free our fellow men, women and children from the abject and dehumanizing conditions of extreme poverty, to which more than one billion of them are currently subjected. We are committed to making the right to development a reality for everyone and to freeing the entire human race from want. (UN 2000)

A total of eight Millennium Development Goals (MDGs) were outlined, consisting of 21 quantifiable targets measured with 60 indicators (Table 2-3). The deadline for achieving those goals was set for the year 2015. Discussion is already under way to design a post-2015 development agenda (UNDP 2014).

Progress toward reaching the MDGs over the past 12 years has been monitored through a series of report cards issued at five years (2005) and 10 years (2010) after the declaration was announced. Plenty of data have been generated

Table 2-3. The Eight UN Millennium Development Goals and 21 Targets

Goals	Description
1	**Eradicate extreme poverty and hunger** • **Target 1A**: Halve, between 1990 and 2015, the proportion of people whose income is less than $1 a day • **Target 1B**: Achieve full and productive employment and decent work for all, including women and young people • **Target 1C**: Halve, between 1990 and 2015, the proportion of people who suffer from hunger
2	**Achieve universal primary education** • **Target 2A**: Ensure that, by 2015, children everywhere, boys and girls alike, will be able to complete a full course of primary schooling
3	**Promote gender equality and empower women** • **Target 3A**: Eliminate gender disparity in primary and secondary education, preferably by 2005, and in all levels of education no later than 2015
4	**Reduce child mortality** • **Target 4A**: Reduce by two-thirds, between 1990 and 2015, the under-five mortality rate
5	**Improve maternal health** • **Target 5A**: Reduce by three-quarters the maternal mortality rate • **Target 5B**: Achieve universal access to reproductive health

Table 2-3. The Eight UN Millennium Development Goals and 21 Targets (*Continued*)

Goals	Description
6	**Combat HIV/AIDS, malaria, and other diseases** • **Target 6A**: Have halted by 2015 and begun to reverse the spread of HIV/AIDS • **Target 6B**: Achieve, by 2010, universal access to treatment for HIV/AIDS for all of those who need it • **Target 6C**: Have halted by 2015 and begun to reverse the incidence of malaria and other major diseases
7	**Ensure environmental sustainability** • **Target 7A**: Integrate the principles of sustainable development into country policies and programs and reverse the loss of environmental resources • **Target 7B**: Reduce biodiversity loss, achieving by 2010 a significant reduction in the rate of loss • **Target 7C**: Halve, by 2015, the proportion of people without sustainable access to safe drinking water and basic sanitation • **Target 7D**: By 2020, have achieved a significant improvement in the lives of at least 100 million slum dwellers
8	**Develop a global partnership for development** • **Target 8A**: Develop further an open, rule-based, predictable, nondiscriminatory trading and financial system (includes a commitment to good governance, development, and poverty reduction—both nationally and internationally) • **Target 8B**: Address the special needs of the least developed countries • **Target 8C**: Address the special needs of landlocked developing countries and small island developing states • **Target 8D**: Deal comprehensively with the debt problems of developing countries • **Target 8E**: In cooperation with pharmaceutical companies, provide access to affordable essential drugs in developing countries • **Target 8F**: In cooperation with the private sector, make available benefits of new technologies, especially information and communication

Note: This table summarizes the goals and targets for each goal. For more specific data, go to <http://www.un.org/millenniumgoals> (Aug. 27, 2012).

by tracking development trends around the world. The progress has been well documented in various reports produced on a yearly basis by organizations such as the World Bank (in the World Development Report series), the UNDP (in the Human Development Report series), other UN agencies, and a wide range of NGOs since 2000. A recent update of progress toward reaching the MDGs and report cards for each MDG for different regions of the world can be found in the latest Millennium Development Goals report (UN 2012) and other documents (WHO 2012, 2013b; UN 2013a). Analysis and visualization tools such as the software tools developed by Gapminder have helped in making sense of the trends observed since 2000.

Moving Development Forward

Clearly development has evolved over the past 50 years. Table 2-4, taken from a UNDP report (2009), shows the evolution of development through four stages (which we could call Development 1.0, 2.0, 3.0, and 4.0) from development aid (promoting dependency) to capacity development (emphasizing empowerment and community participation). It also lists the assumptions, practices, and results associated with the different stages of development.

The jury is still open as to whether development over the past 50 years has been successful. After all, people today live longer and better than ever before. Even though millions have been lifted out of the poverty trap, criticisms of development abound. Easterly (2006) and others go as far as saying that development in the past has had a limited effect for the monetary investment, $2.3 trillion spent over the past 50 years on international development, with questionable results. A good example illustrating and supporting such an opinion is the failure of the international and national development communities to secure safe water, sanitation, and hygiene for all. A report by the Water Supply and Sanitation Collaborative Council (WSSCC 2004) attributes such failure over the past 50 years to a conservative mindset characterized by (1) no willingness on the part of those who have contributed to failed water, sanitation, and hygiene (WASH) projects to learn from or even be accountable for their failures; (2) a commitment to business as usual with no incentives to change the mindset and incapacitating bureaucracy; (3) a lack of considering community empowerment in decision making; and (4) a myopic expert–elite approach to solving problems using the same rational and predictable methods, applied mostly on large projects. Such criticisms are haunting the development industry, not only in the field of water and sanitation, but in many others as well. Recent assessments of World Bank and Asian Development Bank projects have shown project success rates ranging only from 50% to 70% (ADB 2012; IEG 2012).

Thinking forward, the question remains as to what the future model of human development is and how it can truly be about guaranteeing healthy and fulfilling lives and, as a result, a better quality of life for all. The future of development is still uncertain because there has been a lot of debate about finding the right balance between economic growth (and the material affluence resulting from it) on one side and quality of life and well-being on the other side (Max-Neef 1991; Huesemann and Huesemann 2011; Kenny 2011; Helliwell et al. 2012).

Table 2-4. Evolution of UNDP's Capacity Development Approach

	The Assumption	The Practice	The Result
FIRST	"Developing countries need money"	**Development aid:** Developed countries lend or grant money to developing countries	• Greater focus on investment and supporting than on results • Mounting debt • Dependence on foreign aid • Projects end when money runs out
THEN	"Developing countries should just model themselves after the developed ones"	**Technical assistance:** Foreign experts come in to operate their own projects, which they expect to yield similar results as those seen in developed countries	• Projects launched but disconnected from local goals or priorities • Assumes few or no resources available locally • Dependence on foreign experts • Expertise not always transformed from foreigners to locals • The externally driven model may ignore local realities • Idea of assistance highlights unequal relationship between developed and developing countries
FOLLOWED BY	"Developing countries should partner with developed ones"	**Technical cooperation:** Greater emphasis on training, transferring knowledge, based on national policies and priorities	• Local expertise enhanced • Projects somewhat more in line with local priorities and goals • Driven by outside forces, opportunities missed to develop local institutions and strengthen local capacities • Expensive
AND CURRENTLY	"Developing countries should own, design, direct, implement, and sustain the process themselves"	**Capacity development:** A focus on empowering and strengthening endogenous capabilities	• Makes the most of local resources—people, skills, technologies, institutions—and builds on these • Takes an inclusive approach in addressing issues of power inequality in relations between rich and poor, mainstream and marginalized (countries, groups, and individuals) • Emphasizes deep, lasting transformations through policy and institutional reforms • Values "best fit" for the content over "best practice" because one size does not fit all

Clearly the attributes of economic growth are as important as growth itself (Peet and Hartwick 2009). More specifically,

- If growth negatively affects the environment, it is not development;
- If growth produces junk products and services, it is not development;
- If growth concentrates wealth in the hands of a few, it is not development;
- If growth is controlled by a powerful few, it is not development; and
- If growth is about consumerism, it is not development.

A need seems to exist for a new mindset of development (development 5.0) that questions the orthodoxy of development as usual. That mindset (1) recognizes the idea that development is about people, their culture, participation, rights, and empowerment and not necessarily material affluence through unlimited growth and advancements in GDP per capita; (2) embraces institutional and individual accountability, quality control, and quality assurance in projects; (3) encourages the effectiveness and efficiency of development programs; (4) accepts the notion that development is a work in progress (adaptive) and is complex where questions are raised, new decisions are made, and lessons are learned along the way; and (5) uses systems tools that better account for the dynamic of human systems than traditional cause-and-effect tools. The ultimate goal here is not to come up with *one* model of development but rather a range of learning, organic, and mutually respectful models of *authentic* human development leading to more stable, equitable, safe, prosperous, and sustainable communities, while learning from the lessons of the past.

Above all, it is necessary to remember that development is about *people* and their right to develop their full potential and lead productive, creative, and valuable lives in accord with their needs and interests (UNDP/HDR 1990). "Development means changing the world for the better.... and it entails economic, social, and cultural progress, including, in the latter senses, finer ethical ideals and moral values" (Peet and Hartwick 2009). It goes well beyond Western ideals and values and beyond economic growth. As remarked by Julius Nyerere (1974), president of Tanzania from 1964 to 1985, "Development brings freedom, provided it is development of people. But people cannot be developed; they can only develop themselves."

Measuring Development

Development has been measured in many different ways and at different scales by different groups, such as the World Bank (2013c) or the UNDP (2013). Among all development indicators, those proposed by the World Bank, such as the gross national product (GNP), the gross domestic product (GDP), and more recently the gross national income (GNI), have been used as metrics of the economic health and well-being of nations.

According to the World Bank, the gross national income of a country "comprises the value of all products and services generated within a country in one year (i.e., its gross domestic product), together with its net income received from other countries (notably interest and dividends)." As shown in Figure 1-1, all 188 World Bank member countries and all other economies with populations of more than 30,000 (214 total) are divided into four groups based on GNI per capita (based on purchasing power parity). The most recent classification based on 2012 GNI data (World Bank 2013a) divides economies into the following: low-income (<$1,035); lower-middle-income ($1,036–$4,085); upper-middle-income ($4,086–$12,615); and high-income (>$12,616). As of 2012, these four groups represent 36 (17%), 48 (22%), 55 (26%), and 75 (35%) of the 214 economies in the world, respectively. Low-income and middle-income economies are regrouped as *developing economies*, with reservations and some intense discussion as to the relationship between income and the status of development.

Starting in the mid-1990s, discussion arose regarding the relationship between economic growth and well-being and how the former leads to the latter. Max-Neef (1995) introduced the concept of "threshold hypothesis," remarking that "for every society there seems to be a period in which economic growth (as conventionally measured) brings about an improvement in the quality of life, but only up to a point—the threshold point—beyond which, if there is more economic growth, quality of life may begin to deteriorate." In other words, beyond a certain point of economic growth measured in terms of GDP, GNP, or GNI, the quality of life needs to become a higher priority in development.

Similar reservation about the limitations of GDP and GNP (GNI was not introduced at the time) as measures of development was voiced in an article entitled "If the GDP Is Up, Why Is America Down?" by Cobb et al. in the October 1995 issue of *The Atlantic*. In that article and subsequent publications (Baker 1999; Cobb et al. 1999), GDP or GNP as a measure of growth has been shown as not being necessarily a measure of progress because it only reflects increased spending of a nation, whether that spending is good or bad. It ignores the social and environmental costs (externalities) associated with growth and therefore can be a misleading measure of progress and well-being. As remarked by Arrow et al. (1995), "economic growth is not a panacea for environmental quality."

An alternative index, called the genuine progress indicator (GPI) was introduced in the 1990s as a new measure of economic well-being of a nation and addressed whether economic growth benefits people. It starts with the same consumption-related data of the GDP and GNP but excludes activities in calculating the index that are harmful to the environment and the people (e.g., resource and natural capital depletion; social issues such as crime, family breakdown, and reduced quality of family life; pollution and its effect on environmental health; erosion of farmland; and loss of wetlands). It also adjusts the contribution of certain activities that lead to more equitable income and resource distribution and the contribution of activities such as household work and volunteer work. Needless to say, the GPI has been a topic of intense discussion among different groups of economists. Furthermore, it has been promoted in a limited number of developed countries only, such as the United States (in some states only), Canada, Australia, Chile, and some European Union countries.

Another metric of development to measure country-specific development that goes beyond the GDP (or GNP, or GNI) and economic growth and accounts for human capabilities and socioeconomic progress (i.e., human development) is the human development index (HDI). It was developed by the UNDP in 1990 through the work of the Pakistani economist, Dr. Mahbub ul Haq, the Indian economist Dr. Amartya Sen, and other economists (UNDP/HDR 1990). HDI recognizes that *people and their capabilities* are key criteria in assessing the level of development of a country. The HDI is calculated as the *geometric mean* of three subindices corresponding to three dimensions of human development: (1) a long and healthy life in terms of life expectancy (I_{life}), (2) access to knowledge and education in terms of literacy rate and school enrollment ($I_{education}$), and (3) a decent standard of living in terms of GDP per capita at purchasing power parity (I_{income}).

$$HDI = \sqrt[3]{I_{life} \times I_{education} \times I_{income}}$$

Each subindex is normalized between 0 and 1 using a min-max rescaling theme as follows:

$$Subindex = \frac{actual\ value - minimum\ value}{maximum\ value - minimum\ value}$$

The maximum and minimum values (2011 data) necessary to calculate the three subindices can be found in Table 2-5. Note that $I_{education}$ is itself a function of two indices, one related to the mean years of schooling and one related to expected years of schooling. An example of calculations for the HDI of Vietnam is shown in Table 2-6.

In general, HDI varies between 0 and 1. The most recent values of the HDI for 187 countries based on 2012 data were released in the 2013 UNDP/HDR report. The HDI values range between a maximum of 0.955 (Norway) and a minimum of 0.304 (Democratic Republic of the Congo and Niger), with a world average of 0.694; the 29 lowest HDI values are for countries in Africa, except for Afghanistan and Haiti. According to the UNDP, countries fall into four categories based on the value of their HDI: very high human development, high human

Table 2-5. Goalposts for the Human Development Index

Dimension	Observed maximum	Minimum
Life expectancy	83.4 (Japan, 2011)	20
Mean years of schooling	13.1 (Czech Republic, 2005)	0
Expected years of schooling	18 (capped at)	0
Combined education index	0.978 (New Zealand, 2010)	0
Per capita income (PPP $)	107,721 (Qatar, 2011)	100

Source: UNDP/HDR (2011), with permission from United Nations Development Programme.

Table 2-6. Calculation of the Human Development Index for Vietnam

Indicator	Value
Life expectancy at birth (years)	75.2
Mean years of schooling (years)	5.5
Expected years of schooling (years)	10.4
NI per capita (PPI $)	2,805

Note: Values are rounded

$$\text{Life expectancy index} = \frac{75.2 - 20}{83.4 - 20} = 0.871$$

$$\text{Mean years of schooling index} = \frac{5.5 - 0}{13.1 - 0} = 0.419$$

$$\text{Expected years of schooling index} = \frac{10.4 - 0}{18 - 0} = 0.578$$

$$\text{Education index} = \frac{\sqrt{0.419 \times 0.578} - 0}{0.978 - 0} = 0.503$$

$$\text{Income index} = \frac{\ln(2,805) - \ln(100)}{\ln(107,721) - \ln(100)} = 0.478$$

$$\text{Human Development Index} = \sqrt[3]{0.871 \times 0.503 \times 0.487} = 0.593$$

Source: UNDP/HDR (2011), with permission from United Nations Development Programme.

development, medium human development, and low human development. The first three groups have 47 countries each and the last group has 46 countries. The last two categories are somewhat folded into a broader category of *developing countries* with an HDI of 0.7 or less.

In the technical note section of the 2011 UNDP/HDR report, three additional indicators were introduced:

- The inequality-adjusted human development index (IHDI), which accounts for inequalities in the three dimensions that enter into the calculation of the HDI;
- The gender inequality index (GII), which is a "composite measure reflecting inequality in achievements between women and men in three dimensions: reproductive health, empowerment and the labor market"; and
- The multidimensional poverty index (MPI), which incorporates three critical dimensions of human development: (1) health (nutrition and child mortality); (2) education (years of schooling and children enrolled), and (3) living conditions (availability of cooking fuel, toilets, water, and electricity; floor type; and assets).

Values of the IHDI, GII, and MPI have been determined for all 187 countries in the 2011 and 2013 UNDP/HDR reports. Detailed calculations of these indices can be found in the technical appendixes of the 2011 report.

An index proposed by the Grameen Foundation called the progress out of poverty index (PPI) aims at measuring the poverty level at the household level in specific countries (Grameen Foundation 2014). For each country, a set of 10 questions is created to determine whether households live below the poverty line for that specific country. The scorecard at the household level is developed from existing national household survey data and "captures a snapshot of poverty levels" that can be monitored over time.

Another alternative index initially proposed by the country of Bhutan is the gross national happiness (GNH) index (Ura et al. 2012). Its four pillars are promotion of sustainable development, preservation and promotion of cultural values, conservation of the natural environment, and the establishment of good governance. Like the GPI, the GNH is about wellness and well-being rather than economic growth and consumption. Since it was first proposed in 1974, the GNH has been the topic of various studies and is an integral part of the model of development of the country of Bhutan's vision for 2020 (Bhutan 1999). Using well-being and happiness as indicators of development (instead of GDP or GNP), the GNH has been a subject of recent discussion at the Rio+20 conference (Helliwell et al. 2012) as an alternate measure of development that goes well beyond the economic growth paradigm.

At the Rio+20 meeting, the International Human Dimensions Programme on Global Environmental Change (IHDP 2012) released a report that analyzes the so-called "inclusive wealth" of 20 nations. The ranking of nations is based on three types of assets: physical capital (e.g., infrastructure and machinery); human capital (measured in terms of skills and education); and natural capital (lands, forests, fossil fuels, and minerals).

Despite proposed alternatives to the GDP, GNP, and GNI indices and various criticisms by some about their relevance to development, these indices still remain the dominant ones used in global policy circles when ranking countries based on their level of development. Unfortunately, the old paradigm of economic growth as the only measure of economic success is still deeply rooted in the minds of policy makers, economists, and development agencies. But as the "World Happiness Report" (Helliwell et al. 2012) indicates, a consensus seems to exist that alternatives to these purely economic development indices need to be considered and that "we need to re-think the economic sources of well-being" in rich and poor countries alike.

Community Development

Let's look more specifically at what comprises the development of a community. The term "community" is seen for the purpose of this discussion as an assembly of interacting households and individuals with a mutual sense of belonging and common interests (see more discussion in Chapter 5). In general, community development is a process with unique characteristics. Craig (2004, 2007) reported, for instance, the recommendations from the 2004 Budapest Conference on "building civil society in Europe though community development." The conference participants called for bottom-up empowerment of communities rooted in giving people a voice and a choice in deciding their destiny through access of knowledge and resources. They acknowledged that community development is about

- *Strengthening civil society* by prioritizing the actions of communities and their perspectives in the development of social, economic, and environmental policy;
- *Empowering local communities*, communities of interest or identity, and communities organizing around specific themes or policy initiatives;
- *Strengthening the capacity* of people as active citizens through their community groups, organizations, and networks and the capacity of institutions and agencies to have dialogues with citizens to shape and determine change in their communities; and
- *Supporting active democratic life* by promoting the autonomous voice of disadvantaged and vulnerable communities.

Community development is often synonymous with *community-based development* or *community-driven development* and is closely linked to the concepts of decentralization and participation (Mansouri and Rao 2012). Such a concept is not new; it was initially shown as an alternative to top-down development in the 1950s and early 1960s by the UN and USAID. This concept was inspired by the humanistic concepts of the village self-rule (Swaraj) movement in India (Gandhi 1962) and Liberation Theology in South America (Freire 2007). It was subsequently abandoned in the mid-to-late 1960s for more modern and larger forms of development, to be revitalized again starting in the early 1990s (Mansouri and Rao 2012) with the concept of development as transformation and freedom (Sen 1999; Easterly 2007) and the rights-based approach to development (CARE 2001).

Since the 1960s, there has been ample discussion about what community development should be compared to development solely based on economic growth and income level and dictated from the top down and/or by outsiders. The discussion has moved from economic growth toward understanding the conditions necessary for people to address their basic human needs and to realize their basic human rights. As remarked by Mansouri and Rao (2012), "community development supports efforts to bring village, urban neighborhoods, or other household groupings into the process of managing development resources without relying on formally constituted governments."

In general, community development can take multiple forms, some more successful than others. For most outsiders in rich countries, it still remains an ideology of archaic political and economic concepts perpetuated by experts from academia, government and nongovernment agencies, and international organizations. Community development is still based predominantly on economic growth, reliance on markets, transformation from the top down, and special interests that benefit a few. Finally, it is based on the misconceived belief that poor countries are not in a position to use aid effectively and cannot develop themselves (UNDP/HDR 2005), thus perpetuating a culture of dependency on assistance from outsiders.

Another misconception of community development is that it is about charity. Very often, well-meaning charity groups use some form of intervention based on technology (e.g., pumps or photovoltaic systems) to alleviate a perceived crisis. It often stops there and success is proclaimed by the aid organizations once the technology is in place. Such a haphazard approach has not been very successful over the past 50 years. It has been characterized by an absence of quality control procedures, accountability, and community participation. The best way to characterize this charity approach is to use a metaphor: the "blind" (outsiders) leading the "blind" (community). The developing world is littered with failed technologies contributed in part by the charity model of community development. A good illustrative case study of failed charity solutions is described in the paper entitled "The Stranger's Eyes" by Carlson (1995).

Finally, community development is often misconceived as being just about technology. Simply providing technology to a community is not enough because development can rarely, if ever, be limited to technical dimensions. This approach fails to recognize that development is multidisciplinary and cross-disciplinary. Technical and nontechnical issues interact, and various stakeholders are involved. A purely technical solution to a community problem may actually do more harm than good if it fails to account for the context in which it will be used.

After discussing what community development is not, what would the characteristics of "authentic" community development be? It encompasses many of the following components:

- Change in a dynamic, complex, and multidisciplinary environment;
- Participation and integration of various disciplines and stakeholders;
- Beneficiaries in the driver's seat defining what development is and is not;
- Justice, equity, equality (social power, income, wealth, opportunity) and human rights protection;
- Freedom to find meaningful solutions;
- Common ground between bottom-up and top-down approaches; and
- Strategies that empower not only the community but also the individual, the private sector, the state (and the public sector), and the household (Stiglitz 1998).

Others see more specific tasks in community development, such as

- Creating partnerships to identify the solutions (technical and nontechnical) that best match the community development level and the community capacity;
- Bringing the community to a higher level of development (education) while spearheading social entrepreneurship, infrastructure, health, and economic growth through capacity building; and at the same time ensuring the respect of human rights (right-based approach); and
- Developing solutions that link technology–health–education–poverty–gender–security–policy–governance and development.

As emphasized by David Silver (personal communication, 2011) with ideas from Korten (1981), community development can also be seen as *transformation* at four interdependent levels: personal, household, community, and community of life.

1. *Personal transformation* is best illustrated by two insights: one by Dr. Ariyaratne, president of the Sarvodaya Shramadana movement in Sri Lanka, who reminds us that in development projects, we do not come to develop others, but ourselves. Likewise, Lilla Watson, a Brisbane-based Indigenous educator and activist in Australia relates transformation to liberation: "If you come here to help me, you are wasting your time … but if you have come here because your liberation is bound up in mine, then let's work together."

2. *Household transformation* is about ensuring that basic human needs at the household level are met. This goal is accomplished when the needs are identified by the insiders and not imposed by the outsiders.

3. *Community transformation* is about respecting the wealth of communities, thus respecting their cultural heritage, diversity, and inclusivity and building self-reliance.
4. *Community of life transformation* is about adopting a seventh-generation approach toward the population and the environment, thus illustrating the sustainable nature of the proposed solutions.

It is noteworthy that the aforementioned humanistic and grassroots components of community development are *not* incompatible with economic growth. For instance, Chang (2010) calls for development to return to its traditional definition of transformation of productive structure and capabilities, while *at the same time* paying greater attention to human development, politics, technological development, institutions, sustainability, and the environment than was done 50 years ago. A new and more holistic mindset is therefore needed that breaks away from (1) the traditional and outdated categorization of the First and Third Worlds (Zoellick 2010) and (2) the belief that development as growth and development as a purely humanitarian endeavor are separate and cannot be reconciled.

2.5 Sustainability and Sustainable Development

Rationale for a New Mindset

Lifting billions of people out of poverty through a hybrid model of industrialization with humanistic principles calls forth the development of innovative strategies and a new mindset using available and limited resources more wisely than ever before. It also calls forth a new model of interaction between a population and the carrying capacity of the environment on which it depends.

The rationale for embracing sustainability and sustainable development can easily be made within the context of *both* the developed world and the developing world. In the developed world, the challenge is to consume less and more intelligently while being more efficient and respectful of natural and human systems. In the developing world, the challenge is to ensure that proposed economic solutions address the basic needs of the people and are good for the environment and the people without duplicating the mistakes made by the developed world over the past 50 years. After all, nothing prevents the developing world from leapfrogging from survival to healthy market economies without repeating those mistakes, while "constructing new development pathways that place much less strain on the global environment" (Ewing et al. 2010). This goal obviously assumes that the developing world has an opportunity to do so, is given the "right to develop," and that conditions are in place for development to take root, such as equity, rule of law, and democracy.

Integrating the concepts of sustainability and sustainable development into development work may represent the mindset necessary for taking that shortcut (UNDP/HDR 2011). This mindset is particularly critical because vast populations in today's emerging markets have a large combined purchasing power. As noted by Winter (2011), "by 2030, the economies of Brazil, Russia, India and China (BRIC countries) are expected to account for 41% of the world's market capitalization, up from just 18% today. As a result, there will be around a billion new middle class consumers demanding products to meet their specific needs." The fate of our planet depends a great deal on the behavior of those new customers, along with the few billion people who will follow in the forthcoming centuries (UNDP/HDR 2013).

Looking at past models of development that have led to the growth of Western economies since World War II, we clearly cannot legitimize using the same approach of heavy consumption of resources combined with high waste production when trying to lift billions of people out of poverty through the development of infrastructure and energy and market systems that are signatures of a global economy. Simply put, a new approach to industrialization is needed. More specifically, when it comes to future technological solutions for the entire planet, the population cannot afford to perpetuate decisions that

- Substantially or totally deplete natural resources;
- Eliminate options for the future of natural and human systems;
- Create inequalities among people and divide them;
- Escalate costs to prohibitive levels that all cannot afford; and
- Increase the probability of catastrophic future disasters, either natural or technological.

Current Production–Consumption Model

Our current method of development in the so-called developed world is not sustainable in the long term. Of great concern are the unintended consequences that are related to the technology inherent in the current method of development and the fact that there are always negative effects of any technology, whether new or a quick fix to an existing one (Huesemann and Huesemann 2011). A goal is therefore to reduce the impact of those effects. Furthermore, regardless of how innovative technology may be, it cannot escape the constraining nature of basic laws of conservation of mass and the two laws of thermodynamics.

Figure 2-5 shows, in a nutshell, a schematic of the production–consumption model that has been the driving force throughout the industrial revolution. Such a *cradle-to-grave* (take–make–waste) model has contributed to the making of a technical wonder world but has also contributed to the making of a technical waste world (Berry 1990). It is resource intensive, generates a lot of waste, and uses essential resources and ecological carrying capacity faster than they can be

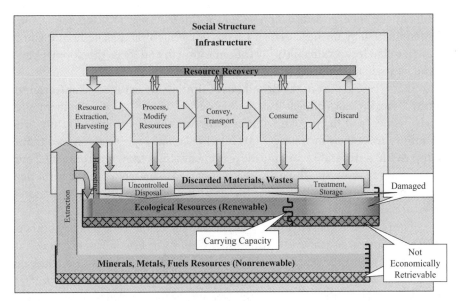

Figure 2-5. Cradle-to-Grave Production Consumption Model: Take–Make–Waste

Source: Wallace (2005), with permission from ACEC.

replaced or replenished. It also (wrongly) assumes that there will be an unlimited source of raw materials and energy available forever and that technology will always come up with appropriate solutions.

Exporting this dysfunctional production–consumption model of economic growth to millions of people in the developing world would jeopardize their growth by trapping them in a downward spiral of ecological and economic decline and dysfunctional social consequences.

In summary, the question is, can we develop a mindset and strategies to (1) stay within the limits of our available resources and carrying capacities and (2) avoid (or minimize) irreversible long-term negative environmental, economic, and social consequences in future production–consumption models that could affect developed and developing countries alike? The limitation and effect of traditional production–consumption systems and the need to adopt a new approach to development was recognized as early as 1980.

The Sustainability Movement

In 1983, the World Commission on Environment and Development (WCED) under the leadership of Gro Harlem Brundtland, then the prime minister of Norway, was asked by the General Assembly of the United Nations to develop "a global agenda for change." The work of the commission was to build on the results of the Independent Commission on International Development Issues, chaired in 1980 by former German Chancellor Willy Brandt. The Brandt Commission called for sustainability in economic growth and social development and in the natural environment (Brandt 1980).

In the 1987 report of the Brundtland Commission, entitled "Our Common Future," the WCED defined "sustainable development" as "development that meets the needs of the present without compromising the ability of future generations to meet their own needs" (WCED 1987). The report emphasized that it is impossible to separate economic development issues from environmental and social issues because the three are interconnected. The report called for a new era of international economic growth, "growth that is forceful and at the same time socially and environmentally sustainable" (WCED 1987).

The recommendation of the WCED led to the Conference on Environment and Development organized by the United Nations in Rio de Janeiro, Brazil, in June 1992. One of the results of the so-called "Earth Summit" was a 600-page summary of 2,500 sustainability issues and recommended solutions, also known as Agenda 21 (UNCED 1992), and the creation of the Commission on Sustainable Development (CSD). The 10 years that followed the Earth Summit focused mostly on developing policy on sustainable development, with limited concrete results because of a lack of political will (Banuri and Najam 2002).

Ten years after the Rio Summit, world leaders met again in Johannesburg in September 2002, a meeting with limited outcome. The world concluded that little progress had been made since 1992, especially in the developing world. It reaffirmed that sustainable development remains a central element of the international agenda and gave new impetus to global action to fight poverty and protect the environment. This idea was echoed by the World Bank in its 2003 World Development Report on balancing transforming institutions, growth, and quality of life in sustainable development for the entire planet.

The 20-year anniversary of the Rio meeting, called the Rio+20, was held in Rio de Janeiro in June 2012. It presented a unique platform for humanity to refresh its commitment to making the world a better place for *all* by merging three powerful initiatives: the sustainable development initiative that originated in 1992; the human development initiative best represented by the Millennium Development Initiative launched in 2000; and the Earth Charter, also introduced in 2000, focusing on the ethical principles necessary to foster sustainable development. Knowledge platforms on sustainability and sustainable development have been developed by multilateral agencies such as the United Nations (2013b) and the World Bank, bilateral in-country agencies (e.g., USAID and U.K. Department for International Development), and other organizations.

It is clear that since its inception in the Western world, sustainable development has been seen as an encouraging and promising solution to the problems created by humans on planet Earth. Over the past 20 years, sustainability has been recognized as the best and most logical way for people to meet their needs while simultaneously nurturing and restoring the environment. It has been embraced, at least *in theory*, by all sectors of the economy—public, private, academia, and government—and has been presented as a radical new platform of opportunities for research, education, technology, and business in the developed world. It has also been acknowledged that actions need to be undertaken worldwide to accelerate progress in a transition toward sustainability in the near future.

Despite that enthusiasm, the sustainability movement has been slow in grounding itself in the real world and has sometimes been a source of controversy in some political and economic circles. It is still perceived today in many different ways by various political and economic groups as (1) a buzzword and academic virtual concept, (2) an environmental luxury for rich and established communities in the developed world, (3) a threat to rapid economic growth in some emerging markets, and/or (4) a necessary evil in some parts of the world. Furthermore, few metrics and indicators of sustainability have been proposed and implemented. For developed countries, sustainable development is still perceived as an option. The problem, however, is that for poor countries, sustainable development is synonymous with long-term survival. It is therefore imperative that sustainable development and human development be addressed simultaneously. As remarked in the Rio+20 meeting, a need exists to develop a concrete and realistic action plan that builds upon the promising development trends observed over the past 20 years.

Four practical questions arise with regard to sustainability and sustainable development: (1) Do we have a clear definition of these two concepts? (2) What represents a sustainable system, structure, or community? (3) Can sustainability and sustainable development ever be quantified, monitored, and evaluated? (4) What is sustainable development within the context of the developing world?

Components of Sustainability

According to dictionary.com, the word "sustainability" comes from "sustain" derived from the Latin word *sustenere,* which means to uphold and prolong, to keep in existence, to endure and withstand. Many definitions for sustainability have been proposed in literature. The next question is *what* do we want to keep, prolong, sustain, or withstand? The answer to that question is obvious: the preservation of activities from which humans can derive their sense of well-being, i.e.,

- The natural environment (air, water, land, biota);
- The human race and its civil fabric (family, individuals, and communities);

- The built environment (facilities and infrastructure systems);
- The systems of production (goods, products, and services); and
- The different forms of capital (resources and knowledge).

Next, it is important to ask over what physical scale (where) and for how long (when) we want to preserve those activities. We can consider two aspects of the scale of sustainability: the spatial scale of sustainability (site, local, state, regional, national, or global footprint) and the temporal scale of sustainability (e.g., today, 1 year, 1–5 years, or 5–10 years).

It is generally agreed upon that sustainability is characterized by harmonizing three basic elements: people, planet, profit (the three Ps) or equity, environment, economics (the three Es). This is often referred to as the *triple-bottom line,* which is often represented as three overlapping circles, one circle for each of the three Es or Ps. The triple-bottom line acknowledges the intimate interaction and balance that exist between society (the anthrosphere); the environment (biosphere, lithosphere, atmosphere, and hydrosphere); and economic and financial (capital and production) systems. The triple-bottom line is captured in the following American Society of Civil Engineers Board approved definition of sustainability:

> *as a set of economic, environmental and social conditions in which all of society has the capacity and opportunity to maintain and improve its quality of life indefinitely, without degrading the quantity, quality or the availability of natural, economic and social resources.* (ASCE 2013)

A comprehensive definition of sustainability, which requires special attention, was proposed by Ben-Eli, where sustainability is seen as an organizing principle, i.e.,

> *A dynamic equilibrium in the processes of interaction between a population and the carrying capacity of an environment such that the population develops to express its full potential without adversely and irreversibly affecting the carrying capacity of the environment upon which it depends.* (Ben-Eli 2011)

Core principles of sustainability were proposed by Ben-Eli in five fundamental domains.

1. The *material domain* (flow of energy and materials) should "contain entropy and ensure that the flow of resources through and within the economy is as nearly non-declining as permitted by physical laws."

2. The *economic domain* (creating and managing wealth) should "adopt an appropriate accounting system, fully aligned with the planet's ecological processes and reflecting true comprehensive biospheric pricing to guide the economy."

3. The *life domain* (biosphere) should "ensure that the essential diversity of all forms of life in the biosphere is maintained."

4. The *social domain* (social interactions) should "maximize degrees of freedom and potential self-realization of all humans without any individual or group adversely affecting others."

5. The *spiritual domain* (code of ethics) should "recognize the seamless dynamic continuum of mystery, wisdom, love, energy, and matter."

As noted by Ben-Eli (2012), these core principles can be summarized as: "contain entropy; account for externalities; maintain diversity; self-actualize benignly; and acknowledge the mystery." For each domain and associated core principle, a set of policy and operational implications have been proposed. The five domains are interrelated because of the systemic nature of the situation being analyzed, i.e., the interaction of a population and its environment.

Since the 1992 Rio de Janeiro Summit, sustainability tools have been proposed by various authors in the Western world to comprehend and model the dynamic of interaction that exists between humans and their environmental surroundings. These tools include the concepts of biomimicry (Benyus 1997), eco-efficiency (DeSimone and Popoff 2000), natural capitalism (Hawken et al. 1999), "natural step" (Nattrass and Altomare 1999), industrial ecology (Graedel and Allenby 2009), earth systems engineering and science (Allenby 2001; Steffen and Tyson 2001), and eco-effectiveness or cradle-to-cradle (McDonough and Braungart 2002). Figure 2-6 shows a cradle-to-cradle alternative to the production–consumption model in Figure 2-5. In this diagram, production–consumption is about increased use of renewable resources and materials, reduced disposal of waste, and increased resource recovery, and preventing reaching the carrying capacity of the natural resource pool. The production-consumption model in Figure 2-6, which more closely follows the regenerative characteristics of life on Earth, calls for a complete redesign of industrial processes and economies (external change) combined with a new mindset (internal change). As an example, Huesemann and Huesemann (2011) recommend discontinuing current industrial and economic practices and emphasize instead (1) sustainable energy generation, (2) use of sustainable materials, and (3) sustainable waste discharge. All three conditions must be respectful of the regenerative (or assimilative) capacity of ecosystems and must not cause environmental damage. Furthermore, a need exists to protect critical natural capital and nature's biochemical and geochemical processes, without which ecosystem services on which we depend on a daily basis would not be provided (e.g., sunlight, photosynthesis, and pollination). It should also be remembered that nature is a 4.7 billion year experiment in sustainability. If it is not sustainable by now, and/or we are not willing to learn from it, why are we still discussing what sustainability is?

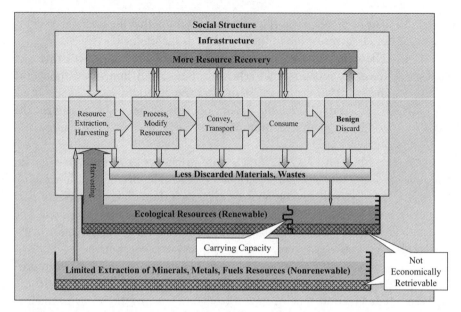

Figure 2-6. A Cradle-to-Cradle Model of Take–Make–Waste

Source: Adapted from Wallace (2005), with permission from ACEC.

The Ecological Footprint

The effect of a population on the environment and its carrying capacity can be measured using the concept of a footprint, as proposed by the Global Footprint Network (Wackernagel and Rees 1996; Chambers et al. 2002). The *ecological footprint* measures how much productive land and water area (biologically productive space) measured in hectares per person a population (an individual, a city, a country, or all of humanity) or an activity requires to produce all the resources it consumes and to absorb all the waste it generates (using prevailing technology). In many ways, it gives a measure of sustainability, i.e., the level of interaction between people and their environment that can be calculated for different physical scales: personal footprint, community footprint, national footprint.

The ecological footprint, which represents the demand on nature, can be compared to the available biologically productive space (the renewable biological capacity). If the hectares of a footprint exceed the hectares of biologically productive space, an ecological deficit is recorded. Such a situation is deemed unsustainable. Statistics on the ecological footprint of the Earth and 150 nations since 1961 have been proposed by the Global Footprint Network. These nations were divided into *biocapacity debtors* and *biocapacity creditors*, as shown in Figure 2-7.

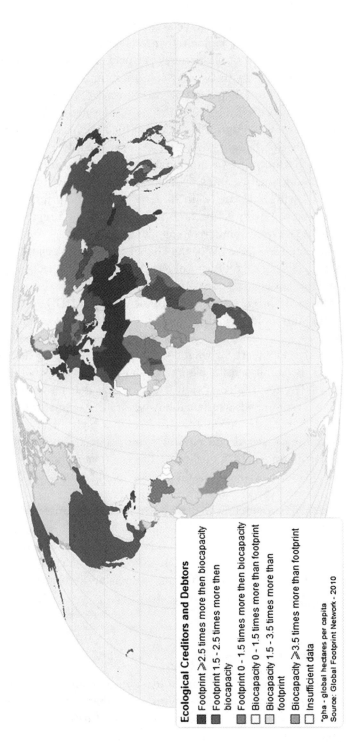

Figure 2-7. The Ecological Wealth of Nations

Note: Based on data from the Global Footprint Atlas (Global Footprint Network 2010).

Source: ChartsBin statistics collector team (2011) <http//chartsbin.com/view/1046> (Oct. 24, 2013); with permission from ChartsBin.

In 2005, for planet Earth as a whole, the average footprint was 2.7 hectares per person. When compared with a biologically productive area of 2.1 hectares per person, the result is a deficit of 0.6 hectares per person, which means that 1.3 planets are needed to support the world's consumption (2.7/2.1 = 1.3). In 2012, that number increased to 1.52 planets needed (Borucke et al. 2012) and is expected to reach 2.0 planets needed by 2030 and 2.5 planets needed by 2050. It is noteworthy that the average footprint is somewhat high because of the over-consumption of certain countries in the world. As remarked by Ewing et al. (2010), "half of the global footprint was attributable to just 10 countries in 2007, with the United States of America and China alone each respectively using 21 and 24 percent of the Earth's biocapacity." In 2002, the average footprint globally was 2.1 hectares per person, with 9.6 hectares per person in the United States compared with 1.0 hectare or less in developing nations.

Based on data gathered since 1980, the Global Footprint Network concluded that an average ecological footprint of less than 1.79 global hectares per person makes the resource demands of a country globally replicable (using 2008 values). When combined with a human development index (HDI) of 0.7 or larger, which is the threshold for high human development according to the UNDP, this gives us a benchmark, or so-called *sustainability quadrant* (sustainable human development) for what could represent successful sustainable development at the country scale. Figure 2-8 shows a plot of global footprints versus their HDIs for 151 countries. It also shows that few countries currently fall into the recommended sustainability quadrant. As remarked in the report by ISI (2012), "most developed countries enjoy a high quality of life but do so by consuming material and natural resources at a rate our planet cannot support." Furthermore, that quality of life is often accompanied by environmental degradation at a larger scale than in developing countries (UNDP/HDR 2011, 2013).

The challenge then becomes how to drive down wealthy countries' ecological footprint while retaining their high HDI values in addition to integrating sustainability and green concepts into new market economies' development as early as possible, a concept suggested in the report entitled "Inclusive Green Growth" by the World Bank (2012a). After all, sustainable development and economic growth are not mutually exclusive in principle.

Sustainable Development Projects

If we see sustainability as a desired end state of dynamic equilibrium between a population and the carrying capacity of its environment, then it can be reached through a *process* called sustainable development. Different definitions of that process have been proposed. Three definitions have been retained for the sake of

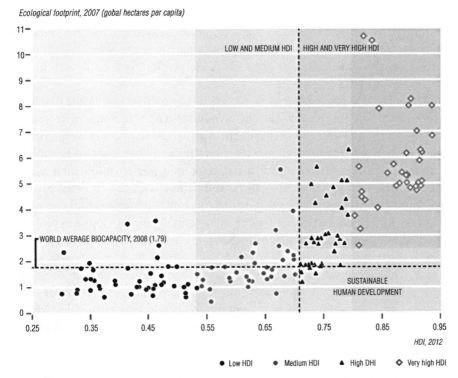

Figure 2-8. Ecological Footprints of 151 Countries vs. the Human Development Index (HDI)

Source: UNDP/HDR (2013), with permission from UNDP.

simplicity among many others (Hawken 1993; Hawken et al. 1999; Huesemann and Huesemann 2011).

- "Development that meets the needs of the present without compromising the ability of future generations to meet their own needs" (WCED 1987). This definition is clearly anthropocentric because it does not include ecosystems, at least not in an explicit way. Furthermore, as remarked by Ben-Eli (2012), the definition is weak in the sense that future generations do not participate in the current decision process.
- "Sustainable development is the challenge of meeting human needs for natural resources, industrial products, food, transportation, shelter, and waste management while conserving and protecting environmental quality and the natural resource base essential for future development" (ASCE 2013).
- "Improving the quality of life while living within the carrying capacity of supporting ecosystems" (Chambers et al. 2002).

In all three definitions, sustainable development can be seen as a process of finding a *balance* between social and economic development and environmental stewardship. These definitions focus on "what" is sustainable development, but not on the "how." So, how are sustainable development projects different from other projects?

To answer that question, we must first recognize that sustainable development projects not only need to be *done right* but also need to be the *right projects* for the community and the environment with which they interact (ISI 2012). The challenge is to integrate sustainable practices and criteria for sustainability in all project phases to account for social, economic, and environmental issues. These criteria and practices need to be compatible with the expected project performance. This goal requires a more holistic project management framework. Below is a list of global characteristics of "right" sustainable development projects (adapted from the characteristics of sustainable businesses proposed by Hawken [1993]). These projects must

- Take responsibility for their effects on the natural world by doing no harm and by not diminishing the diversity of the Earth's systems;
- Create structures and systems of durability and long-term utility whose ultimate use or disposition will not be harmful to current and future generations;
- Change the conversation by educating all stakeholders involved;
- Deliver efficient and resource-conserving solutions that reduce consumption, energy use, distribution costs, economic concentration, soil erosion, atmospheric pollution, and other forms of environmental damage;
- Consider what they take, make, and waste in designing solutions; and
- Deliver solutions that work in harmony with the assimilative and regenerative capacity of the Earth's systems.

Doing the right projects while doing the projects right requires a new mindset in the way projects are conducted. Some components of that new mindset are summarized in Table 2-7 and are compared with the old mindset. It should be noted that the new mindset encompasses more than the three pillars of sustainability, i.e., people, planet, and profit. It includes a fourth pillar of "global ethical and spiritual consciousness" (Clugston 2011).

The aforementioned recommendations apply to all projects, small or large, whether they are carried out in developed or developing countries. Another unique feature of these projects is that they require full participation of community stakeholders in decision making, project design, execution, operation, and maintenance. Finally, these projects take time and require patience.

2.6 Frameworks for Sustainability

Talking about sustainability is one thing; implementing sustainable development projects where the aforementioned attributes are included, at least partially, is

Table 2-7. Attributes of Old and New Mindsets in Sustainable Development

Old Mindset	New Mindset
• Linear	• Cyclical and systemic
• Earth made for humans	• Humans made for Earth
• Violent (brute force)	• Caring and restoring
• Control nature	• Emulate nature
• Short time frames	• Longer time frames
• Earth is a limitless sink for waste	• Earth is a finite sink for waste
• Technology is omnipotent	• Technology as a solution
• Doing well	• Doing well by doing good
• Extractive processes	• Renewable processes
• Waste as waste	• Waste as resources
• Externalize externalities	• Internalize externalities
• Benefits a few	• Benefits all
• Creates waste	• Creates value

another. Because it is usually easier to manage what we measure, a challenge in carrying out such projects is measuring sustainability using metrics and indicators. In project sustainability management, indicators help monitor how well sustainable practices are implemented and whether sustainability criteria are met. This monitoring also helps in developing remedial measures. It is recommended that such indicators be put in place as early as possible in project design and that those responsible for project management understand sustainability and are committed to making it work.

Indicators of project sustainability can be qualitative or quantitative. They need not only to measure the economic, social, and environmental components of sustainability individually but also be able to measure how those components interact. According to the Sustainable Measures company website (2013), effective sustainability indicators need to be (1) relevant to what is being addressed, (2) easy to understand by various stakeholders, (3) reliable and trustworthy, and (4) based on accessible and reliable data. Another way of looking at effective indicators is to use the acronym SMART, which is often referred to in the logical framework of development projects (see Chapter 8). According to the acronym, indicators need to be specific, measurable, achievable or attainable, relevant or realistic, and time-bound or time-limited.

In general, frameworks that include sustainability metrics at the project level in the developed world are limited. In the field of engineering and architecture, several frameworks for project sustainability management have been proposed. A good summary of them can be found in a paper by Talbot and Venkataraman (2011). They include the project sustainability management

(PSM) guidelines proposed by the International Federation of Consulting Engineers (FIDIC 2013), the Global Reporting Initiative (2013), and the Facility Reporting Project (FRP) of Ceres, a Boston-based sustainability advocacy organization, among others. Following is a more recent framework called the Envision Framework (ISI 2012), which has promising application for projects in the developed and developing world.

In the developing world, it should be noted that no such extensive frameworks exist. However, over the past 15 years, there has been ample discussion about what constitutes sustainable service delivery in the fields of rural water supply and sanitation, for instance. This discussion has been driven by the observation that in the developing world, many water supply and sanitation projects do not function well or at all, often shortly after installation. As we discuss further in Chapter 15, proxies other than "number of built and functioning systems" have been proposed to define the sustainability of such projects, but this is still a work in progress.

Envision Framework

The Envision Framework is being developed in collaboration between the Zofnass Program for Sustainable Infrastructure at Harvard University Graduate School of Design and the Institute for Sustainable Infrastructure. Its overall purpose is to evaluate and rate infrastructure projects over their entire life cycles. More specifically, the framework aims to improve the project's sustainable aspects from two points of view: "performance" (how to do projects right) and the "social, environmental, and economic perspective" (how to do right projects). The framework focuses on certain types of infrastructure in Western countries, such as "roads, pipelines, bridges, railways, airports, dams, levees, landfills, water treatment systems, and other civil infrastructure that make up the built environment.... It does not include buildings and facilities." The latest version of the framework (ISI 2012) focuses on assessment of the sustainable aspects of infrastructure projects in the planning and design phases only.

The current version of the framework introduces an infrastructure rating system consisting of five categories that cut across the three dimensions (people, planet, and profit) of sustainability. Taken directly from the ISI (2012) report, the categories include the following:

- *Quality of life* relates to a "project's impact on surrounding communities, from the health and well-being of individuals to the well-being of the larger social fabric as a whole. These impacts may be physical, economic, or social. Quality of Life particularly focuses on assessing whether infrastructure projects are in line with community goals, incorporated into existing community networks, and will benefit the community long-term. For that purpose, community involvement should be sought by infrastructure owners. Community members (both users and non-users) affected by the project should be considered important stakeholders in the decision-making process (during design as well as during operations)."

- *Leadership* relates to how project teams "communicate and collaborate early on, involve a wide variety of people in creating ideas for the project, and understand the long-term, holistic view of the project and its life cycle" and how "collaborative leadership produces a truly sustainable project that contributes positively to the world around it."

- *Resource allocation* refers to "the quantity, source, and characteristics of these resources and their impacts on the overall sustainability of the project. Resources addressed in this rating system include physical materials, both those that are consumed and that leave the project, energy for construction, operation, and maintenance, and water use. Each of these materials is finite in its source and should be treated as an asset to use respectfully."

- *Natural world* relates to how to "understand and minimize negative impacts while considering ways in which the infrastructure can interact with natural systems in a synergistic, positive way."

- *Climate and risks* refers to how to "minimize emissions that may contribute to increased short and long-term risks and to ensure infrastructure projects are resilient to short-term hazards or alter long-term future conditions."

Each category is divided into subcategories that are assigned a rating (credit) depending on a sustainability achievement level that the infrastructure project designers wish to attain: improved, enhanced, superior, conserving, or restorative. Each subcategory has intent and a definite metric. More detailed information can be found in the ISI (2012) report about the specifics in defining each subcategory: (1) the meaning of each level of achievement, (2) recommendations on how to advance to a higher level of achievement, (3) evaluation criteria that are used to determine the credit value, (4) explanation of sustainability issues and practices associated with each credit, and (5) related credits.

A unique aspect of the Envision Framework is that it is about engineering projects and their relation to the environment and people. But it is also about participation, integration of various seemingly independent disciplines, and innovation. The framework is about looking at infrastructure projects over time and allowing for changes in the projects themselves, the environment, and those affected by the infrastructure. Finally, many of the categories and subcategories in the Envision Framework apply to developed and developing communities alike, even though the rating values and their corresponding evaluation criteria in each category of achievement might be different.

2.7 Progress in Human Development

A key question arises as to whether the world is getting better or worse for populations in the developing world. The answer to that question is "it depends." First, it depends largely on who we ask, i.e., intellectuals in the developed world or individuals in developing communities who experience poverty on a daily basis. As we have seen earlier in this chapter in our discussion on poverty, because the latter are actually never consulted, most of the trends have been drawn from various studies sponsored by development agencies that tend to be biased toward reporting success in whatever types of projects they do.

Second, the answer also depends on the scale used: country, community, household, and individual. A success story at one scale does not always translate to another scale. Finally, the answer depends on what "getting better or worse" truly means and what are the indicators of measuring progress. For instance, Kenny (2011) noted that things are getting better as far as the quality of life for more people on our planet is concerned and that the improvement is not necessarily correlated with an increase in GDP per capita and income. Better quality of life results from access to improved goods and services and effective technologies at reduced prices and access to better education and communication systems. Of course, clearly income growth should not be ignored, but it is not an end in and of itself.

On *average*, a general consensus exists that the past 20 years have seen consistent gains in human development that started in the 1990s with increased life expectancy, better health, more access to education, and more democratic governance around the world. Global average life expectancy in 2000 was 66 years, compared with 31 years in 1900 (Kenny 2011). The 2011 UNDP/HDR report mentioned that the "world's average HDI increased 18% between 1990 and 2010 (41% since 1970)." The 2013 UNDP/HDR report identified "more than 40 developing countries that have done better than expected in human development in recent decades, with their progress accelerating markedly over the past ten years." The magazine *Foreign Policy* in its September–October 2010 issue (Kenny 2010) called the first 10 years of the 21st century "humanity's finest … even for the world's bottom billion." They concluded that over that 10-year period, "more people lived lives of greater freedom, security, longevity, and wealth than ever before."

We have also seen growing optimism as countries in Asia and South America that were classified as developing just five or 10 years ago now are fast moving emerging markets. Robert Zoellick (former President of the World Bank group) remarked in his April 14, 2010, lecture that we are seeing the end of what was known as the Third World, the same way that the fall of the Berlin wall in 1989 witnessed the end of the Second World.

In March 2012, more good news was published when the World Bank announced that its estimates for 2010 seemed to indicate that the overall percentage of people living on less than $1.25 per day was about half of what it was in 1990, thus meeting Target 1.A for Millennium Development Goal (MDG) 1, five years early (Chen and Ravallion 2012). This success can be attributed in part to the rapid economic growth observed in China since 2000, where the population that lived below $1.25 a day fell from 77% in the early 1980s to 14% in 2008. The 2012 World Bank announcement also praised a rapid positive change in Africa, Latin America, and Central Asia between 2005 and 2008 despite the economic crisis that affected the developed world. A recent World Bank research paper by Demombynes and Trommlerova (2012) reported optimistic news in a huge reduction in infant and under-five mortality rates across Africa. Likewise, Liu et al. (2012) reported that "between 2000 and 2010, the global burden of death in children younger than 5 years decreased by 2 million, of which pneumonia, measles, and diarrhea contributed the most to the overall reduction."

A report published in the spring of 2012 announced that the number of people without access to safe drinking water has been reduced by 50% since 1990 (WHO/UNICEF 2012). The report concluded that in 2010, 89% of the world's population was using an improved water source, compared with 76% in 1990. Over that period, 2 billion people gained access to improved water sources (see Table 15-1), 50% of them in India and China, still leaving 780 million without improved water sources today. Regarding access to sanitation, the 2012 report concluded that the MDG of reducing the number of people without access to sanitation in half will not be achieved despite the promising trend that between 1990 and 2010, 1.8 billion people gained access to improved sanitation facilities (40% of them in India and China), still leaving 2.5 billion people (more than 50% of them in India and China) without improved sanitation facilities today. It has been estimated that 63% of the world's population use improved sanitation facilities, up from 49% in 1990. In addition, the WHO/UNICEF report concluded that for the 59 countries analyzed, more users of improved sanitation are likely to use improved water sources than the other way around. Overall, it was found that half of the 59 countries' populations use both.

In the fall of 2012, the Food and Agriculture Organization (FAO) released its report entitled *State of Food Insecurity in the World*. The report showed that the number of hungry (undernourished) people had decreased from 1 billion (18.6% of the world's population) in 1990–1992 to 870 million people (12.5% of the world's population) in 2010–2012, with most of the improvement taking place before 2006. The report acknowledged that one in eight people in the world today is undernourished and that this number is still unacceptably high. It also showed considerable disparities in hunger statistics at the regional level. For instance, in Africa, the number of hungry people increased in sub-Saharan Africa and North Africa. In Asia, it increased in western and southern Asia. In comparison, large declines were observed in the number of undernourished people in Southeast Asia, East Asia, and South America.

Finally, in November 2012, the World Health Organization released positive statistics on the health-related MDGs (MDGs 4–7). They also noted that "while some countries have made impressive gains in achieving health-related targets, others are falling behind. Often the countries making the least progress are those affected by high levels of HIV/AIDS, economic hardship or conflict" (WHO 2012).

There is indeed reason to be enthusiastic about recent trends. At the same time, the world is still facing poverty and its consequences, meaning that additional global challenges lie ahead. As remarked by Zoellick (2010), "poverty remains and must be addressed. Failed states remain a major issue. Global challenges are intensifying and must be addressed." In the 2011 United Nations report on the MDGs, it was noted that

> Although many countries have demonstrated that progress is possible, efforts need to be intensified. They must also target the hardest to reach: the poorest of the poor and those disadvantaged because of their sex, age, ethnicity or disability. Disparities in progress between urban and rural areas remain daunting. (UN 2011)

The World Bank update by Chen and Ravallion (2012) also showed that the progress in poverty reduction was mostly for the poorest of the poor. This is not the case for those living between \$1.25 and \$2 a day, whose number increased from 648 million in 1981 to 1.18 billion in 2008, who are still very vulnerable, with limited opportunities to climb further out of poverty. On a global scale, things may look better for the developing world but with success that is not equally distributed. Even at the local level, the success stories are uneven and vary from country to country and even within regions of a given country. In some cases such as sub-Saharan Africa, regress in human development has been observed, especially in the area of health (UNDP/HDR 2011).

Various statistics have been proposed by organization such as the United Nations (UN 2013c) and the U.S. Central Intelligence Agency (CIA 2013). There are still 18 countries in the world (out of 194) with a life expectancy of 50 or lower, the same as it was in the United States and the Western world in the early part of the 20th century. Life expectancy varies between a high of more than 80 years in Japan and some countries in Europe to a low of about 40-plus in many countries of Africa and in Afghanistan, with world average of about 67 years. Even in wealthier countries, huge disparities exist in life expectancy from one region to the next. In the United States, for instance, life expectancy in different counties in 2007 ranged from 65.9 to 81.1 years for men and 73.5 to 86.0 years for women (Kulkarni et al. 2011).

Similar regional variations were reported by the WHO/UNICEF (2013) report regarding water and sanitation as shown in Figures 2-9 and 2-10, respectively. Additional disparities have been found between rural and urban areas (and within parts of urban areas based on residence and wealth), with more disparity for sanitation services than for drinking water supply. The situation is even grimmer in the 48 countries designated as least developed by the UN. It has been estimated that both sub-Saharan Africa and Oceania will not be able to meet the MDG drinking water target.

Today, extreme poverty is still a main issue in many parts of the developing world, and even in poor enclaves in the developed world. It is a disheartening reality for the world's young people, who make up half of the world's unemployed (WDR 2007). Those at the bottom of the income pyramid are still vulnerable to crises, and in many cases, crises that are well beyond their control. On any given day, it has been estimated (Andrew Yager, personal communication, 2012) that about 25,000–30,000 people (about 15% of all daily deaths) in the world die for reasons that are purely preventable, including 5,000 from indoor air pollution, 5,000 from poor water and sanitation, 5,000 from malaria, more than 5,000 from tuberculosis, and more than 5,000 from HIV/AIDS. This fact translates to roughly 9 million people who die per year, or about 200,000 people who die every week, about the same number estimated to have died in the 2010 Haiti earthquake or the 2004 tsunami. As noted by Kenny (2011), "every three seconds, a child dies before the age of five in the developing world."

In summary, paraphrasing Charles Dickens, on our planet today it is "the best of times and the worst of times," all at the same time. The answer to the question of whether the world is getting better or worse depends on who is answering, who is asking the question, and who is serving as a proxy for the voices of the poor. Simply put, we have the resources and the know-how to address the basic needs of all people on our planet. Many constraints are holding back progress in all areas (e.g., WASH, education, and energy). As a matter of fact, these constraints are not technical but rather political, institutional, and financial.

2.8 Concluding Remarks

This chapter presented the foundation for why it is imperative to focus on sustainable community development in both the developed and developing world and why it is necessary and urgent for emerging markets and nations to embrace principles of sustainable development while climbing the growth ladder. It also emphasized the importance of investing in communities and that "people are the real wealth of a nation," as remarked in the 1990 UNDP/HDR report.

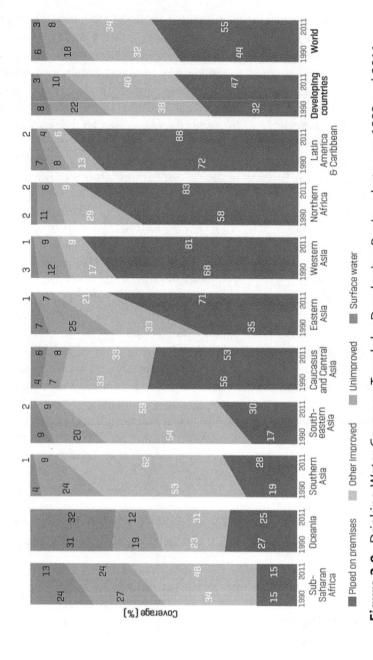

Figure 2-9. Drinking Water Coverage Trends by Developing Regions between 1990 and 2011

Source: WHO/UNICEF (2013), with permission from UNICEF.

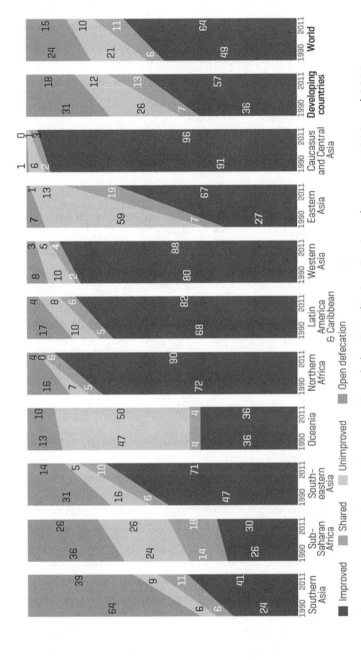

Figure 2-10. Sanitation Coverage Trends by Developing Regions between 1990 and 2011

Source: WHO/UNICEF (2013), with permission from UNICEF.

Development is about the right of people to develop their full potential and to lead productive, creative, and valuable lives in accord with their needs and interests. Development starts at the household level and expands to communities, regions, and eventually entire countries. The overarching goal of sustainable community development is to build communities with the potential to become more stable, equitable, secure, prosperous, and above all, more resilient and healthy over time. In turn, such attributes translate into more opportunities for economic development. For a community to express its full potential, the dynamic equilibrium between the community and its environment needs to be nurtured, i.e., the community needs to be able to operate sustainably from a socioeconomic and environmental point of view.

Based on the principles detailed in this chapter, key characteristics and attributes of sustainable communities can be outlined. They

- Allow *all* members of the community to enjoy a quality of life and well-being where basic human needs, freedoms, rights, and meaningful work are fulfilled in a safe and secure environment;
- Have equitable access to resources and knowledge, thus being capable of sustaining themselves economically, socially, and environmentally;
- Are places where individuals and households have the opportunity to express their full potential without adversely and irreversibly affecting the carrying capacity of the environment upon which they depend;
- Are parts of a system where rule of law and good governance are the norm; and
- Ensure sustainable livelihood opportunities for future generations.

When successfully combined, these five key characteristics contribute to an overall increased level of livelihood, security, and well-being in the basic economic and social units that form the community, i.e., the households, as discussed earlier in this chapter. Because "good livelihoods make good communities" (Banuri and Najam 2002), an increased level of livelihood security and well-being can be seen as one possible indicator of progress in sustainable community development. By being more secure, communities are less vulnerable to adverse events: internal or external, small or large, routine or exceptional, natural or nonnatural, isolated or interrelated. As a result, communities possess more (1) *inherent* (or coping) resilience in their ability to withstand the effect of adverse events and (2) *adaptive* resilience in their ability to adapt to the consequences of those events and recover from their effects. Both definitions of resilience were adapted from those used in the hazard and disaster resilience literature by Paton and Johnston (2006) and Tierney (2009).

In turn, an increased level of household livelihood security and well-being, combined with good governance and economic development, results in more peaceful and stable communities, which translate into more stabilized nations and the reduced likelihood of failed nations. This dynamic is captured in the overall framework for stabilization and reconstruction of countries proposed by the U.S. Institute of Peace (USIP 2009). The framework is based on five interconnected end states:

- *Safe and secure environment* where people have the ability to carry out their daily lives without fear of violence;
- *Rule of law* where people have the ability to access just law and a legitimate justice system based on accountability, human rights protection, safety, and security;
- *Stable governance* where people have the ability to "share, access or compete for power through non-violent political processes and to enjoy the collective benefits and services of the state" (USIP 2009);
- *Sustainable economy* where people have access to opportunities for their livelihoods; and
- *Social well-being* where people have basic needs addressed and can "coexist peacefully in communities with opportunities for advancement" (USIP 2009).

These end states illustrate that sustainable human development is a multifaceted process whose outcome is the realization of full socioeconomic and human potential for all. That process can only occur when minimum standards safeguarding human dignity are in place and policies to ensure the adoption and preservation of human rights (social, economic, and cultural) are adopted, promoted, and enforced through good governance. As remarked by Abdellatif (2003), good governance can mean different things to different agencies and people. It can be seen as "a process by which power [authority] is exercised" in the management of a country's affairs. Since the 1948 Universal Declaration of Human Rights (UN 1948), the international community has promoted the links among human rights, governance, and sustainable human development through a series of declarations, such as the Declaration on the Right to Development, the Millennium Declaration, and others (OHCHR 2007).

Finally, it should be remembered that sustainable community development is a process that requires community participation and empowerment. It can only be defined by the beneficiaries based on a vision *they* have for their community, or it will not work. It may benefit from the input of outsiders, but ultimately the community is the driving force in guaranteeing a quality of life for its households and the members of those households over time. As mentioned earlier in this chapter, people need to be given a chance to develop themselves and the resources and skills to do so.

References

Abdellatif, A. M. (2003). "Good governance and its relationship to democracy and economic development." *Global forum III on fighting corruption and safeguarding integrity*, Seoul. <http://www.arab-hdr.org/resources/publications.aspx?tid= 981> (May 12, 2012).

Adam, E. A., and Agib, A. R. A. (2001). *Compressed stabilized earth block manufacture in Sudan*, UNESCO, Paris. <http://unesdoc.unesco.org/images/0012/001282/128236e.pdf> (Sept. 13, 2012).

Allenby, B. (2001). "Earth systems engineering and management." *IEEE Technology and Society Magazine*, Winter 2000/2001, 10–24.

Altaf, S. (2011). "The end of population growth." *Nation of Change*, Oct. 31. <http://www.nationofchange.org/end-population-growth-1320069591> (Dec. 20, 2012).

American Society of Civil Engineers (ASCE). (2013). "The role of the civil engineer in sustainable development." ASCE Policy Statement 418. <http://www.asce.org/uploadedFiles/Government_Relations/State_Government_Relations/Sustainability%20State%20Issue%20Brief.pdf> (Feb. 10, 2014).

Arrow, K., et al. (1995). "Economic growth, carrying capacity, and the environment." *Ecological Economics*, 15(2), 91–95.

Asian Development Bank (ADB). (2012). *Development effectiveness review: 2011 report*, Asian Development Bank, Manila, Philippines. <http://www.adb.org/sites/default/files/defr-2011.pdf> (March 15, 2013).

Baker, L. (1999). "Real wealth: Use of gross domestic product figures as economic indicator may no longer be valid." *The Environmental Magazine*, May 1, <http://www.thefreelibrary.com/realwealth.-a054623303> (June 10, 2013).

Banuri, T., and Najam, A. (2002). *Civic entrepreneurship: A civil society perspective on sustainable development*, Vol. 1. Gandhara Academy Press, Islamabad, Pakistan.

Barnes, D. F., van der Plas, R., and Floor, W. (1997). "Tackling the rural energy problem in developing countries." *Finance and Development*, 34(2), 11–15.

Ben-Eli, M. (2011). "The five core principles of sustainability." <http://www.sustainabilitylabs.org/page/sustainability-five-core-principles> (Jan. 10, 2013).

Ben-Eli, M. (2012). "The cybernetics of sustainability: Definition and underlying principles." *Enough for all forever: A handbook for learning about sustainability*, J. Murray, G. Cawthorne, C. Dey, and C. Andrew, eds., Common Ground Publishers, Champaign, IL.

Benyus, J. (1997). *Biomimicry, innovation inspired by nature*, Quill, William Morrow, New York.

Berry, T. (1990). *The dream of the Earth*, Sierra Club Books, San Francisco.

Bhutan. (1999). *Bhutan 2020: A vision of peace, prosperity and happiness*, Planning Commission, Royal Government of Bhutan, Thimphu.

Birley, M. (2011). *Health impact assessment: Principles and practice*, Earthscan, London.

Bista, D. B. (2001). *Fatalism and development*, Orient Longman Ltd., Hyderabad, India.

Bornstein, D. (2005). "So you want to change the world." The Hart House Lectures Series, Toronto, Canada. <http://davidbornstein.files.wordpress.com/2008/12/hart-house-lecture-final1.pdf> (Sept. 10, 2013).

Bornstein, D. (2007). *How to change the world*, Oxford University Press, New York.

Borucke, M., et al. (2012). *The national footprint account, 2011 edition*, Global Footprint Network, Oakland, CA. <http://www.footprintnetwork.org/images/uploads/NFA_2011_Edition.pdf> (Jan 24, 2013).

Brandt, W. (1980). *North-South: A program for survival*, The MIT Press, Cambridge, MA.

Caldwell, R. (2002). *Project design handbook*, Cooperative for Assistance and Relief Everywhere (CARE), Atlanta.

Carlson, J. (1995). "The stranger's eyes." *Notes on Anthropology and Intercultural Community Work*, 20, 34–38.

Central Intelligence Agency (CIA). (2013). "The world factbook." <https://www.cia.gov/library/publications/the-world-factbook/> (Oct. 1, 2013).

Chambers, N., Simmons, C., and Wackernagel, M. (2002). *Sharing nature's interest*, Earthscan Publications Ltd., London.

Chambers, R. (1983). *Rural development: Putting the last first*, Pearson Prentice Hall, London.

Chang, H.-J. (2010). "Hamlet without the prince of Denmark: How development has disappeared from today's development's discourse." *Towards new developmentalism: Market as means rather than master*, S. Khan and J. Christiansen, eds., Routledge, Abingdon, U.K.

ChartsBin statistics collector team. (2011). "Ecological creditors and debtors." *ChartsBin.com*, <http://chartsbin.com/view/1046> (Oct. 16, 2013).

Chen, S., and Ravallion, M. (2012). "An update to the World Bank's estimates of consumption poverty in the developing world." <http://siteresources.worldbank.org/intpovcalnet/resources/global_poverty_update_2012_02-29-12.pdf> (Feb. 12, 2013).

Clugston, R. (2011). "Ethical framework for a sustainable world: Earth Charter plus 10 conference and follow-up." *J. Education for Sustainable Development*, 5(2), 173–176.

Cobb, C., Goodman, G. S., and Wackernagel, M. (1999). *Why bigger isn't better: The genuine progress indicator—1999 update*, Redefining Progress, San Francisco. <http://rprogress.org/publications/1999/gpi1999.pdf> (March 10, 2012).

Cobb, C., Halstead, T., and Rowe, J. (1995). "If the GDP is up, why is America down?" *The Atlantic*, October, <http://www.theatlantic.com/past/politics/ecbig/gdp.htm> (Oct. 5, 2011).

Collier, P. (2007). *The bottom billion: Why the poorest countries are failing and what can be done about it*, Oxford University Press, New York.

Cooperative for Assistance and Relief Everywhere (CARE). (2001). *Benefits–harms handbook*, Cooperative for Assistance and Relief Everywhere, Atlanta. <http://www.care.org/getinvolved/advocacy/policypapers/handbook.pdf> (Oct. 25, 2011).

Craig, G. (2004). "The Budapest declaration: Building European civil society through community development." *Community Development J.*, 39(4), 423–429.

Craig, G. (2007). "Community capacity building: Something old, something new?" *Critical Social Policy*, 27, 335–359.

Cuny, F. (1983). *Disasters and development*, Oxford University Press, New York.

Demombynes, G., and Trommlerova, S. K. (2012). "What has driven the decline of infant mortality in Kenya." Research paper 6057. World Bank, Africa Region, Washington, DC. <https://openknowledge.worldbank.org/bitstream/handle/10986/6580/WPS6057.pdf?sequence=1> (Feb. 22, 2013).

DeNavas-Walt, C., Proctor, B. D., and Smith, J. C. (2012). *Income, poverty and health insurance coverage in the United States 2011*, U.S. Department of Commerce, Washington, DC. <http://www.census.gov/prod/2012pubs/p60-243.pdf> (July 23, 2013).

DeSimone, L. D., and Popoff, F. (2000). *Eco-efficiency: The business link to sustainable development*, MIT Press, Cambridge, MA.

de Soto, H. (2003). *The mystery of capital: Why capitalism triumphs in the West and fails everywhere else*, Basic Books, New York.

Dodd, R., and Munck, L. (2002). *Dying for change*, World Health Organization, Geneva. <http://www.who.int/hdp/publications/dying_change.pdf> (Nov. 12, 2011).

Easterly, W. (2005). "What did structural adjustment adjust? The association of policies and growth with repeated IMF and World Bank adjustment loans." *J. Development Economics*, 76, 1–22.

Easterly, W. (2006). *The white man's burden: Why the West's efforts to aid the rest have done so much ill and so little good*, Penguin Press, New York.

Easterly, W. (2007). "The ideology of development." *Foreign Policy*, July/August, <http://dri.fas.nyu.edu/docs/IO/11786/IdeologyofDevelopment.pdf> (Aug. 12, 2012).

Economist. (2005). "The hidden wealth of the poor." Nov. 5. <http://www.economist.com/node/5079324?story_id=5079324> (Aug. 12, 2012).

Ewing, B., et al. (2010). *The ecological footprint atlas 2010*, Global Footprint Network, Oakland, CA.

Federation Internationale des Ingenieurs Conseils (FIDIC). (2013). *Project sustainability management: Application manual*, Federation Internationale des Ingenieurs Conseils, Paris, France.

<http://fidic.org/books/project-sustainability-management-applications-manual-2nd-edition-2013> (Nov. 30, 2013).

Florida, F. (2005). "The world is spiky." *The Atlantic Monthly*, October, 48–51.

Food and Agriculture Organization (FAO). (2012). *State of food insecurity in the world 2012*, Food and Agriculture Organization of the United Nations, Rome. <http://www.fao.org/news/story/en/item/161819/icode/> (Oct. 1, 2012).

Forbes, S. (2009). "Sustainable development extension plan (SUDEX)." Doctoral dissertation, University of Texas at El Paso.

Frankenberger, T. R., Drinkwater, M., and Maxwell, D. (2000). *Operationalizing household livelihood security: A holistic approach for addressing poverty and vulnerability*, CARE, Atlanta. <http://pqdl.care.org/Practice/HLS%20-%20Operationalizing%20HLS%20-%20A%20Holistic%20Approach.pdf> (Nov. 15, 2011).

Freire, P. (2007). *Pedagogy of the oppressed*, Continuum International Publishing, New York. First published in 1970 by Seabury Press (same title).

Friedman, T. (2005). *The world is flat: A brief history of the 21st century*, Farrar, Straus and Giroux, New York.

Gandhi, M. (1962). *Village Swaraj*, Navijan Press, Ahmedabad, India.

Global Adaptation Institute. (2011). "Global Adaptation Index (GaIn): Measuring what matters." Global Adaptation Institute, Washington, DC. <http://index.gain.org/> (March 15, 2012).

Global Footprint Network. (2010). "Ecological footprint atlas 2010." <http://www.footprintnetwork.org/en/index.php/GFN/page/ecological_footprint_atlas_2010/> (Oct. 16, 2013).

Global Issues. (2013). "Poverty facts and stats." <http://www.globalissues.org/article/26/poverty-facts-and-stats> (Oct. 3, 2013).

Global Reporting Initiative. (2013). <https://www.globalreporting.org/Pages/default.aspx> (Oct. 1, 2013).

Goldman, D. P. (2011). *How civilizations die*, Regnery Publishing, Washington, DC.

Graedel, T. E. H., and Allenby, B. R. (2009). *Industrial ecology and sustainable engineering*, Prentice Hall, Upper Saddle River, NJ.

Grameen Foundation (2014). "Progress out of poverty." <http://www.progressoutofpoverty.org/> (March 15, 2014).

Hassall, C., and Sherratt, T. N. (2011). "Statistical inference and spatial patterns in correlates of IQ." *Intelligence*, 39, 303–310.

Haughton, J., and Khandker, S. R. (2009). *Handbook on poverty and inequality*, World Bank, Washington, DC. <http://siteresources.worldbank.org/INTPA/Resources/429966-1259774805724/Poverty_Inequality_Handbook_FrontMatter.pdf> (July 3, 2012).

Hawken, P. (1993). *The ecology of commerce*, Harper Business, New York.

Hawken, P., Lovins, A., and Lovins, L. H. (1999). *Natural capitalism*, Little, Brown and Company, Boston.

Helliwell, J., Layard, R., and Sachs, J., eds. (2012). *The world happiness report*, <http://www.earthinstitute.columbia.edu/sitefiles/file/Sachs%20Writing/2012/World%20Happiness%20Report.pdf> (March 13, 2013).

Houben, H., and Guillaud, H. (2005). *Earth construction: A comprehensive guide*, Practical Action, London.

Huesemann, M., and Huesemann, J. (2011). *Techno-fix: Why technology won't save us or the environment*, New Society Publishers, Gabriola Islands, BC, Canada.

Independent Evaluation Group (IEG). (2012). "World Bank projects performance ratings: Projects completed in period 1981–2010." <http://ieg.worldbankgroup.org/content/ieg/en/home/ratings.html> (April 4, 2013).

Institut du Developpement Durable et des Relations Internationales (IDDRI). (2010). "The state of environmental migration 2010." Institut du Developpement Durable et des Relations Internationales, Paris. <http://www.iddri.org/Publications/Collections/Analyses/STUDY0711_SEM%202010_web.pdf> (Feb 15, 2012).

Institute for Sustainable Infrastructure (ISI). (2012). *Envision: A rating system for sustainable infrastructure* (Version 2.0), Institute for Sustainable Infrastructure, Washington, DC. <http://www.sustainableinfrastructure.org/portal/workbook/GuidanceManual.pdf> (March 13, 2013).

International Human Dimensions Programme on Global Environmental Change (IHDP). (2012). "Inclusive wealth report: Measuring progress toward sustainability." <http://www.ihdp.unu.edu/article/iwr> (Mar. 1, 2013).

Kenny, C. (2010). "Best. Decade. Ever. The first 10 years of the 21st century were humanity's finest— Even for the world's bottom billion." *Foreign Policy*, Sept./Oct., 28–29. <http://www.foreignpolicy.com/articles/2010/08/16/best_decade_ever?wp_login_redirect=0> (Feb, 10, 2011).

Kenny, C. (2011). *Getting better: Why global development is succeeding and how we can improve the world even more*, Basic Books, New York.

Korten, D. (1981). "The management of social transformation." *Public Administration Review*, 41(6), 609–618.

Korten, D. (2006). *The great turning: From empire to Earth community*, Berrett-Koehler, San Francisco.

Kulkarni, S. C., et al. (2011). "Falling behind: Life expectancy in U.S. counties from 2000 to 2007 in an international context." *Population Health Metrics*, 9:16, doi 10.1186/1478-7954.

Liu, L., et al. (2012). "Global, regional, and national causes of child mortality: An updated systematic analysis for 2010 with time trends since 2000." *The Lancet*, 379(9832), 2151–2161.

Mansouri, G., and Rao, V. (2012). *Localizing development: Does participation work?* World Bank Publications, Washington, DC.

Maslow, A. H. (1943). "A theory of human motivation." *Psychological Review*, 50, 370–396.

Max-Neef, M. (1991). *Human scale development: Conception, application and further reflection*, Apex Press, Muscat, Oman.

Max-Neef, M. (1995). "Economic growth and quality of life: A threshold hypothesis." *Ecological Economics*, 15, 115–118.

McDonough, W., and Braungart, M. (2002). *Cradle to cradle*, North Point Press, New York.

Narayan, D., and Petesch, P., eds. (2002). *From many lands*, Vol. 3 in Voices of the Poor, International Bank for Reconstruction and Development/World Bank, Washington, DC. <http://siteresources.worldbank.org/INTPOVERTY/Resources/335642-1124115102975/1555199-1124115210798/lantoc.pdf> (Dec. 10, 2012).

Narayan, D., et al. (1999). *Can anyone hear us? Voices from 47 countries*, Vol. 1 in Voices of the Poor, World Bank, Washington, DC. <http://siteresources.worldbank.org/INTPOVERTY/Resources/335642-1124115102975/1555199-1124115187705/vol1.pdf> (Dec. 10, 2012).

Narayan, D., et al. (2000). *Crying out for change*, Vol. 2 in Voices of the Poor, Oxford University Press, New York. <http://siteresources.worldbank.org/INTPOVERTY/Resources/335642-1124115102975/1555199-1124115201387/cry.pdf> (Dec. 10, 2012).

Nattrass, B., and Altomare, M. (1999). *The natural step for business: Wealth, ecology and the evolutionary corporation*, New Society Publishers, Gabriola Islands, BC, Canada.

Nolan, R. (2002). *Development anthropology: Encounters in the real world*, Westview Press, Boulder, CO.

Nolan, R. (2011). "What do we know, and what can we do with what we know? Anthropology, international development, and U.S. higher education." Keynote lecture at Collective Motion 2.0, Princeton University, Princeton, NJ, Nov. 12.

Norwegian Ministry of Foreign Affairs. (2004). "Peacebuilding: A development perspective." Norwegian Ministry of Foreign Affairs, Oslo, Norway. <http://www.regjeringen.no/upload/UD/Vedlegg/Utvikling/peace-engelsk.pdf> (May 25, 2012).

Nyerere, J. (1974). *Freedom and development—Uhuru na Maendeleo: A selection from writings and speeches 1968–1973*, Oxford University Press, New York, p 58.

Office of the High Commissioner for Human Rights (OHCHR). (2007). *Good governance practices for the protection of human rights*, Office of the High Commissioner for Human Rights, New York. <http://www.ohchr.org/Documents/Publications/GoodGovernance.pdf> (Sept 1, 2013).

Organisation for Economic Co-operation and Development (OECD). (2008). *Environmental Outlook to 2030*, OECD Publishing, Paris. <http://www.oecd.org/environment/environmentalindicatorsmodellingandoutlooks/oecdenvironmentaloutlookto2030.htm> (Oct. 2, 2012).

Oxfam. (2013). "The cost of inequality: How wealth and income extremes hurt us all." Oxfam media briefing, Jan. 18. <http://www.oxfam.org/sites/www.oxfam.org/files/cost-of-inequality-oxfam-mb180113.pdf> (May 15, 2013).

Paton, D., and Johnston, D. (2006). *Disaster resilience: An integrated approach*, Charles C. Thomas Publishing, Springfield, IL.

Peet, R., and Hartwick, E. (2009). *Theories of development: Contentions, arguments, alternatives*, 2nd Ed. Guilford Press, New York.

Poverty–Environment Partnership Joint Agency (PEP). (2008). "Poverty, health & environment: Placing environmental health on countries' development agendas." <http://www.unpei.org/PDF/Pov-Health-Env-CRA.pdf> (May 2, 2012).

Prahalad, C. K. (2006). *The fortune at the bottom of the pyramid: Eradicating poverty through profits*, Wharton School Publishing, Upper Saddle River, NJ.

Prahalad, C. K., and Hart, S. L. (2002). "The fortune at the bottom of the pyramid." *Strategy+Business*, Jan. 10, 26.

Ratterman, W. (2007). "Solar electricity for the developing world." *Home Power*, 119, 96–100.

Ricigliano, R. (2012). *Making peace last: A toolbox for sustainable peacebuilding*, Paradigm Publishers, Boulder, CO.

Rodrik, D. (2006). "Goodbye Washington Consensus, hello Washington confusion? A review of the World Bank's economic growth in the 1990s: Learning from a decade of reform." *J. Economic Literature*, 44, 973–987.

Rostow, W. W. (1960). *The stages of economic growth: A non-communist manifesto*, Cambridge University Press, Cambridge, U.K.

Sachs, J. (2006). *The end of poverty: Economic possibilities for our time*, Penguin Press, New York.

Schumacher, E. F. (1973). *Small is beautiful*, Harper Perennial, New York.

Sen, A. (1999). *Development as freedom*, Anchor Books, New York.

Shah, A. (2013). "Poverty facts and stats." *Global Issues*, Jan. 7. <http://www.globalissues.org/article/26/poverty-facts-and-stats> (Oct. 30, 2013).

Sphere Project. (2011). *Humanitarian charter and minimum standards in humanitarian response*, Practical Action Publishing, Rugby, U.K.

Steffen, W., and Tyson, P., eds. (2001). "Earth system science: An integrated approach." *Environment*, 43(8), 21–27.

Stiglitz, J. E. (1998). "Towards a new paradigm for development: Strategies, policies, and processes." Ninth Raul Prebisch Lecture at UNCTAD, Geneva. <http://unctad.org/en/docs/prebisch9th.en.pdf> (Sept. 5, 2013).

Sustainable Measures. (2013). "Characteristics of effective indicators." <http://www.sustainablemeasures.com/node/92> (Mar. 8, 2013).

Talbot, J., and Venkataraman, R. (2011). "Integration of sustainability principles into project baselines using a comprehensive indicator set." *Int. Business and Economics Research J.*, 10(9), 29–40.

Tearfund. (2011). *Reducing risks of disasters in our community*, Tearfund, Teddington, U.K. <http://tilz.tearfund.org/Publications/ROOTS/Reducing+risk+of+disaster+in+our+communities.htm> (Oct. 22, 2012).

Third World Academy of Sciences (TWAS). (2002). *Safe drinking water: The need, the problem, solutions and an action plan*, Third World Academy of Sciences, Trieste, Italy. <http://twas.ictp.it/publications/twas-reports/safedrinkingwater.pdf> (April 15, 2012).

Tierney, K. (2009). "Disaster response: Research findings and their implications for resilience measures." Research report 6, Community and Regional Resilience Institute (CARRI), Oak Ridge, TN. <http://resilientus.wp.in10sity.net/publications/research-reports/#> (Jan. 2, 2012).

United Nations (UN). (1948). "Universal declaration of human rights." <http://www.un.org/en/documents/udhr/> (March 15, 2014).

United Nations (UN). (1992). *An agenda for peace: Preventive diplomacy, peacemaking and peacekeeping*, Report of the Secretary-General pursuant to the statement adopted by the Summit Meeting of the Security Council on Jan. 31, 1992. <http://unrol.org/files/A_47_277.pdf> (March 16, 2013).

United Nations (UN). (1998). "Statement of commitment for action to eradicate poverty adopted by administrative committee on coordination." <http://www.unesco.org/most/acc4pov.htm> (Sept. 15, 2012).

United Nations (UN). (2000). "United Nations Millennium Declaration 55/2." <http://www.un.org/millennium/declaration/ares552e.htm> (Dec. 5, 2012).

United Nations (UN). (2011). *The Millennium Development Goals Report 2011*, United Nations, New York.

United Nations (UN). (2012). *The Millennium Development Goals Report 2012*, United Nations, New York.

United Nations (UN). (2013a). "Millennium development goals reports." <http://www.un.org/millenniumgoals/reports.shtml> (June 30, 2013).

United Nations (UN). (2013b). "Sustainable knowledge platform." <http://sustainabledevelopment.un.org/> (Oct. 1, 2013).

United Nations (UN). (2013c). *World population prospects: The 2010 revision, demographic profiles (population studies)*, United Nations, New York.

United Nations Children's Fund (UNICEF). (2005). *The state of the world's children, 2005: Childhood under threat*, UNICEF, New York. <http://www.unicef.org/publications/index_24432.html> (Dec. 3, 2012).

United Nations Children's Fund (UNICEF). (2013). "Progress in saving children's lives." <http://www.unicefusa.org/work/> (Oct. 3, 2013).

United Nations Conference on Environment and Development (UNCED). (1992). "Agenda 21." <http://www.un.org/esa/sustdev/documents/agenda21/english/Agenda21.pdf> (Dec. 5, 2012).

United Nations Development Programme (UNDP). (2009). *Capacity development: A UNDP primer*, United Nations Development Programme, New York. <http://www.undp.org/content/dam/aplaws/publication/en/publications/environment-energy/www-ee-library/climate-change/capacity-development-a-undp-primer/CDG_A%20UNDP%20Primer.pdf> (March 24, 2012).

United Nations Development Programme (UNDP). (2013). "International human development indicators." <http://hdrstats.undp.org/en/indicators/> (Oct. 1, 2013).

United Nations Development Programme (UNDP). (2014). "Post-2015 development agenda." <http://www.undp.org/content/undp/en/home/mdgoverview/mdg_goals/post-2015-development-agenda/> (Feb. 15, 2014).

United Nations Development Programme Human Development Report (UNDP/HDR). (1990). *Concepts and measurement of human development*, United Nations Development Programme, New York.

United Nations Development Programme Human Development Report (UNDP/HDR). (2005). *International cooperation at a crossroads: Aid, trade and security in an unequal world*, United Nations Development Programme, New York.

United Nations Development Programme Human Development Report (UNDP/HDR). (2006). *Beyond scarcity: Power, poverty, and the global water crisis*, United Nations Development Progamme, New York.

United Nations Development Programme Human Development Report (UNDP/HDR). (2011). *Sustainability and equity, a better future for all*, United Nations Development Programme, New York. <http://hdr.undp.org/en/media/HDR_2011_EN_Complete.pdf> (April 22, 2012).

United Nations Development Programme Human Development Report (UNDP/HDR). (2013). *The rise of the south: Human progress in a diverse world*, United Nations Development Programme, New York. <http://www.undp.org/content/dam/undp/library/corporate/HDR/2013Global HDR/English/HDR2013%20Report%20English.pdf> (Dec. 5, 2013).

United Nations Educational, Scientific and Cultural Organization (UNESCO). (2003). *Small is working: Technology for poverty reduction*, UNESCO/ITDG/TVE, Paris. <http://portal .unesco.org/science/en/ev.php-url_id=3180&url_do=do_topic&url_section=201.html> (Oct. 7, 2011).

United Nations HABITAT (UN-HABITAT). (2006). "Urbanization facts and figures." <http:// www.unhabitat.org/cdrom/docs/WUF1.pdf> (Oct. 15, 2012).

United Nations Population Fund (UNFPA). (2011). *State of the world population 2011: People and possibilities in a world of 7 billion*, UNFPA, New York. <http://www.unfpa.org/webdav/ site/global/shared/documents/publications/2011/EN-SWOP2011-FINAL.pdf> (March 15, 2012).

United Nations WOMEN (UN-WOMEN). (2013). "Women, poverty & economics." <http:// www.unifem.org/gender_issues/women_poverty_economics/> (Oct. 2, 2013).

United States Census Bureau. (2011). "Income, poverty and health insurance coverage in the United States: 2010." U.S. Department of Commerce, Washington, DC. <http://www.census.gov/ newsroom/releases/archives/income_wealth/cb11-157.html> (Dec. 7, 2011).

United States Institute of Peace (USIP). (2009). *Guiding principles for stabilization and reconstruction*, United States Institute of Peace Press, Washington, DC. <http://www.usip.org/sites/default/ files/resources/guiding_principles_full.pdf> (March 22, 2012).

Ura, K., Alkire, S., Zangmo, T., and Wangdi, K. (2012). *A short guide to gross national happiness index*, Center of Bhutan Studies, Thimphu, Bhutan. <http://www.grossnationalhappiness.com/ wp-content/uploads/2012/04/Short-GNH-Index-edited.pdf> (March 17, 2013).

Valadez, J., and Bamberger, M. (1994). *Monitoring and evaluating social programs in developing countries: A handbook for policymakers, managers, and researchers*, World Bank Publications, Washington, DC. <http://www.pol.ulaval.ca/perfeval/upload/publication_192.pdf> (May 22, 2012).

van der Haak, B., Director. (2006). *Lagos wide & close: An interactive journey into an explosive city*, DVD. Distributed by Ideabooks and DE Filmfreak Distributie, Amsterdam, Holland.

Wackernagel, M., and Rees, W. (1996). *Our ecological footprint—Reducing human impact on Earth*, New Society Publishers, Gabriola Islands, BC, Canada.

Wallace, B. (2005). *Becoming part of the solution: The engineer's guide to sustainable development*, American Council of Engineering Companies (ACEC), Washington, DC.

Water Supply and Sanitation Collaborative Council (WSSCC). (2004). "Listening—To those working with communities in Africa, Asia, and Latin America to achieve the UN goals for water and sanitation." Water Supply and Sanitation Collaborative Council, Geneva. <http://www.comminit.com/?q=africa/node/189342> (June 2, 2012).

Williamson, J. A. (2004). "A short history of the Washington Consensus." <http://www.iie.com/publications/papers/williamson0904-2.pdf> (Oct. 15, 2011).

Winter, A. (2011). "Engineering's new frontier." *Mechanical Engineering Magazine*, November, 31.

World Bank. (2008). "2008 World development indicators: Poverty data." International Bank for Reconstruction and Development/World Bank, Washington, DC. <http://siteresources.worldbank.org/Datastatistics/Resources/WDI08supplement1216.pdf> (Sept. 24, 2011).

World Bank. (2011). "How we classify countries." World Bank, Washington, DC. http://data.worldbank.org/about/country-classifications (Jan. 22, 2012).

World Bank. (2012a). "Inclusive green growth: A pathway to sustainable development." World Bank, Washington, DC. <http://siteresources.worldbank.org/EXTSDNET/Resources/Inclusive_Green_Growth_May_2012.pdf> (Oct. 3, 2013).

World Bank. (2012b). "Turn down the heat: Why a 4°C warmer world must be avoided." World Bank, Washington, DC. <http://climatechange.worldbank.org/sites/default/files/Turn_Down_the_heat_Why_a_4_degree_centigrade_warmer_world_must_be_avoided.pdf> (March 1, 2013).

World Bank. (2012c). "What a waste: a global review of solid waste management." World Bank, Washington, DC. <http://web.worldbank.org/wbsite/external/topics/exturbandevelopment/0,,contentmdk:23172887~pagepk:210058~pipk:210062~thesitepk:337178,00.html> (March 1, 2013).

World Bank. (2013a). "How we classify countries." World Bank, Washington, DC. <http://data.worldbank.org/about/country-classifications> (Oct. 1, 2013).

World Bank. (2013b). *On thin ice: How cutting pollution can slow warming and save lives*, World Bank and the International Cryosphere Climate Initiative. World Bank, Washington, DC. <http://www-wds.worldbank.org/external/default/WDSContentServer/WDSP/IB/2013/11/01/000456286_20131101103946/Rendered/PDF/824090WP0v10EN00Box379869B00PUBLIC0.pdf> (Dec. 2, 2013).

World Bank. (2013c). "World development indicators." World Bank, Washington, DC. <http://data.worldbank.org/data-catalog/world-development-indicators> (Oct. 1, 2013).

World Commission on Environment and Development (WCED). (1987). "Our common future." *World Commission on Environment and Development*, Oxford University Press, New York.

World Development Report (WDR). (2003). *Sustainable development in a dynamic world: Transforming institutions, growth and quality of life*, World Bank and Oxford University Press, New York.

World Development Report (WDR). (2007). *Development and the next generation*, World Bank, Washington, DC.

World Development Report (WDR). (2012). *World Development Report—Gender equality and development*, World Bank, Washington, DC.

World Health Organization (WHO). (2001). *Non-communicable diseases and mental health*, World Health Organization, Geneva. <http://www.who.int/mip2001/files/2008/NCDDiseaseBurden.pdf> (July 10, 2011).

World Health Organization (WHO). (2012). "Millennium development goals." <http://www.who.int/mediacentre/factsheets/fs290/en/> (Apr. 10, 2013).

World Health Organization (WHO). (2013a). "Children: Reducing mortality." <http://www.who.int/mediacentre/factsheets/fs178/en/index.html>, (Sept. 5, 2013).

World Health Organization (WHO). (2013b). "The top 10 causes of death." <http://www.who.int/mediacentre/factsheets/fs310/en/> (July 12 2013).

World Health Organization (WHO/UNICEF). (2012). *Progress on drinking water and sanitation: Joint Monitoring Programme Update*, World Health Organization, Geneva. <http://www.who.int/water_sanitation_health/publications/2012/jmp_report/en/index.html> (July 12, 2013).

World Health Organization (WHO/UNICEF). (2013). *Progress on sanitation and drinking water— 2013 update*, WHO/UNICEF Joint Monitoring Programme for Water Supply and Sanitation, World Health Organization, Geneva. <http://www.childinfo.org/files/JMP2013Final_Eng.pdf> (Dec. 10, 2013).

World Resources Institute (WRI). (2005). *World resources 2005: The wealth of the poor*, World Resources Institute, Washington, DC.

World Water Assessment Program (WWDR). (2009). *Water in a changing world*, UNESCO Publishing, Paris. <http://www.unesco.org/new/en/natural-sciences/environment/water/wwap/wwdr/wwdr3-2009/> (Feb. 3, 2013).

Zoellick, R. B. (2010). "The end of the third world? Modernizing multilateralism for a multipolar world." Woodrow Wilson Center for International Scholars Lecture, April 14, Washington, DC. <http://web.worldbank.org/WBSITE/EXTERNAL/NEWS/0,,contentMDK:22541126~pagePK:34370~piPK:42770~theSitePK:4607,00.html> (March 14, 2014).

3

Engineers and Development

3.1 Context

The work of engineers has contributed enormously to economic development, prosperity, and improved quality of life over the past 200 years but mostly in the developed world. Surprisingly, the engineering profession has shown limited interest and competency in addressing development issues for the developing world. In an article in the May 2011 issue of the *Professional Engineers* (PE) magazine, Cathy Leslie, executive director of Engineers Without Borders–USA (EWB–USA) emphasized that a lack of competency did in fact exist and that it was nearly impossible to respond positively to the question "Are engineers competent enough to undertake development work?" It is widely agreed that today's engineers do not have the skills, tools, nor the education to address the global problems that our planet is facing or will be facing within the next 20 years.

> *Development work is not equal to large scale, developed-world projects. Engineers typically are not educated in working in different cultures and the developing world. Nor are they educated in the skills necessary to identify the correct problem, design the appropriate solution, and ensure that a community has the ability to maintain the solution in the long term. These are important gaps of knowledge in addressing critical development world problems.* (Leslie 2011)

The limitation facing engineers when conducting development work was further echoed by Nolan (2011) in reference to the inability of (U.S.) students to function in foreign cultures. He remarked that today, students lack the "ability to look beyond facts and figures to uncover meanings and patterns, to learn in unfamiliar surroundings, and to gain entry into the cultural world of others." It has been the experience of the author that not only are engineering students ill prepared to work in foreign cultures, but so are the faculty members who are responsible for their education.

In reality, most engineering solutions have been developed for the 1–2 billion richest people on the planet, those at the top of the economic pyramid.

The solutions for that group of customers tend to assume that anything bigger, faster, stronger, and more complex is a better solution to any given problem. As remarked by Radjou et al. (2012), "Western engineers have come to equate complexity with progress ... and 'low cost' with 'poor quality.'" Traditional engineering practice is *not* interested in small projects because they are not perceived as financially rewarding. As long as the Western world sees development as economic growth (and unlimited growth), the engineering profession is bound to just meet the demands and serve the most powerful segments of the world's population.

As a result, this bias toward large-scale engineering for rich societies has left developing nations without adequate facilities and infrastructure to build sustainable economies, especially in rural areas. When developing countries acquire adequate infrastructure, i.e., the utilities, facilities, and systems upon which society depends for its normal functions, the projects are more likely to be in cities and imply large financial and physical capital costs. Furthermore, the infrastructure is likely to resemble that found in Western countries, which benefits an upper class, does not respond to the needs of the masses, and often creates environmental degradation or other unintended consequences, which are then ignored altogether. Although these effects have been criticized by society for at least three decades, the engineering community has continued to be unresponsive to the needs of those at the bottom of the economic pyramid. Furthermore, engineering expertise is rarely sought out in community-based development projects.

In today's world, what is needed is a new form of engineering *project delivery*, one that addresses the technical and nontechnical challenges involved in working in developing communities and at the same time delivers socially and economically appropriate and sustainable solutions to all. Paraphrasing Schumacher (1973), we need to develop solutions (and not just technology) with a human face: solutions that are simpler, socially appropriate, designed at the human scale, not requiring large financial and physical capital, and decentralized, and solutions that create jobs and opportunities that are satisfying and meaningful (Huesemann and Huesemann 2011).

This new form of project delivery is indeed a *disruptive concept* to the status quo of traditional engineering education and practice and, by association, to the underlying industrial and economic systems that rely on science and technology. A simple question arises: Do today's engineering graduates and engineers have the skills and tools to address the global problems that our planet and humans are facing today or will be facing within the next 20 years?

Because the answer to that question has been demonstrated in the previous chapters to be negative and we cannot solve tomorrow's problems with yesterday's tools and skills, a *new epistemology* of engineering practice and education is needed, one that is based on the idea of reflective and adaptive practice, systems thinking, engagement, and a holistic approach to global problems. This new form of engineering education and practice must be designed to cover a wide range of technical and nontechnical issues to train *global citizen engineers* and *whole persons*, capable of operating in a multicultural world, not just narrow-minded technical experts (Nolan 2011).

3.2 Engineers Indispensable to Development

It is broadly acknowledged that technology and development are closely related (Romer 1990; Fagerberg 1994) and that engineers have contributed enormously to society in terms of economic development and quality of life (e.g., better hygiene, nutrition, water supply, waste disposal, and housing). As noted by Weingardt (1998), "engineers are probably the single most indispensable group needed for maintaining and expanding the world's economic well-being and its standard of living." Well grounded in the basic sciences, engineers design and build wide-ranging combinations of systems, structures, processes, and machinery, all for the purpose of meeting societal needs safely and economically.

Although the work of engineers can be traced back to ancient times (Sprague de Camp 1963), the bulk of the engineering advances happened after the Enlightenment period of the 18th century. Inventions like James Watt's steam engine transformed how work was done and sparked the Industrial Revolution. Continuing discoveries in electricity, mechanics, materials, processes, and testing have resulted in thousands of new products and services, all contributing to increased levels of health, comfort, and productivity that were previously unthinkable. Moreover, over the past 50 years, these advances have been accomplished with surprising ease, to such a degree that for any new problem or need, people, specifically those living in the developed world, have been conditioned to expect solutions to appear immediately.

In some ways, that expectation is valid. The fields of biology, computers and information technology, materials, nanotechnology, and others continue to advance rapidly. New discoveries seem to appear almost every day, lending a sense of boundlessness to how far society might advance. However, in other ways this expectation is not realistic. Many of these advances have come at a cost, one that is not generally recognized or factored into societal forecasting, planning, and accounting (e.g., externalities). Despite the benefits to society associated with technological progress, society has also questioned how engineers have been applying technology, noting substantial and unforeseen effects on human and natural systems (Huesemann and Huesemann 2011).

The first serious questions that challenged the panacea of science and technology were raised about 50 years ago by Rachel Carson (1962) in her book, *Silent Spring*. In this landmark book, Carson challenged the notion that DDT and other pesticides (then in common use) were safe and brought into question the effects of synthetic chemicals on the environment. These pesticides were used widely without any consideration whatsoever of the harm they might cause, mostly out of ignorance. Later, more comprehensive studies on these effects were initiated across the world, questioning not only the effects of chemicals on the environment but also the relationship and importance of the environment to global economic growth and development.

Following Carson's work, and as discussed in the previous chapter, the sustainability movement arose in the late 1980s with the publication of "Our Common Future" (WCED 1987) by the Brundtland Commission. It pointed out that society's current methods for exploiting the environment and natural resources could not sustain society's current form of development. Documents such as "Our Common Future" and "Agenda 21" from the United Nations Conference on Environment and Development (UNCED 1992), also known as the Rio Summit, and the Millennium Development Goals (see Section 2.7) launched in 2000 have all emphasized the importance of science and technology in addressing the global problems faced by our planet.

All these aforementioned events have called for the engineering profession to change its mindset and consider a new mission statement—to contribute to the building of a more sustainable, stable, and equitable world, not only in the developed countries but also in the countries in various stages of development. The engineering profession is now challenged to build on its great achievements of the 20th century (Wulf 2000; NAE 2003, 2008) and expand such achievements to all humans on the planet, not just to a limited few in the developed world. Meeting that challenge is still a work in progress.

3.3 Engineering and Society

The effect of engineering on society has often been debated over the past two centuries. Although it is clear to most of us today that engineering and society are closely related, this was not often the case. At times, that link was controversial.

For the first part of the 19th century, the dominant view was that engineering should develop apart from society and that technology was nothing more than applied science and economics (NAE 1991), an approach that was called "internalist" or "determinist" by science history experts (Hughes 1991). The second half of the 19th century and the first half of the 20th century changed the dynamic between technology and society. The 100-year period from 1850 to 1950 is often referred to as the first Golden Age of engineering, where the importance of engineers and their contribution to society were seen as unquestionable (Florman 1987). Technology in the Western world meant material progress at all costs and "maximization of profits for owners and stakeholders" (Huesemann and Huesemann 2011). As remarked by Schön (1983), the epistemology of engineering practice and education at that time was based on *technical rationality*, solving well-defined problems following a positivist attitude that originated at the end of the 19th century. That epistemology helped define "the proper division of labor between the university and the professions" in the Western world, and the "split between research and practice." It also helped craft the concept of the "expert," focusing on narrowly technical practice. However, these experts were not supposed to include values in their technical decisions.

Since then, our quality of life has been built upon a complex and highly productive set of technological, industrial, and municipal systems and structures. Technological advances and their corresponding engineering applications have led to continuous improvements and services: better performing materials; more efficient extraction methods; better and faster communication tools; and new, more effective production techniques; among other accomplishments. All of these things occurred in no small way thanks to the contributions of engineers and scientists working to improve the built environment.

Yet ironically, these technical successes have also contributed to problems: unplanned, deleterious, or undesirable effects of technology on natural and human systems. Such unintended consequences have been objects of criticism by society, such as during the antitechnology movement of the 1970s. These criticisms are still valid today. As discussed by Bugliarello (1991), Hollomon (1991), and various authors in the fields of industrial ecology and earth systems engineering (NAE 2000), these effects have forced the engineering profession to acknowledge its limitations and revisit its assumptions and guiding principles, some of which are listed here:

- Many engineering decisions cannot be made independently of the surrounding natural and human-made systems.
- Our ability to cause planetary change through technology is growing faster than our ability to understand and manage the nontechnical consequences of such change (Leopold 1992).
- The quality of engineering decisions in society directly affects the quality of life of humans and natural systems today and in the future.
- All technologies have negative effects that may backfire. The traditional approach that engineering is only a process to devise and implement a chosen solution amid several purely technical options must be challenged.
- The positivist approach to technology of the 19th century and for most of the 20th century has failed to recognize that real-world problems are often "messes incapable of technical solution." Such a traditional approach is unable to handle "situations of uncertainty, instability, uniqueness, and value conflict" and has contributed to create "a gap between professional knowledge and the demands of real-world practice" (Schön 1983).

- A more holistic approach to engineering requires an understanding of interactions between engineered and nonengineered systems, inclusion of nontechnical issues, and a systems approach to comprehend such interactions. A challenge is reconciling the linear models of engineering with the chaotic (nonlinear), open, diverse, dissipative, uncertain, and adaptive nature of natural and human systems. Unpredictability requires integrating a *logic of failure* (Dörner 1997) and a *reflective practice* (Schön 1983) into project planning and design. It also implies emphasizing adaptation in projects instead of specification and satisficing instead of optimizing (Simon 1972).

- Preparing engineers to become facilitators of sustainable development, appropriate technology, and social and economic changes is one of the greatest challenges faced by the engineering profession today. Meeting that challenge may provide a unique opportunity for renewing the leadership of the engineering profession around the world.

- The compartmentalized 19th century model of engineering education no longer fits the needs of society. Unfortunately, a disconnect still exists today among (1) the magnitude of the problems in our global economy and what is expected of young engineers in engineering firms; (2) the recommendation for general education suggested by accreditation boards; and (3) the limited skills and tools traditionally taught in engineering programs in U.S. universities.

- Engineers must become more engaged in major societal leadership positions at the corporate and governmental levels. Figure 3-1 shows, for instance, the most common professions for politicians worldwide (*Economist* 2009); engineering comes last!

These recommendations are particularly important because modern engineering systems have the power to significantly affect social, economic, and environmental systems far into the future. The effect of such projects on systems outside their technical boundaries is particularly important on larger projects. Because the life span of infrastructure projects varies a lot (e.g., 30–75 years for bridges; 20–50 years for highways; 30–75 years for coal power stations; 50–100 years for commercial buildings; and 50–100 years for housing, railways, and dams), their positive and negative effects can be long lasting and require the adoption of policies early in the design and planning phases to minimize future effects, an issue that is explicitly addressed in the operational manual on environment and social safeguard policies of the World Bank (2013).

Even if all of the aforementioned recommendations are considered by the engineering profession, a need still exists for a reality check because no matter how advanced and well thought out technology might be, it always has negative effects. The unavoidability of the unintended consequences of technology needs to be accepted by the engineering and scientific profession and requires vigilance.

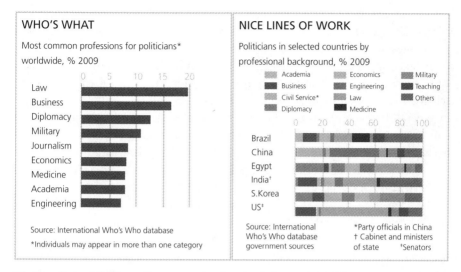

Figure 3-1. (a) Most Common Professions for Politicians; (b) Distribution of Politicians in Selected Countries

Source: *The Economist* 2009, reproduced with permission.

The goal is not to stop those negative effects but rather to minimize them and anticipate them. As remarked in a book by Huesemann and Huesemann (2011), one cannot escape the practical consequences of conservation of mass and the two laws of thermodynamics in designing new technologies and fixes to existing technologies.

A quantum leap in the awareness of the link between technology and society took place in the 1960s and 1970s in association with the environmental and sustainability movement (Wallace 2005). According to Nattrass and Alto-mare (1999), the history of that movement can be divided into four eras as part of an overall learning curve with which the industry has been faced: compliance in the 1970s, beyond compliance in the 1980s, eco-efficiency in the 1990s, and *some* endorsement of sustainability in the 2000s and beyond. For each era, the corporate response and the industry goals have been different and have evolved from being unprepared before the 1970s, to reactive in the 1970s, to anticipatory in the 1980s, to proactive in the 1990s, and to more holistic in the 2000s. A question now arises as to what will be the new era of the second and third decades of the 21st century and how industry will respond to a new set of global challenges related to climate change, security, population growth (especially in cities), and socioeconomic challenges.

A major benchmark in steering the role of engineering in society was the 1992 United Nations Conference on Environment and Development, also known as the Rio Summit. In many ways, that conference helped pave the way for many issues that are still of concern today in the areas of economic development, climate change, and poverty reduction. A major outcome of the Rio Summit was the publication of "Agenda 21" (UNCED 1992), which was a blueprint for actions to be taken in all aspects of human activities and consisted of 40 chapters with 120 action programs. The Rio Summit exemplified the critical role played by science and technology in shaping future societies and the health of the planet in general. It also emphasized the critical role of society in shaping technology and economic decisions. Finally, it called for scientists and engineers to stop being "value neutral" and to be responsible for shaping new ideas.

The 1992 Rio Summit and "Agenda 21" created a major revival and a renewed sense of purpose for the engineering profession, something comparable to the "can-do" attitude that drove the engineering pioneers of the 100-year Golden Age (1850–1950) and the post–World War II era. Several engineering committees on sustainability were established, and conferences and workshops were held to address the Agenda 21 recommendations; a good review of those initiatives can be found in AAES (1994), WFEO (1997, 1999, 2000), and Roberts (2002). The 1990s and early years of the 21st century were fueled by that enthusiasm, which helped elevate the image of the engineer among the youth. It also helped reassert the fact that "our economic and social health depends directly on the health of the engineering endeavor, and the health of engineering depends, in turn, on the support of society" (NAE 1985). It became clear, as suggested by Maurice Strong, secretary general of the 1992 Rio Summit, that "sustainable development would be impossible without full input by the engineering profession" (UNCED 1992).

Sustainable development represents a formidable challenge and opportunities for the engineering profession and society in general.

> *Moving toward sustainability will require more or less a complete overhaul of the world's infrastructure, replacing or refurbishing existing systems with new, cleaner and more efficient processes, systems and technologies. As such, new world markets for sustainable engineering services are being created as industries and governments alike begin the changeover to more sustainable practices. (Wallace 2005)*

Sustainable development is not just about technical fixes. Rather, it is an integrated approach that requires a new mindset that balances technology and society and the three basic elements of sustainability discussed in Chapter 2: environment, equity, and economics (the 3 Es) or people, planet, and profit (the 3 Ps). Preparing for climatic changes and mitigating their effects are reasons for the human race to revisit the connection between technology and society. As noted by Wallace (2005), the 1992 Rio Summit, the Earth Charter launched in 2000, the subsequent 2002 Johannesburg summit, and the Millennium Development Goals all show that there are "no miracle cures available. Solutions would come only through practical and sustained efforts." Sustainable development is a journey in which engineers and scientists are called to play a critical role in various fields, which according to Roberts (2002), include the following:

- Developing and extracting resources,
- Processing and modifying resources,
- Designing and building transportation infrastructure,
- Meeting the needs of consumers,
- Recovering and reusing resources,
- Environmental restoration, and
- Producing and distributing energy.

Therefore, it is critical for the engineering profession to become familiar with the problems facing the entire world, not just the needs and wants of the developed world. At the same time, the engineering profession needs new tools, strategies, and policies to reach the goal of sustainability at a planetary scale. A new mindset must be born that acknowledges the breadth, complexity, and systemic nature of the problems at hand. The challenge of creating a sustainable world demands a new and holistic look at the role of engineering in society. Providing value-free quick technical fixes to societal problems is part of the old mindset and is no longer an option.

The new mindset provides unique opportunities for the engineering profession to place itself in a position to (1) renew leadership in research and development, (2) contribute to economic development at the global scale, and (3) be a key player in international security and peacemaking. As remarked by Bugliarello (1991), "today, technology has an unprecedented opportunity to exercise leadership in showing how technology can offer the means for creating a better world, out of the ashes of collapsing or obsolete political and economic systems."

This statement, written in 1991, is more relevant today than ever. It also provides a unique vision and a powerful marketing tool to help bring more young people into engineering and reverse a declining engineering enrollment trend over the past 10 years (Moskal et al. 2008).

3.4 Engineering for Sustainable Human Development

International development has traditionally been conducted by expert professionals in multilateral and bilateral in-country agencies (e.g., the United Nations (UN), the World Bank, and the U.S. Agency for International Development) or individuals working for nongovernmental organizations (NGOs). The former group tends to focus on large programs with multiple projects and operates under top-down management. Project success is often measured in terms of number of projects delivered and people served and not necessarily in terms of their long-term performance. Such an approach has had limited success in the past. Yet it is still advocated by some large funding agencies and organizations, which note the absence of well established alternatives.

The NGO approach to development, which is more a bottom-up one, is not without flaws either. It focuses mostly on small projects, is often a piecemeal approach, and lacks technical and systemic input. It is not my intent to criticize the work of NGOs. However, it has been my observation that more often than not, their projects are often narrowly focused, lack planning and accountability, and lack monitoring and evaluation tactics. Having said that, some pioneering NGOs and international NGOs (INGOs) are delivering excellent projects.

Surprisingly, engineers are not often involved in small-scale community development projects, an area in which the interaction of technology with society is critical. In the next two decades, a population growth estimated at 1.5 billion (97% in the developing world) will create unprecedented demands for energy, food, land, water, transportation, materials, waste disposal, earth moving, public health care, environmental cleanup, telecommunication, and infrastructure. Clearly the involvement of engineers will be critical in fulfilling those demands at various scales, ranging from remote small communities to large urban areas (megacities), mostly in the developing world. This fact raises a larger question: What needs to be done now and in the near future to allow *all* humans to enjoy a quality of life where basic needs of water, sanitation, nutrition, health, safety, and meaningful work are fulfilled? A secondary question is this: How do engineering practice and education need to change to address such global demands?

The role of science and engineering in addressing basic human needs and poverty reduction became a reinforced topic of discussion in the early part of the 21st century (Juma and Yee-Cheong 2005) after the announcement of the UN Millennium Development Goals in 2000 (UNDP/HDR 2001). Another call for the engagement of science and technology (and engineering) in achieving these goals was brought to the attention of the UN General Assembly in September 2005 (ICSU 2013). It was signed by representatives of major scientific, engineering, and medical organizations and reads as follows:

> Stronger worldwide capacities in science and technology are necessary to allow humanity to achieve the UN Millennium Development Goals. A concerted global effort among the world's scientists, engineers, and medical experts is needed to identify successful strategies and to help implement effective programs. Sustained progress in reducing poverty and related problems will require strengthened institutions for science, technology, and innovation throughout the world, including in each developing nation. (ICSU 2013)

Another emphasis on the role that engineers and other built environment professions (architects, surveyors, and planners) play in disaster risk reduction was made by Lloyd-Jones et al. (2009), where the authors describe in detail the specific role of these professions in addressing the needs of communities in the emergency and recovery phase after a disaster. It includes risk and vulnerability assessment, disaster risk and mitigation, disaster preparedness and predisaster planning, emergency relief, early recovery and transition, and reconstruction and postreconstruction development and review.

Over the past 10 years, we have witnessed a growing interest by engineering students and professionals in direct, "hands-on" participation in international community-based development projects through the Engineers Without Borders movement. For example, Engineers Without Borders–USA (EWB–USA) has grown rapidly since it was organized 10 years ago. EWB–USA is a U.S.-based nongovernmental organization with a threefold mission: (1) partnering with disadvantaged communities to improve quality of life, (2) implementing environmentally and economically sustainable engineering projects, and (3) developing internationally responsible engineers and engineering students. EWB–USA's mission is unique in that it focuses on improving the quality of life of developing communities through projects delivered by teams of engineering and nonengineering students, supported by professional engineers and academics, and in participation with local communities. Virtually all who participate in these projects call the experience "transformational."

What started as a small project in Belize in 2000 by the author and a handful of undergraduate engineering students has quickly grown into an organization of more than 14,000 professional and student members making up more than 300 chapters in the United States and working on about 400 projects in 45 countries. About 43% of all EWB–USA (student and professional) members are women, which is especially noteworthy considering that women comprise fewer than 11% of the engineering workforce in the United States (NSF 2008). The same enthusiasm for international projects has been observed by the author in other countries as well, as testified by the equally rapid growth of the EWB–International network.

EWB–USA is made up of individuals who are willing to donate their time and expertise for the purpose of community capacity building, predominantly at the grassroots level. The projects are smaller in size and funding than traditional engineering projects and do not compete with those conducted by large engineering firms. The work is usually done on a voluntary basis over the course of one or more academic years. Students work on site in the field with mentors (professional and/or academic) once or twice per year and work on their respective projects between trips. They are responsible for the design (reviewed by the mentors), fund raising, assessment, project management, implementation, and monitoring and evaluation. As per the guidelines suggested by EWB–USA, the chapter must show a five-year commitment to each community it is serving. Over the past 10 years, the training that EWB–USA volunteers acquire while on projects in-country has been of special interest to the engineering profession in the United States, which has been looking for the next generation of leaders and change makers. It has been my observation that the engineering profession is indeed in good hands.

The work of EWB–USA falls into what Bugliarello refers to as *engineering for development* or *development engineering*, an interdisciplinary thrust in engineering that

> responds to the global need for engineers who understand the problems of development and sustainability, can bring to bear on them their engineering knowledge, are motivated by a sense of the future, and are able to interact with other disciplines, with communities and with political leaders to design and implement solutions. Bugliarello (2008)

Because engineering for development applies equally well to both developed and developing countries, *Engineering for Sustainable Human Development* (the title of this book) is better suited to denote a field that addresses problems specific to the developing world (such as those addressed by the eight Millennium Development Goals) while integrating the principles of sustainability.

3.5 Design for Sustainable Human Development

In his book, *Out of Poverty*, Polak (2008) makes an important remark about the design of products and services used by populations on our planet today: "The majority of the world's designers focus all their efforts on developing products and services exclusively for the richest 10% of the world's customers; nothing less than a revolution in design is needed to reach the other 90%."

The revolution that Polak is calling for, design for (or with) the other 90%, has the potential for (1) renewing the leadership of the engineering profession at the international level; (2) enrolling more young people in the many fields of science and engineering; (3) fostering technical innovation in research and development, and more importantly; (4) developing solutions that are good for the planet, appropriate to the communities being served, affordable, equitable, and capable of spearheading social entrepreneurship and economic growth for the world's poor.

It is noteworthy that design for (or with) the other 90% can be low-tech or high-tech design. It first appears that such design requires *at least* the same level of ingenuity and quality control as the design for the elite 10%. But it requires more than that because it needs to account for such things as social appropriateness, affordability, accessibility, availability, sustainability, scale, distribution, and operation and maintenance. As noted by Bornstein (2007), "social entrepreneurs have to reach far more people with far less money, so they have to be especially innovative to advance solutions at scale."

The developing world is indeed in a better position to deliver such technologies than the developed world, if it has the resources (knowledge and skills) to do so. Design for (or with) the other 90% can be seen as *disruptive* to all the economic, industrial, and social models of the Western world, a good and necessary disruption indeed. It is no longer a dream; it has become reality, especially in emerging markets, and it has the potential for under-cutting the well-being of the Western business world that has had so far a monopoly in advocating and enforcing expensive and complex technologies (see Chapter 13).

An excellent analysis of technical innovation in emerging markets, designed for clients in those markets, can be found in an article titled "The World Turned Upside Down" (*Economist* 2010). The article introduces key terms, such as "reverse" innovation, "constrained-based" innovation, and "frugal" innovation to characterize a revolutionary bottom-up development and management approach to innovation for the world's poor. Innovations range from the $800 handheld electrocardiogram machine, capable of delivering tests at $1 per patient, to the Tata Swach water filter, which costs $22 and produces 3,000 L of clean water for 200 days for a family of five. Other examples of affordable solutions can be found in the book by Radjou et al. (2012).

Polak (2008) also implies that it is challenging to transplant engineering solutions from the developed world to the developing world without first making appropriate changes that account for local cultural, social, economic, and other issues. For instance, at times complex water and energy systems would be out of context in the developing world. Some areas in the world will never (and do not need to) receive large water-treatment and wastewater-treatment systems or be connected to a power grid. In such cases, smaller, decentralized systems are more appropriate and realistic. In cities, larger systems are likely to be more appropriate and the only option to respond to increased demands. The jury is still out with regard to the merit of small or large systems. For instance, recent advances in wastewater-treatment technologies, coupled with a reevaluation of large-system life-cycle cost, seem to suggest that smaller and more distributed systems are a better option.

The bottom line is that the engineering profession needs to develop a repertoire of solutions and associated practices that could be used in different settings and situations. Engineering practices that can be used at different community scales that range from large cities to slums, refugee camps, and remote villages and communities are desperately needed. A variety of texts (and not enough of them) is available in the literature on engineering practices within the context of developing world projects. They include, among others, *Field Engineering* by Stern (1998), which addresses simple engineering work in rural areas; *Sustainable Wastewater Management in Developing Countries* by Laugesen and Fryd (2009); *Field Guide to Environmental Engineering for Development*

Workers by Mihelcic et al. (2009); and *Rural Energy Services* by Anderson et al. (1999).

We also need engineering practices that can be deployed at different time scales, such as those that are needed in the prevention, rapid response, recovery, and development phases of disasters associated with natural and nonnatural hazards (Lloyd-Jones et al. 2009). Another area that requires special skills relates to engineering practice in conflict or imminent conflict situations (Muscat 2011). Such a repertoire of engineering practices is slow to develop by the engineering profession in the Western world. In comparison, as discussed by Radjou et al. (2012), strong science and technology initiatives are being developed in emerging markets (such as India and China) that are likely to change the world for the better and that, when combined with social entrepreneurship, could have lasting effects. This fact raises the question, "How are engineers trained to design solutions for (and with) the other 90%, in addition to continuing to address the needs of the top 10%?"

3.6 Engineering Education for Sustainable Human Development

Today's engineers need to be able to show a high level of adaptation and flexibility to address more global problems in a dynamic environment where multidisciplinary approaches are the norm. As remarked by Reynolds (2013), today's global problems "do not respect national boundaries and require cooperation in science and engineering to address them successfully." Today's engineers must be able to address the formidable challenges associated with the interaction of adaptive technical systems with societal and environmental systems.

A need exists to create *global engineers* to address global problems (Bourn and Neal 2008). The world calls for it, the engineering profession calls for it, and today's students desire a more meaningful education. Yet engineering education and those who dispense that education are still hesitant in responding to the call. There is indeed a need to question the orthodoxy of traditional education in general (Palmer and Zajonc 2010) because it works out of context with the problems society is facing today and what its customers (students and professionals) are demanding. As in engineering practice, there is indeed a need for *disruptive* engineering education.

Today, engineering education is still subject to conflicting demands. On one side, a need exists for more specialization in depth, or *I*-type of education (Figure 3-2a) in narrower and more abstract fields of study to respond to ever-increasing knowledge and complexity. On the other side, a need exists for more general education so that engineers can work in global areas in interdisciplinary teams, which requires exposure to more general education and more breadth of

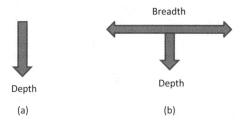

Figure 3-2. Two Different Types of Engineering Education:
(a) Traditional *I*-Type Education, (b) More Holistic *T*-Type Education

education beyond just the technical stuff. Both depth and breadth, or *T*-type education (Figure 3-2b) needs to be integrated into a four-year engineering curriculum (Grasso and Burkins 2009). The balance between the two types of education continues to be a topic of major discussion in the engineering profession today. Furthermore, both depth and breadth need to be supplemented with the acquisition of lifelong learning skills. Clearly our current traditional virtual education of engineers is not enough to respond to the world's problems! As remarked by my colleague (Bud Ahearn, personal communication, 2012) "engineering education is too important to be left [only] in the hands of academics."

Regrettably, mainstream academia today is still slow in realizing that its 19th century compartmentalized model of engineering education does not address the needs of a 21st century world. As remarked by Palmer and Zajonc (2010), "the ways in which we educate students today are, in large part, a reflection of our worldview, which itself is an image of nineteenth-century science." In academia today, there is still a large disconnect among (1) the magnitude of the problems in our global economy and what is expected of young engineers in engineering firms, (2) the recommendation for general education suggested by accreditation boards, and (3) the limited skills and tools traditionally taught in engineering programs in U.S. universities.

Academia does not adequately prepare students to address problems that may occur down the road (Orr 1998), specifically when they will be in managerial positions 10–20 years from now. It does not realize that students attending college now will live to see a 20–30% increase in world population and the consequences of that growth, including climate change and major losses in biological and cultural diversity on Earth. Today's engineers, who currently manage projects around the world, graduated about 15–20 years ago and were taught values, knowledge, and skills that were appropriate 15–20 years ago when fewer than 6 billion people lived on the planet and when communication and computational tools were primitive, compared with today's. As remarked by Nolan (2011), "we have somehow managed, in our universities, to construct institutions in which the connections are largely missing between some of our best minds and some of the world's most pressing problems." Academia has also failed to recognize that today's global problems cut across disciplinary boundaries and cannot be addressed separately.

Despite the aforementioned barriers to integrating global issues into engineering education, we have observed over the past 10 years several independent efforts worldwide to integrate service learning, civic engagement, and outreach into higher education curricula. This trend reflects a growing consensus among a limited number of pioneering engineering faculty, practicing engineers, and university administrators that the current system of engineering education is not adequate to create skilled engineers who are able to address complex geopolitical and economic problems. Such concerns have been voiced by various authors (NAE 2005; Downey et al. 2006; Duderstadt 2008) and addressed at various conferences. The Engineering Education for Sustainable Development (EESD) Conference in 2004 and the EESD Observatory, led by UNESCO and three European technical universities, developed the Barcelona Declaration, a declaration that provides a unique set of recommendations for engineering for sustainable development that encompasses human development as well. The role of the observatory has been to monitor advances in the declaration since 2004 and progress during the 2005–2015 decade in education for sustainable development (ESD) spearheaded by the United Nations.

In 2004, the Barcelona Declaration (EESD 2004) stated that "today's engineers must be able to do the following:

- understand how their work interacts with society and the environment, locally and globally, in order to identify potential challenges, risks and impacts;
- understand the contribution of their work in different cultural, social and political contexts and take those differences into account;
- work in multidisciplinary teams, in order to adapt current technology to the demands imposed by sustainable lifestyles, resource efficiency, pollution prevention and waste management;
- apply a holistic and systemic approach to solving problems and the ability to move beyond the tradition of breaking reality down into disconnected parts;
- participate actively in the discussion and definition of economic, social and technological policies, to help redirect society towards more sustainable development;
- apply professional knowledge according to deontological principles and universal values and ethics; and
- listen closely to the demands of citizens and other stakeholders and let them have a say in the development of new technologies and infrastructures."

Furthermore, the Barcelona Declaration recommended that educational institutions, in participation with the engineering and scientific community must:

- Have an integrated approach to knowledge, attitudes, skills, and values in teaching;
- Incorporate the disciplines of the social sciences and humanities;
- Promote multidisciplinary teamwork;
- Stimulate creativity and critical thinking;
- Foster reflection and self-learning;
- Strengthen systemic thinking with a holistic approach;
- Train people who are motivated to participate and who are able to make responsible decisions; and
- Raise awareness for the challenges posed by globalization.

In the United States, the Accreditation Board for Engineering and Technology (ABET) and professional organizations such as the American Society of Civil Engineers (ASCE) have been spearheading major changes in engineering practice and education. These groups recognize that the traditional engineering method, largely based on education in the engineering sciences and mathematics applied to artificially well defined and closed problems, is no longer adequate in preparing young people to enter the world and deal with the complex problems our society faces (Downey et al. 2006). ABET outcomes (2012) and the ASCE Body of Knowledge Committee (2004, 2nd Ed. 2008) criteria have been designed to overcome those limitations.

The following ABET (2012) outcomes (under Criterion 3) are particularly relevant to engineering education for sustainable human development:

- An ability to design a system, component, or process to meet desired needs within realistic constraints, such as economic, environmental, social, political, ethical, health and safety, manufacturability, and sustainability;
- An ability to function on multidisciplinary teams;
- An ability to identify, formulate, and solve engineering problems;
- An understanding of professional and ethical responsibility;
- An ability to communicate effectively;
- A broad education necessary to understand the effect of engineering solutions in a global, economic, environmental, and societal context; and
- An ability to use the techniques, skills, and modern engineering tools necessary for engineering practice.

The second edition of the ASCE Body of Knowledge Committee (2008) offers in-depth criteria relevant to *civil engineering*, which are especially critical in development work, such as the following:

(1) Formulating and solving an ill-defined engineering problem appropriate to civil engineering;

(2) Evaluating the design of a complex system, component, or process, and assessing compliance;

(3) Analyzing systems of engineered works, whether traditional or emergent, for sustainable performance;

(4) Analyzing the effect of historical and contemporary issues on the identification, formulation, and solution of engineering problems;

(5) Analyzing the effect of engineering solutions on the economy, environment, political landscape, and society;

(6) Evaluating the design of a complex system or evaluating the validity of newly created knowledge or technologies;

(7) Analyzing engineering works and services to function at a basic level in a global context; and

(8) Justifying a solution to an engineering problem based on professional and ethical standards and assessing personal, professional, and ethical development.

Additional initiatives on engineering education and sustainable development have been spearheaded by the U.S. National Academy of Engineering (NAE 2005). ASCE has been active in promoting sustainability and capacity building for development in a series of policy papers and standards and in its code of ethics. It has also been successful at articulating the need for engineers who understand the problems of development and sustainability in a series of reports such as "The Vision for Civil Engineering in 2025" (ASCE 2007) and through the work of its technical advisory committee on sustainability.

In the related field of service learning and social engagement, there has also been a strong push toward integrating changes in engineering education and the overall university mission (Lima and Oakes 2006). In 2005, several universities launched the Talloires network (Tufts 2013). Convened by Tufts University's president in the United States, the Talloires network consists of an international collection of individuals and institutions devoted to strengthening the civic roles and social responsibilities of universities in all parts of the world. It acknowledges the fact that academic institutions do not exist in isolation from society and that they have a commitment to do social good. It also acknowledges that (1) no dichotomy exists between civil engagement and excellence, (2) the university's mandate is to educate and train responsible and dedicated citizens, and (3) civic engagement should be a priority within research and scholarship.

An example of civic engagement and a service learning program in the United States that emphasizes the human nature of engineering is the Engineering Projects in Community Service (EPICS) program, started at Purdue University in Indiana (Coyle et al. 2005; Lima and Oakes 2006). Founded in 1995, it is described as "a unique program in which teams of undergraduates are designing, building, and deploying real systems to solve engineering-based problems for local community service and education organizations." Furthermore, "students earn one or two academic credits each semester and may register for up to four years. Projects may last several years, so tasks of significant size and impact can be tackled."

Presented below is a program developed at the University of Colorado called Engineering for Developing Communities (EDC). The program is offered through the Mortenson Center in Engineering for Developing Communities, and it advocates integrated and participatory solutions to humanitarian development by (1) educating globally responsible engineering students and professionals, (2) promoting research on developing community issues, and (3) reaching out to build local capacity and resiliency in developing communities worldwide.

3.7 The Making of the Global Engineer

A need exists for training a new generation of engineers who can better meet the challenges and address the demands of the developing world and those of the developed world. The challenge is the education of *global engineers* who (1) have the skills and tools appropriate to address issues that our planet is facing today and is likely to face within the next 20 years; (2) are particularly knowledgeable in sustainability principles; (3) are aware of the needs of the developing world; (4) contribute to the relief of the endemic problems afflicting developing communities worldwide by providing knowledge, resources, and appropriate and sustainable solutions; and (5) are able to solve complex, uncertain, and ill-defined problems often associated with sustainable development and human development issues. A more succinct definition of the engineer of the future proposed at a workshop on earth systems organized by the author at the University of Colorado in 2001 is the following: "the engineer of the future applies scientific analysis and holistic synthesis to develop sustainable solutions that integrate social, environmental, cultural, and economic systems."

The making of such a new generation of engineers requires adopting a new mindset in engineering education where the education of engineers is not limited to using value-neutral and theoretical tools introduced in the classroom environment designed to solve well defined problems. These tools need to be supplemented with new ones that enable engineers to acquire attitudes and values, and knowledge and skills, in two categories: (1) *technical* to be able to handle uncertainty and complexity and (2) *nontechnical* to deal with the socio-environmental and economic aspects of communities. In the field of sustainable human development, global engineers are expected to be able to work hand-in-hand with local communities and interact with nontechnical professionals in fields such as social sciences, economics, business, human rights, and professionals in nongovernment and international organizations. Table 3-1 lists several possible concepts that constitute the breadth of the education of global engineers, according to Bourn and Neal (2008).

Table 3-1. Suggested Concepts Related to the Global Dimension of Engineering Education

Sustainability	Cross-cultural capability
Development education	Diversity
Global ethics	Inclusivity
Human rights	Gender/Race/Ethnicity
International relations	Nationality/Disability
Political analysis	Business responsibility
Justice and equality	Citizenship

Source: Bourn and Neal (2008), with permission from Institute of Education, London.

How do global engineers gain the supplementary skills that differentiate them from regular engineers? This can be done in many complementary ways, such as (1) exposing students to real team projects in the classroom, (2) integrating concepts of sustainable development and human development in traditional engineering courses, (3) requiring students to participate in internships in the industry, (4) offering them field experience where they learn the fundamentals of project management (Mintz et al. 2014), and (5) engaging them in voluntary work and experiential learning activities, such as service learning, active learning, project based learning, or EWB–USA. It has been recognized that the benefits of exposing students to EWB types of projects are manifold. The projects

- Give the students an opportunity to experience all aspects of engineering: problem identification, assessment, design, implementation, and monitoring;
- Give the students an opportunity to work with professional mentors during their school year, develop good contacts within the industry, and learn by doing;
- Provide the students with a direct, hands-on engineering educational experience in a new and safe environment;
- Give students the opportunity to work in teams on larger projects as opposed to the discipline-specific concrete canoe (in civil engineering) or the new flushing toilet (in mechanical engineering);
- Demonstrate to students that engineering problems can be complex and not always well defined, can be solved in more ways than one, and often require working effectively with people who think differently (including engineers and nonengineers) and have different cultural backgrounds;
- Teach students how to interact with different cultures and think "outside the box" with limited tools; and

- Train students to develop awareness of professional ethics and the role that engineering plays in addressing community needs.

These seven sets of characteristics match perfectly well (and even go beyond) what is expected of engineering students by ASCE (2008) and ABET (2012). They also match the recommendations outlined in the 2004 Barcelona Declaration mentioned earlier.

Above all, EWB-type projects give students a global outlook, a sense of belonging and engagement, a way of expressing passion and empathy, and a societal context for their engineering work. It also gives them an opportunity to reflect on themselves, develop values, take action on things about which they are passionate, become good listeners, work with other professions, and ultimately "think globally and act locally." It is noteworthy that before participating in EWB-type projects students need to be trained and prepared (preparatory phase) to face issues they are likely to encounter in the field (doing phase). This phase is followed by a reflecting phase upon their return from the field. As remarked by Scott et al. (2011), these three phases are necessary for students to learn more effectively in the field. Scott et al. (2011) also concluded, based on their experience in environmental sciences, that there is a "growing body of evidence that fieldwork is an important way of enhancing undergraduate learning."

Unfortunately, it has been the experience of the author that many schools do not try very hard to incorporate the EWB experience into their engineering curricula. There are many reasons for this. One of them is that many academics have limited field experience and don't look into acquiring any. Another reason is that such work is not rewarded in the traditional promotion and tenure mindset in U.S. universities. Third, it is much easier for faculty to replace field-work with "virtual computer-generated exercises," as remarked by Scott et al. (2011). Furthermore, in many instances, EWB activities are still perceived by many deans of engineering, chairs of engineering departments, and faculty members as extracurricular clubs that should be run by students. It is unfortunate that university administrators and faculty members are not capable of leveraging the student enthusiasm as a recruiting tool and a springboard for new modes of education. They fail to recognize that today's youth demand meaningful engineering education and not the same regurgitated traditional educational experience. As I often mention in my lectures to student groups and faculty, students will go where they will get the education that best matches their global interests and needs of engagement. Short of changing their myopic view of engineering education, many schools will be left with empty classrooms in the near future if they do not begin to change curricula to match what students are asking for and if they do not adopt pedagogical methods to which students can relate.

Global Engineering at CU–Boulder

After the creation of EWB–USA in 2001 and the subsequent student interest nationwide, it became clear to several faculty members at the University of Colorado at Boulder (CU–Boulder), including the author, that a need existed to supplement the education of the CU–Boulder engineering students before their participation in EWB–USA projects. In 2004, the College of Engineering and Applied Science at CU–Boulder started a new program called the Engineering for Developing Communities (EDC) Program in the Civil, Environmental, and Architectural Engineering (CEAE) Department. The program was renamed the Mortenson Center in Engineering for Developing Communities (MCEDC) in 2009 after an endowment from the Mortenson Construction Company and the Mortenson family based in Minneapolis.

MCEDC presents a unique opportunity for a new generation of engineers to combine education, research, and outreach in providing sustainable and appropriate solutions to the endemic problems faced by the people on our planet who are most in need. MCEDC does not purposely train development experts. It is based first on the fundamental principle that *all* engineers need to be exposed, as part of their education, to human development issues. They also need to have a contextual understanding of how technical decisions interact with cultural issues and society. Simply put, engineers need to have both depth and breadth—depth in their engineering skills and breadth in their exposure to various skills and disciplines that enter into engineering work—a *T*-type of education, as mentioned before (Figure 3-2b). This philosophy applies to engineers who will be working in developed and developing countries alike.

Whereas EWB–USA focuses on extracurricular outreach projects, MCEDC addresses education, research and development, and service and outreach, and more importantly the relationship among those three components within the context of community development and capacity building. MCEDC has three objectives:

- Education: create awareness of developing world issues and provide the knowledge and tools for engineering professionals in recovery and development;
- Development research: coordinate research, design, and development of holistic solutions for developing communities that account for technical and nontechnical issues; and
- Capacity: build local community capacity through collaboration with organizations working in developing communities worldwide.

Since 2010, MCEDC has offered a Graduate Certificate in Engineering for Developing Communities. Its body of knowledge covers five major areas: engineering and technology, public health, social entrepreneurship, public policy and governance, and security and vulnerability. Students are trained to use an integrative and participatory framework for development projects in communities in developing countries (as discussed in Chapter 4) on real case studies during the academic year and through a 4–6 week-long practicum in country during the summer after their course work. The MCEDC body of knowledge is divided into three categories: mastery, competence, and exposure.

- *Mastery* of a topic results in the ability to analyze, synthesize, and evaluate situations encountered. Students are able to adapt their responses to situations in ways that align with these concepts and to explain the importance of following these methodologies to attain the desired outcome. Within the context of EDC, students are expected to gain mastery in field readiness and experience, systems thinking, knowledge in integrative and participatory frameworks for development projects, appropriate and sustainable technologies, sustainability, and sustainable development.
- *Competence* in a topic results in the ability to apply the concepts and techniques learned to solve problems in a different way than examples presented. Students value the importance of these topics in a way that consistently exerts influence on their behavior. Within the context of EDC, students are expected to gain competence in public health, team work, leadership, cultural sensitivity, and behavior change communication.
- *Exposure* to a topic results in the ability to recognize terminology and comprehend the concepts being presented. Students learn to value issues but are not expected to demonstrate higher level cognitive skills in this area. Within the context of EDC, students are expected to gain exposure to social entrepreneurship, security issues, and public policy and governance.

The graduate certificate consists of four courses taught over a period of 12 months:

- *Sustainable Community Development I* (SCD 1—fall) focuses on the fundamental tools necessary to address sustainable community development projects in low-income communities (LICs). During the first part of the semester, the focus is on human development and sustainable development issues. This part is followed by the presentation of an integrative and participatory framework for development projects in LICs that incorporates several project phases: appraisal, analysis, design, implementation, monitoring and evaluation, planning an exit strategy, and ensuring long-term benefits of implemented solutions. The ADIME-E framework, presented in Chapter 4, consists of a combination of development and engineering project management tools. Use of the framework is illustrated through case studies and student-driven team projects.

- *Sustainable Community Development II* (SCD 2—spring) covers the principles, practices, and strategies of appropriate technology as part of an integrated and systems approach to community-based development. Course content areas include technical issues in development, environmental health and communicable disease, appropriate and sustainable technologies with hands-on workshops, and global cooperation in development. Use of the framework is illustrated through case studies and student-driven team projects.
- *Sustainable Community Development Field Practicum* (SCD 3—summer) provides an opportunity for students to (1) gain insight into the field of international development, (2) experience the reality of working in developing communities, and (3) apply theoretical foundations of SCD classes to real-world experiences. Field-based experiences are an important component of the EDC program. A true understanding of humanitarian engineering requires students to actively engage in a significant field-based experience in a developing community. The course encompasses humanitarian engineering fieldwork and analytical reporting.
- *Life Cycle Analysis of Civil Infrastructure Systems* (spring) addresses the philosophical and analytical issues for lifetime design and operation of civil systems; optimization tradeoffs of construction, management, and sustainability; and utility of operation and service, including present-value economic analysis. Students also learn about decision-making alternatives of safety and performance.

The graduate certificate has been integrated into four graduate tracks that offer M.S. and Ph.D. degrees in civil engineering at CU–Boulder. The tracks are environmental engineering, civil systems engineering, building systems, and construction engineering and management. The aforementioned body of knowledge has been developed to give engineering students after graduation the opportunity to provide technical expertise to development agencies or various engineering firms in a manner that recognizes the various technical and nontechnical facets of community development that lead to sustainable solutions.

MCEDC Core Values and Principles

During their education, MCEDC students are presented with core values and principles that are expected to help them deal with the challenges they are likely to encounter when interacting with communities on future development projects in their professional careers. In addition to acquiring their much needed engineering skills, they are taught that:

- The planning, management, and implementation of projects in developing communities cannot be done using the same blueprint approach as in the developed world. Engineers need to be able to make decisions in a flexible, learning, and adaptive environment that changes rapidly, is characterized by complexity and uncertainty, and involves community partnerships.

- Doing a project right is necessary but not sufficient because it focuses more on project performance. Doing the right project is equally, if not more, important because it focuses more on whether the project is in balance with the societal, economic, and environmental systems with which it is interacting.
- Students need to abide by a strict professional code of ethics with regard to behavior, accountability, quality control and quality assurance, and the delivery of projects whether in a developed or developing world context.
- They need to be conscious of their mission, vision, values, and approach to development, and those of the intervening organization that employs them.
- They must take responsibility for educating themselves and acquiring the necessary skills before conducting projects.
- A participatory and integrated approach must be in place for all projects, with special attention to capacity building toward long-term sustainability.
- Project innovation is driven by the needs of the users and their capacity, and not outsiders' needs to introduce new and fancy solutions. This situation often means that the most effective and appropriate solutions could well be modifications to existing ones at the community level. It also means that unless the solutions are appropriate and likely to work, these solutions should not be introduced.
- Collaboration with various internal and external stakeholders is a strong part of development project work. Teamwork in this context includes not only working with a culturally and intellectually diverse group but also continual mentoring of future local leaders and change makers.
- Interventions (time and tasks) must be designed to maximize the direct response to community needs and desires, while considering the resources available and the project phase.
- A long-term authentic commitment to the community is vital for sustained success.
- Monitoring and evaluation are needed to assess the progress of a project right from the start and to allow students to decide on necessary changes and respond to identified shortcomings.
- An exit strategy and a long-term sustainability (benefits) plan need to be in place at the start of a project or shortly thereafter.

In summary, the MCEDC curriculum has been designed to serve as a blueprint for the education of engineers in the 21st century who are called to play a critical role in contributing to peace and security in an increasingly challenged world. As of the fall semester of 2012, the program is capped at 60 graduate students to focus on quality of education over quantity.

3.8 Chapter Summary

It is broadly acknowledged that engineers have contributed to society in terms of economic development and quality of life. However, their effect has mostly been limited to the richest segments of the world's population. Today, the engineering profession is called to address a larger challenge, which is to contribute to the building of a more sustainable, stable, and equitable world, not only in developed countries but also in countries in various stages of development. This larger vision calls for a demystification of engineering practice, which traditionally provides value-neutral technical solutions to well defined problems, irrespective of any social context. Instead, engineers are called to be change makers, peacemakers, social entrepreneurs, and facilitators of sustainable development.

Sustainable human development also calls for a new epistemology of engineering practice and education, a new literacy, based on the idea of reflective and adaptive practice, systems thinking, engagement, and fieldwork. It requires global engineers who are able to look at problems in a more holistic way and interact with a wide range of technical and nontechnical stakeholders from various disciplines and walks of life rather than remaining in their traditional silos of expertise. The challenge faced by academia is how to adequately educate engineers to

- Have the skills (technical and nontechnical) and tools appropriate to address critical world issues;
- Think across disciplines and interact with others (e.g., in health, economics, and business);
- Be trained to assess, design, implement, and monitor projects in developing communities;
- Be flexible and resourceful enough to deal with unfamiliar equipment and approaches;
- Become systems thinkers who are able to consider the unintended consequences of their "solutions"; and
- Have access to contemporary technology but have the humility to understand that the "developed world" doesn't have all of the answers.

Not only do engineers have to be proficient in their craft, i.e., doing their *projects right*, but they also have to deliver the *right projects* that are good for the environment and the communities that interact with that environment. This is best described in the following recommendation by Martin et al.:

> engineers ... are responsible not only for the safety, technical and economic performance of their activities, but they also have responsibilities to use resources sustainably; to

minimize the environmental impact of projects, wastes and emissions; and to use their influence to ensure that their work brings social benefits which are equitably distributed.
(Martin et al. 2005)

Therefore, engineers need to be able to combine principles of sustainable development into community development. Finally, human development requires a humanization of the engineering profession and the realization that engineering in that context is above all—as it has always been—about people.

References

Accreditation Board for Engineering and Technology (ABET). (2012). *Criteria for accrediting engineering programs, effective for evaluations during the 2013–2014 accreditation cycle*, Accreditation Board for Engineering and Technology, Baltimore.

American Association of Engineering Societies (AAES). (1994). *The role of engineering in sustainable development*, American Association of Engineering Societies, Washington, DC.

American Society of Civil Engineers (ASCE). (2004). *Civil engineering body of knowledge for the 21st century: Preparing the civil engineer for the future*, Body of Knowledge Committee of the Committee on Academic Prerequisites for Professional Practice, American Society of Civil Engineers, Reston, VA.

American Society of Civil Engineers (ASCE). (2007). "The vision for civil engineering in 2025." *Civil Engineering*, 77(8), 66–71.

American Society of Civil Engineers (ASCE). (2008). *Civil engineering body of knowledge for the 21st century: Preparing the civil engineer for the future*, 2nd Ed., Body of Knowledge Committee of the Committee on Academic Prerequisites for Professional Practice, American Society of Civil Engineers, Reston, VA.

Anderson, T., et al. (1999). *Rural energy services: A handbook for sustainable energy development*, Intermediate Technology Publications Ltd., London.

Bornstein, D. (2007). *How to change the world*, Oxford University Press, New York.

Bourn, D., and Neal, I. (2008). *The global engineer: Incorporating global skills within UK higher education of engineers*, Institute of Education, London. <http://engineersagainstpoverty.org/_db/_documents/WEBGlobalEngineer_Linked_Aug_08_Update.pdf> (Oct. 12, 2012).

Bugliarello, G. (1991). "The social function of engineering: A current assessment." *Engineering as a social enterprise*, National Academies Press, Washington, DC, 73–88.

Bugliarello, G. (2008). "Engineering: Emerging and future challenges." *Engineering: Issues and challenges for development*, UNESCO, Paris.

Carson, R. (1962). *Silent spring*, Houghton Mifflin Company, Boston.

Coyle, E. J., Jamieson, L. H., and Oakes, W. C. (2005). "EPICS engineering projects in community service." *International J. Engineering Education*, 21(1), 139–150.

Dörner, D. (1997). *The logic of failure: Recognizing and avoiding error in complex situations*, Perseus Books, Cambridge, MA.

Downey, G. L., Lucena, J. C., Moskal, B. M., Parkhurst, R., et al. (2006). "The globally competent engineer: Working effectively with people who define problems differently." *J. Eng. Education*, 95(2), 107–122.

Duderstadt, J. J. (2008). *Engineering for a changing world: A roadmap to the future of engineering practice, research, and education*, University of Michigan, Ann Arbor. <http://milproj

.ummu.umich.edu/publications/EngFlex%20report/download/EngFlex%20Report.pdf> (Oct. 20, 2012).

Economist. (2009). "There was a lawyer, an engineer, and a politician." Apr. 16. <http://www.economist.com/node/13496638> (Nov. 1, 2012).

Economist. (2010). "The world turned upside down: A special report on innovation in emerging markets." Apr. 17. <http://www.economist.com/node/15879369> (May 5, 2010).

Engineering Education for Sustainable Development (EESD). (2004). "The Barcelona declaration." <https://www.upc.edu/eesd-observatory/who/declaration-of-barcelona/BCN%20 Declaration%20EESD_english.pdf> (Jul. 10, 2012).

Fagerberg, J. (1994). "Technology and international differences in growth rates." *J. Economic Literature*, 32(3), 1147–1175.

Florman, S. C. (1987). *The existential pleasures of engineering*, St. Martin's Griffin, New York.

Grasso, D., and Burkins, M., eds. (2009). *Holistic engineering education: Beyond technology*, Springer, New York.

Hollomon, J. H. (1991). "Engineering's great challenge: The 1960s." *Engineering as a social enterprise*, National Academies Press, Washington, DC, 104–110.

Huesemann, M., and Huesemann, J. (2011). *Techno-fix: Why technology won't save us or the environment*, New Society Publishers, Gabriola Islands, BC.

Hughes, T. P. (1991). "From deterministic dynamos to seamless-web systems." *Engineering as a social enterprise*, National Academies Press, Washington, DC, 7–25.

International Council for Science (ICSU). (2013). "Science, technology and innovation for achieving the UN Millennium Development Goals." <http://www.icsu.org/publications/icsu-position-statements/sci-tech-un-mdg/> (Oct. 3, 2013).

Juma, C., and Yee-Cheong, L. (2005). *Innovation: Applying knowledge in development*, Earthscan, London.

Laugesen, C., and Fryd, O. (2009). *Sustainable wastewater management in developing countries: New paradigms and case studies from the field*, ASCE Press, Reston, VA.

Leopold, A. (1992). "Engineering and conservation." *The river of the mother of god: And other essays by Aldo Leopold*, B. Callicott and S. Falder, eds., University of Wisconsin Press, Madison, first published in 1938.

Leslie, C. (2011). "In developing world, special needs call for special attention." *Professional Engineers (PE)*, May, 8–9.

Lima, M., and Oakes, W. (2006). *Service learning: Engineering in your community*, Oxford University Press, New York.

Lloyd-Jones, T., et al., eds. (2009). *The built environment professions in disaster risk reduction and response: A guide for humanitarian agencies*, Max Lock Center, London. <http://www.preventionweb.net/english/professional/publications/v.php?id=10390> (Dec. 5, 2012).

Martin, S., Brannigan, J., and Hall, A. (2005). "Sustainability, systems thinking and professional practice." *Journal of Geography in Higher Education*, 29(1), 79–89.

Mihelcic, J. R., et al. (2009). *Field guide to environmental engineering for development workers: Water, sanitation, and indoor air*, ASCE Press, Reston, VA.

Mintz, K., et al. (2014). "Integrating sustainable development into a service-learning engineering course." *J. Prof. Issues Eng. Educ. Pract.*, 140(1), doi: 10.1061/(ASCE)EI.1943-5541.0000169.

Moskal, B. M., Skokan, C., Muñoz, D., and Gosink, J. (2008). "Humanitarian engineering: Global impacts and sustainability of a curricular effort." *International J. Engineering Education*, 24(10), 162–174.

Muscat, R. (2011). "Peace and conflict: Engineering responsibilities and opportunities." *Global Services USA Newsletter*, 13(1), September.

National Academy of Engineering (NAE). (1985). *Engineering in society*, National Academies Press, Washington, DC.

National Academy of Engineering (NAE). (1991). *Engineering as a social enterprise*, National Academies Press, Washington, DC.

National Academy of Engineering (NAE). (2000). "Earth systems engineering exploratory workshop." Unpublished report. Washington, DC.

National Academy of Engineering (NAE). (2003). *Century of innovation: Twenty engineering achievements that transformed our lives*, National Academies Press, Washington, DC.

National Academy of Engineering (NAE). (2005). *Educating the engineer of 2020: Adapting engineering education to the new century*, National Academies Press, Washington, DC.

National Academy of Engineering (NAE). (2008). *Grand challenges for engineering*, National Academies Press, Washington, DC.

National Science Foundation (NSF). (2008). *Women, minorities, and persons with disabilities in science and engineering*, Publication 07-315, National Science Foundation, Washington, DC. <http://www.nsf.gov/statistics/wmpd/> (Dec. 10, 2012).

Nattrass, B., and Altomare, M. (1999). *The Natural Step for business: Wealth, ecology and the evolutionary corporation*, New Society Publishers, Gabriola Island, BC, Canada.

Nolan, R. (2011). "What do we know, and what can we do with what we know? Anthropology, international development, and U.S. higher education." Keynote lecture at Collective Motion 2.0, November 12, Princeton University, Princeton, NJ.

Orr, D. (1998). "Transformation or irrelevance: The challenge of academic planning for environmental education in the 21st century." In *Proceedings of the 1998 Sanibel Symposium*, P. B. Corcoran, J. L. Elder, and R. Tehen, eds., North American Education for Environmental Education, Washington, DC, 17–35.

Palmer, P. J., and Zajonc, A. (2010). *The heart of higher education: A call to renewal*, Jossey-Bass Publishers, San Francisco.

Polak, P. (2008). *Out of poverty: What works when traditional approaches fail?* Berrett-Koehler Publishers, San Francisco.

Radjou, N., Prabhu, J., and Ahuja, S. (2012). *Jugaad innovation*, Jossey-Bass Publishers, San Francisco.

Reynolds, A. (2013). "Diplomacy, development, defense, and engineering." *Mechanical Engineering*, March, 18.

Roberts, D. (2002). *Engineers and sustainable development*, World Federation of Engineering Organizations ComTech Committee, Washington, DC.

Romer, P. M. (1990). "Endogenous technological change." *J. Political Economy*, 98(5), S71–S102.

Schön, D. A. (1983). *The reflective practitioner: How professionals think in action*, Basic Books, New York.

Schumacher, E. F. (1973). *Small is beautiful*, Harper Perennial, New York.

Scott, G. W., et al. (2011). "The value of fieldwork in life and environmental sciences in the context of higher education: A case study in learning about biodiversity." *J. Sci. Educ. Technol.*, 21(1), 11-21.

Simon, H. A. (1972). "Theories of bounded rationality." *Decision and organization*, C. B. McGuire and R. Radner, eds., 161–176, North-Holland Pub., Amsterdam, Netherlands.

Sprague de Camp, L. (1963). *The ancient engineers*, Ballantine Books, New York.

Stern, P., ed. (1998). *Field engineering: An introduction to development work and construction in rural areas*, ITDG Publishing, London.

Tufts University. (2013). "The Talloires network." <http://talloiresnetwork.tufts.edu/> (Jan. 2, 2013).

United Nations Conference on Environment and Development (UNCED). (1992). "Agenda 21." <http://www.un.org/esa/sustdev/documents/agenda21/english/Agenda21.pdf> (Dec. 5, 2012).

United Nations Development Programme Human Development Report (UNDP/HDR). (2001). "Making new technologies work for human development." United Nations Development Programme, New York.

Wallace, W. (2005). *Becoming part of the solution: The engineer's guide to sustainable development,* American Council of Engineering Companies, Washington, DC.

Weingardt, R. (1998). *Forks in the road,* Palamar, Denver.

World Bank. (2013). "Safeguard policies." <http://go.worldbank.org/L0WZ82PW60>. (Oct. 1, 2013).

World Commission on Environment and Development (WCED). (1987). "Our common future." WCED, Oxford University Press, New York.

World Federation of Engineering Organizations (WFEO). (1997). *The engineer's response to sustainable development,* World Federation of Engineering Organizations, Washington, DC.

World Federation of Engineering Organizations (WFEO). (1999). *Production efficiencies: The engineer's report,* World Federation of Engineering Organizations, Washington, DC.

World Federation of Engineering Organizations (WFEO). (2000). *Sustainable agricultural and natural resources management engineering practices: The engineer's report,* World Federation of Engineering Organizations, Washington, DC.

Wulf, W. A. (2000). "Great achievements and grand challenges." *The Bridge,* 30(3–4), 5–10.

4

Development Project Frameworks

4.1 Guiding Principles

This book is about guidelines to conduct small-scale community-based engineering projects in developing countries. It does not address large-sector programs consisting of multiple interventions and projects at the country level, nor does it address traditional projects conducted by engineering companies in the developed world. The book uses the definition of a project proposed by the Project Management Institute (PMI) in their *A Guide to the Project Management Body of Knowledge* (PMI 2008) as "a temporary endeavor undertaken to produce a unique product, service, or result." Another relevant definition is that used by Lewis (2007) where a project is "a problem scheduled for a solution" and a problem is seen as "a gap between where you are and where you want to be, with obstacles existing that prevent easy movement to close the gap."

Well executed projects, whether in the developed world or in the developing world, require following a methodology, a framework, and a management structure. They also demand trained and competent project leaders and managers. However, both of these characteristics do not necessarily guarantee successful projects because projects may not perform as planned for many reasons, even in "ideal" conditions. In the field of international development project management (IDPM), it is commonly accepted that "many internal and external, visible and invisible factors ... influence the environment and create a high amount of risk in accomplishing the project objectives" (Kwak 2002).

The factors that are deemed critical to the success of a project, called critical success factors (CSFs), vary with each project type and how project success is perceived by agencies or organizations that undertake and implement the project. For instance, Kwak (2002) lists 10 CSFs that could affect the outcome of a project: political, legal, cultural, technical, managerial or organizational, economic, environmental, social, corruption, and physical. Belassi and Tukel (1996) regroup these factors into four categories: factors related to the project, factors related to the project team and project manager, factors related to the agency or organization, and factors related to the project environment. More recently, Ika et al. (2012) showed a good correlation between the success of World Bank projects and five CSFs related to project monitoring, coordination, design, training, and institutional environment. It is important to note that the CSFs are interrelated because of the systemic nature of development projects.

For small-scale community projects in the developing world, the fundamentals of management might be the same as in Western countries, but the approaches used in project planning, design, and execution are likely to differ substantially because of the uncertain and complex context in which the projects unfold. For example, absent in developing community projects are predefined detailed blueprints that ensure control and predictability, which are found in large engineering projects. So development project managers have to be able to manage challenging and sometimes seemingly competing tasks that are not often found in major engineering projects.

Such challenges arise, for instance, in the way project managers ensure that work is completed "on time, within budget and scope, and at the correct performance level," according to Lewis (2007). In development projects, the four project constraints of performance, cost, time, and scope (Figure 4-1), defined as PCTS, need to be considered within the context of a different culture. Communities change, evolve, and reconfigure themselves over time. These conditions may induce changes in the PCTS space of a project. A change of any one side of the triangle in Figure 4-1 results in changes in the other two sides, once a given triangle area, i.e., a project scope (S), has been selected. Likewise, changing the scope of the project (area of the triangle) may change the length of some sides of the triangle. As remarked by Lewis (2007), in projects all three sides [of the triangle] and the area cannot be arbitrarily assigned. Only three constraints can be selected, and the fourth one is dictated by the other three. As a result, as noted by Laufer (2012), in adaptive environments (such as those in the developing world) where problems abound because of inherent complexity and uncertainty,

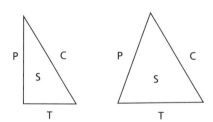

Figure 4-1. Project Performance, Cost, Time, and Scope (PCTS)

Source: Lewis (2007), with permission from AMACOM Books.

"the project manager must be willing and able to make significant changes and to challenge the status quo."

In small-scale community development projects, managers need to possess unique skills because they have to be aware of the context in which projects are conducted, the multitude of nontechnical issues that could derail projects, and the complexity inherent in those projects. They must also recognize that the tools used to understand complex systems are quite different from those for simple and complicated systems (Snowden and Boone 2007). Unlike in *simple* and *complicated* systems, which rely on order, are fact based, and have clean patterns that connect cause and effect, there are no definite answers to problems faced by *complex* systems, which involve disorder and require pattern recognition. Therefore, as noted by Narayan (1993), development project managers must be able to deliver project quality control and assurance, manage unpredictability, and reduce the project unknowns to an acceptable level without prematurely imposing inappropriate solutions. Even in a perfect world, addressing such competing constraints represents for project managers a balancing act between what's possible and what the sponsor and clients want.

A framework for the management of small-scale community development projects in developing countries is therefore necessary. In general, project management consists of a sequence of processes, such as project definition and initiation, planning, execution, monitoring and evaluation, and a closeout phase. In the developed world, such a framework is routinely used and promoted (PMI 2008). However, for projects in developing countries, there is not always the same range of practices and quality control or quality assurance as in the developed world, especially when projects are conducted by (but not limited to) charity groups and/or inexperienced nongovernmental organizations (NGOs).

In the overall spectrum of small-scale community development projects are those that follow a so-called *ready–fire–aim* approach, where the projects are conducted without much design, preparation, and planning (if any) and are expected to be done quickly, at a minimum cost, and of high quality. Needless to say, many of those projects set themselves up for failure. They do not perform well and are often left with large reworking costs that far exceed the cost the projects would have incurred if they had followed proper planning right from the start (Lewis 2007). Projects that follow a ready–fire–aim approach fail to recognize that all projects change over time, need planning, and are constrained by various parameters such as performance, cost, scope, quality, schedule, resources, customer satisfaction, and risks (PMI 2008). As a result, the project outcome may differ substantially from what the local beneficiaries expected at the start of the project.

Development projects with outcomes different from intended ones are everywhere and are not limited to those handled by inexperienced NGOs and charity groups. Even international development projects conducted by agencies that have in place strict management guidelines have shown poor performance. This fact was reported recently by the Independent Evaluation Group (IEG 2012) that reviewed almost 10,000 World Bank projects. They found an overall project success rate of 57%; the highest rate was 69% for transportation projects, and the lowest rate was 42% for health projects. Large variations were also observed across regions. Another organization, Water.org (2012) is often quoted as reporting that "over 50 percent of all water projects fail and less than five percent of projects are visited, and far less than one percent has any longer-term monitoring." Often, the systems fail shortly after installation. Baumann (2006) noted that in water projects "handpump failure rates can be anywhere from 15 to 50 percent, averaging 30 percent in Africa."

The aforementioned statistics do not reflect well on how various development agencies (big and small) manage projects. In principle, the core components of project management in the developing world should not be any different from those in traditional project management used in Western countries. They should lead to the same quality of projects. Questions arise about why this situation is currently not the case, what should be done differently, and how project management practices borrowed from those in the developed world can be adapted to the developing world. Last but not least, another question arises about how such practices can be used on different time scales, such as those in the prevention, rapid response, recovery, and development phases of disasters associated with human and natural hazards. A range of good practices addressing the management of community projects in developing countries from crisis to development is therefore needed.

Ten guiding principles are recommended when considering a framework for the management of small-scale community projects in developing countries.

- The first is to ensure that the framework accounts for the context in which the projects are conducted. As remarked by Nolan (1998) in his book *Projects that Work*, "other things being equal, projects which fit with their surroundings will work—those which do not fail." Too often, projects that are successful in one context are imposed into another context with limited or no success.
- A second principle is that the projects have to be done right from a performance and technical point of view. They must also be the right projects for the community (beneficiaries) and the environment that interact with the projects. Sustainable practices are an integral part of community projects.
- A third principle, related to the first one, is that of community participation, i.e., stakeholder engagement must be included in all stages of the framework. According to Barton (1997), stakeholders "include all persons and groups who have the capacity to make or influence

decisions that have impact on project design or implementation." Stakeholders also include those who do not have a voice and who will be affected by the project. Hence, we need to ensure that the process used in the framework helps community users "generate information to solve problems they have identified, using methods that increase their capacity to solve similar problems in the future" (Narayan 1993). Participation is not just about getting involved in the different phases of project management. It is also a way for the beneficiaries to be able to have power in influencing project outcomes. As remarked by Narayan (1993, 1995) in relation to water, sanitation, and hygiene (WASH) projects, "experience has demonstrated that involving users in decision making, goal setting, design and management increases the chances that water and sanitation facilities will be financed, used fully and looked after properly."

- A fourth guiding principle is to use an integrated approach to project design that recognizes the multidisciplinary, dynamic, and complex nature of development projects. For engineers reading this book, development needs to be understood well beyond providing just value-free technical solutions, something of a difficult concept to accept by engineers. Engineers interested in development projects need to become acquainted with nontechnical issues and need to have been exposed to a T-type education, which includes depth (technical) and breadth (nontechnical), as discussed in Chapter 3.

- A fifth guiding principle is the need to include and follow as best as possible a project logic rather than a random or piecemeal approach. The strategy consists of a combination of steps that follow a cause–effect hierarchy. Projects have an overall effect that depends on reaching goals, which themselves require meeting objectives by carrying out activities, which in turn necessitate different forms of input and resources. These various steps have measurable indicators, modes of verification, and specific targets and benchmarks. They also assume that some assumptions and preconditions are met. When not met, they have the potential for putting projects at risk.

- A sixth guiding principle is the need to adopt an adaptive and reflective practice (Schön 1983) as projects unfold to arrive at satisfactory solutions and not necessarily the best solutions. In development projects, rationality is bounded by complexity and uncertainty. As noted by Simon (1972), "complexity and uncertainty make global rationality impossible." An adaptive and reflective practice through learning by doing contributes to making sounder management decisions as the project (or an activity) is happening (using reflection in action) or after it has been completed (using reflection on action). That practice must also consider the lessons learned from completed projects, whether successful or not, and how such projects have performed over time.

- A seventh guiding principle relates to using appreciative inquiry to identify and learn from the positive deviants in the community, i.e., those individuals and households who "succeed against all odds" and yet are exposed to the same conditions and constraints as everyone else (Pascale et al. 2010). These changemakers represent leverage points in the community who, after interacting with the rest of the community, can accelerate change though participation. The solutions of the positive deviants are already proven within the context of the community and are easier to scale up across the community than solutions from external experts.

- An eighth guiding principle is about the sustainability (i.e., the long-term benefits) of the results that the framework generates and the inclusion of rights-based issues; inclusiveness; and respect for human dignity, diversity, and equity. Too often, projects fail in the long term because they have not been designed accordingly right from the start. In other instances, they tend to divide people because they become entangled into geopolitical issues (even local issues) that benefit one group or individual at the expense of others. In other cases, project assessment (monitoring and evaluation) and an exit strategy have not been incorporated into the project from the beginning and are seen as an afterthought.

- A ninth guiding principles deals with the characteristics of the solutions that projects create. In the developing world, the solutions need to be accessible, affordable, available, sustainable, reliable, and scalable. They also need to be appropriate, contextual, and equitable.

- Finally, a tenth guiding principle is about using frameworks of project planning and management that are results driven rather than activity driven. Simply put, there is more to projects than a list of technologies and activities and how many pumps, photovoltaic panels, and other artifacts have been installed. Projects are defined by the quality of the solutions that unfold from their implementation and not by the nature and quantity of technical stuff.

At the outset, this chapter reviews the main components of project management as defined by the Project Management Institute (PMI). This section is followed by a review of project life-cycle tools and frameworks used by some agencies involved in development work. It then highlights the components of the framework used in the rest of this book. The chapter concludes with a discussion on rights-based issues and a design-as-you-go (adaptive) approach to project design and management.

4.2 Project Life-Cycle Management

Whether addressing problems that communities face in the developed world or in the developing world, project management needs to be a key factor in discussions to secure the delivery of meaningful and high-quality projects. Project management involves many technical and nontechnical (e.g., socioeconomic and political) issues that ensure project success. These issues are particularly critical in community development projects. Even though this book focuses on small-scale community projects (and not large-sector programs), the steps of traditional project management still need to be followed as closely as possible. The steps can, however, be adapted to the smaller scale, depending on the context of the projects under consideration. In many instances, they do not have to be as sophisticated and demanding in context as many larger projects are. In fact, simpler (but not simplistic) strategies are likely to be more beneficial to the community and are more likely to deliver tangible solutions. Having said that, project recipients, regardless of where they live, should not expect anything less than high-quality project delivery from outsiders. Project management cannot and should not be ignored.

PMI (2008) defines project management as "the application of knowledge, skills, tools and techniques to a broad range of activities in order to meet the requirements of a particular project." According to PMI, project management is usually done through the integration of processes divided into five main groups: initiating, planning, executing, monitoring and controlling, and closing. A brief definition of each process group is given here:

- *Initiating* focuses on defining the project by determining the nature and scope of the project, how many phases it contains, the project environment, the stakeholders, the risks, the context, and whether the project can realistically be completed. The project scope includes goals, budgets, and time lines. A vision and mission for the project are developed. At the end of the project initiation phase, a *project charter* (or project hypothesis) is established.

- *Planning* is about strategy, tactics, and logistics. This step is where a project plan is outlined that includes the work to be performed, goals, procedures, budget, schedule, resources needed, risk analysis, planning team, deliverables, work breakdown, and activities needed to achieve the deliverables. A specific list of things to be done to meet goals and objectives within specific boundaries is created. The various tasks and their logic are also selected. The output of this phase is a *project management plan.*

- *Executing* corresponds to putting the project management plan to work. This phase involves coordinating resources and people and also integrating and performing project activities in accordance with

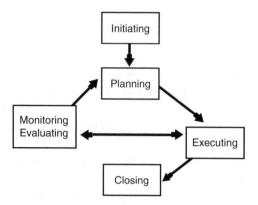

Figure 4-2. Interaction Among Project Management Processes

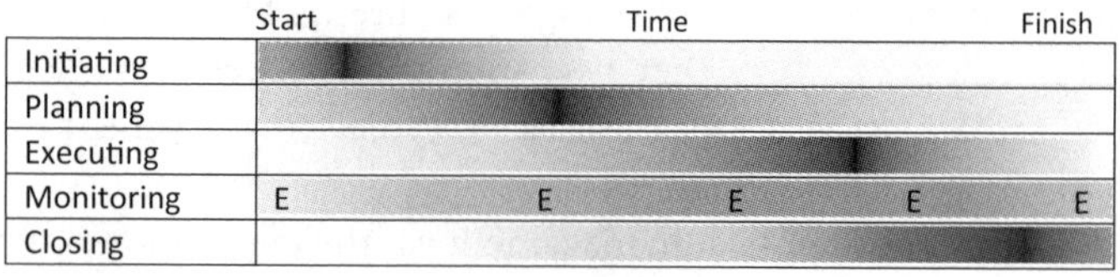

	Start	Time	Finish
Initiating			
Planning			
Executing			
Monitoring	E	E	E	E	E
Closing			

Figure 4-3. Overlap of Different Processes in a Project Life Cycle

Note: E = evaluating.

the project plan. The output of this phase consists of completing *deliverables*.

- *Monitoring and controlling* is the step that involves keeping track of and evaluating the various phases of the project, its operation, how the tasks are executed, how the outputs compare with the plan, and monitoring and evaluating the main project variables, i.e., performance, cost, risks, quality, schedule, resources, and scope, among many others.
- *Closing* is the last phase that depends on the satisfaction of the client. It also includes the reflective component of looking at lessons learned (what went well and what could be improved for the next project). The output of this phase consists of *archived project documents*.

All five process groups of project management often are iterative during the project life cycle because of the potential for change as projects unfold. Figure 4-2 shows the interaction among these five processes, and Figure 4-3 illustrates how they are likely to overlap during the lifetime of a project.

Change is inevitable in project management, and it can be managed in different ways depending on the type of project and how it has been planned (Nolan 1998). It may vary from no change permitted (*directive* or blueprint planning) to using an adaptive approach (*interactive* or learning-process planning). In general, project management entails a progressive structure where each process group input depends on what was accomplished in the previous one or what was learned in previous project iterations. At times, a directive approach is better suited to complete some phases of a project that are well defined. At other times, more flexibility is necessary to handle uncertainty and complexity, and an adaptive approach is more appropriate.

The five process groups mentioned may correspond to one major phase of a project. If the project consists of several phases as in large-sector programs, each with specific deliverables, the five processes are repeated for each phase. The phases can be sequential, where the closing part of one phase becomes the initiation for the next phase of the project. The phases can also overlap in the design and construction phases of a project.

Readers interested in more detailed components of project management ought to consult *A Guide to Project Management Body of Knowledge* (PMI 2008). They will find that the Project Management Institute suggests nine knowledge areas in their management training toward certification: integration management, scope management, time management, cost management, quality management, human resources management, communication management, risk management, and procurement management.

4.3 Project Design

Project design can mean different things to different people in different industries. As best defined by Simon (1972), "design is concerned with the discovery and elaboration of alternatives." Exploring these alternatives is integral to the five processes of project management mentioned earlier. The alternatives emerge from the interaction of analytical tools, creativity, and imagination.

For some projects, project design starts even before the initiation process of a project when the client brings forth a problem to a consultant. For other projects, project design starts after the initiation phase and requires first the gathering, analysis, and synthesis of information that helps identify problems.

The engineering and development communities look at design in different ways. According to the Accreditation Board for Engineering and Technology, engineering project design is defined as

> the process of devising a system, component or process to meet desired needs. It is a decision-making process (often iterative), in which the basic sciences, mathematics, and engineering sciences are applied to convert resources optimally to meet a stated objective. Among fundamental elements of the design process are the establishment of objectives and criteria, synthesis, analysis, construction, testing and evaluation. (ABET 2012)

This definition of design has an open-ended structure and acknowledges that the design process must follow a logic model. It encompasses two aspects: the *conceptual* part of design, which addresses what solutions are needed, and the *detailed* part of design, which looks more specifically at the details of the solutions and their implementation. The ABET definition of engineering design is content based rather than context based.

In the development literature, project design has a more holistic meaning. For instance, the Cooperative for Assistance and Relief Everywhere (CARE) defines project design as

> the collaborative and systematic identification and prioritization of problems and opportunities and the planning of solutions and ways of assessing project outcomes, which together will promote fundamental and sustainable change in target populations and institutions. (Caldwell 2002)

This definition is more contextual in nature and emphasizes developing "solutions" rather than delivering "technical fixes." In this book, the CARE definition of project design is adopted, and the need to include both content and context in developing alternative solutions to a problem is emphasized. The development approach to design is supplemented with additional components that are used in engineering practice as needed.

4.4 Project Life-Cycle Frameworks

Comprehensive frameworks for the management of development projects in developing countries have been proposed by various development agencies, such as CARE, Mercy Corps, the United Nations Development Programme, Europe-Aid, U.K.'s Department for International Development (DFID), Oxfam, and USAID. In fact, all these agencies have some form of project life-cycle framework, some more elaborate than others. In all cases, the content of the frameworks is driven by a strong desire of the agencies, in collaboration with their respective partners, to do the following:

- Provide and deliver high-quality programs and projects that improve people's lives and give them healthy choices and opportunities,
- Enable the measurements of project outcomes and effects,
- Provide documentation for future projects and develop a database of projects,
- Ensure accountability to donors, and
- Educate their respective staffs.

A common feature of these frameworks is that they all emphasize the need to include community participation or mobilization in the project phases. Another common feature is that all the frameworks recognize the cyclical nature of projects; their sequential and hierarchical structure; and the need to have a coherent information system in place in project planning, management, monitoring, and evaluation. Projects are broken down into phases whose duration and importance vary with each project. The terminology used to describe those phases may vary from one framework to the next. In all cases, each phase implies activities where decisions need to be made, monitored, and evaluated; reporting is required; and specific responsibilities are assigned. In general, for projects in developing communities that involve technical decisions, the engineering profession has had limited input into the development of existing frameworks.

Project Logic

Existing major development frameworks recognize that good project management delivery depends on adopting a strategic combination of steps that follow a cause–effect hierarchy, i.e., a project logic, which provides clear definitions of vision, mission, goals, and objectives in a project. When combined, these key components yield a clearer road map in addressing identified problems. The project logic is a model that consists of several targets and uses verifiable indicators (measurements) to qualify and quantify the progress of development projects and addresses the assumptions and risks involved in all project steps. As remarked in the *Logic Model Workbook* by the Innovation Network (2013), a logical model can support many project activities, including planning, management, communication, consensus building, and fund raising. It must be remembered though, that a model is always an abstraction of reality.

In many current development frameworks, a logical framework approach (LFA) (or logframe) is used to describe the logic involved in project planning and management. LFA asks project planners and managers to see a solution to a problem as emerging from a strategic combination and logical progression of identified inputs (resources) that are necessary for conducting various activities. These activities deliver outputs and help meet specific objectives. These objectives, in turn, produce effects and reach goals that ultimately have an overall impact (or outcome, overarching goal). To be meaningful, these components of the framework need to have indicators, modes of verification, and targets and are all subject to assumptions and preconditions. As shown in a literature review of 18 agencies by Bakewell and Garbutt (2005), LFA has become standard in development projects and is often required by donors. It also provides a common platform of understanding and communication among project stakeholders.

In general, LFA is used in project design and assessment (monitoring and evaluation) and in addressing the long-term sustainability of projects. Table 4-1 shows the basic components of the LFA in the form of a logical framework matrix. The matrix can be interpreted from the bottom up and/or the top down (vertical logic). In all cases, the "impact" or "outcome" represents the end state

Table 4-1. Typical Components of the Logical Framework Matrix

Element	Explanation	Examples
Impact (outcome/aim)	Long-term tangible changes in human well-being, organizations, and systems resulting from meeting goals	• Improved health and well-being • Increased gender equity
Effects (goals/ purposes)	Short-term and intermediate changes in human behaviors and systems resulting from meeting objectives	• Safe behaviors • Improved health care • Improved services (WASH, energy, etc.)
Outputs (objectives)	Deliverables, products, and services created by conducting project activities	• Infrastructure • Trained personnel • Better institutions
Activities	Processes, technology, tools, and actions necessary to convert inputs into outputs and meeting objectives	• Construction • Equipment install • Recruiting/training • Curriculum develop • Manufacturing
Inputs	Resources necessary to undertake activities	• Funds • Materials equipment • Personnel

and the overall changes the project is expected to make, i.e., tangible development changes. It often includes the type of improvement in human conditions after the project has been completed, the identification and number of beneficiaries, and an estimation of when change is expected to occur.

A summary of the LFA can be expressed in the form of a causal hypothesis statement:

> *this set of INPUTS and ACTIVITIES will result in these products and services [OUTPUTs], which will facilitate these changes in the population [EFFECTS], which will contribute to the desired IMPACT.* (RHRC 2004)

As an example, "external funding and expertise will be used to train governmental representatives to provide health and hygiene education of community members and training in the installation, operation, and maintenance of water pumps. This will result in better health and the supply of clean and reliable water sources. In turn, this will lead to an improvement of community well-being and economic development."

The LFA represents the *strategic* component of project planning. Once in place, it provides the necessary information to develop the project *logistics* and

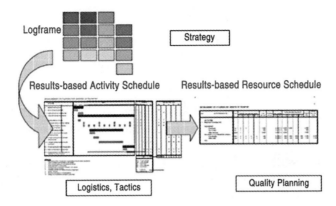

Figure 4-4. From Strategy to Logistics, Tactics, and Quality Planning

Source: EuropeAid (2002), with permission from Particip GmbH.

tactics, as shown in Figure 4-4. They define what activities and resources are needed for the project and the corresponding time frames of activity and resource delivery.

It is noteworthy that the terminology used to describe the key components in the project logic (i.e., inputs, activities, outputs, effects, and impact/outcome) can differ from one development agency to another (Mercy Corps 2005). Despite those differences, the underlying concept is to always have in place a structured approach to a project that is clearly articulated and can be understood by all project stakeholders. Furthermore, it is an approach that can be communicated to partners and donors. More discussion on LFA and how to use the logical framework matrix in development projects can be found in Chapter 8.

Alternatives to the LFA have been proposed by some development agencies. They include the Results Framework Approach used by USAID (2010), the Causal Pathway Framework (RHRC 2004), and the Theory-Based Evaluation Approach (Weiss 1997; World Bank 2002), which will not be discussed herein. In other instances, some groups and agencies use a project logic that makes sense to them, based on previous field experience.

4.5 Review of Major Development Frameworks

All major development agencies have frameworks in place for the planning and management of international projects. Even though the terminology, time line, methodology, and reporting used in the frameworks vary from country to country, they all recognize a need for structure and consistency in addressing projects and in dealing with partners and various stakeholders. Below, several of these frameworks are summarized. A modified version of the CARE framework is then used as the backbone framework for the rest of this book.

CARE Framework

The Cooperative for Assistance and Relief Everywhere (CARE) has developed a strong framework for facilitating the planning and execution of its development projects worldwide. The framework, called design, monitoring, and evaluation (DME), is best described in a CARE publication entitled *Project Design Handbook* written by Caldwell in 2002. Other supportive material to the framework can also be found on the CARE website.

The CARE framework essentially consists of five sequential steps: holistic appraisal, analysis and synthesis, focused strategy using a logical framework, coherent information systems with monitoring and evaluation, and reflective practice. The framework provides a logical and cyclical progression of project initiation, execution, monitoring and evaluation, and eventual conclusion.

Fully embedded into the CARE framework is the idea of community participation in all steps of the process and a rights-based approach (RBA). The latter ensures that people have, at the very minimum, conditions of living in dignity and empowerment to claim their rights and exercise their responsibilities.

Mercy Corps Framework

The *Design, Monitoring and Evaluation Guidebook* by the Mercy Corps (2005) provides an overall framework for conducting projects in nonemergency and nonconflict situations. It is supplemented with additional resources available on the Mercy Corps website. Eight steps are recommended for Mercy Corps project design, which can be regrouped into four categories: (1) assessment; (2) a detailed logical framework analysis; (3) a "reasonable test," which requires an outside review of project design; and (4) the completion of a work plan around the logical framework components and project management. Like the CARE framework, the Mercy Corps framework emphasizes the participatory nature of projects. It also emphasizes the needs for monitoring and evaluation and the use of indicators.

United Nations Development Programme Framework

According to the United Nations Development Programme (UNDP 2009), the delivery of high-performance projects with demonstrable results requires an integrated approach among project planning, monitoring, and evaluation. In the project framework, these three components depend on each other and can be modified in an iterative manner. UNDP (2009) offers clear, result-focused definitions of these three phases, which are outlined here:

- *Project planning* is "the process of settings goals, developing strategies, outlining the implementation arrangements and allocating resources to achieve those goals."
- *Project monitoring* is "the ongoing process by which [project-related] stakeholders obtain regular feedback on the progress being made toward achieving their goals and objectives." Monitoring is about "reviewing progress against achieving goals" and not just progress in implementing actions.
- *Project evaluation* is "a rigorous and independent assessment of either completed or ongoing activities to determine the extent to which they are achieving stated objectives and contributing to decision making." Evaluation can apply to projects and such elements as activities, program, and strategy.

The overall strategy in managing these three components in an integrated manner is called results-based management (RBM). The rationale behind this concept is that a poorly planned project is likely to be difficult to monitor and evaluate. Likewise, a poorly monitored project cannot be evaluated and requires subsequent changes in planning. As summarized by the UNDP (2009), RBM focuses on achieving results and ensures (1) better management of risks and opportunities, (2) better accountability and responsibility, (3) a process to make corrective actions and improve the execution of projects, (4) learning from experience, and (5) a way to make informed decisions.

EuropeAid Project Cycle Management

The project cycle management (PCM) framework was developed by EuropeAid in 2001 for people interested in the planning, management, and evaluation of projects funded by the European Commission's external aid programs. The *Project Cycle Management Handbook*, published in 2002, presents a detailed description of six phases in project cycle management: programming, identification, appraisal, financing, implementation, and evaluation. Figure 4-5 shows how the six phases are related, the documents that are expected as deliverables for each phase, and the decision process reached after each phase.

The PCM framework is quite comprehensive and extensively uses the LFA in all project phases. It is also very structured, with clear deliverables, indicators, modes of verification, and criteria for success. In PCM, principles of programming are clearly outlined.

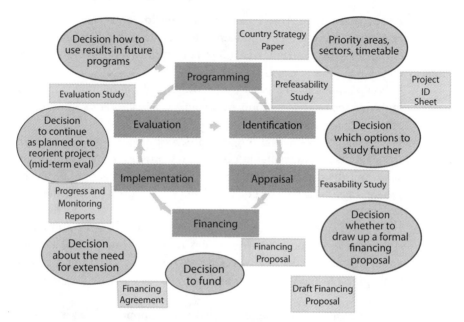

Figure 4-5. The Project Cycle: Phases, Major Documents, and Decisions

Source: EuropeAid (2002), with permission from Particip GmbH.

Department for International Development Framework

The Department for International Development (DFID) in the United Kingdom proposed in 2002 a series of tools and techniques to "undertake development activities and interventions of any kind." They include identification, design, start-up, implementation, monitoring, review, impact evaluation, and building partnerships. The guide entitled *Tools for Development* by DFID (2002) addresses the same building blocks as the other frameworks. Although more general in its description of the various components of the project cycles than the other frameworks, it offers detailed recommendations on issues such as risk management, building partnerships, and teamwork. Use of the framework is illustrated in the DFID guide with various development case studies.

U.S. Agency for International Development Framework

The U.S. Agency for International Development (USAID 2010) uses a results framework approach in its project logic instead of the traditional logical framework approach (LFA). The terminology in the USAID framework is similar to that used in the other frameworks. For instance, it uses a strategic objective (also called assistance objective), which is equivalent to the impact or outcome in the LFA. For the strategic objective to be met, several intermediate results need to be achieved. They are equivalent to the goals or effects in the LFA. The intermediate results have connections and links that need to be identified. They also involve various partners that are critical for the results to be reached and require that some critical assumptions be made. The immediate results require outputs and inputs, as for the LFA. Likewise, performance indicators and targets, along with a monitoring and evaluation plan, are integrated into the framework.

Health Communication Partnership

The Health Communication Partnership developed a framework that emphasizes community mobilization in addressing health issues and social change (Howard-Grabman and Snetro 2003). The framework consists of a community action cycle with seven well defined logical steps, plus another to make it a cycle: (1) prepare to mobilize, (2) organize the community for action, (3) explore health issues and set priorities, (4) plan together, (5) act together, (6) evaluate together, and (7) prepare to scale up, and prepare to mobilize again (step 1).

Even though it focuses strictly on health issues, the framework is actually generic and compares well with the other frameworks described above. As a result, it is adaptable to a variety of settings and conditions. This framework is well established, well documented, and easy to use, and it has been tested in countless situations.

4.6 Proposed Framework

The previous section reviewed the characteristics of several major frameworks used by agencies involved in development projects. Clearly development agencies have different ways of managing project cycles: different guidelines, best practices, processes, and policies regarding the selection and implementation of their projects. No single framework is superior to the others.

It is not the intent of this book to propose a new framework but rather to build on one major framework and develop a *toolbox* that can be used by engineers in development projects. The CARE project design framework (Caldwell 2002) was selected as the core framework. When appropriate, it is supplemented with tools used by other agencies (UNDP, Mercy Corps, and EuropeAid) and analysis tools more commonly used in engineering practice. They include risk analysis, capacity analysis, resilience analysis, and other systems tools for project planning (Delp et al. 1977). A system dynamics approach to analyze community development projects and a design-as-you-go (adaptive and reflective) approach for their implementation are included in the proposed framework as a way of accounting for the complexity and different forms of uncertainty inherent in development projects. Throughout the rest of this book, proper reference is given to the CARE model and other models, as appropriate.

It is noteworthy that the DME acronym used by CARE does not entirely provide an accurate representation of the steps used in its framework. For instance, an "A" should be included in the acronym to include the *appraisal* phase of the framework. The CARE framework also misses entirely the implementation and intervention (I) phase of the project (like many frameworks), which aligns well with engineering practices. Finally, it does not expand on the exit (E) phase (or closeout phase) of the project, an important component that needs to be decided at the outset of any project. The long-term benefits of projects must be addressed well beyond the exit phase.

In view of these limitations, the proposed framework is referred to throughout the rest of this book as ADIME-E: appraisal, design, implementation, monitoring and evaluation, and exit strategy. The framework is shown in Figure 4-6. The project planning phase (strategy, logistics, and tactics); behavior change communication; and project quality planning are included in step 3. Risk analysis, capacity analysis, and resilience analysis have been combined into step 4 of the framework. Step 0 (referred to as project initiation or identification) has been added to regroup all the activities that need to be in place before conducting community appraisal. Step 5 corresponds to project execution and assessment (monitoring and evaluation). Step 6 is about how to plan for project closure, long-term benefits (sustainability), and scaling up. Finally, step 7 is about including a reflective practice upon project completion.

Following is a brief summary of the different steps in the ADIME-E framework. All steps (except steps 0 and 7) are expanded further as separate chapters in this book (Chapters 5–12) and are presented in a sequential manner with illustrative examples.

It should be noted that the terminology used in the management of projects by developing agencies may differ from that used in traditional engineering projects in the developed world. Table 4-2 makes an attempt at comparing the terminology used in the ADIME-E framework with that used in traditional engineering management.

Initiation and Identification

It is not always clear how a project is initiated and identified at the outset. In the case of the small-scale projects in this book, a project can originate from a request by in-country individuals or groups, existing NGOs, or governmental organizations. Figure 4-7 from Nolan (2002) shows possible sources of project ideas. In all cases, some form of request starts the process. Before the project can move ahead, a prefeasibility or preappraisal phase must be carried out. According to EuropeAid (2002), this step is done to "help identify, select or investigate specific ideas, and to define what further studies may be needed to formulate a project or action."

0. Project Initiation and Identification
Prefeasibility and Preappraisal

1. Community Appraisal (Chapter 5)
Purpose – Develop a comprehensive
community baseline profile
Key Steps/Tools:
- Creating an appraisal team
- Preparing for appraisal
- Participatory action research
- Primary and secondary data collection
- Stakeholder and beneficiary analysis
- Partner analysis
- Gender analysis
- Assessment of rights
- Capacity appraisal
- Vulnerability appraisal
Output: Problem identification and ranking

2. Project Hypothesis (Chapter 7)
Purpose – Prioritize problems and develop
feasibility of solutions
Key Steps/Tools:
- System perspective (**Chapter 6**)
- Causal analysis
- Problem and solution trees
- Hierarchical analysis of solutions
- Multicriteria utility assessment matrix
Outcome: Problem statement, preliminary
design, and work plan

3. Strategy and Planning (Chapter 8)
Purpose - Identify project interventions,
project design, and develop an
operational plan
Key Steps/Tools:
- Logical framework model
- Project logistics and tactics
- Planning of project activities
- Planning of management activities
- Planning of project quality
- Predicting project impact
- Behavior change communication
- Selection of appropriate and
  sustainable technologies
  (**Chapter 13**)

7. Reflective Practice
Purpose – Reflect on action and
identify lessons learned
Key Steps/Tools:
- Encourage reflective practice
- Incorporate changes
- Explore areas of improvement
- Promote life-long learning
Outcome: Better practices and more
qualified practitioners

Improved Community Livelihood through Participation

**6. Exit Strategy, Sustainability,
and Scaling Up (Chapter 12)**
Purpose – Plan for project closure and
ensure long-term project benefits for an
"extended" period of time
Key Steps/Tools:
- Define exit strategy
- Ensure long-term benefits
- Develop criteria for sustainability
- Measures and processes are in place
- Develop critera for scaling-up
Outcome: Refined project work plan,
execution, closing, sustainability, and
scaling up

**5. Project Execution and
Assessment (Chapter 12)**
Purpose – Deliver sound and long-
lasting solutions with assessment of
project progress and impact
Key Steps/Tools:
- Develop monitoring and
  evaluation plan
- Identify targets, benchmarks,
  and
  indicators
- Develop contingency plans
- Ensure rights-based approach
Outcome: Refined project work plan
and execution

**4. Capacity Analysis, Risk
Analysis, and Resilience
Analysis
(Chapters 9, 10, and 11)**
Purpose – Refine the project
work plan and identify enabling,
constraining, and risk
environment
Key Steps/Tools:
- Analysis of community's
  capacity and resilience to
  carry out the work plan
- Risk identification and
  management
- Project impact assessment
Outcome: Refined project work
plan

Figure 4-6. ADIME-E Framework Used in this Book

The initiation and identification phase is used to establish a "rough project description" and to decide whether the project will receive a green light to proceed. If it does, that phase also serves to prepare the community for action. According to Howard-Grabman and Snetro (2003), this phase is about orienting the community, informing the community about the project and inviting participation, building trust and relationships, and identifying a core group to represent the community through the life of the project.

Based on preliminary interviews with the stakeholders and those requesting the project, combined with possible site visits and data gathering, and past experience with similar projects, a decision is made regarding whether the project

Table 4-2. Terminology Used to Describe the Steps in Managing a Project

Proposed Framework	Traditional
Appraisal and Diagnosis	Definition and Initiation (Defining a problem determines how to solve it) Define problems, develop vision, identify mission
Analysis and Synthesis	Develop Solution Options

Strategy and Planning
Capacity, Risk, and Resilience
Quality Analysis
	Planning
	(Strategy, logistics, and tactics)
	Who, what, when, why, how much, how long, tasks,
	    schedule, resources, risk management,
	    contingency, quality, and management activities

Execution	Execution
Monitoring and Evaluation	Monitoring, Control, and Evaluation (Assessment of quality and quantity of work)
Exit & Reflective Practice	Closeout (Feedback, lessons learned review)

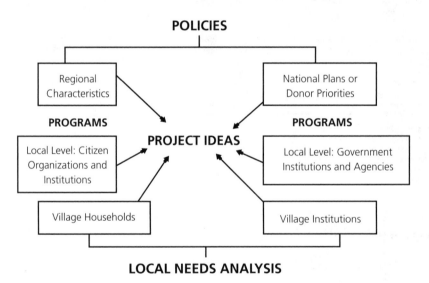

Figure 4-7. Possible Sources of Project Ideas

Source: Nolan (2002), with permission from Westview Press.

is viable and can move into the appraisal phase or if the project should be rejected. In this evaluation phase, great care must be taken to assess whether the outside organization that will intervene in the project has the capacity to manage and complete the project or if it needs to bring in other partners (subcontracts) to supplement that capacity. EuropeAid (2002) uses an extensive list of relevance and feasibility criteria to decide whether a project is realistic. Nolan (2002) cites five major decision criteria: (1) relevance (importance to community, urgency of problem at stake, effect of solution, and organizational capacity to take the project); (2) associated changes expected if projects succeed or do not succeed; (3) potential for learning from the project; (4) project complexity and its associated level of effort; and (5) how manageable the project is with the existing community. Other decision criteria may include project sustainability (results and effects) and scalability.

If the decision is made to move forward, documents are then drafted. They may include elements such as the terms of reference (TORs), contracts, a memorandum of understanding (MOU), and a scope of work (SOW). These documents are specific to the agencies conducting the study.

Community Appraisal

As for all engineering projects, a need exists to assess the community and identify its strengths, weaknesses, challenges, threats, capacity, vulnerability, resources, and the hazards (adverse events) to which it might be exposed. The main purpose of this step is to learn about the context in which the project will be conducted. In general, community appraisal goes further in depth than the initiation or identification phase. Another purpose of community appraisal is to develop a community baseline, i.e., a way to assess the current situation of the community and the main issues it is facing. Appraisal is first and foremost a sociocultural appraisal, where using participatory research and participation action research (PAR) tools, primary and secondary and qualitative and quantitative data are collected and analyzed. In addition to sociocultural issues, other community attributes are observed: e.g., environmental, economic, technical, and human resources. The results of the appraisal are presented in various forms of analysis: stakeholder and beneficiary, gender, partnership, capacity, vulnerability and vulnerable groups, social network, and uncertainty.

Problems identified by the community are acknowledged and ranked according to various criteria. For instance, a need exists to examine the added value in solving each problem in terms of improving people's quality of life compared with associated costs (cost–benefit analysis). Often, the cost of an activity and whether it is justifiable for the expected project outcomes must be addressed. Other criteria may include assessing the level of local support available to solve the problems and existing comparative advantages.

Project Hypothesis

Once the data have been analyzed and the problems identified and ranked, there is a need to further analyze each problem in terms of cause and effect. *Causal analysis* helps map the root causes and consequences (effects) of each problem identified and leads to a comprehensive picture of each problem in a hierarchical manner and a view of how different problems might be related to each other. *Problem trees* are used to represent how causes and effects relate to each problem and to each other in a systemic way. The same tree analogy is used to show how addressing the various causes and effects contributes to a solution to the problem. *Solution trees,* also called objective trees, provide a comprehensive picture of the solution, including the various activities needed to reach the solution and how these activities are interrelated in a systemic way.

In this step, a range of potential and alternative solutions and a preliminary action plan are presented. Various criteria are introduced to reduce those solutions to a more appropriate list. At the end of this step, a *project hypothesis* is developed that summarizes for each problem (1) the problem being addressed, (2) the anticipated outcomes in solving the problem, (3) the critical causes of the problem, (4) the relationships among the problem's causes and its effects, and (5) the effects of possible interventions. Assumptions and preconditions necessary to support the project hypothesis are also outlined, and the risks involved if the assumptions and preconditions are not met. At this stage of the framework, the project hypothesis can be analyzed further in terms of relevance, feasibility, and sustainability. This stage can be seen as a project feasibility study.

Strategy and Planning

A comprehensive work plan is now developed that includes project strategy, logistics, and tactics. A work plan is "a collection of documents that communicates essential information about a project to everyone who is involved in the project" (CH2M HILL 1996). *Strategy* refers to the overall game plan or overall method that will be followed to conduct the work. Tactics and logistics are related to the implementation part. *Tactics* refer to the who, what, when, where, and how of a project. *Logistics* refer to activity and resource scheduling and procurement.

Strategy is expressed in terms of a logical framework approach (LFA), which summarizes the structure of the project and its internal logic in terms of impact, goals, objectives (outputs), activities, and inputs. This approach translates into a practical operational (implementation) plan ready to be executed with well defined indicators, modes of verification of success, targets, and taking into account assumptions and preconditions that may create risk. The plan contains detailed information about project tactics and logistics, including activity and resource scheduling, responsibility charts, work breakdown structures, budgets, resource plans, and Gantt charts (used to describe the human, material, and financial means necessary to undertake the activities). Multiple project life-cycle costs are considered and may include initial, installation, energy, operational, maintenance and repair, downtime, environmental, decommissioning, and disposal costs, depending on the nature of the project.

In addition to planning the project activities, a need also exists to plan the management of project activities, such as quality control, reporting, budget control, and staff. Similar to the project activities, human, physical, and financial resources necessary to undertake the management activities need to be outlined, procured, and mobilized.

In this phase of the project cycle, the planning of project quality needs to be addressed as part of a quality management plan. This plan includes defining quality standards and the characteristics of those standards. This plan helps define a strategy for quality assurance, quality control, and quality improvement. Another aspect of project planning has to do with assessing the impact (local, regional, and global) that the project activities and solutions may create on people's health and well-being and on the environment. Issues may include noise, land, water and air pollution, deforestation, and reduced biodiversity. Special precautions must be taken to ensure that local and/or national regulations are respected. A third aspect of project planning is taking care of zoning issues and permits that are necessary before project execution.

Success of the operational plan does not rely only on engineering solutions. More often than not, the success of implementing the components of the action plan depends on necessary changes in community behavior. A behavior change communication (BCC) strategy may need to be introduced to promote positive behaviors and create a supportive environment so that the behaviors are sustainable in the long term and eventually become habits.

Complementary Analyses

Additional forms of analysis can be included in the framework, such as capacity analysis, risk analysis, and resilience analysis. They help refine and confirm the decisions and solutions made in the work plan. For instance, capacity analysis helps (1) ensure that the community has the capacity to move forward with the proposed action, (2) ensure that the solutions in the work plan match the level of community development and therefore refine what technologies and solutions are most appropriate, (3) identify the weakest links in the community, and (4) determine the necessary steps in eliminating those weak links through community capacity building so that the community can achieve a higher level of development and success over time. In addition, risk analysis explores the possibility of undesired outcomes (or absence of desired outcomes) in the different phases of the implementation plan that may result from the complex and uncertain nature of development projects. Risk management contributes to the identification, prioritization, resolution, and monitoring of risks and their management. Finally, resiliency analysis explores how communities may be resilient to various forms of adverse events and what should be done to increase their resilience.

Project Execution and Assessment

Once the work plan is in place and has been endorsed by all project stakeholders, the work plan is finally executed. Throughout the execution phase, a need exists to manage change and assess progress toward achieving the goals and objectives outlined in the logical framework starting from the community baseline developed at the end of the appraisal phase. Indicators, targets, and benchmarks are included to measure progress. Project assessment through monitoring and evaluation over the project duration helps (1) make informed decisions, (2) ensure effective and efficient use of resources, (3) assess impact, (4) define progress and project reporting, (5) ensure the collection of and use of information in a timely and systemic manner, and (6) contribute to better future projects based on lessons learned. Monitoring and evaluation help in addressing project quality assurance and quality control. Finally, they also help decide whether scaling up of the project is feasible.

Monitoring and evaluation may also help develop corrective measures and project updates, as shown in Figure 4-2, if the project deviates substantially from its objectives and goals, which may jeopardize its impact. As part of project quality improvement, the work plan is updated to reflect issues that may arise with project implementation (resources, activities, and schedules) or with issues related to project management.

Exit Strategy, Sustainability, and Scaling Up

A project closure plan is critical to successful project delivery (CH2M HILL 1996). It is an ongoing process, with activities that span the entire project life cycle starting in the identification and initiation phase.

As the project comes to an end (planned or unplanned), lessons learned must be evaluated and a final evaluation of the project must occur. Project sustainability must be ensured once the project has ended, and long-term project performance (postevaluation). Sustainability must be ensured, meaning that the project continues to deliver benefits to the target community and that measures and processes are in place should problems arise and decisions need to be made.

Criteria for project sustainability are mostly subjective. According to Nolan (2002), project sustainability depends on how the project is "compatible with its surroundings." Good design, along with community participation, sound financial support and economic environment, and fitting policy are great attributes to sustainable projects. According to the World Wildlife Fund (Gawler 2005), "a project can be said sustainable when it continues to deliver benefits for an extended period, after the main part of external support has been completed."

Finally, for long-term success, projects should be evaluated for scaling up, i.e., expanding the project scope and implementation toward a greater impact within the community or other communities.

Reflective Practice

It is important for community development practitioners to reflect on a project once it has been completed. This reflection on action process (Schön 1983) represents a valuable learning (debriefing) exercise in identifying what has worked and not worked in a project. It helps incorporate changes in future projects and explore areas of improvement. Reflective practice is also a valuable tool for the practitioners because it promotes self-learning and self-motivation, enhances skills and knowledge, increases confidence and understanding, and supports professionalism. In general, a continuum of reflection occurs on all projects, starting with project assessment (monitoring and evaluation), which is more reflection in action, all the way to the postproject reflection on action. Reflection on action can also be incorporated in the evaluation phases during project execution.

4.7 Rights-Based Approach

No matter how well planned and well executed projects are, they always cause some impact on human lives and community livelihood. Negative or unexpected impacts and consequences of projects that affect human dignity and fulfillment of human potential should be of concern. Examples include increased tensions among components of a community; marginalization of existing groups; and increased vulnerability, isolation, and gender inequalities. As discussed earlier, the uncertainty of community dynamics is likely to create unintentional problems leading to unexpected consequences. In some cases, the projects may exacerbate existing tensions in the community. Having said that, uncertainty does not legitimize the existence of problems and failed projects. Ultimately, development agencies are held accountable for the overall impact of their projects.

An example of rights-based analysis was proposed by CARE (2001) called benefits–harms analysis. Based on the human rights field and the concept of "do no harm," the proposed approach consists of nine tools. Three main categories of rights are considered: political, security, and socioeconomic. For each category of rights, unintended effects can manifest from one or several of three reasons (O'Brien 2002):

- Incomplete knowledge of the context in which the project takes place or will take place,
- Incomplete analysis of causes and effects that lead to possible effects, and
- Failure to take a course of action to reduce unintended harms or improve previously unforeseen benefits.

4.8 Uncertainty in Development Projects

Uncertainty is an integral part of small-scale community development projects. As noted by Morgan (1998), "the process of community development is inherently unpredictable and un-programmable. It depends critically on constant learning and adaptation to be effective." Uncertainty can arise from many sources, including, among others, the community itself; the data collected in the community appraisal; and the models used by outsiders to represent the community and its components. As remarked by Knight (1921), uncertainty differs from risk: "if you don't know for sure what will happen, but you know the odds, that's risk … if you don't even know the odds, that's uncertainty."

Even if uncertainty could be fully managed, additional surprises and unintended consequences would likely arise as projects are being implemented because of unpredictability. In many ways, one can say with a high level of certainty that "uncertainty is a constant" in small-scale community development projects. Thus, as suggested by Thompson and MacMillan (2010) within the context of social enterprise in the developing world, the challenge is to transform the realm of "anything possible can happen" (uncertainty) to the realm of "plausible, probable, and plannable," by taking actions that incorporate reflection, adaptation, monitoring and evaluation, and feedback mechanisms. As best summarized by Narayan (1993) for water and sanitation projects, the challenge for project managers "becomes managing unpredictability by reducing the unknown to acceptable levels without prematurely imposing inappropriate structures." In other words, project managers must be able to combine leadership roles (envision, align, deploy, and learn) and management roles (plan, do, check, and act) (CH2M HILL 1996).

The structure of each community is unique, the behavior of each community is unpredictable, and each community cannot be known in its entirety. No two communities are alike, and there is rarely a way to gather complete information about them. Uncertainty is linked to the fact that communities are complex systems rather than predictable ones. As a result, as discussed in Chapter 6, communities can be better described as systems consisting of subsystems, which themselves interact with other dependent systems in a hierarchical manner. In those systems, humans play a critical role and bring with them a limited potential for perceiving and processing information toward making decisions. In human systems (Simon 1957, 1972; Slovic 1987; Callebaut 2007; C-Change 2010),

- People interpret and make meaning of information based on their own context;
- Cultures, norms, and networks influence people's behavior;
- People cannot always control the issues that create their behavior;
- People are not always rational in deciding what is best for their well-being and health; and
- People are more likely to rely on intuitive risk perception and heuristic strategies than on actual facts.

Furthermore, as remarked by Simon on the principle of bounded rationality of human judgment,

> ... *the capacity of the ... mind for formulating and solving complex problems is very small compared with the size of the problems whose solution is required for objectively rational behavior in the real world—or even for a reasonable approximation to such objective rationality.* Simon (1957)

The inherent uncertainty at the community level, combined with the bounded rationality and unpredictability of human beings in processing information and acting upon it (e.g., heuristics), has serious implications for how decisions are made by insiders and outsiders on community projects.

- Decisions are not likely to be fully rational. They depend somewhat on the context and on many other factors, such as emotions, feelings, subconscious, and nonrational factors that are ill defined.
- Decisions are made on simplified versions of real problems and based on perceptions, perspectives, beliefs, experience, and habits. The problems are rarely analyzed in full, and all problems and associated solutions cannot be foreseen in their entirety.
- Optimal decisions are impossible to make because of incomplete knowledge of the problems at stake, and therefore decision makers must account for margins of errors. Engineering in complex and uncertain situations is more about finding "good enough" alternative solutions (satisficing) rather than the best solutions (optimizing) as remarked by Simon (1972). It is about getting to maybe instead of yes or no (Westley et al. 2007).

When engaging in community development, these implications will affect (1) how community appraisals are conducted (see challenges and biases discussed in Section 5.4), (2) how the information and data obtained by the appraisal are analyzed and presented to the community, (3) how the vulnerability of the community and associated risks are perceived and interpreted, and (4) how the recommendations from outsiders are received and understood by the community. In turn, these implications control how projects are designed, planned, executed, monitored, and evaluated in the ADIME-E framework and how predictable the outcome of those projects might be. Finally, they affect the sustainability of projects well beyond the completion phase.

Uncertainty also implies that "as complexity rises, precise statements lose meaning and meaningful statements lose precision," according to the law of incompatibility formulated by the mathematician Lofti A. Zadeh. Simply put, predictable and rational answers to uncertain problems are not possible. The system dynamics tools discussed in Chapter 6 allow for exploring variability and the effect of uncertainty on system behavior by exploring "what if" with mental models. But even mental models of systems have their own limitations and uncertainty and cannot generate definite answers because they depend on the perspective of those who build the models and their decisions in defining the system boundaries. As a result, in any model, things are included by decision makers because they are deemed important whereas other issues are ignored and made unimportant. This method supports the mathematician Box's observation in his book with N. Draper (1987) that "essentially all models are wrong, but some are useful."

In summary, inherent community uncertainty cannot be eliminated because of the inherent complexity and unpredictability of human systems and the other systems with which they interact. Therefore, it must be embraced and accounted for accordingly by using appropriate decision-making tools. One positive aspect of uncertainty is that it forces project managers to make mindful decisions and pay attention while working on projects.

Project Failure

As mentioned several times in this book, a lot of criticism exists in the literature about the delivery modes of small-scale community development projects, and concerns have been raised about their high rates of failure. Aside from failure resulting from poorly executed projects conducted by inexperienced groups or decision makers who have no skills in making those decisions (which is often the case), there are many instances where the projects have failed despite good planning, management, and execution. In many ways, the previous discussion on uncertainty can shed some light on that issue.

In his book, *The Logic of Failure*, Dörner (1997) proposed four reasons for why things fail in complex situations despite good initial planning:

- Slowness in human thinking: we feel obliged to economize and simplify;
- Slow speed in absorbing new material: we don't think about problems we don't have;
- Self-protection: we need to make things easier and have things under control to preserve our expectation of success; and
- A limited understanding of systems: we have a hard time comprehending complexity, we make hypotheses, and we are ignorant of what we don't know.

Dörner concludes that all four reasons, when combined, may lead to unexpected behavior, and he cites several real-life examples illustrating this conclusion. In one example regarding the development of the fictitious country of Tanaland, it is shown that, despite good intentions, failure to keep asking questions in the decision process in a project may lead to its rapid failure because decisions are made using nonupdated information.

Poor decision making is not the only thing that leads to unintended consequences. They may also arise from poor group dynamics. According to Diamond (2005), failure may happen when groups fail to

- Anticipate a problem before the problem actually arrives,
- Perceive a problem that has actually arrived,
- Solve a problem once it has been perceived, and
- Solve a problem after trying to do something about it.

There are plenty of reasons why projects fail in any of the phases shown in Figure 4-2, and the literature on the subject is quite extensive. Nolan (2002) lists various causes of project failure in the framing, management, and assessment stages of a project. At the *framing stage*, a project may fail because various aspects of the project have been overlooked or because wrong or incomplete decisions have been made from the appraisal data. At the *management stage*, project failure may result from bad or inadequate implementation of the project, or parts of it, or from failure to adapt to changes as the project unfolds. Finally, at the *assessment stage* (monitoring and evaluation), a project may fail because results are overlooked or lessons from the past have not been transferred into future project activities.

Clarkson (2013) lists the top 10 reasons why projects fail in general as the following:

1. Poor sponsorship,
2. Unclear requirements,
3. Unrealistic timescales or budgets,
4. Poor risk management,
5. Poor process and documentation,
6. Poor estimating,

    7. Poor communication and stakeholder engagement,

    8. Poor business cases,

    9. Inadequate and incorrectly skilled resources, and

   10. Scope creep.

In practice, and especially in small-scale community development projects, it is hard to avoid failure. In fact, we can say with a high level of certainty that failure is a possible option. Among the aforementioned reasons, it has been the experience of the author that scope creep is more a rule than an exception because it is indeed not uncommon that an intervention fails because the situation has become something other than the one planned for a project. A common scenario is for a project to spin off into subprojects that require additional planning, management, and resources. For instance, it is not uncommon for a "drinking water supply only" water project to branch out into sanitation, hygiene, health, food, and energy subprojects. Project spin-off may quickly become a major challenge for project managers who start out having to deliver high-quality projects with strict constraints. The experience of project managers, combined with monitoring and evaluation of projects as early as possible in project planning and management, is critical to reduce the effect of scope creep and to avoid unexpected surprises.

Finally, there can be positive aspects to project failure. Failure leads to valuable lessons learned, especially from projects that had all the components for success. Unfortunately, detailed documentation of past projects with positive and negative outcomes is rarely found in the development industry (Valadez and Bamberger 1994). It is indeed rare for organizations to report failed projects; and if they do, no effort is placed on trying to link negative outcomes to project design and management. The literature is mostly biased toward reporting good outcomes (often superficially) rather than failed projects. It is time to replace "failure" by "lessons learned" and start learning from past case studies.

4.9 Project Delivery in Complex Systems

One difficulty in using any of the project cycle management frameworks mentioned previously is that there is a need to accommodate the uncertainty, complexity, and unpredictability inherent in developing community projects in all stages of the project cycle. Absent in such projects are the blueprints that guarantee control and predictability, as in projects in the developed world. This context requires changing the role that managers and practitioners play in projects, their responsibilities, and attitudes when encountering challenges. As remarked by Narayan (1993), project managers need to serve as "managers of change rather than as overseers of [project] schedules." Furthermore, "their central task is to design a learning and problem-solving environment characterized by facilitative leadership, goals and a vision that are shared by users, systems for two-way knowledge generation, resource generation, conflict resolution, and generally accepted rules and regulations."

More often than not, in developing community projects, practitioners are confronted with setting the problems before addressing them, a difficult task indeed when the situation is ill defined at the beginning (Schön 1983). Once the problems have been set, solving them requires integrating adaptive change, variability, and flexibility and introducing contingency plans and options in project planning and execution. As remarked by Narayan, practitioners need to balance what is expected of them in traditional problem solving by being able to maneuver in an unpredictable learning and changing environment. The strategy used depends greatly on the nature of the projects, the community, and who the project decision makers and stakeholders are. In recovery and development projects, for instance, where time lines are less stringent and more time can be used for reflective practice (and practical reflection), uncertainty can be managed (or at least reduced) using an adaptive approach that allows for a certain amount of flexibility.

Because uncertainty in development projects arises from the complexity of communities and human systems, the tools used in planning such projects must reflect those unique features. Practitioners must also recognize that participatory community development takes time, is culture and context dependent, and cannot be imposed over a rigid time frame nor using a top-down approach only. Therefore, using traditional management tools developed in the Western world that expect nothing less than predictability is unlikely to be of any use in the context of projects in developing communities. In Chapter 6, we discuss how system dynamics tools are better suited to address the multidisciplinary, cross-disciplinary, multistakeholder, and uncertain nature of development projects.

Project management in uncertain conditions is not new. In fact, it is more a rule than an exception in engineering practice, even in the developed world, where uncertainty can be better handled through a combination of objective and subjective decision-making tools (Elms and Brown 2012). In his book *Projects that Work*, Nolan (1998) divides project planning methods into two groups of methods: *interactive* methods and *directive* methods (Table 4-3). Interactive methods are used when "the elements of the project evolve as time goes on, and as new learning occurs." Schön (1983) calls this approach *reflective practice*, which is more in line with the intervention of *self-reflective practitioners* than experts (Caldwell 2002).

Interactive and reflective methods account better for uncertainty, are more flexible and adaptive, and require preplanned adaptability and more subjective decision making. They are better suited for a learning environment. These methods are an integral part of what Patton (2011) calls *developmental evaluation*, which is an approach recommended to evaluate progress and make

Table 4-3. Aspects of Directive versus Interactive Project Planning

Directive Planning	Project Features	Interactive Planning
The impetus for the project comes from above.	Origin of the project	The impetus for the project comes from below.
Interventions are temporary.	Nature of the intervention	Involvement is long term.
The environment is stable and familiar.	The environment	The environment is unstable or unfamiliar.
Projects center on things rather than people.	Focus of the project	Projects emphasize growth in human capacity rather than material things alone.
Detailed knowledge of techniques, outcomes, and contingencies is assumed to exist at the start of the project.	Role of existing knowledge	Incomplete knowledge is assumed; learning about what to do becomes a major project goal.
Little learning or new knowledge is assumed to be necessary to make the project work.	Role of new knowledge	Learning and new knowledge are seen as central to the success of the project.
Overall strategies and objectives are spelled out in advance.	Strategies and objectives	Objectives and strategies emerge gradually from on-site study of the situation.
The research, decision-making, and action functions in the project are separated and done by different groups.	Integration of effort	Research, decision making, and action are combined and done by essentially the same group of people.
All resources, activities, and timetables are spelled out in advance.	Choice of resources, activities, and timetables	Resources, activities, and timetables are determined as the project proceeds on the basis of experience gained in this field.
Project decisions are relatively "pure" and can be made in terms of a few controllable variables, preferably of a quantitative nature.	Decision making	Project decisions are "impure" and are made in terms of shifting often qualitative factors.
Implementation is routine and involves the application of prespecified solutions. Tasks are somewhat routine and repetitive.	Implementation tasks	Implementation is creative and experimental and changes as the project evolves. Tasks are not routine, but may need to be done differently at different times.

Table 4-3. Aspects of Directive Versus Interactive Project Planning (*Continued*)

Directive Planning	Project Features	Interactive Planning
Few modifications of the project plan are possible at later stages.	Modifications of plans	Continual modification of the project plan is necessary to take account of new learning.
Little local initiative or participation is required.	Local input	Local participation is necessary to shape the project.

Source: Nolan (1998), with permission from Riall W. Nolan.

decisions in complex and uncertain settings in social innovation. The approach is about exploring the parameters of an innovation and, as it takes shape, changing the intervention as needed (and if needed), adapting it to changed circumstances, and altering tactics based on emergent conditions.

Interactive methods differ from more traditional *directive planning* methods, which are more rigid and linear, require predetermined accurate information and objective decision making, and rely on the input of experts. Most civil engineering projects (e.g., building a bridge) that deal with manufactured materials rely on directive planning or *blueprint planning*.

In a recent paper, Elms and Brown (2012) discuss the decision-making process necessary to address complex situations. They argue that even though the engineering field prides itself on its rational objective approach to decision making, there are many complex situations in which a subjective (or intuitive) approach is better suited with or without the use of additional rational decision tools. That combination of approaches can be included in all components of decision making, i.e., defining the project aim, deciding on a plan of action and alternative actions, assessing the project constraints, and accounting for the context in which the project is taking place. Elms and Brown remark that using dominantly subjective methods seems to lead to better decisions "for problems at the interface between straightforward technics—the traditional province of engineers—and the environments (natural, social, economic, political and so on) surrounding them," which are the complex problems of interest herein.

In the framework proposed in this book, an interactive planning approach called the design-as-you-go method is recommended. It has been used, for instance, in geotechnical engineering for tunnel project management over the past 40 years. Geological conditions in which tunnels are to be excavated and human systems in a community setting have something in common: they are not completely known at the outset of the project. More information and data about the ground conditions and the community emerge as the project moves along from appraisal through design, implementation, and closing. Thus, in both instances, there is a need to adjust project decision making accordingly and allow for reflective practice, flexibility, and preplanned adaptability. In some instances, a need exists to introduce simple (but not simplistic) solutions to handle the complexity. The decision process becomes a combination of objective decisions based on solid engineering principles and subjective decisions based on the encountered field conditions. Contingency plans are introduced to account for the complex and uncertain nature of the projects at stake. In tunnel construction, the design-as-you-go approach called the new Austrian tunneling method, or NATM (2013), has three main attributes in terms of project planning and decision making.

- All parties involved in the project (clients, engineers, and contractors) are required to agree beforehand on a joint approach to handling changes in geologic conditions, a preplanned adaptability. Because tunnel design is about designing its support system, all three parties agree on various support methods based on the most likely geologic conditions to be encountered. Upon ground excavation, all parties meet to check the conditions and agree on selecting a type of support system until new ground data become available.

- At all times, the tunnel is instrumented and the ground conditions are monitored. The performance of the tunnel is evaluated continuously. Any unexpected change results automatically in a change in project design.

- Contract documents reflect the multiparty agreement in the decision-making process and have been approved by all parties. The litigious nature of the project is therefore reduced through relationship and trust.

How does the NATM translate into community development projects? Its first attribute relates to participatory planning, where all parties and stakeholders (insiders, outsiders, and government) involved in the development project agree on a range of action plans depending on the community conditions encountered. The second attribute expresses itself in the form of having a project monitoring and evaluation structure in place as soon as possible as a way of assessing the progress of the projects in a participatory manner. The third aspect deals with the contractual practices of development agencies and the need to include alternative dispute resolution agreements in contractual documents to handle issues and disagreements between stakeholders as they arise (Nolan 1998). It also relates to the fact that collaborative projects are easier to handle, and more likely to succeed, than those that are confrontational.

In recovery and development projects, interactive planning approaches such as the design-as-you-go approach are better suited to the situation because more time is allowed to make decisions and adapt to changes in project conditions. In rapid response projects (refugee camps and other crisis situations), directive planning makes more sense because the response time is constrained, there is less tolerance for uncertainty (in a very uncertain environment), and more rigorous and timely decisions need to be made quickly in relation to saving lives and protecting populations.

In general, practitioners involved in small-scale development projects need to be cognizant of both interactive and directive planning approaches and use the appropriate ones as needs arise. They must recognize that each project is unique and requires a specific approach. Failing to recognize the uniqueness in project planning and execution by using the same tools irrespective of the project context may create more harm than good.

4.10 Chapter Summary

In this chapter, the basic components of the ADIME-E framework for small-scale projects in developing communities were introduced. A more robust description of each component of the framework is provided in the forthcoming chapters. The main attributes of the proposed framework can be summarized as follows:

- Its components are borrowed from existing development frameworks.
- It includes logic used in engineering project management.
- It emphasizes the contextual and participatory nature of development projects.
- It is centered on household livelihood, i.e., the smallest basic human community unit.
- It acknowledges and accounts for the uncertain nature of development projects in all stages of the project cycle.
- It accounts for the integrated and multidisciplinary characteristics of development projects where technical and nontechnical issues are intertwined.
- It recognizes that depending on the project phase and the type of project, adaptive planning methods (with subjective decision making dominant) or directive planning methods (with objective planning methods dominant) need to be used.
- It recognizes that development projects involve human systems and interaction with other systems, and therefore a need exists for using system dynamics tools rather than linear cause–effect tools in project planning and design.

The overarching goal of the proposed framework is to provide the conditions necessary for sustainable community development and social change leading to communities that are more stable, resilient, equitable, secure, healthy, and prosperous. In general, such communities have the capacity (or ability) through resources and knowledge to (1) address their own problems, (2) be self-sustaining, (3) cope with and adapt to various forms of stress and shocks,

(4) satisfy their own basic needs, and (5) demonstrate livelihood security for current and forthcoming generations.

Finally, it must be recognized that the proposed ADIME-E framework is not a magic bullet that transforms a community overnight and guarantees its transformation. Development takes time and hopefully consists of more steps forward than steps backward on the road to community self-reliance.

References

Accreditation Board for Engineering and Technology (ABET). (2012). *Criteria for accrediting engineering programs, effective for evaluations during the 2013–2014 accreditation cycle*, Accreditation Board for Engineering and Technology, Baltimore.

Bakewell, O., and Garbutt, A. (2005). "The use and abuse of the logical framework approach." Swedish International Development Cooperation Agency, Stockholm, Sweden. <http://www.intrac.org/data/files/resources/518/The-Use-and-Abuse-of-the-Logical-Framework-Approach.pdf> (Jan. 10, 2013).

Barton, T. (1997). *Guidelines for monitoring and evaluation: How are we doing?* CARE International in Uganda, Monitoring and Evaluation Task Force, Kampala, Uganda. <http://portals.wdi.wur.nl/files/docs/ppme/Guidelines_to_monitoring_and_evaluation%5B1%5D.pdf> (Jan. 12, 2012).

Baumann, E. (2006). "Do operation and maintenance pay?" *Waterlines*, 25(1), 10–12.

Belassi, W., and Tukel, O. I. (1996). "A new framework for determining critical success/failure factors in projects." *International J. Project Management*, 14(3), 141–151.

Box, G. E. P., and Draper, N. R. (1987). *Empirical model-building and response surfaces*, Wiley, New York.

Caldwell, R. (2002). *Project design handbook*, Cooperative for Assistance and Relief Everywhere (CARE), Atlanta.

Callebaut, W. (2007). "Herbert Simon's silent revolution." *Biological Theory*, 2(1), 76–86.

C-Change. (2010). "Communication for change: A short guide to social and behavior change (SBCC) theory and models." <http://www.slideshare.net/CChangeProgram/communication-for-change-a-short-guide-to-social-and-behavior-change-sbcc-theory-and-models> (March 15, 2014).

CH2M HILL. (1996). *Project delivery system: A system and process for benchmark performances*, CH2M HILL and Work Systems Associates, Inc., Denver.

Clarkson, I. (2013). "Top ten reasons why projects fail." <http://pmstudent.com/top-ten-reasons-why-projects-fail/> (Oct. 3, 2013).

Cooperative for Assistance and Relief Everywhere (CARE). (2001). *Benefits–harms handbook*, Cooperative for Assistance and Relief Everywhere, Atlanta. <http://www.care.org/getinvolved/advocacy/policypapers/handbook.pdf> (Oct. 10, 2012).

Delp, P., et al. (1977). *Systems tools for project planning*, International Development Institute, Bloomington, IN.

Department for International Development (DFID). (2002). *Tools for development*, version 15, Department for International Development, London.

Diamond, J. (2005). *Collapse: How societies choose to fail or succeed*, Penguin Group, New York.

Dörner, D. (1997). *The logic of failure: Recognizing and avoiding error in complex situations*, Perseus Books, Cambridge, MA.

Elms, D. G., and Brown, C. B. (2012). "Decisions in a complex context: A new formalism?" *Proc. International Forum on Engineering Decision Making*, 6th International Forum on Engineering Decision Making, Lake Louise, AB, Canada.

EuropeAid. (2002). *Project cycle management handbook*, version 2.0, PARTICIP GmbH, Fribourg, Germany. <http://www.sle-berlin.de/files/sletraining/PCM_Train_Handbook_EN-March2002.pdf> (March 10, 2012).

Gawler, M. (2005). *WWF introductory course: Project design in the context of project cycle management*, Artemis Services, Prévessin-Moëns, France <http://www.artemis-services.com/downloads/sourcebook_0502.pdf> (March 15, 2012).

Howard-Grabman, L., and Snetro, G. (2003). *How to mobilize communities for health and social change*, Center for Communication Programs, Baltimore. <http://www.jhuccp.org/resource_center/publications/field_guides_tools/how-mobilize-communities-health-and-social-change-20> (March 22, 2012).

Ika, L. A., Diallo, A., and Thuillier, D. (2012). "Critical success factors for World Bank projects: An empirical investigation." *International J. Project Management*, 30, 105–116.

Independent Evaluation Group (IEG). (2012). *World Bank projects performance ratings: Projects completed in period 1981–2010*, World Bank, Washington, DC, <https://databox.worldbank.org/> (April 4, 2013).

Innovation Network. (2013). "Logic model workbook." <http://www.innonet.org/client_docs/File/logic_model_workbook.pdf> (Oct. 1, 2013).

Knight, F. H. (1921). *Risk, uncertainty, and profit*, Reprinted in 2009 by Signalman Publishing, Kissimmee, FL.

Kwak, Y. H. (2002). "Critical success factors in international development project management." *Proc. CIV 10th Int. Symp. on Construction Innovation & Global Competitiveness*, B. O. Uwakwhe and I. A. Minkarah, eds., CRC Press, Boca Raton, FL.

Laufer, A. (2012). *Mastering the leadership role in project management*, FT Press, Upper Saddle River, NJ.

Lewis, J. P. (2007). *Fundamentals of project management*, 3rd Ed., Amacom, New York.

Mercy Corps. (2005). *Design, monitoring and evaluation guidebook*, Mercy Corps, Portland, OR, <http://www.mercycorps.org/sites/default/files/file1157150018.pdf> (March 15, 2012).

Morgan, P. (1998). *Capacity and capacity development—Some strategies*, Canadian International Development Agency (CIDA), Policy Branch. CIDA, Guatineau, Quebec <http://portals.wi.wur.nl/files/docs/SPICAD/14.%20Capacity%20and%20capacity%20development_some%20strategies%20(SIDA).pdf> (March 15, 2012).

Narayan, D. (1993). *Participatory evaluation: Tools for managing change in water and sanitation*, World Bank, Washington, DC, <http://www.chs.ubc.ca/archives/files/Participatory%20Evaluation%20Tools%20for%20Managing%20Change%20in%20Water%20and%20Sanitation.pdf> (Dec. 10, 2012).

Narayan, D. (1995). *The contribution of people's participation: Evidence from 121 rural water supply projects*, World Bank, Washington, DC, <http://www-wds.worldbank.org/external/default/WDSContentServer/WDSP/IB/2006/12/29/000011823_20061229153803/Rendered/PDF/38294.pdf> (Dec. 10, 2012).

New Austrian Tunneling Method (NATM). (2013). <http://en.wikipedia.org/wiki/New_Austrian_Tunnelling_method> (March 22, 2013).

Nolan, R. (1998). "Projects that work: Context-based planning for community change." Unpublished report. Department of Anthropology, Purdue University, West Lafayette, IN.

Nolan, R. (2002). *Development anthropology: Encounters in the real world*, Westview Press, Boulder, CO.

O'Brien, P. (2002). "Benefits–harms analysis: A rights-based tool developed by CARE International." *Humanitarian Exchange Magazine*, March, 20.

Pascale, R., Sternin, J., and Sternin, M. (2010). *The power of positive deviance: How unlikely innovators solve the world's toughest problems*, Harvard University Review Press, Cambridge, MA.

Patton, M. Q. (2011). *Developmental evaluation: Applying complexity concepts to enhance innovation and use*, Guilford Press, New York.

Project Management Institute (PMI). (2008). *A guide to the project management body of knowledge*, PMBOK guide, 4th ed., Project Management Institute, Newtown Square, PA.

Reproductive Health Response in Crises Consortium (RHRC). (2004). "Consortium monitoring and evaluation toolkit." <http://www.rhrc.org/resources/general_fieldtools/toolkit/index .htm> (March 30, 2013).

Schön, D. A. (1983). *The reflective practitioner: How professionals think in action*, Basic Books, New York.

Simon, H. A. (1957). *Models of man: Social and rational*, Wiley, New York.

Simon, H. A. (1972). "Theories of bounded rationality." *Decision and organization*, C. B. McGuire and R. Radner, eds., 161–176, North-Holland Pub., Amsterdam, Netherlands.

Slovic, P. (1987). "Perception of risk." *Science*, 236(4799), 280–285.

Snowden, D., and Boone, M. (2007). "A leader's framework for decision making." *Harvard Business Review*, November, 69–76. <http://www.mpiweb.org/CMS/uploadedFiles/Article%20for%20 Marketing%20-%20Mary%20Boone.pdf> (June 10, 2013).

Thompson, J. D., and MacMillan, I. C. (2010). "Business models: Creating new markets and societal wealth." *Long Range Planning*, 43, 291–307.

United Nations Development Programme (UNDP). (2009). *Handbook on planning, monitoring and evaluating for development results*. United Nations Development Programme, New York. <http://web.undp.org/evaluation/handbook/documents/english/pme-handbook.pdf> (May 22, 2012).

U.S. Agency for International Development (USAID). (2010). *Performance monitoring and evaluation: Building a results framework*, 2nd Ed. U.S. Agency for International Development, Washington, DC, <http://pdf.usaid.gov/pdf_docs/PNADW113.pdf> (Jan. 22, 2013).

Valadez, J., and Bamberger, M. (1994). *Monitoring and evaluating social programs in developing countries: A handbook for policymakers, managers, and researchers*, World Bank Publications, Washington, DC.

Water.Org. (2012). <http://water.org/water-crisis/water-facts/water/> (March 22, 2012).

Weiss, C. H. (1997). "Theory-based evaluation: Past, present, and future." *New Directions for Evaluation*, 76, 41–55.

Westley, F., Zimmerman, B., and Patton, M. Q. (2007). *Getting to maybe: How the world is changed*, Vintage Canada.

World Bank. (2002). *Monitoring and evaluation: Some tools, methods and approaches*, World Bank, Washington, DC, <http://www.worldbank.org/oed/ecd/tools/> (April 4, 2012).

5

Defining and Appraising the Community

5.1 About Appraisal

Defining and appraising a community are critical steps in project design. Even though this work may only represent 10–20% of an entire project's effort, failure to appraise a community properly right from the beginning may have lasting consequences throughout the life of the project. In general, having bad information about the context in which a project takes place, combined with good intentions gives bad results, i.e., *garbage in, garbage out*. Hence, community appraisal cannot be ignored. Its importance is best stated by Schumacher (1973): "find what people do and know (knowledge and their resources) and teach them to do it better." In other words, before change can be proposed to a community, its operating environment (enabling and constraining) and that of its basic units, i.e., the households, must be understood. This information is both qualitative and quantitative. A community *baseline profile* must be established and defined within geographical boundaries (area of influence) and time boundaries.

It should be noted that the term *assessment* is sometimes substituted in the development literature for *appraisal*. In this book, a clear distinction is made between these two terms. Assessment is used within the context of monitoring and evaluation, as discussed in Chapter 12. Appraisal is about learning about the community through the collection of data, the transformation of data into useful information, and the analysis of that information.

Development agencies use various methods and tools to carry out community appraisal. Figure 5-1 shows for instance the various steps used by the Namsaling Community Development Center (NCDC 2013) in Nepal in their community appraisal. Regardless of the method selected, community appraisal takes time, requires patience, and can be costly. At the same time, it has to remain realistic and significantly appropriate without going overboard; unnecessary and excessive details are only detrimental in the long run. Knowing when to stop spending time and effort in collecting data represents a major decision from the

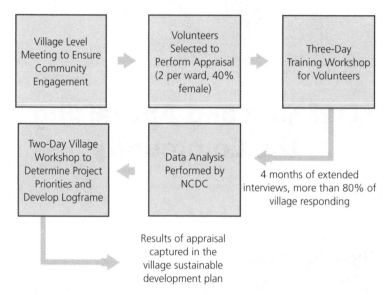

Figure 5-1. Methodology Used by the NCDC in Conducting Community Appraisal in Mabu, Nepal

Source: Courtesy of Barry Bialek.

data collection team because it is easy to fall into paralysis in analysis. Finally, it is assumed that those conducting community appraisal have appropriate skills and experience and are aware of the biases they bring into the process.

The overall approach discussed herein is similar to the approach used in geotechnical engineering (a subset of civil engineering focusing on the interaction among soils, rocks, and other geomaterials with engineering structures) projects, where uncertainty is the norm. Site exploration and investigation of geological formations in engineering projects always reveals the variability of the geology at the project site. We cannot get an accurate picture of the site geology, only information at selected locations. A limited number of boreholes, trenches, and other techniques are often used to get an idea of the geology in place, but never a complete one. Hence, extrapolation and professional judgment must be applied when designing and constructing infrastructure (e.g., foundations, dams, and tunnels) that interacts with geological formations (soils and rocks). The same is true when working with a community. A community is a system that cannot be known in its entirety; there will always be some "information deficit," as noted by Nolan (1998). As discussed in Section 4.8, the goal is to reduce the uncertainty in the participatory decision process and convert uncertainty (not quantifiable) into risk (quantifiable). As discussed in the previous chapter, uncertainty can be integrated into project design and implementation by integrating reflective practice, interactive planning, developmental evaluation, and a design-as-you-go approach. Failure to ignore uncertainty may result in unpleasant surprises as projects unfold.

In the overall ADIME-E framework shown in Figure 4-6, the results of the appraisal discussed in this chapter feed directly into the second step in the framework, which consists of causal analysis and synthesis, e.g., the identification and prioritization of problems and the identification of priorities and feasible solutions. This chapter discusses participatory research and participatory action research (PAR), a methodology and a broad set of ethnographic tools used in social and health sciences, agriculture, and development that try to answer questions such as the following: What is participation? Why participate? Who participates? How do people participate? When do they participate? Where do they participate? What are the components of participation? This questioning is followed by a discussion on what represents a community.

After the participatory research and community discussions, the reader is presented with the various steps in the appraisal phase, which include (1) creating an appraisal team, (2) preparing for the appraisal, (3) designing and carrying out the appraisal, and (4) analyzing and presenting the appraisal results to the stakeholders in the community. The chapter ends with recommendations on how to identify and rank the problems that emerge from the results of the community appraisal. In the present context and according to Caldwell (2002), a problem is seen as "a condition or set of conditions that affect people in a negative way e.g. death, infectious diseases, poverty, low income, low agricultural production, inadequate housing."

5.2 Appraisal Outcome

Starting with the end in mind, the main goal of the appraisal phase is to learn as much as possible about the community through the collection of data, the transformation of data into useful information, and the analysis of that information. It provides a *local context* of the community's operating environment, its cultural setting, and its level of development.

The appraisal also provides information about the more *global context* of the country and region in which the community resides: (1) the country's level of development (e.g., its gross domestic product or its human development index) and its global governance policies with regard to demographics, health, education, gender equality, employment, water and sanitation, and land usage and ownership; (2) its geopolitical structure; (3) its various types of industry; and (4) its strengths, weaknesses, and priorities. In general, a successful appraisal outcome depends largely on building trust and developing good relations with the community members and its leaders through *participation*. At the end of the appraisal phase, a detailed *community baseline* is established.

Core Information

For a given appraisal type (e.g., of health; the environment; energy; water, sanitation, and hygiene (WASH); or food and nutrition) or a combination of appraisal types, the goal is to have at the end of the appraisal phase a clearer idea of (1) the community's needs, vulnerability, constraints, capacity, and livelihood security; (2) the community structure, dynamic, and variations (differentiation and disaggregation); (3) the individuals and groups that have an interest in the project outcome, i.e., the project stakeholders and partners (at the local, regional, and national levels); (4) the range of issues (problems, threats, needs, and challenges) of concern to the community and their rankings; and (5) opportunities for solving key issues.

More specifically, the appraisal phase provides core information (qualitative and quantitative) about the following:

- The community itself: location; demographic, geographic, socioeconomic, political, cultural, and environmental factors; health; education; beliefs and practices; attitudes; feelings; human rights; power distribution; forms of behavior; and positive deviance;
- The community dynamic, including social groups, vulnerable groups, government, institution, the decision process and leadership roles, marginalized groups, rights assessment, gender equality, support groups, connection and social networks, community vision, and priorities;
- How people with different identities (because of tradition, gender, patriarchy or matriarchy, ethnicity, race, caste, childhood, aging, or disability) experience poverty, violence, or oppression;
- The positive deviant individuals or groups (changemakers) in the community who do things differently and successfully using uncommon behavior and attitude;
- The range of stakeholders and groups in the community (through stakeholder and partner analysis) and their interests, resources, and levels of influence (positive or negative);
- The community resources, skills, strengths, and capacity (institutional, human resources, technical, economic and financial, energy, environmental, social, and cultural) and the quality, quantity, and state of those resources and skills;
- The range of adverse events (small, medium, and large) the community has experienced in the past (the events need to be mapped in terms of type, location or extent, intensity, severity, duration, surprise effect, probability of occurrence, risk drivers, and how they affected the community);
- The community concerns, priorities, sense of vulnerability, and risks (real and perceived) that could harm people, property, services, livelihoods, and the environment on which people depend;

- The community needs that, according to Caldwell (2002), can be broken down into three types: (1) *normative needs* based on "desirable conditions" defined by national and/or international standards; (2) *felt needs* based on the perceptions by community members; and (3) *relative needs* based on comparing one group (community, household, or individuals) with another;
- The community dynamic across various seasons (in rural areas, seasons define community and household activities); and
- In-country governance, policy, and social, political, and economic issues at the regional and national levels that the community needs to consider in its development; regional and national policies in public health and sanitation, education, job creation, shelter, transportation, energy, poverty reduction, and other areas need to be identified because they may facilitate community development in some cases or create impediments to development in other cases.

Table 5-1 gives a list of possible relevant questions that may be asked when collecting the aforementioned information. Examples of detailed baseline survey questionnaires can be found in Forbes (2009) and Zahnd (2012).

Specific Information

In addition to core information about the community, the appraisal phase also contributes information about the capacity of the community to deliver more specific services to its members related to energy, WASH, health, shelter, education, food, or transportation. Using appropriate indicators, the quality and quantity of existing services can be appraised and compared with existing standards to identify service gaps. Having a strong capacity appraisal in place for various types of service delivery is critical in selecting correcting options, implementing appropriate solutions, and monitoring and evaluating the long-term well-being of the community. More specifically, the appraisal needs to identify for each type of existing service what works well, what does not work well, and what could be improved.

For energy projects, it is important to assess and understand what sources of energy are currently used, how they are collected, and by whom. Figure 5-2 shows an example of how energy is used by different groups of stakeholders in a village in Mali (Johnson and Bryden 2012). For renewable energy projects in developing countries, Zahnd (2012) suggests paying particular attention to the following issues during community appraisal: (1) how and where the energy will be used; (2) how much energy is needed; (3) what sources of renewable energy (wind, water, solar, biomass) are available at various locations and seasons; (4) what the agricultural practices in the community are and how they vary over a year; (5) what the seasonal living patterns of the community are; (6) what the

Table 5-1. Suggested Questions in Community Appraisal

- **Sociocultural Factors and Conditions**
  - How do the community members cook their meals?
  - What type of roles do women play in local society?
- **Technical Factors and Conditions**
  - Does the community have access to electricity? If so, where does it come from?
  - What type of technology is present (i.e., computers, cars)?
- **Economic Factors and Conditions**
  - How do the community members make a living and earn money?
  - How do they prioritize typical expenditures (i.e., food and water over entertainment)?
- **Political Factors and Conditions**
  - What type of relationship does the community have with the local, regional, and national branches of the government?
  - Is there some sort of social hierarchy?
  - What does the power and authority structure within the community look like?
  - Is there a local chief, mayor, sheik, or committee of elders?
- **Environmental Factors and Conditions**
  - What happens to wastewater?
  - What is the local geography like (e.g., by a river, in a desert, or on a mountain)?
- **Community Health Factors and Conditions**
  - How often are members of the community sick? What are common symptoms?
  - Do they have access to clinics and/or doctors?
- **Institutional Factors and Conditions**
  - Have there been or are there now any NGOs or other service-based organizations active in the area and/or community?
  - Is there a potential for any partnerships between the team, the community, and any local or regional NGOs?
- **Educational Factors and Conditions**
  - What kinds of schooling do children receive (i.e., public education, home schooling)?
  - What levels of education do children typically complete (i.e., elementary school, high school, college)?

Source: Zweig (2011), with permission from Will Zweig.

social, economic, cultural, and religious practices are; and (7) how possessions are distributed across the community and within each household. Detailed guidelines on how to conduct an energy service survey at the village level can be found in the book entitled *Rural Energy Services* by Anderson et al. (1999). Community energy appraisal requires collecting energy data in an anecdotal way (by using questionnaires and/or focus groups) and hard scientific data, such as solar radiation, wind power, water power, and amount and calorific value of collected biomass.

Likewise, in water supply projects, issues of water availability (quantity), water quality, and the functioning of existing water supply systems (pumps,

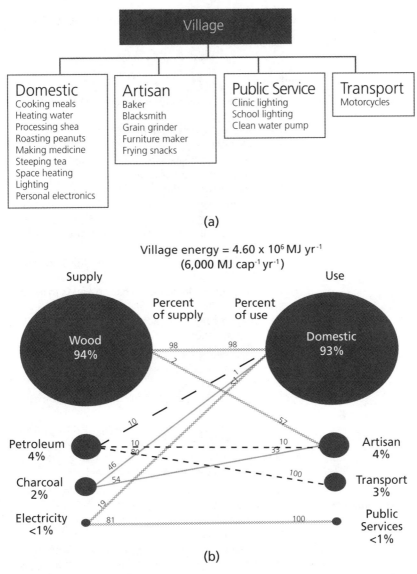

Figure 5-2. Survey of Energy Use in Village in Mali by (a) Group of Stakeholders and (b) Energy Type Used

Source: Adapted from Johnson and Bryden (2012), with permission from Elsevier.

storage, distribution, and treatment systems) need to be appraised. Table 5-2 lists various indicators that could be of help to an appraisal team in developing a baseline survey before deciding on a water management strategy (Barton et al. 1997) that may decide whether to improve existing water systems or build new ones. In addition, project teams may want to add information related to existing

Table 5-2. Indicators of Water Availability, Water Quality, and Functioning of Water Supplies and Distribution Systems at the Local Level

Topic	Indicators
Water availability	• Percentage of households with safe domestic water sources (e.g., wells or taps) • Average walking time from house to source of safe, drinkable water • Average number of liters per capita available in the household in different seasons • Hourly capacity of springs in different seasons • Depth of the water table as measured in different seasons in a sample of wells • Regularity of stream flow (e.g., overflow after rainfall or dry in summer) • Average time spent daily for watering cattle in different seasons • Percentage of productive units having access to irrigation systems • Surface of irrigated land plots • Length of irrigation systems
Water quality (human use only)	• Turbidity, chemical pollution, and bacterial pollution • Number of fecal coliforms per ml in different water sources and seasons • Salt concentration per mL in different water sources and seasons • Frequency and appropriateness of chlorinization of wells, tanks, and piped-water systems • Percentage of households satisfied with the taste and appearance of water in different seasons
Functioning of water supply systems	• Number of days per year in which household wells or taps are not functioning • Percentage of households relying on domestic water harvesting systems • Presence and function of local water committee controlling maintenance and support for water supply • Number and availability of local mechanics with training to repair wells or taps • Liters lost per minute because of major leakages in the supply system • Seasonal differences in depth of water table

Source: Barton et al. (1997), with permission from International Union for Conservation of Nature (IUCN).

water sources (springs, wells, or aquifers) and water catchment systems; the seasonality of surface water and groundwater; annual rainfall; existing vegetation cover; erosion; current practices (where, when, how, and who) in water collection and wastewater disposal; existing sources of water contamination; drainage and infiltration systems; and how water sources are shared within the community and with other communities. The energy sources that are used to pump, store, distribute, and treat the water need to be identified as well. A more detailed list of questions related to the survey of existing water systems can be found in a paper by Butts (2010).

For WASH projects, Narayan (1993) proposed indicators that can be used to monitor project performance starting from a WASH baseline survey during community appraisal. Among the various indicators are those that measure the following:

- The sustainability and reliability of WASH systems and related human capacity development, local institutional capacity, cost sharing and unit costs, and collaboration among organizations; and
- The effective use of WASH facilities expressed in terms of optimal use, hygienic use, and consistent use.

Additional information related to existing sanitation systems (e.g., open defecation, latrines, or septic tanks), their location (with respect to water sources), use, functioning, and maintenance needs to be collected as well. Hygiene practices, behavior, cultural habits, and the existence of any form of hygiene education at the community level need to be mapped to identify existing practices of hand washing, bathing, and transmission channels of virus and bacteria. The health effect of existing WASH conditions must also be assessed. A detailed participatory approach and methodology on how to gather and analyze information regarding rural water supply and sanitation projects and programs in developing countries can be found in Mukherjee and van Wijk (2003) and Ockelford and Reed (2006).

In regard to community health, various strategies have been proposed in the literature to develop the health profile of a community and conduct a health needs assessment (Conant and Fadem 2008; Birley 2011). For example, according to Barton et al. (1997), this health profile can be determined using three sets of indicators that define the quality and quantity of health:

1. Type and location of health risks:
   - The range of health problems that the community is facing and has faced over the past 6 to 12 months and who those problems affect;
   - The distribution and location of health problems in the community;
   - The risk factors that may contribute to health problems, such as poor sanitation and hygiene, crowded living conditions, and location of communities relative to known hazards and environmental issues.

2. Child nutrition.
3. The capacity and vulnerability of existing health services.

In poor communities in urban and peri-urban areas, a need also exists to understand how health is affected by factors related to the built environment (e.g., roads, transportation, pollution, gas emission, climate, lack of sidewalks, infrastructure, water supply, and sewerage). This need is particularly important as cities become more densely populated. The publication entitled "Understanding the Relationship between Public Health and the Built Environment" by the LEED-ND core committee (2006) gives an excellent survey of the various forms of ill health related to infrastructure and reviews the data that need to be collected.

With regard to food security, the following issues may need to be appraised (Korf 2003):

- Food availability (quality, quantity, seasonability) over a one-year period;
- Food access and distribution in terms of household purchasing power; and
- Food usage and preparation in terms of nutritional and health aspects.

Also related to food, the appraisal of farming practices and land usage (including landless farming and subsistence farming) and types of infrastructure involved in farming (e.g., irrigation, water storage, roads, wells, nurseries, and greenhouses) needs to be evaluated.

Finally, related to both food and health is the issue of cooking practices: the types of fuel and cookstoves used for cooking and heating (e.g., biomass or liquefied petroleum gas) and the level of air pollution and exposure in the household environment. Johnson and Bryden (2012) present an excellent case study of gathering and analyzing information regarding the use of wood collection, consumption, and cookstove use in an isolated village in Africa.

Another related issue is that of solid waste, more specifically, the amount generated, type, and methods of collection and disposal (or lack thereof). As noted in a report by the World Bank (2012), "uncollected solid waste contributes to flooding, air pollution, and public health impacts such as respiratory ailments, diarrhea, and dengue fever." The book entitled *A Community Guide to Environmental Health* by Conant and Fadem (2008) emphasizes the importance of appraising how the environment around a community influences its health.

Community Baseline Profile

In general, at the end of the appraisal phase, a *baseline profile* of the community is established. It can be seen as a series of snapshots about a day in the life of the community, its households, and members. The baseline defines the overall behavior of the community and its corresponding structure. That baseline is usually quite comprehensive and is likely to cover many interrelated areas.

Overall, it should define the community as it sees itself (not as outsiders see it) today through its strengths, weaknesses, opportunities, and challenges. In summary, the baseline profile helps identify the enabling and constraining factors in the community.

It is noteworthy that the appraisal team's perception of the baseline profile is likely to change during the entire appraisal phase. It may be limited in scope at the start when few data are available and become more comprehensive as the appraisal proceeds and the community members are more trusting of outsiders. Furthermore, its refinement does not stop at the end of the appraisal phase and continues well into the project execution, and even later in its monitoring and evaluation phase. As remarked by Nolan (2002), "gaining an insider's view of another culture takes time and effort, as patterns fall into place one piece at a time." However, even the best appraisal is never complete because there is always some form of uncertainty about the community. From a practical viewpoint, the results of the appraisal phase are usually presented in matrix (or tabular) form or by other means of data representation (e.g., sketches, drawings, or videos).

5.3 Community Diagnostic Tools

About Community

Community appraisal requires that a community or communities be selected for project work. Etymologically, "community" comes from the Latin and means "the gift of being together." As pointed out by Craig (2007), a community can mean different things to different people. It can be

- A *geographical community*, e.g., a population sharing a defined physical space (such as a neighborhood, camp, village, or city);
- A *community of identity*, e.g., defined by common interests and needs; or
- An *issue-based community* around some cause or issue (e.g., a club or political party).

In all cases, a community can be seen as a place of social interaction, connection, and participation; a grouping of individuals who feel a mutual sense of belonging; and a system designed to attain some goals (Hillery 1964). In general, we all belong to and interact with many communities of importance in our daily lives. In general, communities have several characteristics, for example, they

- Consist of various components (e.g., social, economic, institutional, or environmental) that all interact in complex ways;
- Consist of interacting units called households;
- Interact with other communities at the local, regional, or global scale;
- Manifest all the attributes of complex dynamic systems (lateral and vertical complexities), e.g., nonlinearity, emergence, synergy, and

uncertainty (synergy may imply that the resilience of the whole can be quite different from that of its components);
- Have capacity (strength), resources, assets (capital), and knowledge;
- Have spirit, engagement, cohesion, and collective action (social capital); and
- Have household security needs and are vulnerable to various hazard events, ranging from everyday issues to large disaster events, each one carrying a certain level of risk for the community.

In other words, communities are complex and dynamic social organizations with their own unique characteristics, security needs, challenges, and potential solutions. We can say with a high level of certainty that no two communities are alike.

Participation

Participation, i.e., the process of taking part and sharing is as old as communities themselves and has been promoted by many cultures across history (Mansouri and Rao 2012). As remarked by Barton et al. (1997), participation is a "condition by which local knowledge, skills, and resources can be mobilized and fully employed." Participation recognizes that there is inherent talent and capacity at the local level and that, as remarked by Schumacher (1973), it can be leveraged to teach people how to do things better, once we know what they can do best. If it takes place at the local level, participation has been shown to add benefits when considering decision making, project sustainability, effectiveness, and efficiency. In the context of this book, the term "participation" can be replaced by "mobilization" (Howard-Grabman and Snetro 2003) or "cocreation" (Prahalad 2006).

The guidelines and methodology presented in this book emphasize a participatory approach to community development. As discussed in Chapter 4, participation is an integral part of the proposed ADIME-E framework and is a key component in each phase of the project cycle. As noted by Pretty (1995), "participation is one of the critical components of success" in development projects. Ultimately, a desirable final product of a community development project consists of solutions that (1) are achieved by mobilizing the community collective action; (2) make the community proud of itself; and (3) the community has the capacity and self-reliance to manage.

Participation can also be seen as a condition by which local knowledge, skills, and resources are mobilized and fully used. It can also be seen as empowerment, and it is more than "taking part" (Cornwall and Jewkes 1995). Participation can take different forms, depending on the dynamic that exists between outsiders and insiders and the social and cultural context in which it takes place. Different cultures look at human interaction differently. For instance, in some cultures, participation is based on building trust, whereas others are more competitive. Some cultures do not promote or even go so far as to discourage participation or limit it to certain genders, castes, or social, political, and economic groups. The reader should realize that ideal democratic participation is a myth and is more often biased than not, whether internally or externally. It can be motivated by an individual or a group of actors (organic participation) or promoted by institutions (induced participation), as noted by Mansouri and Rao (2012).

Biggs (1989) distinguished four modes of participation in agriculture-related projects that can be applied to other types of projects as well. Each mode of participation implies a certain dynamic between insiders and outsiders, which is best summarized as follows by Cornwall and Jewkes (1995):

- *Contractual*: insiders are contracted into projects of outsiders, where outside experts have preconceived solutions to problems as understood by them;
- *Consultative*: insiders are asked for their opinions concerning issues within the community or about possible solutions to those issues;
- *Collaborative:* insiders and outsiders work together on projects initiated and managed by outsiders (the project team); and
- *Collegial* (or collegiate): there is mutual learning, participation, skill and decision sharing and making, and insiders have control over the process.

It should be noted that other classifications of participation have been proposed in the literature. For instance, Pretty (1995) differentiates between manipulative, passive, consultative, material incentives, functional, interactive, and self-mobilization forms of participation, in order of increased level of effect and community empowerment. Likewise, Chambers (2005a) uses the terms nominal, extractive, induced, consultative, partnership, transformative, and self-mobilizing in increasing order of community ownership.

As noted by Biggs, the type of participation depends on the context of the project. For instance, if the emphasis of the project is oriented toward technology testing, the contractual or consultative approach might be more appropriate. At other times, when identifying problems and coming up with solutions, a collaborative or collegial approach is more appropriate. Rather than being specific about the mode of participation, one should recognize that it is a process that evolves over time from contractual to collegial.

The different forms of participation listed above form a ladder of control (or a participation ladder) where decisions go from being made mostly by others (outsiders) to finally being made by the people themselves (insiders). It is noteworthy that transitioning from a contractual mode to a collegial mode with a given community takes time (expressed in years) and relies a lot on building relationships and trust with that community. For that reason, participation has been dominantly contractual and consultative in development projects.

Participatory community development using a collegial approach to participation represents a relatively recent mindset where the beneficiaries are seen as "sitting in the driver's seat" and are not subjected to "precooked" expert recipes. The beneficiaries are involved in the assessment and analysis of the problems they have identified and are active contributors in the design of the solutions. Their knowledge is critical in that process. In general, a collegial approach to participation is more likely to translate into skills, confidence, equity, gender equality, transparency, accountability, and efficiency through ownership.

One of the challenges in community appraisal is to identify what forms of participation best match the capacity of the community. Another piece of useful information before deciding on that issue is to identify the history of participation in the community and what has worked and not worked.

Participatory Research

Over the past 25 years, the concept of participatory research has been proposed in various fields of health, social science, agriculture, and development as an approach or strategy (rather than a suite of techniques) for outsiders to interact with a community and their members and address issues and problems they are facing. It can encompass various aspects of a project, such as appraisal, planning, monitoring, and evaluation. As remarked by Park (1999), the goal of participatory research is to generate knowledge (research) and produce action, thus leading to action-oriented research. It emphasizes collaborative and collegial participation, where the researcher is more a facilitator or catalyst than a director. Participatory research falls into a larger group of methods called participatory action research or PAR. It should be noted that in the rest of this section, no distinction is made between participatory research and PAR, and the two terms are used interchangeably. Readers interested in the various types of PAR methods and their advantages and limitations can find more information in several references: Fals-Borda and Rahman (1991), Cornwall and Jewkes (1995), Park (1999), Scheyvens and Storey (2003), and Chambers (2005a), among others.

PAR is a recognized form of experimental research that focuses on the effects of the researcher's direct actions of practice within a participatory community with the goal of improving the performance quality of the community itself or an area of concern. It is therefore a bottom-up approach to community development projects, emphasizing that learning is best when it is done in the field. Because PAR always begins with the people's problems and indigenous knowledge, is motivated by collective action, and ends in action, it can also be interpreted as people's action research (Park 1999).

As discussed more extensively by Fals-Borda and Rahman (1991), Cornwall and Jewkes (1995), and Park (1999), PAR originated from grassroots work in

South America, India, and the United States in the 1970s. It was strongly inspired by the work in education by Freire (2007) in the 1970s in South America toward giving rights to and empowering marginalized groups to transform their lives and create revolutionary changes. The original ideas associated with PAR at that time included such concepts as authentic participation and commitment, social transformation, creating people's power, collective research, valuing and applying folk culture, social activism, and liberation by one's own consciousness and knowledge. The basic ideology behind PAR according to Fals-Borda and Rahman (1991) was that "a self-conscious people, those who are currently poor and oppressed, will progressively transform their environment by their own praxis. In this process, others may play a catalytic and supportive role but will not dominate."

Cornwall and Jewkes (1995) see PAR as "respecting and understanding the people with and for whom researchers work. It is about developing a realization that local people are knowledgeable and that they, together with researchers, can work toward analyses and solutions."

The PAR approach has several collegial attributes:

- Its methodologies are reflexive, flexible and iterative, and nonlinear.
- It creates knowledge for those (outsiders and insiders) who engage in it.
- It is a catalyst that engages people as active contributors and promotes communication.
- It uses local knowledge and perception, and outside experts have a limited contribution, if any.
- It creates space and confidence.
- It is about respecting and understanding people.
- It implies participation in the appraisal, design, monitoring, and evaluation phases.

Various participatory research approaches have been proposed in the literature since the 1970s. These various approaches have been identified by acronyms such as PRA, RAP, RRA, and PLA and can be viewed as a broader part of PAR (Cornwall and Jewkes 1995; Barton et al. 1997; Pretty et al. 1997; Mukherjee 2003). All these techniques tend to generate meaningful qualitative data about the community using different sets of tools and for different contexts.

Among all participatory approaches proposed in the literature, participatory rural appraisal (PRA) has been used extensively since the early 1980s by nongovernmental organizations (NGOs) and other agencies involved in development (Chambers 1983, 1994a, 1994b). As one of the early pioneers of PAR, Chambers defines PRA as "a family of approaches and methods to enable rural people to share, enhance, and analyze their knowledge of life and conditions, to plan and to act." PRA has also been referred to in the literature as participatory reflection and action and participatory learning and action (Mukherjee 2003;

Chambers 2005b). Multiple tools and techniques are available in PRA (Cavestro 2003) and include direct observation, participatory mapping, transect walks, interviewing, time lines, participatory diagramming, wealth and well-being rankings, and questionnaires.

The PRA approach evolved from an earlier method in the 1970s and 1980s called rapid rural appraisal (RRA) but differs from it because it is more focused on local people's empowerment. According to Chambers (1994b), "in RRA, information is more elicited and extracted by outsiders as part of a process of data gathering; in PRA it is more generated, analyzed, owned and shared by local people as part of a process of their empowerment." PRA field methods and tools are multiple (Mukherjee 2003) and have been regrouped (Chambers 1994a) into three major categories:

- Group and team dynamics (e.g., learning contracts, role reversals, and feedback sessions);
- Sampling and interviewing (e.g., transect walks, wealth ranking, social mapping, focus group discussions, semistructured interviews, and triangulation); and
- Visualization and mapping (e.g., Venn diagrams, matrix scoring, and time lines).

In general, the PRA approach generates both qualitative and quantitative information. It has been used in a wide range of applications dealing with natural resource management, agriculture, poverty and social programs, health and food security, among others (Chambers 1994a). In general, PRA tends to be more consultative or collaborative rather than collegiate (Cornwall and Jewkes 1995). More discussion on the pros and cons of PRA and RRA can be found in a special issue of "Participatory Learning and Action" (*PLA Notes* 1995), the International Institute for Environment and Development (IIED) website, and the Oxfam *Handbook of Development and Relief* (2008).

Another method called rapid assessment process (RAP) is an "intensive team-based *qualitative* inquiry using triangulation, iterative data analysis, and additional data collection to *quickly* develop a preliminary understanding of a situation from the insider's perspective" (Beebe 2001). It belongs to a group of rapid methods such as rapid appraisal, rapid assessment, and rapid rural appraisal mentioned previously. A common attribute of all the rapid methods is their emphasis on local knowledge and understanding the "big picture" in areas that are uncertain and complex by using rapid survey and analysis methods. They generate quick and inexpensive snapshots of the situation at hand (social organization, for instance) rather than detailed statistical analysis. The quick nature of the rapid methods makes them more contractual and consultative than collaborative or collegiate. Furthermore, they are not very rigorous and informative for making decisions. The process is usually conducted by a team of at least two individuals over a period ranging between four days and six weeks, much less than with the aforementioned methods.

Other diagnostic tools that may complement the PAR approach include SWOT (SWOL, SWOC) analysis to determine the internal strengths, weaknesses, opportunities, and threats (limitations or challenges) of resources (human or physical) at different scales in the community. Additional analyses discussed in the following sections include stakeholder analysis; partnership and relationship analysis; gender analysis; and resource, assets, capacity, risk, and service analysis.

5.4 A Reality Check: Challenges and Biases

The participatory approaches and tools presented are not meant to be academic exercises but rather to be tested in the real environment of a community. They have often been sources of intense discussion and debate by sociocultural anthropologists, policy makers, development makers, and others, especially those in the academic world. As remarked by Guijt and Cornwall (1995), "the use of participatory methods alone does not guarantee participation in setting development agendas. Nor does it necessarily lead to empowerment, despite the claims sometimes made." According to Cornwall and Jewkes (1995), several challenges are likely to arise when using the tools in practice and need to be accounted for by practitioners:

- Working with local populations is not easy because of many factors (e.g., cultural, religious, economic, and gender) and preconceived ideas that populations may have about outsiders. Likewise, outsiders may not realize the communication boundaries associated with their "unconscious cultural programming" (Braucher 2013). Gaps in communication are likely to arise between the individualistic cultures traditionally providing development solutions and the more collectivistic cultures receiving these solutions.
- Communication can be difficult, especially in a language that is foreign to the appraisal team, even when translators are available because they may bring their own biases.
- Participation takes time and resources and requires patience and sensitivity. Not all cultures see participation in the same way and not necessarily as perceived in Western democratic political systems (Barton et al. 1997).
- Fieldwork is about finding a "fine balance between rigidity and flexibility" (Scheyvens and Storey 2003), the right amount of data and information, and the proper use of qualitative and quantitative tools (Barton 1997).

- Only the voices of a few special groups (with mostly male members) in the communities are often heard (Cornwall 2003). The voices of women and children, minorities, and vulnerable groups, are rarely heard. Special efforts need to be made to hear those voices.
- Participation is often jeopardized by dysfunctional forms of human behavior driven by corruption and fraud; lack of oversight regulations and their enforcement; and capture by local elites, literate, and financially well off, powerful individuals (Mansouri and Rao 2012).
- Only some communities in a region may be selected for a project. As observed by Bamberger et al. (2004), "the project may cover all of the poorest communities, or it has selected all of the most dynamic communities, or is only being organized in districts where there is strong political support and a commitment of local government funds."
- Participation is time consuming. An underlying assumption in PAR and other appraisal tools is that people in the community are willing and can afford to invest time and energy in the process. This is not always possible. As noted by Scheyvens and Storey (2003), "good information takes time and patience to gather and requires information checking, evaluating and cross checking."
- Communities may not always be well defined geographically, or politically. Outsiders tend to see the community as consisting of components with which they are familiar. In general, reality depends on what project teams choose to measure, or more specifically, which set of lenses (perspective) people choose to look through. The children's book *Zoom* by Banyai (1998) is an excellent tool to understand how different conclusions can be reached by considering an issue with a different perspective.
- Communities may want to emphasize their "wants" rather than their "needs."
- Community appraisal is never neutral. The "outsiders" conducting the appraisal are never completely objective in their approaches because they carry their own perspectives, biases, beliefs, and prejudices into their work. Even if they do not want to disturb the community, their mere presence disturbs it.
- The type of appraisal and its length and extent may be dictated by a specific donor or agency, and/or budgetary and time constraints.
- We always have incomplete information about a community, its needs, its vulnerability, and its capacity. As discussed in Chapter 4, this problem, along with the heuristic approach of humans to complexity and change, affects the decision process and the selection of project alternatives.

- There are also cases when participation does not work and/or when groups and individuals are unable to reach a common middle ground.

When not properly accounted for, these challenges can create biases in community appraisal. According to Chambers (1983), biases can be categorized into six different groups.

1. Spatial: Data collection takes place in areas that are easily accessible to the interviewers.

2. Project: Not all projects are given the same priority; failed projects are often ignored, and successful projects, which are more likely to be featured to the donors, receive more attention.

3. Person: More educated, better off people, and potential users and adopters of solutions in the community are likely to be interviewed, leaving others unable to express themselves.

4. Season: Appraisal is done during a season that is more conducive to travel and reaching out to people, thus ignoring times when the community is in its worst shape.

5. Diplomacy: People interviewed are more likely to say yes than no, when asked. Also, unusual questions are not asked, and certain issues are not raised as a matter of courtesy.

6. Professional: Discussion is often a one-way street from outsiders to community members and involves better educated and more well-off community members.

Short of sounding repetitive, these challenges and biases essentially guarantee a gap between what outsiders conclude they see (and interpret) in the community and the reality of the community itself. That gap could have a strong effect later in the project design and may lead to inappropriate projects and unexpected additional costs. The challenges and biases affect not only how outsiders perceive a community in the appraisal phase but also how they interpret (1) the success of a project during its monitoring and evaluation phase, (2) when to close a project, and (3) whether to scale up the project.

Finally, a last reality check is that data collection requires substantial preparation from those who conduct the techniques in the field. A good review of recommendations on how to (1) prepare for fieldwork, (2) conduct fieldwork, (3) leave the fieldwork, and (4) return from the field can be found in the book *Development Fieldwork: A Practical Guide* by Scheyvens and Storey (2003). Furthermore, several tools are available in the literature to prepare team members to encounter other cultures and deal with intercultural differences. The book *Figuring Foreigners Out* by Storti (1999) and the application CultureGPS developed for smart phones, based on the comparative intercultural research of Hofstede (Hofstede et al. 2010), can be useful for that purpose.

5.5 Building a Support Team

An appraisal starts with creating a professional support team that matches the appraisal type to be conducted and has proper qualifications in PAR methods. Team members need to be selected based on their sensitivity to culture, technical expertise, experience in community appraisals, personal attributes, and the unique skills they bring to the group. They must include the right balance of men and women. The team, as a whole, needs to demonstrate a variety of expertise and interest to match the type of appraisal.

Team members usually receive training from sponsoring agencies before going to the field where they are exposed to the knowledge, skills, expected behavior, and attitudes necessary for the task at stake. Figure 5-1 shows an example of how such training is integrated into the overall appraisal part of community projects conducted by the NCDC in Nepal. The team members should also be made aware of the biases and challenges mentioned previously and the dos and don'ts of good facilitation among themselves and with community members (Table 5-3).

The team may consist of representatives of various organizations such as NGOs or international NGOs (INGOs), governmental organizations (GOs), and community-based organizations (CBOs). In its *Tools for Development Handbook*, the Department for International Development (DFID 2002) in the United Kingdom emphasizes the importance of teamwork in development projects and provides extensive recommendations for creating effective teams, including strategies for team development and management, team player empowerment and training, and leadership in team dynamics.

The team dynamic needs to be understood and cultivated. The Myers–Briggs test (HumanMetrics 2013), or other personality tests, combined with a

Table 5-3. Dos and Don'ts of Facilitation

Do	Don't
– allow time for introductions and explanations	– teach
– show respect for all views and opinions	– rush through process at speed
– watch, listen, learn, and show interest	– lecture
– be sensitive to feelings and culture	– criticize contributions
– be prepared, but flexible	– interrupt speakers
– be creative	– dominate discussions
– show humor	– look bored
– be willing to allow community members to take the lead	– ignore cultural norms
– finish well with thanks to everyone	– laugh at people's ideas
– have agreement on the next step of process	– use mobile phones

Source: Tearfund (2011), reproduced with permission.

SWOC (strengths, weaknesses, opportunities, and challenges) analysis on team members, for instance, may be used to identify the strengths, limitations, and modes of learning and interaction of each team member. Team members must also be exposed to team dynamics tools to help them face issues that may arise in their collaborative work, such as conflict resolution.

The end result is to create a balanced team of 5–10 people in terms of expertise and personality. Specific tasks need to be assigned to team members, such as team leader, note takers, facilitators, translators, and logistics coordinators. Plans must also be included for team members to meet on a regular basis to review and evaluate the appraisal progress, success stories, and various roadblocks that may arise during the appraisal phase.

5.6 Data Collection

Community data are necessary to understand the project environment. They provide the pool necessary to extract information about the community. Table 5-4, taken from Nolan (2002), lists various sources of information that may be collected in community mapping. In the sustainable development extension method, Forbes (2009) lists more than 200 parameters that can be mapped at the community level. It should be noted that the collection of data and their analyses do not stop once the project environment has been identified. As we see in Chapter 12, during project implementation a continuous need exists to monitor and evaluate that environment and assess how well the project is doing as it unfolds. The methods of data collection presented in this chapter on appraisal apply to project monitoring and evaluation as well. Several examples of data collection and community participation in the monitoring and evaluation of water sanitation projects can be found in Narayan (1993) and Mukherjee and van Wijk (2003).

Quantitative and Qualitative Methods

As discussed by Scheyvens and Storey (2003), community data can be collected using a combination of ethnographic quantitative and qualitative methods. Regardless of the methods used, a key priority in data collection is to make sure that the data are authentic, valid, appropriate, meaningful, inclusive, truthful, and accurate, in other words, that we have confidence in using them to draw conclusions about the community. According to Barton (1997), good quality data and information must show the following attributes: accuracy, relevance, timeliness, credibility, attribution, significance, and representativeness. To that list, Frankel and Gage (2007) add coverage, completeness, accessibility, and power of sample size.

Table 5-4. Sources of Information in Community Mapping

Aspects	Examples of Information Needed
People	Who lives in the area? What is their structure and composition? What divisions exist? What is the basic profile in terms of things like health, education, employment, income, and so forth? What are the patterns of leadership? What aspects of their belief systems, values, and practices seem important? Do some groups have more power or influence than others?
Environment	Where are the physical and social boundaries of the community? What aspects of climate, topography, natural resources, or seasonal variations seem important? What outstanding natural features mark the area? How is environment connected with livelihood?
Infrastructure	What institutions, organizations, facilities, or services exist? What is their relationship to local populations, now and in the past? What is likely to change in the future?
Resources	What important assets does this community possess or have access to? These might include financial resources, intellectual resources, human resources, and informational resources. How are these assets held and managed? What rules govern their use?
Modes of livelihood	What are the principal bases of the economy? How are people organized for work? How are they connected and/or differentiated? Are there extremes of wealth and poverty? What are current economic trends? How are resources and benefits distributed? How is time patterned?
Issues and concerns	What things have engaged the time, thought, and energy of people here? What are people's main concerns or issues? How do they see these issues? Are there differences of opinion regarding these? What sorts of options are seen as acceptable or workable for dealing with them?
Principal constraints	What factors or conditions lying largely outside the control or prediction of the community are important for understanding what is happening inside the community itself? How do people see these things? Have they changed over time?

Source: Nolan (2002), with permission from Westview Press.

In general, quantitative and qualitative data collection methods differ in terms of types of data collected, methods used, skills required by those collecting the data, and the scope and scale of data collection.

- *Quantitative* data (hard data) tend to be objective, representative, and, importantly, specified in numbers and can be verified and replicated. Caldwell and Sprechmann (1997b) define them as "data that can be analyzed using measures and techniques that can summarize and describe information into usable numbers (percentage, ratios, mean, average, range, size)." The methods used in quantitative data collection can be *participatory* (surveys, structured interviews, questionnaires) or *nonparticipatory* (structured surveys, measurements, observations). They require a clear understanding of sampling methods (probability and nonprobability) and their limitations. Data collection takes time and must follow a strict protocol; it requires proper skills and experience from those collecting the data. To be meaningful, quantitative data are analyzed using proper statistical methods. Overall, quantitative data are good for interpreting *what* is the current situation and predicting *what* will happen.

- *Qualitative* data (soft data) focus more on the "*why* things occur" and meaning but cannot be described with hard, secure facts. According to Caldwell and Sprechmann (1997b), the methods used in qualitative data collection "result in information which can best be described as words." Qualitative data are about "descriptions of situations, events, people, interactions and observed behaviors; direct quotations from people, and excerpts or entire passages from documents, correspondence, records, and case studies." Methods used to collect qualitative data can be *participatory* (interviews, focus groups, conversation and discourse, life and oral histories, stakeholder observations, mapping, ranking) or *nonparticipatory* (personal observations, fieldwork diaries, media documents). Data collection takes less time than for the quantitative data. The PAR and other methods discussed fall in this category of techniques.

A detailed description of the participatory and nonparticipatory methods for data can be found in Barton (1997) and Caldwell and Sprechmann (1997a, 1997b). In practice, having a combination of both quantitative and qualitative methods is best as they offer different perspectives and allow for the cross-checking (e.g., triangulation) of information.

Primary and Secondary Data

In the social, health, and agricultural science literature, community data are divided into two major categories: I data and II data. Both sets of data can be qualitative or quantitative. *Primary* (I) data are new data obtained directly by the appraisal team from the community and stakeholders, whereas *secondary data* (II) are those that previously exist about the community, the region, or the country, and were collected by someone else; they are sometimes referred to as "secondhand" data. They are available in various forms and indirect sources, such as articles, reports, websites, maps, censuses, individuals who may have visited the community in the past, and previous studies. Some additional data may also be obtained from in-country governmental agencies. Finally, secondary data may also originate from studies that have nothing to do with the current project but give insights into the community. All data are useful in some way, as long as they are framed in a cultural context (Nolan 2002).

Secondary data are often the main data available at the start of the appraisal phase, and they can be used to help draft a first and rough baseline of the community. They can be *aggregated* (from groups of observations around a defined criterion) or *disaggregated* (broken down into gender, age, and ethnicity).

Recommendations have been suggested in the social and health science and the development literature to ensure the quality of secondary data. They include cross-analyzing the data for accuracy through triangulation and explaining sources of discrepancies (McCaston 2005) and investigating who collected the data, why they were collected, and how they were collected (Nolan 1998). In general, secondary data complement but do not replace primary data.

In the prefeasibility phase of a project (initiation and identification in Chapter 4), existing secondary data can be supplemented by a first round of interviews of influential people in the community (e.g., community leaders and social mobilizers). A subset of the appraisal team may want to participate in that preliminary visit. At the end of that visit, insights about the community and its vision for what it wants to be and what it is looking for may emerge. From that interview, hypotheses can also be drawn about potential problems and issues currently faced by the community, thus helping the appraisal team home in on the type of appraisal that needs to be carried out. Finally, information gaps and needs to clarify existing data may also become apparent.

The preliminary visit is usually coordinated with a few community leaders (or a third party who knows the leaders), who serve first as entry points into the community. They also help by acting as the focal points in building trust and long-term relationships among the team members and the rest of the community as the project unfolds over time. Finally, the preliminary visit helps to start communication with community members and to explain how the main appraisal is to be conducted in terms of schedule and logistical arrangements.

5.7 Designing and Carrying Out the Appraisal

While interacting with community members and collecting primary qualitative and/or quantitative data, key questions and concerns arise. The insights gained through the analysis of secondary data should help the appraisal team decide on topics to be investigated, the individuals and groups who are likely to have a stake in addressing those topics, and the PAR tools and methods that best match the concerns. This step may lead to creating an action research matrix, which lists various topics to be investigated and possible stakeholders and participants who are more likely to be interested in each topic and/or to be affected by the project

Table 5-5. Example of Action Research Matrix (Namsaling Project)

What Is to Be Investigated?	Who Could Participate in the Study (Who Has Interest)?
Health and nutrition of children	Mothers
Where supplies can be attained and how goods can be sold (potential buyer contacts)	Women craft workers
People's perception of hiring unmarried women, method of training for jobs, and ability of family to support unmarried women	Unmarried women and potential employers
Potential for additional training, supplies for teachers, tele-education, electricity, water, and sanitation	Teachers
Potential for prenatal supplies, additional training, tele-medicine and funding mechanism	Health care workers
Sustainable agricultural practices and income-generation techniques	Subsistence farmers
Sustainable agricultural practices, income-generation techniques and fair trade agreements	Cash crop farmers
Potential for decentralized water filtration and latrines	All residents

intervention. Table 5-5 shows an example of an action research matrix for a project in Namsaling, Nepal. The matrix represents the first steps in mapping community participation.

One of the biggest assumptions and misconceptions at the start of appraisal is that people in the community are willing to and can afford to invest time and energy in the process. In practice, this is rarely the case because community members have other things to do than be interviewed by outsiders. This is particularly true if the community leaders have not approved the appraisal. A second assumption is that the appraisal team is willing to step aside, ask questions, observe, listen, and more importantly, is aware of the potential biases and challenges listed in Section 5.4.

In addition to the community leaders who introduce the appraisal team to the community, a wide range of individuals can be interviewed by the appraisal team, for example, local authority figures, key informants, religious leaders, elders, children, health clinic staff, teachers, women, and young adults. As noted by Barton et al. (1997), interviews can be open-ended and one-on-one or involve spontaneous natural groups that are encountered during the appraisal, or in the form of focus groups specifically convened by the appraisal team. Closed-ended interviews that focus on specific questions with limited possible answers can be carried out as well.

More specifically, focus groups are groups of individuals (5–10 people) in the community who share something in common (e.g., men, women, teenagers, interest groups, or farmers) about a specific topic of interest to them. The discussion is usually about a small number of open-ended questions that are predetermined by the appraisal team. The interview is confidential and needs to take place in an environment that is conducive to trust and openness. The questions create an organic process of interaction among the focus group participants in which perceptions, attitudes, and needs may be revealed. The discussion requires facilitation skills on the part of the interviewers and recording and observation skills from the recorders and note takers.

A summary of what has been learned in the various interviews can be presented in the form of an interest group matrix, as shown in Table 5-6. It lists the key interests of the various groups that have been interviewed and the locations where the interviews have been carried out. In addition to interviews, the appraisal team should have a variety of quantitative and qualitative PAR tools in its toolbox. These include administering questionnaires; conducting transect walks, instrumentation, participatory mapping, and modeling (homes, natural resources, infrastructure, and social facilities); developing a historical (space and time) perspective of the community; establishing a wish list; conducting a census; mapping health issues (risks, child nutrition); and reviewing health records. Role plays and games may also provide valuable information. In all cases, the tools

Table 5-6. Example of Interest Group Matrix (Namsaling Project)

Groups and Members	Meeting Places	Key Interests
Mothers	Their homes	Nutrition and health of children
Women craft workers	Homes or market	Income generation
Unmarried women	Their homes	Jobs and family support
Teachers	Schools	School supplies, income for school, more teachers, dependable electricity, water and sanitation at schools, tele-education
Health care workers	Their homes	Tele-medicine, medical supplies, funding mechanism, more education
Subsistence farmers	Their fields or homes	Landslides (deforestation), agricultural education, income generation
Cash crop farmers	Their fields or market	Landslides (deforestation), empowerment for fair trade, agricultural education, income generation

must be selected to be appropriate to the participants (they must match writing and reading skills, for instance) and to match the context in which the data are collected and the time and energy participants have available.

It should be noted that the aforementioned participatory tools can be used at different stages in the ADIME-E framework. For instance, some of the tools used for community appraisal, such as interviews, can also be used in assessing the quality of some project interventions, the monitoring and evaluation phase of the project or of specific activities in that project, and/or generating ideas on particular issues as the project unfolds.

5.8 Analysis and Presentation of Data

Data analysis is carried out to make sense of the raw data collected and to convert data into useful information. In general, the data are used to reveal the following:

- The most significant issues, concerns, needs (not just wants) that the community is facing, and their prioritization;
- Perceived core problems and cause–effect relationships for each problem;
- The community's available resources and assets (natural, human, social, economic, and infrastructure capital);
- Issues important to different groups and different areas of service: what works (or has worked) well, what does not work well, what could be improved, what are current roadblocks to improvement;
- Ranking and importance of issues based on such things as gender, age, employed vs. unemployed, caste, belief systems, and married vs. single individuals; and
- Areas where the appraisal team needs to come back and address issues that require more information and/or clarification. This iterative process needs to continue until a general consensus is reached.

Barton (1997) suggests six steps in analyzing the collected data.

1. Data analysis should be developed within the context of the community being addressed, the nature of the project being developed, and the reason the study is being conducted (appraisal, monitoring, evaluation, or diagnosis).
2. The data must be validated and confirmed using triangulation methods to support forthcoming intervention decisions.
3. The data need to be organized logically and checked for completeness and documentation.
4. Data analysis is an ongoing dynamic process that may change as new data become available.
5. Data analysis needs to be able to convince potential users that the data are reliable, credible, accurate, relevant, and timely.

6. Those who conduct the data analysis need to be able to show that conclusions and recommendations can be drawn from the analysis and convince the users that the data are indeed useful.

To obtain a profile of the community, the analysis of the data can be broken down into several categories: stakeholder, partnership, gender, capacity (resources, assets, and services), vulnerability and vulnerable groups, and social networks. The results of the analysis can be shown using descriptive statistical methods for the quantitative data and anecdotal summaries for the qualitative information. The data can be presented in tabular (matrix) forms. SWOT (SWOL or SWOC) analyses are often included in those tables. Table 5-7 gives an example of a SWOT analysis that tallies the strengths, weaknesses, opportunities, and threats for six types of services existing in a community in Guatemala.

Maps are also useful in data analysis. They may include social maps, historical maps, maps of agricultural practices, hydrologic maps, and cross sections. Maps can be generated on site, on the ground, or on flip charts. They can also be created using modern geographic information system (GIS) software that is useful in showing the distribution of geographical information related to the environmental, demographic, health, social, and economic features of a given area. In some cases, smart phone applications have been used by various NGOs to collect field data more accurately and faster. Examples in regard to water, sanitation, hygiene, and health include FLOW (Field Level Operations Watch) developed by Water for People, Water Point Mapper developed by WaterAid), and Episurveyor (now called Magpi). The data are then presented in the form of GIS maps.

Figure 5-3 shows the results of a GIS survey conducted in nine wards in Namsaling, Nepal. In this case study, a door-to-door survey was conducted to identify the number of houses without a toilet and the occurrence of diarrhea in each household over a period of two weeks preceding the survey. Using that information and the location of sources of water, specific interventions can be designed and prioritized.

In all cases, the results of the data analysis need to be shared and confirmed with the community and presented in a format that can be understood by the community in terms of content and process. It also needs to be presented in a way that is easy for dissemination to other groups of stakeholders.

Stakeholder Analysis

The stakeholder analysis seeks to determine individuals, groups, organizations, and institutions that (1) have a direct or indirect interest in the community, (2) are likely to have some form of involvement or level of influence (positive or negative) in project activities, or (3) could be positively or negatively affected by the project. Furthermore, stakeholders may be able to identify risks, issues,

Table 5-7. SWOT Analysis of Existing Community Services (Guatemala Project)

Topics to Investigate	Strengths	Weaknesses	Opportunities	Threats
Water infrastructure	– Highly accessible	– Often broken	– Consistency – Maintenance	– Population growth, people cheat the system – Terrain doesn't allow for certain technologies
Food security	– Land is available for farming – Food is available in the stores and market	– Low crop in poor communities (located in arid areas and on steep hillsides) – Lack of purchasing power	– Increase production – Improve nutrition	– Crops vulnerable to drought and heavy rain storms
Public transportation	– It exists	– Only runs into town once a week	– More access to sell and trade goods	– Steep terrain and unreliable roads
Sanitation infrastructure	– Water reuse – Access to latrines	– Poor solid waste management	– Infrastructure – Community attitude	– Socioeconomic inequality – Lack of education – Maintenance
Natural disaster planning	– Some attention from NGOs	– Lack of understanding	– Understand the risk through education – Adoption of mitigation techniques	– Science conflicting with religion – More immediate concerns/low priority for community
Health care	– Basic first aid is accessible	– Difficult to reach hospitals	– Family planning – Health education	– Lack of transportation – Habit change to improve health conflicting with culture, religion, and values

Source: McAfee et al. (2011), reproduced with permission.

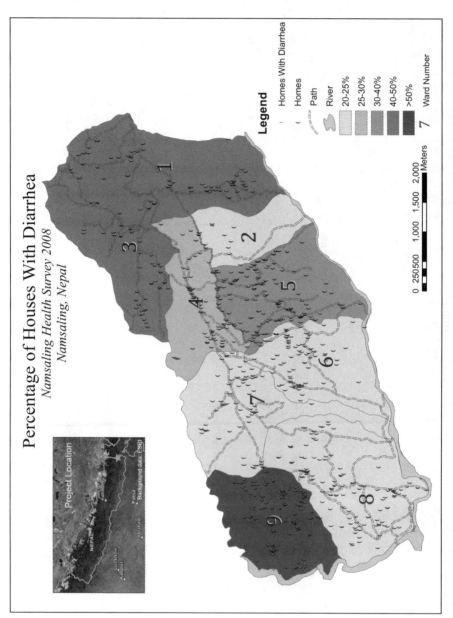

Figure 5-3. GIS Representation of Results of Health Survey Conducted in Namsaling, Nepal

Source: Courtesy of B. Bialek.

problems, and constraints that already exist in the community and rank them accordingly. Stakeholders come from different groups, such as

- Members of the target community and its households: leaders, vulnerable groups, social groups, teachers, religious leaders, and extension workers;
- District, regional, and national leaders: officials, politicians, and service providers; and
- Outside organizations such as NGOs, INGOs, GOs, volunteers, and media.

Table 5-8 shows an example of a stakeholder analysis matrix for the Namsaling project. In this example, three groups of stakeholders were identified with their respective strengths, weaknesses, and levels of influence for a WASH project.

The stakeholder analysis also reveals relationships (collaboration and conflict issues) that exist among various stakeholders and their respective levels of participation, e.g., *key* stakeholders (strong influence); *primary* stakeholders

Table 5-8. Stakeholder Analysis (Namsaling Project)

Group	Strengths	Weaknesses	Influence for/by Project
NCDC	• Based locally • Experience with local projects • Cultural knowledge • Connected locally and internationally	• Communication ability is slow	• Mentors/coworkers/ facilitators for implementing project • Resources of knowledge
Mothers	• Influence on youth and individual families • Can implement change in household	• Not educated • Unemployed • Can only influence within household	• Can pass on information to husband and children • Can gain training in common household health practices • Can assist in implementation of project
Subsistence farmers	• Knowledge of local agricultural practices and problems (e.g., deforestation)	• Uneducated	• Can help in implementation of project • Can gain education about sustainable farming practices (crop planting times, rotation and harvesting techniques)

(affected); and *secondary* or indirect stakeholders (with limited stake and influence), and their respective levels of participation. Other stakeholders that may need to be identified include, according to the World Wildlife Fund (Gawler 2005), opposition stakeholders and marginalized stakeholders. A *gender analysis* should be added for each stakeholder type. It specifically appraises how men and women currently conduct their socioeconomic activities, their specific needs and areas of interest, and their areas of empowerment and participation.

In general, all development agencies have their own criteria to define project stakeholders. For instance, the DFID (2002) suggests using a 2 × 2 matrix to show the influence (power) and importance of various project stakeholders. An example is shown in Figure 5-4 for an appraisal conducted in the Khanalthok village development committee (VDC) in Nepal. In all cases, it should be noted that not all stakeholders have the same influence on a project.

The stakeholder analysis also seeks to answer questions such as the following: (1) Who depends on the project? (2) Who is interested in the outcome of the project? (3) Who could influence the project? (4) Who will be affected by the project? (5) Who could be against the project (threat)? and (6) Who may be left behind in the project? It also tries to link core problems with various groups. The analysis may be further refined by looking at gender, vulnerable and

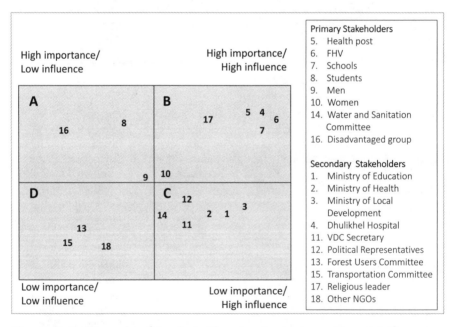

Figure 5-4. Example of Analysis Showing Importance Versus Influence of Various Stakeholders

Source: Khanalthok VDC (Nepal) project conducted in summer 2013.

marginalized groups, and levels of employment. Stakeholder analysis is also critical in developing the project action because it is necessary to link various stakeholders with specific stages in the project planning (design, implementation, monitoring and evaluation, and project sustainability).

EuropeAid (2002) suggests conducting a beneficiary analysis in addition to the stakeholder analysis. Beneficiaries are defined as "those who benefit in whatever way from the implementation of the project" and may include target groups and more at-large beneficiaries (e.g., society, children, and women). Ideally, all stakeholders should benefit from the project. But the benefits will likely differ, and some will benefit more than others.

Partnership Analysis

Partnerships are critical in community development projects and can take multiple forms (e.g., strategic, supportive, logistical, subcontractual, or network). However, they need to be clearly identified before any collaborative work is conducted. Poorly defined partnerships may result in more harm than good, poorly coordinated and executed projects, wasted time and resources, unnecessary duplication, and ultimately confusion at the local level. Main questions that need to be addressed when contemplating and establishing partnerships include the following:

- Who are the partners and their values?
- What level and type of experience and expertise can be expected of each partner, including strengths, weaknesses, opportunities, and challenges?
- How is a working partnership likely to take shape? (1) roles and responsibilities of different partners in proposed projects, (2) information sharing, (3) joint project planning, (4) resource sharing, (5) dealing with conflict, and (6) defining phase-in and phase-out strategies for all partners.
- What are the guiding principles to create efficient and effective partnerships, and what should the performance indicators for that partnership be?

A *partner matrix* may be constructed to summarize for all partner groups their strengths, weaknesses, levels of influence and importance, and the participation level that can be expected. It can be combined with the *stakeholder matrix*.

Capacity Appraisal

In the appraisal phase of the ADIME-E framework, a question arises as to what resources, assets, and services are already available in the community to address current and pending issues. Essentially, this question helps define what strengths

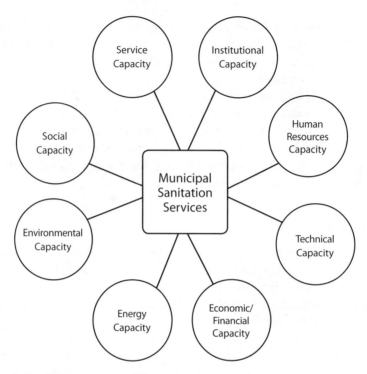

Figure 5-5. Components of Capacity Appraisal

Source: Bouabib (2004), with permission from University of Virginia.

already exist and under what conditions (e.g., quantity, quality, functionality). Capacity appraisal is also about identifying what works, how well it works, and the constraints preventing things from working better.

In general, resources and assets at the community level are multiple and can be divided into subgroups. Figure 5-5 shows a diagram adapted from Garrick Louis and coworkers at the University of Virginia (Ahmad 2004; Bouabib 2004; Louis and Bouabib 2004), which can be used to summarize the *overall capacity* of a community.

This type of resource and asset appraisal requires the mapping of seven categories (indicators) of capacity for a given type of service (e.g., WASH, energy, shelter, education). The categories include institutional, human resources, technical, economic and financial, energy, environmental, and social and cultural. As discussed in more depth in Chapter 9, each capacity category is itself broken down into basic components (subindicators) that are then rated based on agreed-upon metrics. Upon rating each form of capacity in a quantitative way, a radial vector diagram is constructed. An example is shown in Figure 5-6 for a project in Guatemala. The capacity appraisal can also be qualitative (high,

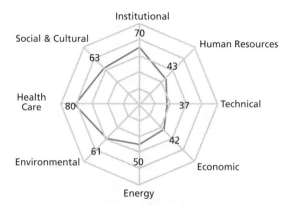

Factor	Score
Institutional	70
Human Resources	43
Technical	37
Economic	42
Energy	50
Environmental	61
Healthcare	80
Social & Cultural	63

Figure 5-6. Example of Capacity Appraisal (Guatemala Project)

Source: McAfee et al. (2011), reproduced with permission.

medium, low) or based on an arbitrary scale ranging from 1 to 5. Each level may be defined by a descriptor such as high (stable and resilient), adequate but not ideal, lacking, severely deficient, or low.

An eighth category of capacity appears in Figure 5-5. Referred to as *service capacity*, it measures the gap between the actual level of a given service provided in the community and some published standard that guarantees an acceptable level of health for the community. For example, in drinking water projects, the level of service capacity depends on the quality and quantity of water available per person, the distance to a water site, and the time it takes to collect the water. According to Howard and Bartram (2003), four service levels can be defined: no access, basic access, intermediate access, and optimal access (Table 15-3). Minimum standards have been proposed by various agencies for different types of context: rapid response, recovery, and development. Those standards are discussed further in Chapter 15.

As is discussed further in Chapter 9, this kind of capacity appraisal reveals the strong points in the community, which development projects to build upon in the future. The radial diagram shown in Figure 5-6 also helps in quickly identifying the weaker components where capacity building is necessary for a given type of service. For instance, in Figure 5-6, it is clear that technical capacity is the weakest. This analysis may be combined with interviews where individuals or groups are asked about what services in the community work well, what is not working well, what can be improved, and what roadblocks could prevent any improvement. In mapping the community's capacity, the PAR team must take note of the categories and subcategories of capacity where reduced or no information is available.

Another type of community asset not shown in Figure 5-5 relates to the resources and services provided to the community by institutions and organizations already interacting with it. They may include community-based organizations (CBOs), community cooperatives, NGOs, INGOs, GOs, service providers, coordinating bodies, and various entities from the private and public sectors. There is a need to analyze the level of contributions from these entities, their past and current performance (scoring), and long-term plans before considering any of them as future project partners. A SWOC analysis may be conducted to map the strengths and weaknesses of those entities, the opportunities they represent, and the foreseeable challenges in working with them. Table 5-9 gives an example of a survey of existing groups in Guatemala that were mapped for a possible municipality project. Table 5-10 summarizes how some of these existing groups could become involved in various community issues and potentially become partners.

Vulnerability and Vulnerable Group Appraisal

Vulnerability reflects the weaknesses of a community in preparing and planning for, absorbing, recovering from, or adapting to actual or potential adverse events. The opposite of vulnerability is security. Within the context of sustainable community development, security refers to the security of households in the areas of food and nutrition, health, habitat and shelter, water and sanitation, education, economy, environment, and civil society. Mapping these various forms of security (or lack thereof) is needed to assess where the community is most vulnerable.

The results of the appraisal can be presented in a radial vector form, shown in Figure 5-7 for a municipality project in Guatemala. The radial diagram helps identify where a community is most vulnerable. Here, vulnerability is in three areas: nutrition (food situation, environmental risks, response capacity); marginalization (exclusion from development); and occupational security (homes with low potential to obtain income or reach minimal levels of consumption of goods and services).

Other forms of vulnerabilities can be included in the radial diagram. For instance, the Tearfund (Tearfund 2011) considers five major groups of vulnerability (and capacity): individual (especially the most vulnerable individuals); social (lack of ties and networks); natural (water, fish, forest, soil); physical (infrastructure); and economic (livelihoods, finances, buying and selling abilities). Even though the Tearfund framework was developed in relation to natural hazards, it can be used for other adverse events as well. In general, when mapping the community's vulnerability, the PAR team must take note of the categories where there is reduced or no information available.

Table 5-9. Existing Support Organizations (Guatemala Project)

	Organizations	Strengths	Weaknesses	Opportunities	Organization Website
International NGOs	USAID	International organization with significant resources	May not have as many connections to community	Could provide support and have U.S. connection	usaid.gov
	Oxfam	Currently working in Guatemala on the issue of hunger among poor farmers	Challenges making a local level connection	Utilize Oxfam knowledge and resources. Connect with the existing program	oxfam.org
	ICRC (International Committee of the Red Cross)	Worldwide connections and support	Focuses on post disaster support; currently low involvement in Guatemala	Utilize resources when natural disaster occurs	icrc.org
In-country NGOs	BPD (Behrhorst Partners for Development)	Strong presence in community, involved in health	Limited resources	Existing partnershp that can continue	behrhorst.org
	ASECSA (Association of Community Health Services)	Works with international aid organizations and disseminates resources on a local level	Dependent on external funding	Potential for partnership with community of Pacoxpon or existing support organizations	
Government organizations	Ministry of Health	Provide project technical support and funding	Not always available or involved	Could have an expanded role	

Source: McAfee et al. (2011), reproduced with permission.

Table 5-10. Linking Problems to Potential External and Internal Partners (Guatemala Project)

Topics to Investigate	External—Government & NGO Support	Internal—Community Support
Water infrastructure	BPD, USAID	Water committee, women responsible for water collection, farmers
Food security	Oxfam, ASECSA	Farmers, mothers
Public transportation		Bus drivers, riders, business owners
Sanitation infrastructure	BPD, USAID, ASECSA	Homeowners, teachers
Natural disaster planning	BPD, Oxfam, ICRC	Farmers
Health care	Ministry of Health, USAID	Midwives, local health care representative

Source: McAfee et al. (2011), reproduced with permission.

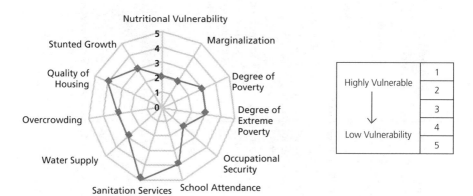

Figure 5-7. Example of Vulnerability Analysis (Guatemala Project)

Source: McAfee et al. (2011), reproduced with permission.

If the community's vulnerability is related to its exposure to natural hazards, these need to be mapped as well. These hazards include such events as earthquakes, floods, drought, landslides, wildfire, coastal storms, tornadoes, hurricanes, and tsunamis. The history of these events must be analyzed, including type, significance, location or extent, intensity, severity, duration, surprise effect, signs (warning), probability of occurrence, trends, risk drivers, and corresponding impact drivers. Some data may be available that measure the characteristics of those events (e.g., water depth and flow, rainfall, wind data, ground acceleration, and earthquake magnitude).

Table 5-11. Risk Assessment (Guatemala Project)

Risk	Causes	Impact	Likelihood	Risk Level
Landslides	Heavy rainfall, poor rainwater control, deforestation	Severe	Unlikely	Medium
Flooding	Heavy rainfall, deforestation	Severe	Likely	High
River overflow	Heavy rainfall, poor solid waste management	Moderate	Somewhat likely	Medium
Forest fires	Carelessness in cooking, arson	Extreme	Very unlikely	Medium
Deforestation	Over-harvesting of trees	Severe	Likely	High
Contamination	Poor waste management	Low	Likely	Medium
Chemical contamination	Overuse of pesticides	Low	Somewhat likely	Low
Earthquakes	Seismic activity	Severe	Somewhat likely	Medium

Impact	Extreme	Medium	Medium	High	High	High
	Severe	Low	Medium	Medium	High	High
	Moderate	Low	Low	Medium	Medium	High
	Low	Low	Low	Low	Medium	Medium
	Negligible	Low	Low	Low	Low	Medium
		Very unlikely	Unlikely	Somewhat likely	Likely	Very likely

Note: Various risks are listed and categorized based on their impact and likelihood.

Source: McAfee et al. (2011), reproduced with permission.

An excellent methodology for the mapping of natural hazards and the identification of associated risks was developed in the United States by FEMA (2001, 2004). The Tearfund (2011) proposed another framework to identify risks because of natural and nonnatural hazards using tools such as community mapping, hazard ranking, time line and history of hazards, Venn diagrams, and transect walks. As an example, Table 5-11 shows a risk map of natural events for a community project in Guatemala. The risk level (low, medium, high) is determined based on the likelihood (very unlikely to very likely) and impact (negligible to extreme) of each risk event and is presented in the form of a so-called risk heat map. A color-code grid helps identify various risk zones. Table 5-11 gives a list of natural risks that exist before project implementation. These risks may affect other risks associated with the project itself and in some cases amplify project-related risks, as discussed in Chapter 10.

Vulnerable groups need to be identified as to the nature of their vulnerability. Of particular focus is the identification of the most vulnerable individuals in a community (e.g., women and children, elderly, sick or physically less able, and marginalized groups).

5.9 Problem Identification and Ranking

At the end of the appraisal phase, the PAR team should be in a position to formulate the different problems that the community is facing in the form of well defined problem statements. Once identified and before proceeding any further, the PAR team needs to confirm and validate those problems with the community. Caldwell (2002) recommends clearly identifying the what, who, and where in the problem statement. The "what" defines "the condition the project is intended to address," and the "who" defines "the population affected by the condition." Finally, the "where" states "the area or location of the population." Examples include the following:

- Acute respiratory problems in 43% of households in rural Ayaviri, Peru;
- Inadequate housing in a (specific) Indian reservation in Montana;
- No access to potable water for 42% of the Loreto, Mexico, community;
- No toilets of minimum hygienic standards available for 70% of the rural population of Loreto, Mexico;
- Electricity available to only 24% of the town of Ayaviri, Peru, and unreliable;
- High incidence of communicable diseases in the northern part of the village of Namsaling, Nepal;
- Flooded latrines during the rainy season in the low-elevation part of the village of Salinas, Mexico.

After identifying the problem, the appraisal needs to list the causes and consequences of existing problems that may have been voiced by community members and other stakeholders and those identified by the appraisal team. According to Caldwell (2002), a suggested approach is to "first identify the subject or the 'who' of the sentence, then state the verb(s) of the sentence. Finally, state the objects of the verb(s), or the subjective completions." Examples include the following:

- Community xyz is exposed to high levels of turbidity and *E. coli* because of a broken water treatment system.
- Children often die of dehydration because of diarrheal diseases.
- Children miss many school days per year because of recurring health problems.

- Farmers cannot bring goods and products to markets because of bad roads.
- The neighborhood is not safe at night because of a lack of security.
- Women are in danger when using the sanitation facilities in the refugee camp at night because of a lack of electricity.

A series of additional questions (UNDP 2009) must be addressed before proceeding further:

- Are the problems real, clear to the stakeholders, and properly worded?
- Are the problems identified the most critical ones needing to be addressed?
- Are the problems identified capturing the concerns of both men and women, marginalized and vulnerable groups?
- Are the problems identified addressing the *current needs* of the community and not its wants and future desires?

Once the problems are formulated, the next thing is to rank them by order of priority. Special attention needs to be placed on how closely related problems might be and their interdependence. One technique to rank problems is to ask various groups of stakeholders to identify their top three or four problems. This ranking can be done at the level of each household or by gender or age group (e.g., married men or women, young men and women, elderly). Ranking exercise matrices are then constructed. Table 5-12 shows an example of such a ranking for the Khanalthok VDC in Nepal, where problems were prioritized as water, roads/access, sanitation/health, and energy.

Another technique, adapted from that used by the UNDP (2009), consists of ranking problems using three criteria. The first is about the *value* that solving the problem would bring to the community and its components (households). This ranking can be combined with cost–benefit and cost-effectiveness analyses (World Bank 2002). The second is based on the amount of *support* that would be available from stakeholders and partners. Finally, the third is about

Table 5-12. Problem Ranking Exercise Matrix

	Rank			
Stakeholders	**1**	**2**	**3**	**4**
Households	Water	Energy	Road	Agriculture
Schools	Water	Sanitation	Access	—
Store owners (3)	Road	Water	Sanitation	Health
Youth	Sanitation	Water	Road	Energy
VDC leaders	Water	Health	Road	Energy
Adult focus group	Road	Water	Agriculture	Sanitation

Source: Khanalthok VDC (Nepal) project conducted in summer 2013.

the *capacity* and *comparative advantage* of the community in partnership with outsiders to work on the problem. Problems that contribute positively to meeting all three criteria have a higher priority because of their overlapping attributes.

5.10 Social Network Analysis

The analysis of community social networks is another tool that can be used in community appraisal. A community's ability to handle daily issues, respond to crisis, or recover from problems ranging from those associated with everyday hazards to disaster event situations is in part dependent on the capacity and effectiveness of its existing social networks. A *social network* is defined as "a group of people or organizations that form a web of relationships" (NRC 2009). It can be represented as a social map consisting of interacting nodes and links (or ties) that define the social fabric or web. An example is shown in Figure 5-8, where circles could represent groups, individuals, or partners of relative importance and the links represent how various stakeholders are interconnected in addressing a specific issue (e.g., water, energy, health, or education). The same representation could also be used to map how decisions are made at the community level, who is involved in these decisions, and what the marginalized groups are.

Members of a community network include the community members and households, the stakeholders, but also the other communities, partners, and organizations that the community is interacting with (NGOs, INGOs, GOs,

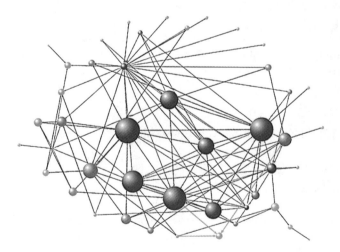

Figure 5-8. Typical Representation of Nodes and Links in Social Network Analysis

Note: The size of each circle corresponds to the level of influence of a component in the network.

private enterprises, government agencies, and other communities). Within the ADIME-E framework, network members play a critical role in their participation in (1) appraising the community needs, (2) developing a community baseline, (3) defining and ranking problems and solutions, and (4) developing and executing an action plan.

Of course, some members in a network are more critical than others in terms of skills, knowledge, resources, and decision making. They can help in developing more effective and efficient solutions. Some can block progress. Others are vulnerable to various forms of hazards. Finally, some members may have skills that others in the community are not aware of and need to be brought into decision making and are clearly change makers. As we see in Chapter 7, knowing the dynamic of social networks can help in project planning and management (e.g., action feasibility matrix, potential partner matrix, participatory planning matrix).

Social network analysis (SNA) is defined as "a process of analyzing a social network and identifying key actors, groups, vulnerabilities, and redundancies as well as the changes in these variables" (NRC 2009). Various SNA tools and software are available in practice to carry out dynamic network analysis using tools for data collection, statistical analysis, and simulation (NRC 2009). In a recent book, Borgatti et al. (2013) show several examples of application of SNA using the software UCINET.

SNA tools and software help not only create a quick visualization of the network and its components (Figure 5-8) but also identify the dynamics (strengths and weaknesses) of interaction, make predictions, conduct simulations, and plan interventions. Although SNA focuses more on the components of social networks and their interactions, it pays less attention to the nature of those interactions that can be handled better by system dynamics analyses, as discussed in Chapter 6.

Advantages of SNA

As discussed in a study conducted by a committee of the NRC in 2009 on SNA and community resilience, SNA as an appraisal tool can add to the understanding of the workings of a community in several ways. It can help

- Study the existing relationship and network communication in the community systems;
- Map how the components of community interact, already work (capacity), don't work (vulnerability), or could work better together in preparation for, response to, and recovery from, an event;
- Identify who makes decisions, who could block decisions, who are key players or threats, and who could be brought into the decision process;
- Map community weaknesses and vulnerable populations;

- Develop a community capacity baseline and adaptive capacity (constantly updated); and
- Visualize how change will progress in the community as its capacity increases.

Challenges of SNA

SNA is not commonly used in development practice. It is often identified in social network platforms and new media. Furthermore, it requires that practitioners be trained and willing to accept new tools. It must also be integrated into community management policy and practice. Finally, incentives for using SNA in practice must be created.

5.11 Chapter Summary

This chapter reviewed the various steps that are necessary in conducting community appraisal. Existing PAR tools commonly used in social and health sciences, agriculture, and development are key to understanding the dynamic of communities and assessing their strengths (capacity), weaknesses (vulnerability), and their potential for building capacity. Engineers are usually not aware of such tools and need to acquire proper training in qualitative and quantitative methods before doing fieldwork. They may not need to become experts in ethnographic studies but should be able to work in partnership with social science experts and comprehend conclusions reached by those experts.

Community appraisal is required. It provides data (and often a lot of data) that need to be processed into useful information. As mentioned in the introduction to this chapter, bad information associated with a sloppy appraisal even though combined with good intentions gives bad results, i.e., garbage in, garbage out. There is no way around it. Furthermore, not recognizing the context in which a project takes place is a recipe for failure.

Failure to include participation in problem identification and ranking is not community development. At the same time, including participation in development projects creates complexity and uncertainty that may produce less than satisfactory projects. This outcome, which is often hard to comprehend for engineers who work in the developed world, requires a higher tolerance by project managers for gaps between reality and what is expected. It also represents unique opportunities for development agencies to improve their methodologies and frameworks for development projects.

References

Ahmad, T. T. (2004). "A classification tool for selecting sanitation service options in lower income communities." MS dissertation, University of Virginia, Charlottesville, VA.

Anderson, T., et al. (1999). *Rural energy services: A handbook for sustainable energy development*, Intermediate Technology Publications Ltd., London.

Bamberger, M., Rugh, J., and Fort, L. (2004). "Shoestring evaluation: Designing impact evaluations under budget, time, and data constraints." *American J. Evaluation*, 25: 5. <http://aje.sagepub.com/content/25/1/5.full.pdf+html> (April 1, 2013).

Banyai, I. (1998). *Zoom*, Puffin Books, New York.

Barton, T. (1997). "Guidelines for monitoring and evaluation: How are we doing?" CARE International in Uganda, Monitoring and Evaluation Task Force, Kampala, Uganda. <http://portals.wdi.wur.nl/files/docs/ppme/Guidelines_to_monitoring_and_evaluation%5B1%5D.pdf> (Jan. 12, 2012).

Barton, T., et al. (1997). "Our people, our resources: Supporting rural communities in participatory action research on population dynamics and the local environment." International Union for Conservation of Nature (IUCN) Publications Services, Gland, Switzerland, and Cambridge, U.K.

Beebe, J. (2001). *Rapid assessment process*, AltaMira Press, Lanham, MD.

Biggs, S. D. (1989). "Resource-poor farmer participation in research: A synthesis of experiences from nine national agricultural research systems." International Service for National Agricultural Research, The Hague, Netherlands.

Birley, M. (2011). *Health impact assessment: Principles and practice*, Earthscan Publications Ltd., London.

Borgatti, S. P., Everett, M. G., and Johnson, J. C. (2013). *Analyzing social networks*, Sage Publications, Thousand Oaks, CA.

Bouabib, A. (2004). "Requirements analysis for sustainable sanitation systems in low-income communities." M.S. dissertation, University of Virginia, Charlottesville, VA.

Braucher, E. (2013). "The blindest blind spot: What effective collaboration lacks." DeVex, Washington, DC, February 4. <https://www.devex.com/en/news/80248/print> (March 15, 2013).

Butts, E. (2010). "Engineering of water systems: Part 3—Reconnaissance of existing water systems." *Water Well Journal*, January, 31–33.

Caldwell, R. (2002). *Project design handbook*, Cooperative for Assistance and Relief Everywhere (CARE), Atlanta.

Caldwell, R., and Sprechmann S. (1997a). *Facilitators' manual*, DM&E Workshop Series: Vol. 2, M&E Workshop Series, CARE International, Atlanta. <http://pqdl.care.org/Practice/DME%20Workshop%20Facilitator%27s%20Manual.pdf> (Nov. 1, 2012).

Caldwell, R., and Sprechmann, S. (1997b). *Handout manual*, DM&E Workshop Series: Vol. 1, M&E Workshop Series, CARE International, Atlanta. <http://pqdl.care.org/Practice/DME%20Workshop%20Handouts.pdf> (Nov. 1, 2012).

Cavestro, L. (2003). "PRA: Participatory rural appraisal concepts, methodologies and techniques." M.S. report, University of Padova, Italy. <http://www.agraria.unipd.it/agraria/master/02-03/participatory%20rural%20appraisal.pdf> (June 20, 2013).

Chambers, R. (1983). *Rural development: Putting the last first*, Pearson Prentice Hall, London.

Chambers, R. (1994a). "The origins and practice of participatory rural appraisal." *World Development*, 22(7), 953–969.

Chambers, R. (1994b). "Participatory rural appraisal (PRA): Analysis of experience." *World Development*, 22(9), 1253–1268.

Chambers, R. (2005a). *Ideas for development*, Institute for Development Studies. Earthscan Publications Ltd., London.

Chambers, R. (2005b). *Participatory workshops: A sourcebook of 21 sets of ideas and activities*, Earthscan Publications Ltd., London.

Conant, J., and Fadem, P. (2008). *A community guide to environmental health*, Hesperian Publ., Berkeley, CA.

Cornwall, A. (2003). "Whose voices? Whose choices? Reflections on gender and participatory development." *World Development*, 31(8), 1325–1342.

Cornwall, A., and Jewkes, R. (1995). "What is participatory research?" *Soc. Sci. Med.*, 41(12), 1667–1676.

Craig, G. (2007). "Community capacity building: Something old, something new?" *Critical Social Policy*, 27, 335–359.

Department for International Development (DFID). (2002). *Tools for development handbook*, version 15, Department for International Development, London.

EuropeAid. (2002). *Project cycle management handbook*, version 2.0, PARTICIP GmbH, Fribourg, Germany. <http://www.sle-berlin.de/files/sletraining/PCM_Train_Handbook_EN-March2002.pdf> (March 10, 2012).

Fals-Borda, O., and Rahman, M. A., eds. (1991). *Action and knowledge: Breaking the monopoly with participatory action research*, The Apex Press, New York.

Federal Emergency Management Agency (FEMA). (2001). *Understanding your risks: Identifying hazards and estimating losses*, Publication No. 386-2, Federal Emergency Management Agency, Washington, DC. <http://www.fema.gov/library/viewRecord.do?id=1880> (Feb. 22, 2012).

Federal Emergency Management Agency (FEMA). (2004). *Using HAZUS-MH for risk assessment: A how-to guide*, Publication No. 433, Federal Emergency Management Agency, Washington, DC. <http://www.fema.gov/pdf/plan/prevent/hazus/fema433.pdf> (Feb. 22, 2012).

Forbes, S. (2009). "Sustainable development extension plan (SUDEX)." Doctoral dissertation, University of Texas at El Paso.

Frankel, N., and Gage, A. (2007). "M&E fundamentals: A self-guided minicourse." U.S. Agency for International Development (USAID), Washington, DC. <http://pdf.usaid.gov/pdf_docs/PNADJ235.pdf> (Nov. 15, 2012).

Freire, P. (2007). *Pedagogy of the oppressed*, Continuum International Publishing, New York. First published in 1970 by Seabury Press (same title).

Gawler, M. (2005). *WWF introductory course: Project design in the context of project cycle management*, Artemis Services, Prévessin-Moëns, France <http://www.artemis-services.com/downloads/sourcebook_0502.pdf> (March 15, 2012).

Guijt, I., and Cornwall, A. (1995). "Critical reflections on the practice of PRA." Participatory Learning and Action Notes, 24. International Institute for Environment and Development, London. <http://pubs.iied.org/pdfs/G01589.pdf> (Feb. 20, 2012).

Hillery, G. A. (1964). "Villages, cities and total institutions." *American Sociological Review*, 28, 32–42.

Hofstede, G., Hofstede, G. J., and Minkov, M. (2010). *Cultures and organizations: Software for the mind*, 3rd Ed., McGraw Hill, New York.

Howard, G., and Bartram, J. (2003). *Domestic water quantity, service level and health*, World Health Organization (WHO), Geneva. <http://www.who.int/water_sanitation_health/diseases/WSH03.02.pdf> (Dec. 10, 2012).

Howard-Grabman, L., and Snetro, G. (2003). *How to mobilize communities for health and social change*, Center for Communication Programs, Baltimore. <http://www.jhuccp.org/resource_center/publications/field_guides_tools/how-mobilize-communities-health-and-social-change-20> (Dec. 5, 2012).

HumanMetrics. (2013). "Jung typology test." <http://www.humanmetrics.com/cgi-win/jtypes2.asp> (Oct. 4, 2013).

Johnson, N. G., and Bryden, K. M. (2012). "Energy supply and use in a rural West African village." *Energy*, 43(1), 283–292.

Korf, B. (2003). *Field guide for participatory needs assessment*, Integrated Food Security Programme, Trincomalee, Sri Lanka.

Leadership in Energy and Environmental Design (LEED-ND). (2006). "Understanding the relationship between public health and the built environment." Leadership in Energy and Environmental Design (LEED-ND), Washington, DC. <http://www.usgbc.org/Docs/Archive/General/Docs1480.pdf> (May 12, 2012).

Louis, G., and Bouabib, M. A. (2004). "Community assessment model for sustainable municipal sanitation services in low income communities." Presented at the 2004 International Council on Systems Engineering, Mid-Atlantic Regional Conference, Arlington, VA, ICSE, San Diego, CA.

Mansouri, G., and Rao, V. (2012). *Localizing development: Does participation work?* World Bank Publications, Washington, DC.

McAfee, K., Sharma, A., Rowley, C., and Van Voast, C. (2011). "Paxocon, Guatemala, term project." CVEN 5929: Sustainable Community Development 2, University of Colorado at Boulder.

McCaston, M. K. (2005). *Tips for collecting, reviewing, and analyzing secondary data*, CARE USA, Atlanta. <http://pqdl.care.org/Practice/DME%20-%20Tips%20for%20Collecting,%20Reviewing%20and%20Analyzing%20Secondary%20Data.pdf> (Nov. 10, 2012).

Mukherjee, N. (2003). *Participatory learning and action: With 100 field methods*, Concept Publ., New Delhi, India.

Mukherjee, N., and van Wijk, C., eds. (2003). *Sustainability planning and monitoring in community water supply and sanitation: A guide on the methodology for participatory assessment (MPA) for community-driven development programs*, World Bank, Washington, DC.

Namsaling Community Development Center (NCDC). (2013). An example of a community appraisal can be found at <http://ncdcilam.org.np/images/uploaded/publication/13-04-12-11-04-17Namsaling%20English%20doc.pdf> (March 10, 2013).

Narayan, D. (1993). *Participatory evaluation: Tools for managing change in water and sanitation*, World Bank, Washington, DC. <http://www.chs.ubc.ca/archives/files/Participatory%20Evaluation%20Tools%20for%20Managing%20Change%20in%20Water%20and%20Sanitation.pdf> (Dec. 10, 2012).

National Research Council (NRC). (2009). *Applications of social network analysis for building community disaster resilience*, National Academies Press, Washington, DC.

Nolan, R. (1998). "Projects that work: Context-based planning for community change." Unpublished report. Department of Anthropology, Purdue University, West Lafayette, IN.

Nolan, R. (2002). *Development anthropology: Encounters in the real world*, Westview Press, Boulder, CO.

Ockelford, J., and Reed, B. (2006). "Participatory planning for integrated rural water supply and sanitation programmes." Water, Engineering and Development Center (WEDC), Leicestershire, U.K. <http://siteresources.worldbank.org/extwss/resources/337301-1147283814231/2532554-1153769449882/poster8-guidelinesforparticipatoryrwss_program planning.pdf> (Dec. 10, 2012).

Oxfam. (2008). *Handbook of development and relief*, Vols. 1 and 2, Oxfam Publ., Cowley, Oxford, U.K.

Park, P. (1999). "People, knowledge and change in participatory research." *Management and Learning*, 30(2), 141–157.

PLA Notes. (1995). "Participatory learning and action." International Institute for Environment and Development, London. <http://www.planotes.org/pla_backissues/24.html> (Nov. 12, 2012).

Prahalad, C. K. (2006). *The fortune at the bottom of the pyramid: Eradicating poverty through profits*, Wharton School Publishing, Upper Saddle River, NJ.

Pretty, J. (1995). "Participatory learning for sustainable agriculture." *World Development*, 23(8), 1247–1263.

Pretty, J., et al. (1997). "A trainer's guide for participatory learning and action." International Institute for Environment and Development, London.

Scheyvens, R., and Storey, D. (2003). *Development fieldwork: A practical guide*, Sage Publications, Thousand Oaks, CA.

Schumacher, E. F. (1973). *Small is beautiful*, Harper Perennial, New York.

Storti, C. (1999). *Figuring foreigners out: A practical guide*, Intercultural Press, Boston.

Tearfund. (2011). "Reducing risks of disasters in our community." Tearfund, Teddington, U.K. <http://tilz.tearfund.org/Publications/ROOTS/Reducing+risk+of+disaster+in+our+communities.htm> (Oct. 22, 2012).

United Nations Development Programme (UNDP). (2009). *Handbook on planning, monitoring and evaluating for development results*, United Nations Development Programme, New York. <http://web.undp.org/evaluation/handbook/documents/english/pme-handbook.pdf> (May 22, 2012).

World Bank. (2002). *Monitoring and evaluation: Some tools, methods and approaches*, World Bank, Washington, DC. <http://www.worldbank.org/oed/ecd/tools/> (April 4, 2012).

World Bank. (2012). *What a waste: a global review of solid waste management*, World Bank, Washington, DC. <http://web.worldbank.org/wbsite/external/topics/exturbandevelopment/0,,contentmdk:23172887~pagepk:210058~pipk:210062~thesitepk:337178,00.html> (March 1, 2013).

Zahnd, A. (2012). "Role of RETS in community development." Lecture notes in CVEN 5929: Sustainable Community Development 2, University of Colorado at Boulder.

Zweig, W. (2011). "Examining and applying the first two phases of the ADME framework." M.S. Thesis, University of Colorado at Boulder.

6

A System Dynamics Approach to Community Development

6.1 Communities as Systems

The data and information obtained from community appraisal reveal that communities are complex systems of subsystems consisting of parts (socioeconomic units, households, individuals), that are related and share, in one way or another, some form of a common vision. Communities tend to manifest in their behavior all the attributes of complex adaptive dynamic (changing over time) systems: e.g., nonlinearity, chaos, emergence, synergy, uncertainty, unpredictability, interdependency, coevolution, and self-organization. Some of these terms are defined in Table 6-1. In particular, synergy implies that the dynamic of a community (the whole) can be quite different from that of its components (e.g., households and community members). As a system, a community is also part of a hierarchy of other systems (e.g., communities, government, and institutions) that the community interacts with at the regional and/or national levels; what happens at one level, policy for instance, may have consequences (intended and unintended) at other levels. Because of their complex nature, communities do not lend themselves to traditional linear tools of analysis based on linear causation. Circular causation, based on cause-and-effect loops, is better suited to model their complex behavior.

In general, communities are human systems interacting with other systems (natural, production, infrastructure, capital) constrained by a wide range of issues (e.g., social, economic, and cultural), as shown in Figure 6-1. These systems can be divided into subsystems. Infrastructure, for instance, includes several components, such as water, telecommunication, shelter, energy, transportation, and waste. It is clear that infrastructure is critical to the well-being of communities and dependency exists within each of its components and across components. Because of the tied connectivity that exists between infrastructure components at the same (intra) level and across (inter) levels, any disruption of any one of the levels could lead to cascading effects, with major consequences to

Table 6-1. General Characteristics of Complex Adaptive Systems

Element	Characteristics
Nonlinearity	Sensitivity to initial conditions; small actions can simulate reactions, [...] in which highly improbable, unpredictable, and unexpected events have huge effects.
Emergence	Patterns emerge from self-organization among interacting agents. What emerges is beyond, outside of, and oblivious to any notion of shared intentionality. Each agent or element pursues its own path but as paths intersect and the elements interact, patterns of interaction emerge and the whole of the interactions becomes greater than the separate parts.
Dynamics	Interactions within, between, and among subsystems and parts within systems are volatile, turbulent, and cascading rapidly and unpredictably.
Adaptation	Interacting elements and agents respond and adapt to each other so that what emerges and evolves is a function of ongoing adaptation among both interacting elements and the responsive relationships interacting agents have with their environment.
Uncertainty	Under conditions of complexity, processes and outcomes are unpredictable, uncontrollable, and unknowable in advance. [...]
Coevolution	As interacting and adaptive agents self-organize, ongoing connections emerge that become coevolutionary as the agents evolve together (coevolve) within or as part of the whole system, over time.

Source: Patton (2011), with permission from Guilford Press.

the well-being of community members. Systems are full of cross-scale correlations and feedback mechanisms. In areas prone to natural disasters, protecting critical infrastructures (lifeline systems) is one of the highest priorities. The same could be said about other major systems shown in Figure 6-1 and the interaction of all the systems. As suggested by Barry Commoner, the first law of ecology is that "everything is connected to everything else."

Communities depend on the systems shown in Figure 6-1 for economic stability and growth, commerce, education, communication, health, energy, and transportation, among other things. The community and its dependence form a sort of metabolism. Community development is not only about looking at how those systems provide separate resources to the community. It is also about looking at how the resources and resource providers interact and depend on each other within that metabolism.

Among all the systems shown in Figure 6-1, human systems are by far the most difficult to comprehend. As remarked in Section 4.8, human systems impose onto other systems unique characteristics, such as "bounded rationality, limited

Figure 6-1. Systems Involved in Community Development

certainty, limited predictability, indeterminate causality, and evolutionary change" (Hjorth and Bagheri 2006); they evolve and adapt as they develop. Therefore, the tools necessary to comprehend, analyze, and predict human systems and the systems they interact with must differ from the reductionist tools traditionally used to arrive at clearly defined answers to clearly defined problems. The tools must embrace adaptiveness, change, variability, and unpredictability. They must also be able to evolve at the same time as the systems themselves, i.e., they must match the dynamic of change.

Because predictable systems are rarely, if ever, found in nature, new tools must be used, and they must be accompanied by a mindset to use them in complex systems, such as communities. The need for a new mindset is quite relevant when considering community resilience to hazard events. From a systems perspective, a resilient community is more than an aggregation of resilient individuals and depends on many interacting components. In many ways, resilience can be seen as an emerging property of strong, evolving, adaptive, and healthy systems and their interdependencies (Gunderson and Holling 2002; Patton 2011; NRC 2012).

A systems approach to development projects has been suggested in the international development literature over the past 10 years but has not gained much popularity and traction. Even though the complexity of such projects has often been acknowledged and a systems approach to those projects would be more appropriate (Valadez and Bamberger 1994; Newman 2010; USAID 2011), development is still driven by a traditional reductionist mindset based on cause and effect that is expected to yield well defined, stable, reliable, and predictable solutions. As discussed in Chapter 2, there is enough of a track record in the history of development over the past 50 years that such an approach has not produced the promised results. The main reason is that the approach does not match well the reality of development projects.

Systems thinking is particularly relevant to the field of engineering for sustainable human development where development workers work as part of multidisciplinary teams, interact with various decision makers, and must be able to integrate and process issues from multiple disciplines, all at the same time. Simply put, they deal and interact with multiple interrelated systems and feedback mechanisms and have to make decisions in a complex and uncertain environment across a variety of issues: economic, social, productivity, political, infrastructure, organization, and natural (Patton 2011). As an example, Table 6-2 shows the relationships and double causality (feedback) that exist among the issues of climate change, poverty, engineering, and globalization. All four issues interact with each other in complex and systemic ways.

A systems approach to community development can be used to do the following:

- Understand the current dynamic of a community and its enabling or constraining environment;
- Explore how certain variables or factors influence that dynamic, including feedback mechanisms, by carrying out various sensitivity studies;
- Explain existing patterns of community behavior over time;
- Identify and explore leverage points (tipping points), i.e., critical places to intervene in the community (Meadows 2008);
- Predict how the community may respond to various constraints and disturbances, and/or strategies of capacity development, thus allowing for reevaluation of decisions and actions;
- Link policy actions that affect communities to the results of development interventions;
- Explore possible unintended consequences of decisions in community development; and
- Monitor and evaluate the performance of development interventions and decide how to make adjustments as projects unfold.

In all cases, using a systems view leads to an enhanced understanding of the dynamic of a community. However, it must be noted that a systems approach is not a panacea because it does not provide definite and predictable answers to how the community responds to specific events.

At the outset, this chapter gives an overview of systems theory. It is not meant to be a comprehensive presentation of systems theory, system dynamics, and dynamic modeling. Instead, it presents the logic behind understanding, analyzing, modeling systems, and simulating simple systems. This is demonstrated using the software package *iThink* and STELLA, developed by ISEE Systems and documented in several books (Richmond 2004; Fisher 2011). We then apply the systems approach to community development projects and explore its advantages and disadvantages.

6.2 Systems and Systems Thinking Basics

Systems science emerged shortly after World War II. Several schools of thought were initiated at that time and are still in use today. They include cybernetics, general systems theory, organizational learning, operations research, total quality management, and system dynamics (Dent 2001). In general, systems science was motivated by a need to handle complex global problems that emerged after World War II.

One branch of systems thinking is the field of system dynamics, which originated at MIT in the late 1950s and early 1960s with the work of Jay Forrester in the study of industrial systems and urban dynamics (Forrester 1961). It is essentially a methodology to study how systems change over time and how the structure of systems affects their behavior, a very important concept in systems thinking. The methodology also allows exploring the consequences of different behavioral forms of system changes, such as growth, decay, overshoot, oscillations, equilibrium, randomness, and chaos. System dynamics has been used since the 1950s as a mental model in a wide range of disciplines, such as engineering, business and economics, health, planning, and management.

The system dynamics approach became more accessible to the public with the publication of *The Fifth Discipline* by Senge (1994a) and *Business Dynamics* by Sterman (2000), who introduced the concept of dynamic modeling. Other interesting texts on various applications of systems thinking and system dynamics include those by Ford (2010) on modeling environmental processes; Hargrove (1998) on health sciences; Robinson (2001) on climate sciences; and Hannon and Ruth (2001a, 2001b), among others. Other books have also been proposed to popularize systems thinking and to make the learning of systems more interesting and fun. Examples include the books by Briggs and Peat (1999), Meadows (2008), Wheatley (1999), Westley et al. (2007), and storybooks designed for children (Sweeney 2001).

In general, systems can be defined as groups of interacting seemingly independent parts linked by exchanges of energy, matter, and/or information. Systems are made of parts and interrelationships (interconnections) among their parts, and they are driven by a common purpose to achieve something (even though the parts may or may not have conflicting purposes). In systems, the whole is greater than the sum of its parts, and systems develop patterns of behavior that

Table 6-2. Double-Causality Relationships between Climate Change, Poverty, Globalization, and Engineering

	Effect of climate change on poverty	**Effect of climate change on globalization**	**Effect of climate change on engineering**
Climate change linkages and effects	Poor hit earliest and hardest with the least capacity to adapt. Climate change may lead to: Loss of habitats and biodiversity; Loss of livelihoods and new opportunities; Increased frequency and severity of natural disasters, flooding, and extreme weather; Water scarcity and desertification; Conflict, civil unrest, and migration; Health effects and food insecurity.	The effects of carbon trading and the shift toward a low-carbon economy, especially in energy, transportation, foodstuffs, manufacturing, construction, and tourism markets. Localization of supply chains and markets because of higher costs. Increased risk, uncertainty, and market volatility. Disruption to agriculture and infrastructure.	New markets and opportunities in renewable energy, alternative fuels, energy conservation, and waste reduction. New research and innovation opportunities. Disaster preparedness and relief and postdisaster reconstruction. Low-carbon economy, especially in energy, infrastructure, and construction markets.
	Effect of poverty on climate change	**Effect of poverty on globalization**	**Effect of poverty on engineering**
Poverty linkages and effects	Farming, energy, transportation, urbanization, and development choices of developing nations are critical if global CO_2 reduction targets are to be met, especially in rapidly industrializing economies (Brazil, Russia, India, and China). Global carbon trading and emissions targets must recognize the needs and rights of the poor and the obligations of industrialized nations.	The responsibility to act ethically, contribute to poverty reduction, and involve poor in decision making is becoming recognized by global corporations. Failure to act responsibly or to address poverty undermines support for (current models of) globalization. Globalization criticized by international development and trade reformers.	Requires low-cost solutions that are appropriate to cultural, political, social, and economic environment. Requires participation of the poor and local knowledge. Developing countries are often high-risk and high-return markets.

Effect of globalization on engineering
Growth in LDC markets, especially in utilities, infrastructure, and the extractive industries.
International supply chains promote technology transfer and standardized systems.
Growth in labor mobility and access to knowledge.

Globalization linkages and effects

Engineering linkages and effects

Effect of globalization on poverty
Social, legal, and environmental safeguards often lower in LDCs.
Offers economic opportunities, especially in natural resources and agriculture, tourism, manufacturing, and fair-trade goods.
LDC economies vulnerable to capital flight and brain drain, trade rules disadvantage LDCs and undermine sovereignty.

Effect of engineering on globalization
Engineering knowledge and innovation especially in transport, energy, manufacturing and ICT are the drivers behind economic integration and globalization.
Sustainability and climate change will force a revised model of engineering and globalization.

Effect of globalization on climate change
International supply chains increase energy and transportation effects.
Reduced production costs increase waste and consumerism, fueling carbon emissions.
Environmental effects displaced to less developed country (LDC) production centers.

Effect of engineering on climate change
Transport, energy, agriculture, infrastructure and manufacturing choices determine effects.
Engineering and innovation key to mitigation and adaptation.
Engineering key to disaster preparedness and reconstruction.

Effect of engineering on poverty
Engineering key to providing pro-poor energy, transport, shelter, health and water products and services.
Platform infrastructure and technologies provide an enabling environment for growth.
Engineering supply chains and technology transfer offer poverty reduction opportunities.

Source: Bourn and Neal (2008), with permission from Institute of Education, London.

may not be predictable by each part's behavior. As remarked by Berry (1990), in systems "nothing is completely itself without everything else."

Systems can be classified in different ways. One way is to look at their mass and energy interaction with what is outside the systems. They can be

- *Isolated*, with boundaries closed to import or export of both mass and energy;
- *Closed*, with boundaries closed to import or export of mass but not energy; or
- *Open*, with exchange of both mass and energy with the surroundings.

Communities are open systems. Another way of looking at systems is based on their complexity.

- *Simple*: Few variables are involved, with limited and easily understandable relationships that are displayed over a short time period. Predictability and certainty are the norm. The process involved in dealing with simple systems consists of sensing, categorizing, and responding to the "known knowns" in the systems and coming up with the correct answer to a problem (Snowden and Boone 2007).
- *Complex*: Many variables are involved, uncertainty is the norm, and complex interactions exist among the variables. The characteristics of such systems include nonlinearity, feedback mechanisms, discontinuities, sensitivity, dynamic equilibrium, emergent behavior, evolution, path dependency, and patterns of behavior. The process in dealing with complex systems is that of probing, sensing, and responding to the "unknown unknowns" in the systems while keeping in mind that no right answer is available. This method should not be confused with the sensing, analyzing, and responding to the "known unknowns" in *complicated* systems, where the correct answer to a problem is still possible (Snowden and Boone 2007).

In addition to being complex, communities can also at times be chaotic systems, in the sense that "they tend to self-organize, preserving their internal equilibrium while retaining a measure of openness to the external world" (Briggs and Peat 1999). In general, chaotic systems have unique characteristics in that

- They are dynamic (change over time) and show self-organization (order may emerge out of chaos);
- Small things can have huge consequences (subtle influences), a concept best illustrated by the "butterfly effect," suggested by mathematician and meteorologist E. N. Lorenz (2014);
- Many coupled negative and positive feedback loops drive the systems;
- They may behave at times in unexpected ways (bifurcation);
- They show nonlinear behavior and processes that may defy expectations leading to unintended consequences and unexpected behavior;
- Their existence and survivability depend on their diversity; and

- They experience synchronicity, where seemingly unrelated events happen at the same time and interact with each other.

All the systems (and the subsystems within them) shown as interacting in Figure 6-1 possess many of the aforementioned characteristics and are far from being predictable entities. They can rapidly change, have ripple effects, and create turbulent actions and cascading effects. These forms of behavior are particularly challenging to traditional engineers when trying to design and implement predictable, stable, reliable structures that interact with uncertain systems, such as natural systems. As noted by Holling and Gunderson (2002), engineering and nature operate differently as far as their resilience to adverse events is concerned. *Engineering resilience*, as defined by Holling and Gunderson, relies on "stability near an equilibrium steady state, where resistance to disturbance and speed of return to the equilibrium are used to measure the property." In comparison, *natural (or ecosystem) resilience* operates "far from any equilibrium steady state, where instabilities can flip a system into another regime of behavior—i.e., to another stability domain."

The process of understanding the interaction between natural systems (biosphere, hydrosphere, atmosphere, and geosphere) and human systems (anthrosphere and built environment) has been of particular interest in the fields of industrial ecology (Graedel and Allenby 2009) and earth systems engineering (ESE). The latter was defined by the U.S. National Academy of Engineering as

a multidisciplinary (engineering, science, social science, and governance) process of solution development that takes a holistic view of natural and human system interactions. The goal of ESE is to better understand complex, nonlinear systems of global importance and to develop the tools necessary to implement that understanding. (NAE 2000)

Consider, for example, the interaction of a concrete dam (a somewhat predictable complicated system) with a river (a nonpredictable complex system). Both have quite different characteristics, yet they interact with each other: the dam affects the environment (biosphere, atmosphere, geosphere, and hydrosphere), and the environment affects the performance of the dam. Predicting the outcome of that interaction depends largely on how the interaction is modeled and what components are included in it. Furthermore, the interaction cannot be understood using traditional linear cause-and-effect tools.

Countless examples exist of engineering projects that have resulted in more harm than good for the environment and people because they failed to recognize the dynamics between infrastructure and environmental systems. With regard to large dams in developing countries, there is still much discussion about their positive or negative social and environmental effects (Birley 2011). Negative effects include, for instance, changes in water regimes leading to

Figure 6-2. A Planned and Linear Model of a Refugee Camp in Port-au-Prince, Haiti

Source: Courtesy of Matthew R. Jelacic.

migration and resettlement and associated changes in the socioeconomic fabric of existing communities (WHO 2013).

In community development, similar issues are likely to arise when designing solutions that do not account for the dynamic of human systems, their culture, diversity, unpredictability, and modes of interaction. As an example, consider Figure 6-2, which shows a refugee camp in Port-au-Prince, Haiti, after the 2010 earthquake. The camp is very linear, and its design is based on a 2,000-year-old Roman and military model of encampment. It has been the experience of the author that when left on their own, local populations rarely adopt the model of Figure 6-2. Instead, their camps are more likely to look like congregations of tents designed organically and driven by a desire for community interaction (e.g., food, communication, and security) and survival.

Systems thinking needs to replace deterministic (Cartesian) thinking in engineering education and practice, including engineering for sustainable human development, because it may help create a higher level of consciousness (Sweeney 2001). It helps us

- See the world around us in wholes instead of snapshots;
- Sense how well parts of systems work together;
- Acknowledge relationships between system components from multiple levels of perspective rather than from cause–effect chains of reaction;
- Understand the dynamic, adaptable, unpredictable, and changing nature of life, including the effect of time and delays;

- Understand how one small event can influence another (positively or negatively) and the associated consequences of such interactions, a process called cross-scale change;
- Understand that what we see happening around us depends on where we are in the system and our attitude and perception toward that system;
- Challenge our own assumptions through mental models (discussed later);
- Become aware of and accept our bounded rationality (Simon 1972), e.g., our need to make decisions without knowing all the facts because of complexity, and our inclination to *satisfice* instead of optimize; and
- Realize that complexity is not an obstacle but an opportunity to step out of the boxes that we have created when describing the world.

All these characteristics create a mindset that is more rewarding for those practicing systems thinking and better for the planet in general. The challenge is to integrate systems thinking as a habit into the body of knowledge and the toolbox of all global engineers (see Chapter 3).

6.3 What Systems Thinking Is Not About

It is important to give a few words of caution regarding what systems thinking is *not* because the concept is often misunderstood or used incorrectly. First, systems thinking is not about analysis, which often involves breaking down a problem into bite-size and manageable pieces with the overall intent to comprehend each part individually. Such a compartmentalized approach originated from Descartes in the seventeenth century and has been the dominant way of linear and deterministic thinking in science and technology over the past 400 years.

Along that line, systems thinking is not about focusing on detailed complexity, which is usually handled by simulating thousands of variables and complex arrays of details. This method often leads to paralysis in analysis, creating intractable overloads of data and information, and distracts from seeing emerging patterns and interrelationships. Such an approach is commonly used in forecasting, planning (engineering included), and business analysis.

Finally, systems thinking is not about making things less complicated, perfect, and simple. It is about embracing the unique attributes of systems listed above, not as obstacles but as opportunities for change, as discussed in the book *Getting to Maybe* by Westley et al. (2007). As remarked by these authors, managing complex issues as if they are complicated or simple may miss the mark and create disasters. Instead, systems thinking is about "acting deliberately and intentionally in a complex uncertain world by virtue of being in and of that world" and welcoming the possibility of change and risk taking.

6.4 Systems Components and Archetypes

Systems Components

Unlike linear systems that consist of serial cause-and-effect relationships (with unidirectional and linear causation), complex systems consist of cause-and-effect loops (with circular causation), some of them positive (or reinforcing) and others negative (balancing). Senge (1994a) defines these two basic building blocks as follows:

- *Reinforcing (+) loops* create a compounding effect and are self-reinforcing processes. Any situation where actions produce results that create more of the same action is an attribute of reinforcing loops. They amplify or add to change, create a snowball effect, and have potential to increase growth or decline.

- *Balancing (−) loops* are those that bring two things into agreement. They are self-correcting processes and prevent unlimited growth or decline. Any situation concerned with achieving a goal or objective is representative of balancing loops. They seek stability and equilibrium. Balancing loops need a goal, a monitoring system, and a corresponding response to change.

In systems, reinforcing and balancing loops interact because there is always a limit to growth. Balancing loops always accompany reinforcing loops to reach dynamic equilibrium and keep them in control and within safe bounds. Delays may also be added to balancing and reinforcing loops to account for the role of time in linking causes and effects.

The interaction between reinforcing and balancing loops can be represented using so-called *mental models* in the form of *causal loop diagrams* or *stock-and-flow diagrams*. It should be remembered though, that mental models are simplifications of actual systems, e.g., abstractions of reality. They are used to understand a particular problem experienced by the system but are not used to model the system itself. Furthermore, it should be remembered that mental models cannot be validated or verified (Sterman 2000, 2006).

Figure 6-3 shows an example of a *causal loop diagram* consisting of two interacting causal loops that define the size of a population. In that diagram, arrows labeled with a + sign link things that move in the same direction. Arrows labeled with a − sign link issues that move in the opposite direction. In this case, births add to the population and deaths decrease it. Also shown in Figure 6-3, the birth and death rates determine how fast the population increases or decreases; a larger population has a greater number of births. Causal loop diagrams such as the one shown in Figure 6-3 are used to visualize what contributes to growth or stability and are mostly used at the project *strategy* level. They are not used to conduct simulations. They help in laying out the different components of a system and how they interact.

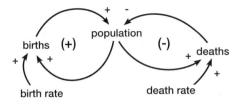

Figure 6-3. Population Causal Loop Diagram

Note: Arrows with a + sign link things that move in the same direction. Arrows with a – sign link issues that move in the opposite direction

Source: Ford (2010), with permission from Island Press.

Another way of describing the dynamics of systems is to use *stock-and-flow diagrams* (Figure 6-4), which consist of combinations of the following basic components:

- Information flows in and out of components, called *stocks*. A stock can be a reservoir and corresponds to an accumulation of something (a delay of some sort) that can be measured at one point in time (e.g., water in a bathtub or behind a dam, population, or wood in the forest). A stock can also be represented as a *conveyor* to account for the time it takes for information to pass through the stock. Stocks can be seen as *state variables*.
- *Flow* (inflow or outflow) is shown in the form of pipelines (with faucet controlling the flow). It refers to activities that cause *change over time* (e.g., number of births per year, inflation rate, flow of a river, cash flow, or carbon emission and sequestration), thus resulting in changes in the stock accumulations. Flows can be seen as *control variables*. Flow may include some form of *delay* or time lag.
- *Clouds* indicate infinite sources or sinks.
- *Connectors* indicate transmission or links of actions and information between components, either stock to flow or flow to flow.

Figure 6-5 shows the stock-and-flow diagram corresponding to the example of Figure 6-3. Such mental models help visualize how things flow and accumulate through reinforcing and/or balancing loops. Using some software (e.g., *iThink* and STELLA), they allow for numerical simulations and parametric or sensitivity studies and can therefore be used at the project *operation* level. In designing such models, the laws of conservation and accumulation must be obeyed.

Systems Archetypes

As remarked by Senge (1994a), some patterns in systems seem to occur over and over again. They are called *systems archetypes* or *generic structures*. Several of these

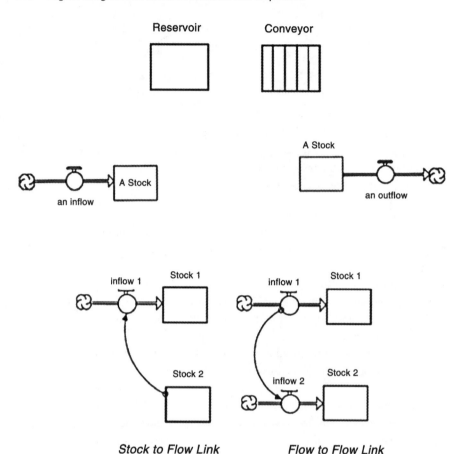

Figure 6-4. Basic Building Blocks of Mental Models: Stocks (Reservoir or Conveyor), Flow (Inflow and Outflow) with Clouds and Connectors

Source: Richmond (2004), adapted with permission from ISEE Systems, Inc.

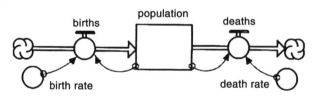

Figure 6-5. Population Stock-and-Flow Diagram

Source: Ford (2010), with permission from Island Press.

archetypes have been identified and are well documented for human systems because they represent common patterns of human behavior. It is likely that in the field of human development, such patterns and the community response to them are similar in different regions of the world, or at least they have similar core characteristics that are supplemented with local characteristics. Archetypes are diagnostic tools to identify such patterns. According to Meadows (2008), archetypes are traps (grooves) forcing a system to produce the same answer under the same conditions; they create *habits*. Recognizing archetypes at play (in a prospective way) is also an opportunity for forcing change "ahead of the game" and creating a way out of the trap or the groove. Archetypes show clearly that the structure of systems in many ways controls their behavior, an important observation mentioned earlier in this chapter. Simply put, *a change in behavior requires a change in structure.*

Systems archetypes are formed by combining fundamental modes of dynamic behavior. According to Sterman (2000), the vast majority of systems archetypes include, in their respective causal loop diagrams, basic modes, such as (1) exponential growth that can be modeled by a single reinforcing loop; (2) goal seeking, which can be modeled using a single balancing loop; (3) oscillation, which occurs when delay is combined with a balancing loop; and (4) delays. Other, higher forms of behavior can be obtained by combining the aforementioned basic modes, such as S-shaped growth (balancing and reinforcing loops), S-shaped growth with overshoot and oscillation (balancing and reinforcing loops with delay), and overshoot and collapse. Other modes of system behavior include equilibrium, random behavior, and chaos.

Senge (1994a) identified and named nine archetypes in human systems: fixes that backfire or fail, limits to growth or success, shifting the burden, eroding goals, escalation, success to the successful, tragedy of the commons, growth and underinvestment, and balancing process with delay. Descriptions of each archetype and examples of their applications to real-world problems can be found in *The Fifth Discipline Field Book* by Senge (1994b) and the systems archetypes toolboxes by Kim (2000).

Among all nine archetypes, the *tragedy of the commons* is often used in the sustainable development literature to explain the diminishing return associated with stakeholders in a group sharing common resources without consideration for the collective good. The concept was originally introduced by Hardin (1968) and emphasizes that without regulations, pursuing one's interest while trying to share resources with others always leads to a tragedy for all, leaving the commons in worse shape than if they had been managed in the first place (Gardner 2005). This archetype emphasizes that the solution to the tragedy of the commons dilemma does not reside at the individual level but rather requires a collective decision process that all stakeholders need to address and upon which they need to agree. This archetype could be used to describe the community sharing of a multitude of resources, including water, food, energy, land, and infrastructure.

6.5 Modeling Systems Dynamics Using *iThink* and STELLA

iThink and STELLA are software tools developed to support dynamic modeling. They were introduced by ISEE Systems in 1983 and originated from the work of Barry Richmond. They provide graphically oriented toolkits for systems simulation using stock-flow representation that is somewhat easy to learn and use. Other software packages are available for modeling dynamic systems (e.g., Vensim, Powersim, and Dynamo), but they are not addressed here.

Regardless of the software package used, system dynamics models are usually built in steps of increasing complexity. Table 6-3 from Ford (2010) lists eight steps to build mental models of systems. The steps are broken down into two groups: qualitative and quantitative. It should be remembered that modeling itself is a highly iterative process that is complex and contains feedback loops.

Guidelines in Systems Modeling

At first glance, systems seem to be intractable and overwhelming. As suggested by Richmond (2004), a few initial steps need to be carried out before building mental models of real-world systems. First, the problem and issues to be addressed and the type of behavior to be modeled need to be framed, contextualized, and described explicitly. In community project management, this step emerges in the project initiation phase (Chapter 4) and at the conclusion of the community

Table 6-3. Steps of System Modeling

Qualitative Modeling	
Step 1.	A is for get Acquainted with the problem.
Step 2.	B is for Be specific about the dynamic problem.
Step 3.	C is for Construct the stock-and-flow diagram.
Step 4.	D is for Draw the causal loop diagram.
Quantitative Modeling	
Step 5.	E is for Estimate the parameters.
Step 6.	R is for Run the model to get the reference mode.
Step 7.	S is for Sensitivity analysis.
Step 8.	T is for Testing the effects of policies.

Source: Ford (2010), with permission from Island Press.

appraisal phase (Chapter 5), where problems have been identified and a causal analysis has been carried out (Chapter 7). Issues that are *endogenous* (i.e., originating from within) and *exogenous* (i.e., originating from without) to the systems are identified. Other issues may need to be excluded, at least during the first step in the iterative modeling process. They may be included later or on an as-needed basis (Ford 2010).

Second, the boundaries of the problem need to be defined in both breadth or extension (horizontal) and depth or intensity (vertical). In the case of sustainable human development projects, the horizontal characteristic deals with the cross-disciplinary nature of the problems at stake (e.g., health, water, sanitation, energy, shelter, or jobs). The vertical component is associated with how deep and detailed it is necessary to go in a reductionist manner into each development discipline. Finally, the time frame (days, months, years) simulated by the mental model must be selected and the system's initial conditions (community baseline).

The system boundaries (spatial and temporal) selected in mental models, do not actually exist in nature, where systems are connected to other systems in space and time. Because it is not possible to include the entire universe and its entire history into mental models, boundaries are necessary and need to be created. In practice, they don't need to be too narrow or too wide. According to Ricigliano and Chigas (USAID 2011), a special effort must be made within the system boundaries to balance comprehensiveness and comprehensibility when developing mental models.

It should be remembered that the conclusions reached in modeling any system depend, to a large extent, on the context and the selected boundaries. Boundaries drive how systems are framed. They differentiate between "what is in and what is out, what is deemed relevant and irrelevant, what is important and unimportant, what is worthwhile and what is not, who benefits and who is disadvantaged" (Williams 2008). As the complexity of the mental model grows, boundaries may need to be expanded to see if conclusions about system behavior change. In sustainable human development projects, the spatial boundaries are likely to be geographical (community, village, household), and the temporal boundaries may include seasonal or yearly activities.

Constructing and Simulating Mental Models

Once the problem has been framed and the boundaries have been selected, it is necessary to follow certain guidelines (Richmond 2004) for constructing mental (virtual) models, such as the one shown in Figure 6-5.

- First, select the basic building blocks (stock, flow, conveyor belts) that need to be included in the model. This is based on (1) what is accumulating (reservoirs), (2) the processes that are flowing in and out of the reservoirs (flow), and (3) the processes that control the flow (reinforcing or balancing). Second, decide about the building blocks that are critical and disregard those that could be set aside (for the time being). Both stocks and flows can be physical or nonphysical.

- Second, decide how the various elements are connected or paired and how they depend on each other, e.g., direct (linear) causality, reciprocal causality, closed-loop causality, and delay.

In constructing mental models, rules must be followed (Richmond 2004). First, care must be taken to respect unit consistency, i.e., each flow in or out of a reservoir must use the same units of measure as the reservoir itself, except that flow is measured in "per unit of time." Likewise, all stocks attached to a given pipeline must use the same unit of measure. The second rule in constructing mental models is that conservation laws of mass and energy must be respected.

Once a mental model is created, simulation models can be run, using simple data to test the model performance and overall stability and compare it with the actual behavior of the real-world problem that is being simulated (model verification and validation). More complex input can then be included once confidence has been built around the model. In general, the process consisting of problem definition, model construction, and simulation is *iterative*. Through the various simulations, the system's overall behavior is studied under various assumptions, leverage points are identified, and possible forms of unintended consequences are explored. Ultimately, a new and better way of understanding the system may emerge through the iterative process (Meadows 2008).

6.6 Systems and Community Development

From the results of the community appraisal discussed in Chapter 5, a systems model and map of a community in its current state (baseline) can be established, showing the range of interconnected issues and problems the community is facing and its subsystems at play. Causal loop diagrams, such as the one shown in Figure 6-3, are ideal to represent such interconnections. The results of the stakeholder, partnership, capacity, vulnerability, and social network analyses described in Chapter 5 can then be interpreted in a more holistic way. Causal loop diagrams can help identify and make sense of current trends, emerging patterns, and interrelationships for specific community issues or for a range of issues.

Stock-and-flow diagrams (like Figure 6-5) supplement causal loop diagrams to simulate the dynamic of communities. More specifically, they are able to predict and help understand how communities are likely to respond and change over time when confronted with day-to-day issues; pushed to limits (hazards); and/or faced with inadequate resources, knowledge, or assets. Because of the inherent uncertainty in communities, systems dynamics tools allow for conducting stochastic simulations, and not just deterministic ones. This ability is particularly critical because the random nature of human and environmental systems and the resulting uncertainty may trigger some unexpected behavior patterns.

Referring again to the equation,

$$\text{Risk} = \text{Hazard} \times \text{Exposure} \times (\text{Vulnerability} - \text{Capacity})$$

presented in Chapter 1, a systems approach to community projects can help better understand (1) the role played by specific components of inherent and adaptive capacity and vulnerability in calculating risk, (2) the interconnectedness among those components, and (3) the dynamic process as some of the components change on their own or are tempered or improved. Different risk scenarios and outcomes can then be simulated to explore the response and behavior of communities and identify leverage points, i.e., places where small changes in the community can have large consequences. This step may help in prioritizing decisions to reach the main goal, i.e., creating communities that are more resilient to hazards, more equitable, more secure, healthier, and/or more prosperous.

In summary, a systems approach has the potential for developing more holistic intervention strategies at the community level than do traditional approaches. It provides an alternative to conventional methods that "tend to produce fragmented, incrementally effective (if not counter-productive) development efforts" (Delp et al. 1977). A systems approach to development projects is also more prone to capturing the unusual characteristics of human systems interacting with other systems. But like everything else, it is not a panacea because it has its own limitations as well. Among those limitations is the fact that a systems approach requires defining boundaries of a system. As remarked earlier, boundaries define "judgments about worth" (Williams 2008). Using a systems approach to address development issues always requires that a choice be made between what is valued and what is not by those who use the approach. Thus, what is inferred from the systems solutions greatly depends on the perspective of those who do the modeling, which could create another form of bias in development work in addition to those discussed in Section 5.4.

System Dynamics and Sustainability

In Chapter 2, we emphasized a definition of sustainability proposed by Ben-Eli (2011) as a dynamic equilibrium in the interaction of human systems and their environment. Figure 6-6 is a simple two-way causal loop representation of that interaction.

Both human and environmental systems are characterized by complexity. Humans create unique characteristics in systems in which they interact,

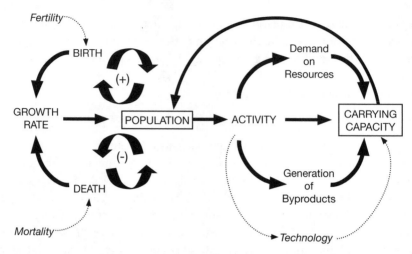

Figure 6-6. Causal Loop Diagram Representing the Interaction between a Population and the Carrying Capacity of the Environment

Source: Ben-Eli (2011), with permission from The Sustainability Laboratory.

as discussed earlier in this chapter. However, natural systems have their own attributes and operating mechanisms. They are diverse, open, dissipative, self-regulating, adaptive and evolving, and self-organized. They are run by a variety of biogeochemical cycles in the atmosphere, biosphere, hydrosphere, and geosphere of which we have limited comprehension.

Using system dynamics tools to model the interaction between human and natural systems can help show sustainability as "a particular system state born by a particular underlying structure" (Ben-Eli 2011). In other words, because structure controls behavior, the state of sustainability results from systems behavior, which itself is controlled by some unique systems structure. The uniqueness comes from the interaction of reinforcing and balancing causal loops capable of holding human, economic, and environmental systems in check. This interaction is done within an adaptive and responsive context that is designed to "optimize benefits and minimize harm" if the process is allowed to operate within the mindset of ecosystem operation rather than that used in controlling engineered systems (Patton 2011).

Among the causal loops that contribute to the sustainability of a community, Hjorth and Bagheri (2006) introduced special ones called *viability loops* as "key loops in the real world, which are responsible for the viability of all ecosystems including human based ecosystems." These loops are based around four main categories: human needs, the economy, the environment, and life-support structures. They are critical in maintaining the balance between humans and their environment if sustainability is to be achieved. They are all balancing feedback loops, which may lead to major unintended consequences if disturbed. An example is what would happen when exceeding the carrying capacity of the environment, as illustrated in the tragedy of the commons example.

As a result, viability loops need to be protected in community development. They are the lifelines of natural systems and human systems. System dynamics can be used to understand the role played by these loops, their interactions, and what happens when one or several of them are disturbed. Causal loop diagrams combined with stock-and-flow diagrams may help in quantifying sustainability.

A question that arises is how broad and detailed system dynamics models should be to truly capture the interaction of human systems and their environment. Models that look at the interaction of human and other systems in a global way are limited because of the sheer complexity at stake. One example is the model called the tool to assess regional and global environmental and health targets for sustainability (TARGETS) developed by Rotmans and de Vries (1997). It consists of five interacting sub-models covering the areas of population and health, energy, land and food, water, and biogeochemical cycles. It is designed to assess global change. Because such large-scale modeling is difficult and models can easily become complex quite rapidly, system dynamics has been more often used to model the dynamic of specific issues, such as health, the environment, water, food, the economy, and in some cases, a combination of several of them.

Modeling Community Health

Newman (2010) describes several examples of applications of system dynamics to development projects in Guatemala, Bolivia, and Peru. His examples illustrate how systems thinking could be used to comprehend the consequences of policy decisions and actions on in-the-field community health issues related to malaria control and chronic malnutrition. Compared with more traditional linear approaches where one outcome is expected from one set policy, a systems approach allows for (1) exploring multiple options; (2) identifying the cost and benefit of each action; and (3) reevaluating and reconsidering the decisions and actions in the form of a reflective practice, as discussed in Section 4.9. Figure 6-7 shows an example of a stock-and-flow diagram used by Newman to model "the processes through which a child becomes chronically malnourished" in Peru.

Other publications illustrating the use of comprehensive system dynamics models of various health-care issues can be found in the papers by Newman et al. (2003), McDonnell et al. (2004), and Rwashana and Williams (2008). Homer and Hirsch (2006) and Jones et al. (2006) discuss, more specifically, the advantages of system dynamics to model chronic disease prevention. In these papers, systems thinking is used to model the interaction of health systems with other

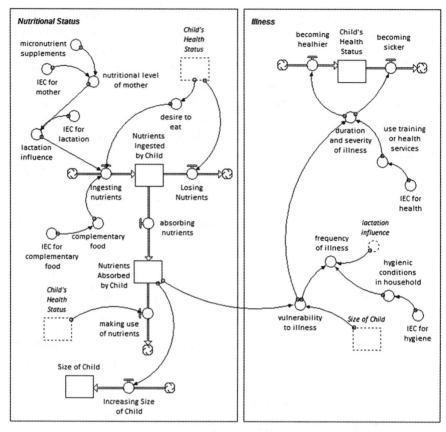

Figure 6-7. Stock-and-Flow Diagram Showing Processes Related to Child Malnutrition in Peru

Note: IEC refers to information, education, and communication programs.

Source: Newman (2010), with permission from ISEE Systems, Inc.

human and social activities (economic, infrastructure development) that contribute indirectly to the health and well-being of communities.

Modeling Water Cycle and Management

The availability of water in terms of quantity, quality, distribution, and water management is a critical issue to all communities, large or small. In general, community water systems consist of natural systems and their associated infrastructure. In water resource management projects, these two systems constantly interact with each other and involve other socioeconomic issues. As reviewed by

Winz et al. (2008), system dynamics has been used over the past 40 years to account for the uncertainties inherent in water resource management.

The WorldWater model is an example of a system dynamics tool to explore strategies of water management (Simonovic 2003). It contains seven interacting components: population, agriculture, nonrenewable resources, economy, pollution, and water quality and quantity. When combined with dynamic modeling tools, the model has been used to gain insight into the dynamic of water use at the global level or at the regional level. An example of application of the model at the country scale (e.g., CanadaWater model) or within regions of a country is described by Simonovic and Rajasekaram (2004).

6.7 Illustrative Example Using *iThink* and STELLA

The software packages *iThink* and STELLA developed by ISEE Systems are flexible and powerful tools used to model sustainable human development. The reader is encouraged to acquire this software and become familiar with its modeling capabilities.

Figure 6-8 shows an example of a stock-and-flow diagram representing how waste can be converted into a commercially viable resource for a community (Teipel 2013, personal communication). The model begins by taking into account the waste generated by the community per capita. Waste can either flow into a landfill or it can be collected and reused. If the waste is collected and sorted, waste accumulates (e.g., conveyor in Figure 6-8) before the reclaimed waste is sent to the sorting reservoir. Once material flows into the preprocessing reservoir, the resulting output material can flow into three new reservoirs. The first option is to accumulate and be processed as a raw bulk material (e.g., empty aluminum cans, wood scrap turned into sawdust). The second option is to further process the material to meet specific commercial applications (e.g., wood powder material used as filler for plastic composites). The third option is to send the preprocessed material back to the landfill. In the last case, the material cannot be sold as raw bulk material on the global market, nor can it be reused to create a new material. In summary, the waste material that is diverted from landfills in the first two options can provide feedstock for a waste-to-resource model and unique commercial opportunities for a community.

6.8 Chapter Summary

Chapter 6 provides the reader with an introduction to system dynamics and its possible applications to sustainable human development. It is clear that a systems approach to development has the potential to provide a greater understanding of the complex interaction that exists among different components of a community and the problems and issues a community is facing. As remarked by

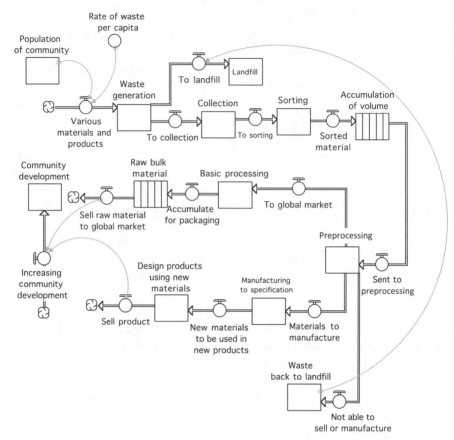

Figure 6-8. Stock-and-Flow Diagram Showing a Waste-to-Resource Community Framework

Source: Courtesy of Elisa Teipel.

Newman (2010), it also has the potential to provide a link between policy actions in development and actual results while evaluating and assessing various possible options, rather than the traditional approach of exploring one option for a predetermined outcome. Systems thinking complements (but does not replace) reductionist thinking. At times, it is indeed necessary to dive deeper into specific issues in the system while keeping the "big picture" in mind.

The examples presented in this chapter have demonstrated the potential use of system dynamics tools to better understand the dynamic of communities and account for their uncertainty and complexity. This chapter did not explore how system dynamics can be combined with other system tools such as those in the fields of cybernetics, operations research, and organizational theory. A discussion on the potential of such interaction can be found in the paper by Basole et al. (2011) in regard to modeling complex enterprise networks in manufacturing and health delivery.

The integration of systems thinking into development is still in its infancy, even though it has been attempted by a limited number of agencies over the past 10 years. It represents an alternative to traditional models that expect predictable and well defined results following well defined actions. We know from past experience that such an approach has had limited results. It is time to endorse a new mindset that follows the true nature of human communities and their interaction with other systems.

References

Basole, R. C., et al. (2011). "Models of complex enterprise networks." *Journal of Enterprise Transformation*, 1, 208–230.

Ben-Eli, M. (2011). "The five core principles of sustainability." <http://www.sustainabilitylabs.org/page/sustainability-five-core-principles> (Jan. 10, 2013).

Berry, T. (1990). *The dream of the earth*, Sierra Club Books, San Francisco.

Birley, M. (2011). *Health impact assessment: Principles and practice*, Earthscan Publications Ltd., London.

Bourn, D., and Neal, I. (2008). *The global engineer: Incorporating global skills within UK higher education of engineers*, Institute of Education, London. <http://engineersagainstpoverty.org/_db/_documents/WEBGlobalEngineer_Linked_Aug_08_Update.pdf> (Oct. 12, 2012).

Briggs, J., and Peat, F. D. (1999). *Seven life lessons of chaos—Spiritual wisdom from the science of change*, Harper Perennial, New York.

Delp, P., et al. (1977). *Systems tools for project planning*, International Development Institute, Bloomington, IN.

Dent, E. B. (2001). "System science traditions: Differing philosophical assumptions." *Systems: J. Transdisciplinary Systems Science*, 6(1–2), 13–30.

Fisher, D. M. (2011). *Modeling dynamic systems: Lessons from a first course*, ISEE Systems, Inc., Lebanon, NH. <http://www.iseesystems.com> (Jan. 10, 2012).

Ford, A. (2010). *Modeling the environment*, Island Press, Washington, DC.

Forrester, J. W. (1961). *Industrial dynamics*, Pegasus Communications, Waltham, MA.

Gardner, G. (2005). "Yours, mine, ours—or nobody's." *WorldWatch*, 18(2), March/April.

Graedel, T. E. H., and Allenby, B. R. (2009). *Industrial ecology and sustainable engineering*, Prentice Hall, Upper Saddle River, NJ.

Gunderson, L. H., and Holling, C. S., eds. (2002). *Panarchy: Understanding transformations in human and natural systems*, Island Press, Washington, DC.

Hannon, B., and Ruth, M. (2001a). *Dynamic modeling*, Springer, New York.

Hannon, B., and Ruth, M. (2001b). *Modeling dynamic biological systems*, Springer, New York.

Hardin, G. (1968). "The tragedy of the commons." *Science*, 152, 1243–1248.

Hargrove, J. L. (1998). *Dynamic modeling in the health sciences*, Springer, New York.

Hjorth, P., and Bagheri, A. (2006). "Navigating towards sustainable development: A system dynamics approach." *Futures*, 38, 74–92.

Holling, C. S., and Gunderson, L. H. (2002). "Resilience and adaptive cycles." *Panarchy: Understanding transformations in human and natural systems*, L. H. Gunderson and C. S. Holling, eds., Island Press, Washington, DC.

Homer, J. B., and Hirsch, G. B. (2006). "System dynamics modeling for public health: Background and opportunities." *American J. Public Health*, 96(3), 452–458.

Jones, A. P., et al. (2006). "Understanding diabetes population dynamics through simulation modeling and experimentation." *American J. Public Health*, 96(3), 488–494.

Kim, D. H. (2000). *Systems archetypes I and II*, Nabu Press, Charleston, SC.

Lorenz, E. N. (2014) "Butterfly effect." Scholarpedia, <http://www.scholarpedia.org/article/Butterfly_effect> (April 20, 2014).

McDonnell, G., Heffernan, M., and Faulkner, A. (2004). "Using system dynamics to analyze health system performance within the WHO framework." 22nd International Conference—System Dynamics Society, Albany, NY <http://www.anysims.com/PDF/Health%20System%20Performance.pdf> (March 2, 2013).

Meadows, D. (2008). *Thinking in systems*, Chelsea Green Publishing, White River Junction, VT.

National Academy of Engineering (NAE). (2000). "Earth systems engineering exploratory workshop." Unpublished report. National Academy of Engineering, Washington, DC.

National Research Council (NRC). (2012). *Disaster resilience: A national imperative*, National Academies Press, Washington, DC.

Newman, J. L. (2010). "Finding system dynamics: An exploration in international development." *Tracing connections: Voices of systems thinkers*, B. Richmond et al., eds., ISEE Systems Inc., Lebanon, NH, 143–165.

Newman, J., et al. (2003). "A system dynamics approach to monitoring and evaluation at the country level." Presented at the 5th Biennial World Bank Conference on Evaluation and Development, Washington DC., Cornell System Dynamics Network, Ithaca, NY. <http://csdnet.dyson.cornell.edu/papers/newman.pdf> (March 5, 2013).

Patton, M. Q. (2011). *Developmental evaluation: Applying complexity concepts to enhance innovation and use*, Guilford Press, New York.

Richmond, B. (2004). *An introduction to systems thinking, STELLA software*, ISEE Systems, Inc., Lebanon, NH.

Robinson, W. A. (2001). *Modeling dynamic climate systems*, Springer, New York.

Rotmans, J., and de Vries, B., eds. (1997). *Perspectives on global change: The TARGETS approach*, Cambridge University Press, Cambridge, U.K.

Rwashana, A. S., and Williams, D. W. (2008). "System dynamics modeling in healthcare: The Uganda immunization system." *Int. J. Computing and ICT Research*, 1(1), 85–98.

Senge, P. (1994a). *The fifth discipline: The art & practice of the learning organization*, Doubleday, New York.

Senge, P. (1994b). *The fifth discipline field book: Strategies and tools for building a learning organization*, Crown Business, New York.

Simon, H. A. (1972). "Theories of bounded rationality." *Decision and organization*, C. B. McGuire and R. Radner, eds., North-Holland Pub., Amsterdam, Netherlands, 161–176.

Simonovic, S. P. (2003). "Assessment of water resources through system dynamics simulation: From global issues to regional solutions." Proc. 36th Hawaii International Conference on System Sciences. Computer Society Press, Washington, DC.

Simonovic, S. P., and Rajasekaram, V. (2004). "Integrated analyses of Canada's water resources: A system dynamic approach." *Canadian Water Resources J.*, 29(4), 223–250.

Snowden, D., and Boone, M. (2007). "A leader's framework for decision making." *Harvard Business Review*, 69–76, November. <http://www.mpiweb.org/CMS/uploadedFiles/Article%20for%20Marketing%20-%20Mary%20Boone.pdf> (March 24, 2013).

Sterman, J. (2000). *Business dynamics: Systems thinking and modeling for a complex world*, Irwin, McGraw Hill, New York.

Sterman, J. (2006). "Learning from evidence in a complex world." *Am. J. Public Health*, 96(3), 505–514.

Sweeney, L. B. (2001). *When a butterfly sneezes: A guide for helping kids explore interconnections in our world through favorite stories*, Pegasus Communications, Waltham, MA.

Teipel, Elisa. (2013). Personal communication.

U.S. Agency for International Development (USAID). (2011). *Complexity event*, U.S. Agency for International Development, Washington, DC. <http://kdid.org/sites/kdid/files/resource/files/Complexity_event_brief.pdf> (Feb. 5, 2013).

Valadez, J., and Bamberger, M. (1994). *Monitoring and evaluating social programs in developing countries: A handbook for policymakers, managers, and researchers*, World Bank Publications, Washington, DC. <http://www.pol.ulaval.ca/perfeval/upload/publication_192.pdf> (May 22, 2012).

Westley, F., Zimmerman, B., and Patton, M. Q. (2007). *Getting to maybe: How the world is changed*, Vintage, Toronto, Canada.

Wheatley, M. J. (1999). *Leadership and the new science*, Berrett-Koehler Publ., San Francisco.

Williams, R. (2008). "Bucking the system." *The Broker*, December 2. Amsterdam, The Netherlands. <http://www.thebrokeronline.eu/Articles/Bucking-the-system> (March 10, 2013).

Winz, I., Brierley, G., and Trowsdale, S. (2008). "The use of system dynamics simulation in water resources management." *Water Resources Management*, 23(7), 1301–1323.

World Health Organization (WHO). (2013). "Human health and dams." World Health Organization, Geneva. <http://www.who.int/hia/evidence/whohia091/en/index.html> (Oct. 15, 2013).

7

From Appraisal to Project Hypothesis

7.1 Preliminary Design

At this stage of the ADIME-E framework (Figure 4-6) and after the analysis of the community appraisal discussed in Chapter 5, the appraisal team has a better understanding about (1) the community, its culture, vulnerability, constraints, capacity, and livelihood security; (2) the community structure, dynamic, and variations (differentiation and disaggregation); (3) the range of issues (problems, threats, needs, challenges) of concern to the community and their rankings; and (4) opportunities for solving key issues. The appraisal team has identified and ranked *core* problems faced by the community, in consultation with the community stakeholders and leaders.

Before addressing those problems and implementing appropriate solutions, a detailed and thorough action plan needs to be outlined. This plan requires additional analysis of the problems, their causes and consequences (effects), and the identification of potential solutions that match, as best as possible, the community capacity and vulnerability. As mentioned in previous chapters, an underlying concept in sustainable community development projects requires that solutions be done right from a technical point of view and that they are the right solutions for the community. Chapters 7 and 8 discuss the steps that are necessary in developing an action plan with these two attributes in mind.

This chapter covers three components of project design that ultimately help frame the project hypothesis (or problem statement) and the what, who, when, where, why, and how of the project. The first component is *causal analysis*, i.e., the systematic identification and prioritization of the consequences and the root causes of each problem identified. The second component addresses the initial identification and ranking of possible solutions to the problems identified. Finally, the chapter explores how to develop project hypotheses in consultation with the community stakeholders and a preliminary action plan to implement those solutions.

From a practical point of view, the steps in the overall ADIME-E framework discussed in this chapter can be considered as comprising the preliminary phase of project design, that is, for a given core problem that has causes and effects, project hypotheses are outlined and *preliminary solutions* are proposed and evaluated. The outcome of this phase is validated again with community stakeholders through participation.

7.2 Causal Analysis: Problem and Solution Trees

According to Caldwell (2002), *causal analysis* (or cause-and-effect analysis) is a process that can be used to determine the underlying causes and consequences of a core problem and explore relationships between these causes and effects. Causal analysis acknowledges the complex cause-and-effect relationships (linear and circular causation) that characterize the dynamic of systems such as communities. These relationships are often the reason why problems exist in the first place and why the problems do not always have easy solutions. It is not uncommon for a problem to actually be the consequence or cause of another problem. Direct and indirect issues with macro- or microlinkages may contribute to a given problem. Being able to comprehend all these connections can be difficult for the human mind, in particular for those who are more comfortable with linear thinking. The system tools presented in Chapter 6 and a variety of visual tools, such as Mind Maps (Buzan and Buzan 1996), may help development workers picture the connections and links.

The results of causal analysis can be presented in different forms (e.g., hierarchical diagrams, fishbone diagrams, tabular format, or causal trees). The analogy of a *problem tree* is used herein (Figure 7-1a). The tree serves as a visual tool to organize information about the factors that interact with a central issue, e.g., a core problem. The core problem is represented by the tree trunk. The consequences (or effects) of the problem are represented by a network of tree branches, the visual part of the tree. Branches may have smaller branches to simulate effects and associated subeffects. The causes, subcauses, and other associated linkages are represented by the tree roots, the hidden part of the tree. Several core problems can be represented by several trees, which in turn can share roots and branches.

Construction of a problem tree is a *participatory* group-dynamic exercise that involves direct interaction with community stakeholders. It can also be constructed separately by the appraisal team during one of its process meetings, then confirmed and improved with the help of the community. The concept of a problem–consequence–root cause logic is simple and universal enough that the problem tree can more than likely be explained to and understood by communities around the world. The best way to build a problem tree is to use sticky notes (e.g., Post-its) of different colors that are placed on a wall. The core problem is

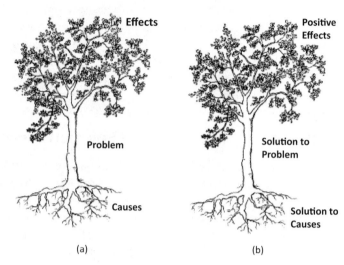

Figure 7-1. Representation of (a) Problem Tree and (b) Solution Tree

written on a note that is placed prominently on a central location on the wall. It is recommended to use green stickers for the consequences (branches) and yellow or brown stickers for the roots.

The problem tree leads to its counterpart; a *solution tree* (Figure 7-1b), also called a *result tree* or an *objective tree* (Delp et al. 1977). Instead of showing negative causes and effects of a major problem, a solution tree has positive roots and optimistic outcomes. The solution to the problem is now at its center and contributes to reaching the project outcome or overarching goal. The solution tree gives a comprehensive picture of the future desired solution in a hierarchical format. As demonstrated in Chapter 8, a strategy in the form of a logical framework consisting of goals, objectives (output), activities, and input needs to be developed to make the outcome or overarching goal a reality.

Figures 7-2 and 7-3 show, respectively, a problem tree and a solution tree for a project in Mabu, Nepal, where one of the problems was not enough water for crops during the dry season, which leads to low crop yield and reduced household living security. Another problem identified in the community (but not shown here) was the lack of electricity in each home. All problems were actually combined, thus resulting in a more holistic solution toward a global outcome and overarching goal, which was to "improve household livelihood security through increased income opportunities and increased food security."

When designing solutions for the core problems, major tracks in the causal analysis need to be identified. In other words, several consequences and root

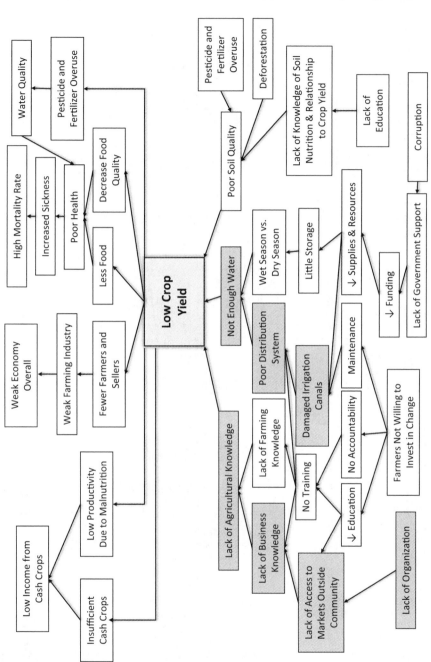

Figure 7-2. Example of Problem Tree for Crop Yield (Mabu Project)

Source: Glover et al. (2011), reproduced with permission.

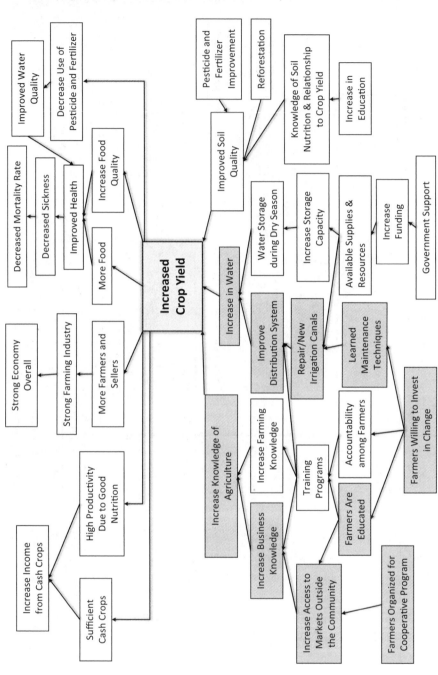

Figure 7-3. Example of Solution Tree for Crop Yield (Mabu Project)

Source: Glover et al. (2011), reproduced with permission.

causes may be linked together and clustered around a common theme or a commonality of purpose, which is often the case in practice (Gawler 2005). A theme may correspond to a *secondary problem* or *subproblem*, for instance. Once identified, each theme may appear as a sequence of sticky notes with the same colors in the problem and solution trees. Stickers can easily be moved during participatory group-dynamic exercises until a theme appears. For the Mabu project, two tracks (water and knowledge in agriculture) were considered in the root causes responsible for low crop yield, as shown by the shaded boxes in Figures 7-2 and 7-3.

Not all root causes and common themes carry the same level of significance in a project. Addressing all root causes and consequences to a given problem is practically impossible because of the inherent complexity. Thus, criteria need to be developed to prioritize issues that need to be addressed first; prioritization in turn helps identify the project objectives. According to Caldwell (2002), to be critical and selected, root causes and themes must meet several key criteria. They need to

- Show good potential to make a significant impact and contribution if eliminated;
- Make sense to community stakeholders;
- Have major impact through synergy, collaboration, and partnering; and
- Be achievable and measurable with the existing skills and resources of locals and outsiders.

Analysis of the appraisal results should help in identifying key root causes that are recurring in the community. If necessary, additional data collection may need to be carried out to further pin down these root causes. The synergy that is required to address the root causes can actually be identified from the results of the social network analysis discussed in Section 5.10. The reader is referred, as an example, to Table 5-10 for the Guatemala project, which shows the links between various community issues and internal and external support. Synergy can also be tested in a qualitative and quantitative way using the causal loop and stock-and-flow diagrams discussed in Chapter 6.

7.3 Preliminary Solutions

Once a core problem and its critical root causes are ready to be addressed and the main project objectives are identified, various design solutions and interventions can be explored. The Department for International Development (DFID 2002) calls this phase *the alternative analysis*, i.e., the identification of possible alternative options, the assessment of their feasibility, and the agreement on a preliminary plan of action. It can also be called *morphological analysis* (Delp et al. 1977) or *feasibility study* (Forbes 2009).

Action Identification Matrix

One of the many challenges is to be able to choose a list of promising alternative solutions that match the various community characteristics identified in the appraisal phase (capacity, vulnerability, resilience). To do this, an action identification matrix is constructed (Barton et al. 1997) that lists potential solutions and interventions for the different causes and identified issues (e.g., water, sanitation, and hygiene; energy; or shelter). Table 7-1 shows an example of such a matrix for the low crop yield component of the Mabu project. At this stage in project design, the solutions and interventions in the matrix are just creative

Table 7-1. Action Identification Matrix for Low Crop Yield
(Mabu Project)

Problems (causes to tackle)	Solutions	Potential actions
Not enough water reaching crops	Provide water for crops during dry season	• Construct new irrigation canals • Train maintenance people to repair old and damaged canals • Implement drip irrigation systems • Construct water storage facilities to provide water during dry season
Limited access to markets	Increase access to markets	• Implement greenhouse farming system to allow off-season crop growth • Switch to herbal intensive farming • Educational campaign for cooperative program
Deforestation leading to erosion	Reforestation	• Implement forestry management system (controlled harvesting and replanting) • Education campaign on importance of forests • Plant trees at field boundaries to create buffer zones for erosion control and wildlife habitat • Implement forest integration farming to reduce deforestation for expansion of agriculture
Poor soil quality	Improve soil quality	• Switch to crops more suited to environment • Implement organic farming methods • Provide education about sustainable agricultural methods • Begin program to use animal and plant waste in place of chemical fertilizers • Implement soil erosion control systems

Source: Glover et al. (2011), reproduced with permission.

alternatives that are listed without consideration for their rankings, feasibility, or appropriateness. Brainstorming is an ideal method for coming up with such a list.

The team of outsiders responsible for developing the action identification matrix must be technically qualified to go through this phase. The team must possess the technical expertise and comparative advantage to conduct the exercise and suggest recommendations that can be technical or nontechnical (behavior and/or policy change, for instance). Additional expertise may be sought, as necessary, from the local communities, government agencies, and other groups and individuals who have experience and have developed best practices in the past.

Multicriteria Utility Assessment (MCUA) Matrix

Ultimately, the various alternative solutions need to be ranked so that more appropriate ones can be selected. As remarked by Caldwell (2002), this can be done on a consensus basis with the stakeholders and partners but preferably using agreed-upon criteria. The field of decision sciences is rich in mathematical methods that can help users make decisions when many variables are at play (Decision Sciences Institute 2013).

A multicriteria utility assessment (MCUA) matrix can be used to carry out the decision process. It is a decision tool that ranks alternative solutions based on their worth for each problem identified. Further description of the MCUA method can be found in Delp et al. (1977). Because not all solutions to a given problem are feasible, they need to be ranked based on their appropriateness. Caldwell (2002) recommends using the following appropriateness attributes or criteria: cost-effectiveness, social acceptability, required management support, community support, sustainability, technical feasibility, political sensitivity, and level of risk. Table 7-2 shows an MCUA matrix for the Mabu project that combines the two issues of availability of water to increase crop yield and electricity production.

Other parameters that may enter into the ranking process include benefit–cost analysis, transportation and delivery costs, operation and maintenance, energy needs, replacement parts and costs, life expectancy, payback period, maintenance, and timing (Forbes 2009). To this list, we can also add social acceptability, political sensitivity, administrative feasibility, community sustainability, community participation, and environmental sustainability, among others.

As shown in Table 7-2, the degree of relevance of each attribute (criterion) to a given alternative solution in the MCUA matrix is recorded using an arbitrary score (1, 2, or 3). Weighting factors (3, 4, or 5) are also introduced to weigh the importance of each attribute or criterion. Multiplying the scores with the weighting factors and adding the products, an overall score (multiple-attribute

Table 7-2. Example of Multicriteria Utility Assessment Matrix (Mabu Project)

Criteria	Weight	Train Maintenance Person for Old Canals Score	Score × Weight	Irrigation Canals Score	Score × Weight	Implement Drip Irrigation Score	Score × Weight	Water Storage Facilities for Year-Round Water Supply Score	Score × Weight	Electrical Transmission Lines from Existing Hydro Plants Score	Score × Weight	Pico-Hydro Plants Score	Score × Weight	Photovoltaic Panels on Individual Homes Score	Score × Weight	Combined Irrigation Canals/Pico-Hydro Systems Score	Score × Weight
Cost-effectiveness	3	3	9	2	6	1	3	1	3	1	3	2	6	2	6	3	9
Social acceptability	5	2	10	3	15	1	5	2	10	2	10	2	10	2	10	3	15
Operations & maintenance feasibility	4	2	8	2	8	1	4	2	8	2	8	2	8	1	4	2	8
Environmental sustainability	5	3	15	1	5	3	15	2	10	3	15	3	15	3	15	2	10
Community participation	4	1	4	3	12	2	8	2	8	3	12	3	12	2	8	3	12
Effect on community health	4	1	4	2	8	2	8	1	4	1	4	2	8	1	4	2	8
Economic impact	3	1	3	3	9	2	6	1	3	2	6	2	6	2	6	3	9
Number of people impacted	4	2	8	2	8	2	8	2	8	3	12	2	8	3	12	3	12
Total			61		71		57		54		70		73		65		83

Source: Glover et al. (2011), reproduced with permission.

value or MAV) is calculated for each alternative solution. In Table 7-2, it varies between 54 and 83. In mathematical form, the MAV value for the ith alternative solution is equal to

$$\text{MAV}_i = \sum_{j=1}^{n} w_j v_{ij}$$

where v_{ij} is the value associated with the j-th attribute (or criterion) for the i-th alternative solution, and w_j is the associated weight. The solution with the greatest overall score (MAV) is defined as the most desirable. The entire rating process must be done, as best as possible, with complete participation of the community. It may also be done in a nonnumerical manner (using high, medium, and low ratings) by the community and translated into numbers by outsiders.

A word of caution about the ratings in the MCUA matrix applies: the aforementioned scores and weighting factors are nothing other than intelligent (intuitive) guesses made by decision makers based on discussion with the community, the expertise of the outsiders, and additional input from others. There is a lot of variability in the ratings. Sensitivity analyses may be integrated into the decision process by using a range for the weights and scores that appear in the MCUA matrix.

At the end of this process, one alternative solution may clearly stand out in the MCUA above the rest. However, this is more an exception than a rule. More often than not, solutions rank close to each other (as shown for some of the water-related solutions to crop yield in Table 7-2). Even solutions that have smaller MAV scores should not be discarded because they may become feasible later if more data become available. After all, it must be remembered that this is still the preliminary phase of project design. Further analysis is therefore needed to narrow down the most appropriate solutions and interventions, as discussed further in Chapters 8, 9, 10, and 11.

It should also be noted that the literature is divided about weight selection when using the MCUA matrix. Existing methods to determine weights include direct entry, as used in Table 7-2; the rank order centroid method; the ratio method; the swing-weighting method; and the analytic hierarchy process pairwise comparison method (Molenaar 2011). Table 7-2 clearly shows that the combined solution "irrigation canals (water) and pico-hydro systems (electricity)" stands stronger than the individual ones.

Project Hypothesis

The preliminary alternative solutions mentioned need to be brought to the attention of the entire community, its stakeholders, and partners and validated through various mechanisms, such as feedback meetings, nominal group process, and prioritization exercises (Delp et al. 1977). DFID (2002) calls this phase *visioning*, i.e., it offers a way for the stakeholders to share a common vision for the future, starting from where they are in the present. It can be conducted in small or large groups using some of the participatory action research methods discussed in Chapter 5. This exercise helps with information sharing, external validation, and building support and acceptance by the community members. From this exercise, certain solutions may emerge as more appropriate to some community members than others. This emergence may confirm the conclusions reached with the MCUA method. In other cases, it may contradict those conclusions and require reexamination of the attributes and criteria used in the initial ranking. Although several alternative solutions may still need to be considered at the end of this selection exercise, there are likely to be fewer options than those listed in the initial MCUA matrix.

This stage of the overall ADIME-E framework is where the focus of the project has shifted from appraisal and identification of community problems to developing a preliminary action plan. A *project hypothesis* (or project statement) can now be laid out, as demonstrated below using the Mabu project as an example, where

- Anticipated outcome (overarching goal)—improve household livelihood security through increased income opportunities and increased food security;
- Problem being addressed—low crop yield caused by lack of water and energy production;
- Critical causes to the problem being addressed:
  - Energy: Lack of technical expertise, inability to transfer energy to all wards, lack of energy infrastructure, lack of technical education, and lack of business education; and
  - Water and crop yield: Lack of agricultural knowledge, not enough water, poor distribution system, lack of business knowledge, damaged water infrastructure, lack of maintenance, lack of education, and lack of organization;
- Relationships between the problem's causes and effects—see problem tree (Figure 7-2);
- Effects of possible interventions—see solution tree (Figure 7-3);
- Rating of the various interventions—see MCUA matrix (Table 7-2); and
- Assumptions and preconditions necessary to support the project hypothesis—see logframe in Chapter 8.

In refining the project hypothesis, several issues and factors are likely to arise. The statements need to be addressed in consultation with the community in the form of participatory planning workshops. One method is to frame them into a series of broad questions that are listed in Table 7-3, which relate to the what, who, when, where, why, and how of the project.

Table 7-3. Possible Questions Used to Refine the Project Hypothesis

What	• What needs to be accomplished and in what order of priorities?
	• What represents success?
	• What external and internal factors in the community could jeopardize success?
	• What are the short-term and long-term benefits to the community?
	• What other institutions or agencies inside and outside in the community need to be involved?
	• What strategies will be used in addressing conflict and disagreement between parties?
	• What are the strengths and weaknesses of the partners?
Who	• Who are the partners and stakeholders?
	• Who will manage the project?
	• Who will be responsible for project strategy, operations, and management?
When	• When would be an appropriate time to carry out the project?
	• When will the various project activities take place and what will be their duration?
Where	• Where will activities take place?
	• Where are the needed resources located?
	• Where is the manpower?
Why	• Why are we conducting the various tasks?
	• Why are we involving this partner or group?
	• Why is it important to have and meet deadlines?
How	• How will the project be funded?
	• How will the project be managed?
	• How will any disagreement with and within the community be handled?

Answers to the questions in Table 7-3 can be regrouped and presented in tabular or matrix form:

- *Action feasibility matrix or participatory planning matrix* showing the actors or players (internal and external) and their contributions, their short- and long-term benefits, and their needs for long-lasting benefits. An example is shown in Table 7-4 for the crop yield component of the Mabu project.
- *Potential partners matrix* showing the partners for each problem and identifying who they are, their strengths, weaknesses, and what they can offer to the partnership (physical and time scale). For the Guatemala project, Tables 5-9 and 5-10 provide that information.

For the Mabu project, that information is presented in the stakeholder/partner role matrix in Chapter 8.

In general, the aforementioned matrices contain enough information to refine the project design hypothesis to a level that is acceptable to all parties

Table 7-4. Action Feasibility Matrix for Crop Yield Issues (Mabu project)

Responsible actors	What contributions should they make?	What benefits will they get in the short run?	What should they do to ensure a long-standing solution?	What benefits will the solution bring to them in the long run?
Interest group	Labor for construction, startup investment, full participation in sustainable agriculture	Increased crop yields, empowerment of farming community	Maintain program and monitor and evaluate components for optimization, continued fund development, identify community members to perform maintenance	Increased sustained crop yields, increased cash crops, improved environment leading to higher food security and better health
Community at large	Support of new farming system funds, input from their perspective, conduct PAR	More food and greater economic base for the community	Assistance with monetary support labor needs where necessary, basic knowledge/ education of agricultural practice	Same as above
Local government agency (VDC)	Initial funding and technical support for design and maintenance, training support	Less dependency from community, community support and partnership	Continued support and regular monitoring and evaluation, some funding, continued technical assistance, follow-up	Community resilience, increased ability to collect donations for other projects due to increased cash crop yields
NGO partner (NCDC)	Support and initial funding, training and support for design and maintenance, training support	Same as above	Same as above	Community capacity for other projects increased

Source: Glover et al. (2011), reproduced with permission.

involved. They provide a sound basis for the selection of an optimum strategy (or strategies) and the development of logistics and tactics in the agreed-upon action plan as discussed in the next chapter.

7.4 Chapter Summary

This chapter presents a series of tools that take the results of the appraisal analysis and convert them into a preliminary action plan. Preliminary solutions to key community problems have been identified and a project hypothesis (statement) has been laid out.

Finally, the end of the preliminary phase of project design provides project managers with the opportunity to decide whether to move forward and develop a full action plan, put the project on hold, or terminate the project. There is no definite list of criteria that can be used to make that decision. It usually depends on the framework used by development agencies and what is deemed important to them. The criteria can be of different types, such as budget, policy priorities, human resources, urgency, social acceptability, and resource availability.

References

Barton, T., et al. (1997). "Our people, our resources: Supporting rural communities in participatory action research on population dynamics and the local environment." International Union for Conservation of Nature (IUCN) Publications Services, Gland, Switzerland, and Cambridge, U.K.

Buzan, T., and Buzan, B. (1996). *The mind map book: How to use radiant thinking to maximize your brain's untapped potential*, Plume, Penguin Group, New York.

Caldwell, R. (2002). *Project design handbook*, Cooperative for Assistance and Relief Everywhere (CARE), Atlanta.

Decision Sciences Institute. (2013). <http://www.decisionsciences.org/>. Oct. 5, 2013.

Delp, P., et al. (1977). *Systems tools for project planning*, International Development Institute, Bloomington, IN.

Department for International Development (DFID). (2002). *Tools for development*, version 15, Department for International Development, London.

Forbes, S. (2009). "Sustainable development extension plan (SUDEX)." Doctoral dissertation, University of Texas at El Paso.

Gawler, M. (2005). "WWF introductory course: Project design in the context of project cycle management." Artemis Services, Prévessin-Moëns, France. <http://www.artemis-services.com/downloads/sourcebook_0502.pdf> (March 15, 2012).

Glover, C., Goodrum, M., Jordan, E., Senesis, C., and Wiggins, J. (2011). "Mabu village term project." CVEN 5929: Sustainable Community Development 2, University of Colorado at Boulder.

Molenaar, K. (2011). "Multi-criteria decision making." Lecture notes, CVEN 5276: Risk and Decision Analysis, University of Colorado at Boulder.

8

Focused Strategy and Planning

8.1 Comprehensive Planning

The alternative solutions and preliminary action plan identified in the previous chapter need to be examined further to arrive at a comprehensive work plan in agreement with the community. Planning is about setting goals, developing strategies, outlining the implementation arrangements, and allocating the resources (physical, human, financial) to achieve the goals. As remarked by Lewis (2007), "planning is not an option" and "you cannot have control unless you have a plan." Of course, the plan can be changed and updated while considering the uncertainty associated with community development projects. Regardless of the uncertainty, a need still exists to have a plan in place. In development projects, having to change plans (sometimes often) is more a rule than an exception. As we see in Chapter 12, changes may be deemed necessary as the project unfolds based on observations made during project monitoring and evaluation.

In the planning phase of the project management cycle, strategy, logistics, and tactics need to be addressed because they play a critical role in project implementation. *Strategy* refers to the overall game plan or overall method that will be followed when conducting the work. Tactics and logistics are part of the implementation process. *Tactics* define and address the why, what, who, when, where, and how of a project. *Logistics* are about ensuring that human resources, materials, and supplies are available at the right time and location (Lewis 2007). Like all phases of the project management cycle, planning requires community participation; it has to be negotiated and agreed upon with the community, which may require several rounds of iteration. In addition to addressing strategy, logistics, and tactics for the different activities included in project intervention, the work plan needs to address how the management activities will be carried out and the associated physical, human, and financial resources that are required.

At the end of this phase of the ADIME-E framework, management activities and project steps (e.g., time lines, resources, and materials) can be recommended and lined up for execution. Many issues and constraints (technical and nontechnical) enter into the planning phase of the project. The results of that phase can be presented in the form of flowcharts, decision trees, decision tables, and other means that are appropriate to execute high-quality projects. These tools must also be understandable by the local community participants. At the end of this phase of the framework, a work plan or action plan that is more comprehensive than the preliminary plan outlined at the end of Chapter 7 exists. However, the work or action plan still needs to be further refined using additional methods of analysis (capacity, risk, and resilience) described in Chapters 9, 10, and 11. Developing the work plan is a work in progress and is part of an iterative process, as shown in Figure 4-2.

8.2 Strategy

Logical Framework Approach

Over the past 25 years, development agencies around the world have debated the need for incorporating rigorous methodologies in the planning and management of community development projects. A methodology called the logical framework approach (LFA), also referred to as logical framework analysis or logframe, initially developed by the U.S. Agency for International Development in the late 1960s, has been adopted by many development agencies as a road map for planning projects. An excellent review of LFA and how it has been used by 18 organizations around the world can be found in Bakewell and Garbutt (2005). As noted by EuropeAid (2002), "LFA is a technique to identify and analyze a given situation and to define objectives and activities which should be undertaken to improve the situation." It can be used in the development of a project action plan and later integrated into project monitoring and evaluation (see Chapter 12).

CARE recommends that the design of all of its projects should follow a logic model called the *project hierarchy* (Caldwell 2002). Within that model, solutions emerge from a "strategic combination" of identified inputs, activities, and outputs (objectives) that produces effects (goals) and impacts or outcomes (Table 4-1). Table 8-1 shows a typical logical framework matrix for the Mabu, Nepal, project where the aforementioned components stack up vertically and relate to each other in the overall project hierarchy and form a *vertical logic*. The matrix also shows a *horizontal logic* describing possible performance indicators and modes of verification to assess how each component progresses and depends on assumptions, preconditions, and risks. The horizontal logic relates to the monitoring of the input, activities, outputs, goals, and overall project impact (see Chapter 12). The logical framework approach (LFA) is the process by which the logical framework matrix is populated through participation.

Table 8-1. Example of Logframe Matrix (Mabu Project)

Outcome/Impact: Improved household livelihood security through increased income opportunities and increased food security.

Project Hierarchy		Indicators	Means of Verification	Assumptions
Goals	1. Improved standard of living of irrigators	1. 5 years after project: 95% of irrigators are able to meet their needs for food and are able to produce a surplus for sale 80% of irrigators have reduced their use of pesticides and fertilizers	1. Community-wide survey in project area 4–5 years after implementation, when system operating at full potential by NCDC	1. Baseline communitywide survey in project area was carried out at beginning of project Irrigated fields produce higher yields
	2. Provide electricity to homes	2. 10 years after project: 30% increase in electricity production in Mabu	2. Count number of houses with electricity available Survey community members and identify trained personnel	2. Community will continue to desire electricity for use in homes and businesses
	3. Support development of small businesses	3. 10 years after project: 6–9 new businesses operating	3. Count operating businesses begun within the past 10 years	3. Support development programs will continue and be a long-term source of data

Subgoals				
1.1	Increased crop yields and irrigator incomes	1.1 Five years after end of project:	1.1 Direct observations of land usage and canals	1.1 Disruptions to irrigation system are caused by poor maintenance practices
1.2	Reduce use of chemical pesticides and fertilizers	25% increase in number of farmers receiving water from irrigation canals	Survey farmers	Surveys result in reliable data as to pesticide usage
1.3	Technical and environmental sustainability	Exceed or achieve project return and timely loan repayment	Soil chemical tests	Able to determine levels of disruption to system before project
		25% reduction in instances of disruptions to the irrigation system	Purchase receipts	
		30% reduction in usage of chemical pesticides and fertilizers		
2.1	Electric lighting available for all homes served by pico-hydro system and other electric appliances supported in subset of homes	2.1 Five years after end of project:	2.1 NCDC surveys and photographs of homes served by system to count number of homes served with various electric appliances	2.1 Electric cooktops will provide incentive for community members to pursue electricity
		Electric lighting available in 75% of homes not previously lit		Pico-hydro system will be properly maintained by community
		Electric amenities available in 50% of homes		
		10 electric blenders available within given community for use in shared applications or private enterprise		

Table 8-1. Example of Logframe Matrix (Mabu Project) (*Continued*)

Outcome/Impact: Improved household livelihood security through increased income opportunities and increased food security.

Project Hierarchy		Indicators	Means of Verification	Assumptions
	3.1 Growth of small businesses using both electricity and mechanical shaft energy	3.1 One small business using mechanical shaft energy at each plant		3.1 Community members will pursue small businesses using the pico-hydro system Users willing to pay electric fees for system maintenance
Outputs	1.1.1 Irrigation canals built	1.1.1 3 canals built at project completion	1.1.1 Observation by NCDC	1.1.1 Records from training programs exist Knowledge of alternatives will result in changed behaviors
	1.1.2 Maintenance workers trained to operate and repair canals and pico-hydro system	1.1.2 12–15 workers completed training program at time canals are completed		
	1.2.1 Behavior change campaign created for pesticide use	1.2.1 3 training workshops held for farmers, and posters and brochures distributed throughout town	1.2.1 Training program records Surveys of farmers	

2.1.1 Pico-hydro system installed	2.1.1 2–3 pico-hydro plants built per canal	2.1.1 NCDC surveys and photographs of plants	2.1.1 Head and flow are sufficient for system
2.1.2 Transmission lines for distribution built	2.1.2 Transmission lines built for each pico- hydro unit	2.1.2 NCDC surveys and photographs	
2.1.3 Behavior change campaign created for electric appliance technology adoption	2.1.3 All houses served received training for use of electric appliances	2.1.3 NCDC surveys and photographs	2.1.3 Community members will discontinue use of kerosene lamps and firewood and use electric lighting and burners, respectively
3.1.1 Business training programs	3.1.1 50–60 people trained in new business development		3.1.1 Community members will invest microloans in new businesses
3.1.2 Microloans for entrepreneurs			

Project Hierarchy			Assumptions
Activities	1.1.1 Irrigation canals:		All data necessary to design pico-hydro and canal system will be collected
	1.1.1.1 Site selection and mapping		
	1.1.1.2 Sizing and demand modeling		
	1.1.1.3 Source labor and materials		
	1.1.1.4 Design and build irrigation canal		

Table 8-1. Example of Logframe Matrix (Mabu Project) (*Continued*)

Project Hierarchy	Assumptions
2.1.1/2.1.2 Pico-hydro system:	Firms or NGOs exist with proper experience for designing and building a canal-based pico-hydro system
2.1.1.1 Site selection and mapping	
2.1.1.2 Sizing and demand modeling	Timely acquisition of funding and materials
2.1.1.3 Select pico-hydro unit	
2.1.1.4 Source labor and materials	
2.1.1.5 Design housing and transmission	
2.1.1.6 Install systems	
2.1.1.7 Build transmission lines	
1.1.2 Maintenance training	A sufficient number of community members will want to train to become maintenance technicians
1.2.1.1 Train community members in canal maintenance practices	
2.1.3.1 Train electric maintenance workers	
1.2.1/2.1.3 Behavior change:	Community members will remain engaged in the program and will participate in behavior change activities
1.2.1.1 Develop messages for change (reduce chemical pesticides and fertilizers)	
1.2.1.2 Hold workshops for farmers	
2.1.3.1 Develop messages for change (electricity adoption)	
2.1.3.2 Develop posters and brochures	
2.1.3.3 Hold workshops for electricity use	
3.1.1. Business training	Community interest in the project remains high
3.1.1.1 New business training for local community	
3.1.2 Microloan program	
3.1.2.1 Develop structure for microloan distribution	

Source: Glover et al. (2011), reproduced with permission.

The logical framework matrix is designed so that each component of the vertical hierarchy in the matrix, starting from the bottom, contributes to the next level up: activities convert input into outputs or deliverables (objectives), outputs cause certain effects and changes to take place (meeting goals), and effects produce impact (outcome). The hierarchy can also be seen as a series of if–then relationships that connect the components of the LFA. For instance, *if* the inputs are met and certain assumptions and/or preconditions are met, *then* activities can be carried out, objectives (outputs) are met, and goals (effects) reached. As noted by Caldwell (2002), the hierarchy can be summarized as follows:

Inputs → Activities → Outputs → Effects → Impact

The whole logical framework can be seen as a coherent information system (a cause–effect logical hierarchy or project logic) that must be clear to the outsiders, community members, and other partners. The UNDP (2009) calls it a *results chain*. It helps clarify in a logical way what the project is expected to achieve, the steps in achieving the desired results, what resources are necessary, and how project success and progress will be measured over time using performance indicators and modes of verification.

The project hierarchy can be interpreted from the bottom up or the top down. The following definitions are used for the components shown in the first column in Table 8-1.

The *outcome/impact* is the ultimate impact and purpose of the project; it is the end state and the big-picture change that the project is expected to make, for instance, in existing systems, the environment, or communities. It can also be called the *overarching goal, aim, impact goal,* or *strategic objective.* In many ways, it can also be seen as the *vision* for the project. According to Caldwell (2002), the outcome/impact must state the type of improvements that can be expected in human conditions after the project has been completed, the identification and number of beneficiaries, and a time when change is expected to occur. The outcome relates directly to the various problem(s) addressed in the causal analysis (trunk of problem and solution trees) discussed in Chapter 7. Usually, an outcome is long-term and big picture and the project under consideration may just be a contributing factor, among many others, to making the outcome a reality. In Table 8-1, the outcome/impact is defined as "improved household livelihood security though increased income opportunities and increased food security." A time frame for that outcome can be added, such as three years, for instance. The beneficiaries include all households in Mabu.

For the outcome/impact to become a reality, several *goals*, also referred to as *effects* or *purposes*, need to be met. The goals describe what must take place and the necessary changes in development conditions (human, economic, civic, environmental) that must be achieved for the outcome/impact to be either short term (immediately or in the near future), long term (over time), or medium term (in between). Goals are defined as endpoints but not as processes. The goals address directly the root causes of the problem and solution trees. For each *critical* root cause, a goal (solution) is established. Goals can also be seen as defining the *mission* of the project and describing what needs to be met to achieve the vision. Goals can be set to affect different scales, such as the individual, household, community, or country levels. In Table 8-1, there are three main goals/effects that contribute to the overall project impact: (1) improve the standard of living of those using irrigation, (2) provide electricity to homes, and (3) support development of small businesses. The goals are then divided into subgoals or targets. In Table 8-1, three subgoals are associated with the first goal and one for the other two goals.

To meet each goal, several objectives or outputs need to be accomplished. They are the deliverables and results and describe what needs to be achieved through the various project activities and interventions. They match the actions outlined in the project hypothesis and can be summarized in the action identification matrix (e.g., Table 7-1).

Activities are detailed actions, tasks, and interventions that are needed to deliver the output. As remarked by the Innovation Network (2013), they may include developing products, providing services, engaging in policy advocacy, and building infrastructure. In general, activities convert input into output.

Inputs represent all the resources that are needed to successfully undertake a set of activities. They may include organizational expertise, physical resources (facilities and technology), assets, financial resources, human resources (stakeholders, consultants, volunteers), and relationships, among others. The resources may already exist or may need to be acquired.

Project teams are reminded of the initial need, the *how* of the project when the framework is read from the bottom to the top. Reading it from the top to the bottom describes *why* the project needs to unfold (Gawler 2005). No matter how the logframe is read, it is put in place for three main reasons: (1) to reach the overarching goal, outcome, or impact; (2) to meet the goals and objectives; and (3) to ensure positive project impact and quality.

It should be noted that the goals and objectives in the project hierarchy need to be meaningful. The acronym SMART is often used to describe the desired attributes of goals and objectives. Paraphrasing Caldwell (2002), the letters in the acronym stand for the following:

- S (specific) in terms of clarity in the what, how, where, and when of change;
- M (measurable), i.e., evidence exists that the goals and objectives are achieved;
- A (area specific) to a geographical area and a given population;
- R (realistic or relevant) in view of existing resources and skills; and

- T (time-bound) in terms of the time frame required to meet the goals and objectives.

At each level of the project hierarchy, indicators, means of verification, and assumptions must be outlined (e.g., second, third, and fourth columns in Table 8-1).

Indicators

Regardless of the terminology used, each step in the framework must show clear, objectively verifiable indicators. According to Caldwell (2002), an indicator is "a variable, measure or criterion used to assist in verifying whether a proposed change has occurred." Indicators are independent measurements used to identify progressive (or regressive) changes related to a project intervention and whether benchmarks have been reached. Meaningful indicators are simply observable changes or events that provide evidence or proof that what has been claimed has actually occurred (Bakewell and Garbutt 2005). Performance criteria are often included with each indicator as a way to assess progress. According to Caldwell (2002), the indicators must have the following eight characteristics: They must be (1) measurable, (2) technically feasible, (3) reliable, (4) valid, (5) relevant, (6) sensitive, (7) cost-effective, and (8) timely. In general, indicators apply to a wide range of project components, including personnel, resources, and funding.

Indicators can be quantitative or qualitative. Quantitative indicators can be a number, percentage, rate, or ratio. Qualitative indicators, however, "reflect people's judgments, opinions, perceptions, and attitudes toward a given solution or subject. They are descriptive observations about changes in sensitivity, satisfaction, influence, awareness, understanding, attitudes, quality, perception, dialogue or sense of well-being" (UNDP 2009). Another way of looking at indicators is in terms of what they actually measure. For instance, Table 5-2 in Chapter 5 shows a range of indicators related to water quality and quantity and water system functionality.

Project indicators are likely to be more meaningful if they are selected by the community members and key stakeholders because they are the ones who will voice what success means to them. Indicators differ and must match each level in the logical framework matrix (i.e., input indicators or activity indicators). Indicators are sometimes divided into *lower level indicators* for the input and activities layers (logistics and tactics) in the logical framework, and *higher level indicators* for the output, goal, and outcome (strategy) layers. In general, lower level indicators are easier to assess than the higher level ones because it takes longer for impact and outcome to become measurable (Barton 1997; Barton et al. 1997).

All the frameworks discussed in Chapter 4 pay special attention to the characteristics of the indicators that enter into their versions of the LFA and monitoring and evaluation plans (see Chapter 12). EuropeAid (2002) uses the SMART acronym mentioned earlier to describe the attributes of good indicators. DFID (2002) uses the acronyms SMARTER (i.e., same as SMART plus enjoyable and rewarding) or SPICED (i.e., subjective, participatory, indirect, cross-checked, empowering, and diverse). The Canadian Aid Agency CIDA suggests using six criteria in selecting indicators: validity, reliability, sensitivity, simplicity, utility, and affordability. The World Bank (2005) recognizes several types of indicators in its projects: leading and process indicators and proxy indicators, in addition to the indicators for the different components of the logframe. Finally, indicators can be divided into impact indicators (measure of impact of solutions), achievement indicators (measure of accomplishment and success), and response indicators (measure of performance), as suggested by Gawler (2005) for World Wildlife Foundation projects.

Means of Verification

In addition to meaningful indicators, the logical framework must include, for each of its components, sources of verification, i.e., means to ensure the success of the different phases of the project hierarchy. This can be done in different ways, e.g., through site visits, interviews, reports, official statistics, or records. The personnel in charge of verification must also be identified and a time line for information delivery.

Assumptions, Preconditions, and Risks

In the logical framework matrix, assumptions, preconditions, and risks behind each action item must be clearly acknowledged because they represent critical factors to ensure success. For instance, several assumptions are required in the inputs to conduct certain activities. Likewise, assumptions or preconditions must be in place to meet certain objectives or deliver outputs, and so on.

Assumptions and preconditions are often beyond the control of the outsiders and depend on the community and the environment. They are part of the external logic of the project because they lie outside the project's accountability. In their project logframe, the World Bank (2005) recommends looking at assumptions in three different ways: generic, specific, and even more specific. In addition to their importance, the risks associated with the assumptions must be determined and can be plotted on an x–y diagram where the x-axis relates to the probability of failure and the y-axis relates to the criticality or importance of each assumption (see an example in Section 10.3). Of special importance are the assumptions with high probability of failure and of high criticality, sometime referred to as "killer assumptions," where risks are very high.

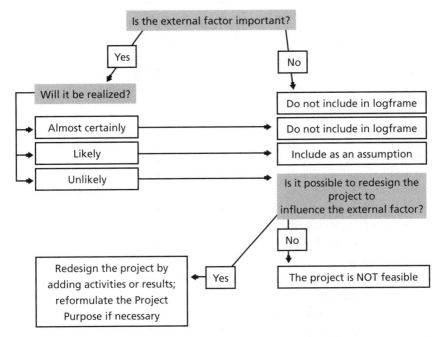

Figure 8-1. Guidelines for Assessing the Importance of Assumptions

Source: EuropeAid (2002), with permission from Particip GmbH.

Assumptions also need to be monitored as the project goes along because of their criticality. If assumptions and/or preconditions are not met, a risk exists that well intentioned actions may lead to negative results and, in some cases, failed projects. Figure 8-1 gives guidelines on how to respectively identify and assess assumptions (external factors) according to EuropeAid (2002). In some extreme cases, the assumptions are so risky and uncertain that the project may not be feasible. In all cases, the risk associated with critical assumptions needs to be integrated into the project risk management plan (see Chapter 10).

Advantages and Disadvantages of LFA

When the logical framework is presented in matrix form, as in Table 8-1, the logic of cause-and-effect can be inferred by directly looking at each vertical column in the matrix (vertical logic). The success or achievement at one level depends on the successes and achievements in the underlying levels. The matrix also offers a horizontal perspective (horizontal logic). For instance, if a goal is not achieved and lower level activities show success, reasons for not achieving the goal may be found in the corresponding indicators, modes of verification, and among the assumptions or preconditions. Finally, the logical framework serves as the hypothesis of the project and forms the basis for subsequent monitoring and evaluation strategies, as discussed in Chapter 12.

The structured nature of the logical framework approach has contributed to making it popular in development organizations around the world. However, it is not universally chosen by all organizations. As discussed by Bakewell and Garbutt (2005), LFA has become the standard in development projects and is often required by donors. Their literature review on the use of LFA by 18 organizations showed that they all recognize a need for project logic combined with a participatory approach to achieve meaningful results. However, there seems to be a wide range of opinions about the effectiveness and practicality of LFA (Bakewell and Garbutt 2005; Oxfam 2008; Jensen 2010). Objections to using project logic using an LFA framework are listed below:

- Too formal and rigid with linear cause–effect (linear causation);
- Does not truly reflect the uncertain and flexible nature of development projects;
- Does not work well with complex situations and unintended consequences;
- Hard to identify meaningful indicators;
- Consumes time and resources to the detriment of the rapidly changing environment itself;
- Hard to change and adjust once in place;
- Culture specific, meaning that it can be hard to implement in some cultures;
- Hard to explain to others and put into practice;
- Often treated as a contract document; and
- Can be used to impose rigid development ideas on communities.

However, proponents of LFA cite several reasons for using it:

- Makes development projects more effective and accountable,
- Provides rigor in all phases of a project,
- Represents a clear way of communicating,
- Is a good road map for setting expectations and reporting on progress and accountability,
- Can be seen as a uniform way of thinking,
- Helps simplify the complexity of projects by providing a rigid structure,
- Represents a consistent way of communication across organizations,
- Forces people to think through the various components that may influence the project, and
- Can easily be combined with a monitoring and evaluation plan as discussed in Chapter 12.

Despite the various arguments for and against the use of LFA in development projects, it still remains the main form of project planning in development organizations where the logical framework is seen as an executive summary of the focused strategy. The information detailed within the logic model provides insight into what the project is expected to achieve, how results will be achieved, which factors are crucial for success, and how success can be measured.

In summary, LFA has strong attributes that warrant its use in project planning and management. A more advanced approach would consider combining LFA with the system dynamics approach discussed in Chapter 6. This combination would reduce the perceived rigidity of LFA while still retaining the added value of its inherent logical structure.

8.3 Operation—Logistics and Tactics

The overall strategy summarized in the logical framework analysis must be followed by the construction of a detailed implementation work plan. This is a step-by-step exercise describing (1) how the goals, objectives, activities, and input are to be implemented; (2) what project and management activities and tasks need to be conducted; (3) how, where, and when (scheduling) they will be conducted; (4) who will conduct the activities and tasks; (5) what resources are necessary; and (6) what contingency plans need to be included. This work plan represents the operational component of the project delivery.

Targets and benchmarks need to be introduced into the logical framework analysis. They represent "expected values or levels of achievement at specific periods of time" starting from an agreed-upon community baseline (Caldwell 2002). In some instances, a target column may be added to the logical framework matrix, showing how achieving the goals, meeting the objectives, conducting the activities, and mobilizing the inputs are expected to progress over time toward achieving the desired results. In general, the logframe leads to a detailed activity schedule, which in turns leads to a resource schedule, as shown in Figure 4-4.

A traditional way for engineers to present project scheduling is to use Gantt charts once a work breakdown structure showing milestones of activities, tasks, subtasks in sequential form, and their respective links has been laid out and agreed upon. This structure allows for the time required to carry out each project or management activity and task to be determined. This phase is usually done without taking much detailed consideration of the resources that are available. Once established, project scheduling is refined to account for resource constraints. Table 8-2 shows an example of a proposed Gantt chart for the Mabu project over a period of 36 months.

In addition to a Gantt chart, a stakeholder–partner role matrix can be created. It shows the key actors and stakeholders in the community likely to be involved in the various activities and tasks, and their expected and/or respective roles and responsibilities. The results of the analysis of the community appraisal

Table 8-2. Gantt Chart for Mabu Project

Tasks	Duration	2011								2012												2013											
		M	J	J	A	S	O	N	D	J	F	M	A	M	J	J	A	S	O	N	D	J	F	M	A	M	J	J	A	S	O	N	D
Assessment																																	
PAR Appraisal	1 mo	░																															
Irrigation Canals																																	
Site Selection/ Mapping	4 mos			░	░																												
Sizing & Demand Modeling	3 mos					░	░																										
Source Labor & Materials	2 mos								░	░																							
Design Canal	4 mos											░																					
Build Canal	3 mos													░	░																		
Train Maintenance Workers	6 mos																░	░	░	░													

Pico-Hydro Systems

Site Selection/Mapping	4 mos
Sizing & Demand Modeling	3 mos
Select Pico-Hydro Unit	2 mos
Source Labor & Materials	2 mos
Design Housing & Transmission	5 mos
Install Systems	3 mos
Build Transmission Lines	5 mos
Train Maintenance Workers	5 mos

Table 8-2. Gantt Chart for Mabu Project (*Continued*)

Tasks	Duration	2011								2012												2013											
		M	J	J	A	S	O	N	D	J	F	M	A	M	J	J	A	S	O	N	D	J	F	M	A	M	J	J	A	S	O	N	D
Behavior Change Campaign																																	
Develop Message for Change	8 mos					■	■	■	■	■	■	■	■																				
Develop Posters and Brochures	3 mos										■	■	■																				
Workshops for Farmers	6 mos														■	■	■	■	■	■													
Workshops on Electricity Use	6 mos																	■	■	■	■	■	■										
Small Business Workshop	6 mos																				■	■	■	■	■	■							
Develop Microloan Program	5 mos																								■	■	■	■	■				
Monitoring & Evaluation																																	
Measure Baseline	3 mos							■	■	■																							
Monitoring	ongoing																																
Evaluation	ongoing																															■	■

Source: Glover et al. (2011), reproduced with permission.

in terms of existing community stakeholders and partners and individuals and groups outside of the community who may participate in project implementation and management activities can be useful in creating the stakeholder–partner role matrix. Table 8-3 shows an example of such a matrix for the Mabu project.

Furthermore, individuals and/or groups among stakeholders and partners can be assigned to different phases of the project, such as decision making, construction, monitoring and evaluation, and maintenance. The second column in Table 8-3 lists possible involvement of stakeholders and partners in the different phases of the Mabu project. If necessary, decisions can be made to train individuals and groups to provide the skills necessary to match the respective assignments and ensure project quality.

Once the activities and tasks are scheduled, the resources (human, physical or material, and financial) necessary to undertake them can now be outlined and allocated, and the associated costs (performance budget) determined. Table 8-4 lists for instance the material and cost estimate for the pico-hydro energy component of the Mabu project.

In general, projects demand many resources (physical, human, financial), but whether these resources exist must be assessed. At this stage of project planning, the overall project schedule may be changed if resources are not available and the project cost becomes too prohibitive. Another option is to keep the schedule as is, consider additional resources to match the schedule, and/or change the project scope, goals, and objectives. As remarked by Lewis (2007), "resource allocation is necessary to determine what kind of schedule is actually achievable. Failure to consider resources almost always leads to a schedule that cannot be met." As discussed in Chapter 4, project management is a balancing act among four components: performance, cost, scope, and time (Figure 4-1). Changing one or several of those four components may derail a project.

Having a range of contingencies and backup plans is important in development projects because of their inherent uncertainties and the difficulty in identifying strict project task deadlines. For instance, it is unlikely that the *critical path* (sequence of activities that determines the minimum total time to complete a project) often used in Gantt charts has any meaning in small-scale community development projects; in most cases, the concept of "strict deadlines" does not even exist, and time can be poorly defined. An alternative to strict deadlines is to use "suggested deadlines" that are not hard and may help with the decision-making process, especially if any uncertainties arise that require alternative solutions. Project managers need to be able to handle the "what if" in small-scale community development projects.

Different plans (i.e., strategies) can be outlined for various levels of expected success (fair, good, excellent). Contingency analysis can take different forms (Delp et al. 1977). It can be developed for a best-estimate scenario (based on the

Table 8-3. Stakeholder–Partner Role Matrix for Mabu Project

Stakeholders	Stage of Project	Interests	Strengths	Weaknesses
Farmers	Design/Ops Maintenance	Increased income Time savings	Knowledge of land/ agriculture Source of labor Access to funds/loans	Resistance to change Equity issues/ conflicts
Electricity Consumers	Design/Ops Maintenance	Access to electricity (higher productivity, education, prestige)	Willingness to pay Early adopter effect	Resistance to change Equity issues/ conflicts
Small Business Owners	Ops & Maintenance	New income or increased opportunities	Innovation Access to funds	Overconsumption Not community oriented
Maintenance Staff	Ops & Maintenance	Job opportunities	Desire to learn	Lack of technical knowledge
Local Govt (VDC)	Design/Ops Maintenance	Reelection Improvement of village	Access to technical experts, government agencies, communication	Limited financial resources Conflict among wards
NCDC	Design/Ops Maintenance	Improvements to Mabu Increased village sustainability	Access to technical experts Resource for community assessment/ planning Access to funds	Not located in community
Current Pico-Hydro System Owners	Design	Preserve customer base, opportunity for expansion	Established operation Practical technical knowledge Local to community	Greed/limited capital/fear of competition
Water Rights Owners	Design	Water abstraction for which they get no benefit	Control of water source	Greed

Source: Glover et al. (2011), reproduced with permission.

Risks/Fears	How Will They Influence the Project?	How to Engage Stakeholders?	Interest Priority	Assumptions
Weather/crop failure	Have to accept the system, rely on them for maintenance	Market the electricity part and reliability parts of project	High	Overestimate how useful technology is
Safety issues	Demand will drive system design	Rely on focus groups, both before and during project	High	People will want electricity in the home
Business failure Rapid growth Exceeding capacity	Rely on businesses as a long-term funding source	Special rate for businesses Seminars on business training	High	Entrepreneurs in Mabu will start businesses
Maintenance gets ignored	Long-term success relies on good maintenance	Provide salaries and training for interested people	High	People want to become maintenance workers and have ability
Corruption	Need their approval, help set up structure for fees	Through meetings, with NCDC as coordinator	High	VDC willing to make time for this project
May prioritize other projects	Need their support (financial and technical) to initiate project	Coordinate through project mentor	High	NCDC supports this project, wants design work done
May oppose project	May use their influence to hinder project Can offer valuable advice and partnership if support	Friendship/ partnership/asking advice	Medium	Systems are still operating/we will be able to overcome any opposition
May withhold water access	May stop project or impose strict regulations	Friendship/ partnership/ determine ways of benefiting owners	High	Will allow water access, will be reasonable

Table 8-4. Material and Cost Estimate for Pico-Hydro Component of Mabu Project

System Component	Required Materials	Required Tools	Labor Source	Estimated Cost
Headwork diversion	Wire mesh, stone cement grout	Hammer, wire cutters, shovels, wheelbarrow	Mason, community members, NCDC surveyor	$4,500
Headwork offtake	Stone masonry	Hammer, chisel, wheelbarrow	Community members	
Transport canal	Packed earth	Shovels, wheelbarrows, lumber for framing, saw, string, stakes	Community members	
Forebay/ settling basin	Stone masonry		Community members	
Penstock	Pipes	Saw, hammer, hot joining plate	Pipe joiner, mason, community members, NCDC surveyor	
Powerhouse	Lumber for framing, corrugated metal, fasteners	Saw, sheet metal cutters, hammer, screwdriver	Carpenter, mason, community members	
Turbine (8.5 kW)	Pelton wheel, nozzle injector	Hammer, screwdriver, wire strippers, wrenches, voltage tester	Electrician, engineer, community members	$3,800
Generator	Three-phase induction motor used as a generator		Electrician, engineer, community member	$600
Voltage regulator	240 V regulator		Electrician, engineer, community member	$700
Transmission lines	Aluminum wire, wooden poles, concrete	Shovels, wheelbarrow, wire cutters, hammers, saws, voltage tester	Electrician, engineer, community member	$5,750

Source: Glover et al. (2011), reproduced with permission.

most likely contingencies) or a worst-case scenario (based on the most adverse contingencies), or for different types of scenarios in between by using a sensitivity analysis. It can also be supplemented by a decision tree analysis that shows various decision options to a problem, their respective probabilities, and their payoffs. As plans are changed for various contingencies, it is necessary that the changes made and the reasons for the changes are documented accordingly.

8.4 Planning of Management Activities

In addition to planning the project activities, plans must be made for the management activities necessary to carry out the project. They are not usually included in the logical framework mentioned. The tasks in the management activities can be presented in the form of a work breakdown structure. In the list below, resources (physical, human, financial) needed to successfully complete these management activities are listed. As discussed by EuropeAid (2002), the management activities may include the following:

- Ensuring the quality control and quality assurance of projects (quality management);
- Information, communication, and reporting (communications management);
- Identifying deadlines, procedures, benchmarks and targets of performance, feedback, control, contractual tasks, decision making (integration, time, and scope management);
- Financial planning in terms of staying within the budget (cost management);
- Management of personnel, including team building, training, coaching, conflict resolution, and staff turnover (human resources management);
- Identification, quantification, analysis, and responding to risk (risk management); and
- Acquisition of goods and services (procurement management).

Another complementary aspect of project management is building and maintaining partnerships with different actors from different sectors at the local, regional, or national levels, which could also be integrated into communications management.

8.5 Project Quality Planning

Quality is an integral part of any engineering project. Along with cost, time, and scope, quality is part of the project plan. As remarked by Rose (2005), "achieving quality in project implementation is not a matter of luck or coincidence; it is a matter of management." The Project Management Institute (PMI 2008) defines quality as "the degree to which a set of inherent characteristics fulfill requirements." Ultimately, project quality management is used to make sure that a project meets the needs originally intended to be met and that the expectations from the project stakeholders are also met. As noted by Rose (2005), everyone on a project is responsible for quality; it includes the organizational management, project managers, project teams, and individual members of those teams. In community development projects, insiders and outsiders have equal contributions to project quality.

In general, projects in developing communities rarely show high levels of quality. Very often, it is because quality is perceived as an elusive concept that is hard to quantify and control. Sometimes, developing a quality plan does not even cross the mind of those who manage the projects. As remarked by Ho (2005) in relation to managing the quality of information technology projects, "quality management is a *'pay me now, or pay me later'* proposition." Poor project quality may create additional unnecessary costs further into the project (that may not have been budgeted), abandonment of projects, and ultimately failed promises to the stakeholders. Including quality management in development projects is not an option; it is an obligation. According to the Project Management Institute (PMI 2008), project quality management consists of three processes: quality planning, quality assurance, and quality control. Rose (2005) also suggests including quality improvement as a fourth component of quality management.

PMI (2008) recommends that quality planning first establish the quality standards relevant to both the delivery (in participation with the community) and management for a specific project. Once this is done, the next step is to develop the characteristics and components of those standards and how to satisfy them. These can be expressed in terms of such things as performance, reliability, suitability, functionality, sustainability, efficiency, level of participation, effectiveness, equity, and duplicability.

A quality plan must be included into the detailed action plan for the project and its management structure. The plan must show how and when the quality standards are satisfied, who is responsible for managing them, the quality metrics used, the costs of the quality plan, and what corrective actions will be taken when no conformance exists between what is being done and the agreed-upon standards. Associated assumptions and preconditions must be identified as well. In many ways, quality planning is included as a preventive measure.

Once quality has been defined, its characteristics outlined, and a plan established, project quality management requires a strategy and methodology for quality assurance, quality control, and quality improvement.

8.6 Refining the Work Plan

At this point in the overall project framework, the following has been completed:

- Appraisal of the community has been conducted, and a community baseline has been established;
- The community capacity and vulnerability have been assessed;
- Data have been collected, analyzed, and processed;
- Problems have been identified (problem tree) and ranked;
- Preliminary solutions have been identified (solution tree) along with potential actions (action identification matrix);
- Solutions have been ranked and weighted based on several criteria (multicriteria assessment matrix);
- Feedback from stakeholders and partners, and prioritization, has been obtained regarding the solutions;
- A project hypothesis and a preliminary action plan have been established where the what, who, where, when, why, and how of the project have been considered;
- An action feasibility matrix has been developed that lists possible contributions of the stakeholders and partners and their short- and long-term benefits;
- A focused strategy and project logic have been considered for selected issues and solutions (logframe);
- The first draft of a work plan or action plan has been outlined, including project and management activities, resource scheduling, role of partners and stakeholders, and contingency planning; and
- Project quality standards for project delivery and management have been outlined.

An additional step in reinforcing and refining the proposed action plan is to compare the proposed solutions with previous case studies where solutions similar to the proposed ones have been implemented in a similar context. This step gives additional insights as to what is likely to work and what may need special attention. As an example, Table 8-5 gives insights gained from three case studies of villages in Nepal where irrigation systems have contributed in the past to improving household livelihood security in farming compared with more traditional rain collection systems. The environmental, gender, and health effects associated with irrigation are also listed.

8.7 Behavior Change Communication

To meet the objectives, reach the goals, achieve the outcome outlined in the logical framework, and ensure quality in the execution of the work plan, changes

Table 8-5. Feedback Gained from Three Villages in Nepal

Case Study: Contribution of Irrigation to Sustaining Rural Livelihoods

Examined the experiences of three villages in Nepal: compared conditions with rain-fed farming to conditions after installation of irrigation system.

Economic Impacts

- Farmers able to grow three seasons of crops
- Farmers able to grow higher-value crops
- Income was found to increase by 25,600 to 70,000 rupees per hectare
- Household yearly income increased to above the poverty line
- Trickle-down income increases were reported by merchants, tailors, blacksmiths, carpenters, etc.

Environmental Impacts

- Use of fertilizer and pesticides increased from around 2% of farmers before irrigation to more than 80% in two of the communities

Gender Equity Impacts

- Women reported higher work loads, most likely due to the increased number of growing seasons and related farm labor

Health Impacts

- Better nutrition as food surpluses became available in place of shortages
- Increased vegetable production and consumption
- Health posts built (unclear on causal link, but possibly due to increased affluence of community)
- Overall family health was reported to have improved

Note: In each village, irrigation systems were used instead of traditional rain catchment systems.

Source: Courtesy of Barry Bialek, personal communication, 2011.

in behavior may be necessary at the community, household, and/or individual levels. Some forms of behavior need to be encouraged and others need to be discouraged, without ever changing the community's culture. Behavior change may be required in one or several general areas, such as agriculture; water; sanitation and hygiene; health (reproductive, infant, HIV/AIDS); family planning, governance and policies; building capacity in various services; community resilience to hazards; or it may involve specific day-to-day activities, such as cooking, heating, or hand washing.

Behavior change communication (BCC) is a communication process that aims to convey changes in behavior to a specific target audience (e.g., children, women, farmers, or castes) regarding issues that are perceived as needing change. These perceptions may originate from outsiders or through collective action with community members and leaders during the project appraisal phase. As discussed in Chapter 6, the behavior of a system depends on its structure. Therefore, BCC can be seen as changing the structure of the community, its households, and its members so that new forms of behavior emerge. It is about creating an environment for an audience to move from action, making a choice, developing judgment, and ultimately having an authentic viewpoint about behavior change. An excellent book on BCC in relation to sanitation is by Kar and Chambers (2008). It demonstrates how a community-led total sanitation strategy based on behavior change, rather than providing technology, can result in positive results. Cairncross et al. (2003) noted that "the promotion of washing one's hands with soap can reduce diarrhea by over 40%."

In general, BCC is not easy. First, it requires outsiders to have good communication skills, such as listening, observation, reflection, and summarizing. These skills must be used throughout the entire project process and are especially critical in the community appraisal phase addressed in Chapter 5. Furthermore, BCC needs to account for a wide variety of factors that have the potential to influence human behavior. According to Booth (2013), they include

- Factors external to an audience, such as access to services and products, policy (laws and regulations), culture, and governance; and
- Factors internal to an audience, such as self-efficacy, learning styles, perceived social norms by gender and age group, perceived consequences, skills, knowledge, attitudes, beliefs, perceived risks, and intentions.

In general, once the target audience and issues faced by the audience have been identified, the next two steps in BCC include developing a persuasive communication strategy and a methodology or plan to promote positive behaviors (and discourage forms of negative behavior) and create a supportive environment so that the behaviors are sustainable and eventually become habits. The field of community-based social marketing (McKenzie-Mohr 2011) provides various proven tools to "foster sustainable behavior."

The behavior change strategy (BCS) necessary to select effective and culturally appropriate means to promote positive behaviors and ensure a supportive environment requires a good understanding of the social, cultural, and economic characteristics of the target audience being addressed and its various forms of vulnerability and capacity. Once a target behavior has been selected, i.e., an end state has been identified (Booth 2013), the following tasks need to be carried out in a pragmatic way:

- Identify the motivators for change in behavior and the barriers that have the potential to prevent or slow down change;
- Review existing forms of behavior, including possible competing ones and their levels of penetration;
- Weigh the benefits of alternative forms of behavior, their effects, and their possible levels of penetration;
- Outline the dominant methods of communication that are most likely to be effective within the target audience and its components and their probability of success; and

- Identify what resources are available and are needed to reach out to the target groups.

According to Fogg (2009), a persuasive strategy for behavior change needs to include three components: sufficient motivation, ability for behavior change, and a triggering mechanism. All three components need to happen at the same time for change in behavior to take place.

The community appraisal described in Chapter 5 should provide insights and data when addressing the aforementioned issues. It should also be used to identify the dominant attitude, skills, and knowledge in the target group that contribute to its current behavior and the cultural dimension in terms of people's concept of self, how context determines their behavior, and their degree of directness. Table 8-6 lists, as an example, areas of change required to develop the proposed action plan in the Mabu project, along with several motivators (benefits) for change.

Once the enabling and constraining environment toward behavior change has been mapped and strategized, a behavior change communication plan (BCCP) or methodology must be considered for each target behavior and must include

- The implementation of methods for behavior change at a small scale (pilot) and how it can be ramped up, more specifically identifying who (is responsible), where (scale and location), when (time frame), and how (the steps);

Table 8-6. Areas of Behavior Change, Motivators (Benefits), and Methods for Change in Agricultural Practice (Mabu Project)

Areas of Change	Motivators for Change
• Farming cooperatives • New markets and transport methods • Less chemical pesticide and fertilizer use • Off-season farming • Water conservation—drip irrigation • Canal maintenance	• Power of many, price stability, etc. • Better long-term soil quality, less pollution • Higher yields • More income, better nutrition • More water available, help environment • Canal will provide lasting benefit • More income at better markets

Methods & Capacity for Change	
• Farmer training workshops • Posters and brochures • School educational component • Demonstration and informational sessions of technology and methods at Mabu yearly fair	

Source: Glover et al. (2011), reproduced with permission.

- The monitoring and evaluation of behavior change; and
- The corrective actions to be used if the target behavior is not in place over a certain physical and/or time scale.

The methods used in BCC need to be specific, measureable, appropriate, realistic, and timely (SMART). Furthermore, the proposed methods need to convince the target audience that the perceived benefits of what is proposed exceed existing perceived barriers; i.e., behavior change is seen as an exchange and a more attractive option. For instance, telling mothers to wash their hands may not be enough. Telling them that washing their hands will result in better health for their children is more meaningful and more conducive for hand washing to become a behavior and eventually a habit. Finally, the approach used in behavior change needs to be participatory and developed from the point of view of the target audience, not from those who only see a need for behavior change. An example was reported by Zahnd (2012) regarding using 17 songs to convey changes in an integrated and contextualized development project in Nepal, in areas such as water, sanitation, hygiene, cooking, energy, and basic education.

According to Booth (2013), a methodology consisting of three successive steps may lead to changes in behavior or habit and form a so-called "habit loop," which can be self-reinforcing. It consists of using tools such as prompts and clues, encouraging commitment, and providing rewards.

- *Prompts and clues* are developed to convey the message to the target audience and remind it of what needs to be done. The communication methods toward behavior change can take different forms and must match the capacity of the community. They can be between individuals (interpersonal), between an individual and multiple people (mass communication), or among or within groups (organizational communication). The prompts and clues may be integrated into a variety of techniques, such as role play, theater, music, film, workshops, or school programs. Signs and posters may be created in areas where changes are expected to occur. A clinic or health posts, for instance, may be good places to display posters that explain infant care or hygiene.
- *Commitment* is important because it ensures long-term sustainability of a BCC initiative. It must be voluntary and not coerced. Actions that may be conducive to enhancing commitment include public commitments versus private commitments and written versus verbal commitments. In general, small commitments are more likely to succeed in changing behavior, rather than large commitments that are difficult to make and sustain. Also, public commitments are stronger than private ones, and written commitments are more effective than oral commitments.

- *Rewards* are in place for small steps because the target audience honors its commitments, thus emphasizing small victories rather than large ones. This success provides further incentive to change behavior.

It should be noted that BCC is more often than not an integral part of the frameworks used by large development agencies. The reader may want to consult the approaches used by the World Bank (Sanitation Marketing Toolkit, WSP 2013) and USAID (Hygiene Improvement Project 2006–2010 and WASHPlus project, USAID 2013), which integrate BCC into the improvement of water, sanitation, and hygiene practices in developing communities. Table 8-7 gives an example of a behavior change communication plan for mothers of children under five in Pakistan (USAID 2007) in four areas: hand washing, purifying water, storing of water safely, and the safe disposal of feces. For each key practice, the table gives who is involved, what is required, and where the practices need to be presented in the community.

A final note about BCC is that it can be integrated into the project logic and the logical framework. This integration is shown in Table 8-1, where among the objectives listed, a BCC campaign about electric appliance technology adoption (e.g., objective 2.1.3) has been included. Also shown in Table 8-1 are the activities and input necessary for meeting that objective.

Finally, the integration of BCC into the project logic is often accompanied with a rights-based approach (RBA), as discussed in Section 4.7, because the two are interconnected. Those involved in developing BCC campaigns at the local level must account for possible positive and negative effects associated with changes in behavior within a given context. In the Mabu project mentioned, special attention was placed on the empowerment of women and members of the lower caste system and how new pico-hydro and irrigation methods shared with those marginal groups would have the potential to empower them.

8.8 Chapter Summary

This chapter started with the preliminary solutions developed in Chapter 7 and provided a methodology (strategy and operation) to convert those solutions into the first draft of a work plan or action plan. In that action plan, it is critical to develop a strategy and a methodology for behavior change, which takes time to implement and bear fruits because of the adaptability, nonlinearity, and evolution of living systems, as discussed in Chapter 6.

As the final work plan is developed, several additional steps need to be carried out to refine and confirm the decisions outlined in the work plan. The first, called capacity analysis, is to ensure that the community has the capacity to move forward with the recommendations and decisions made in the work plan. More specifically, the community's weakest links must be evaluated to determine the areas of capacity building. The second analysis, risk analysis, reviews the risks

Table 8-7. Behavior Change Communication Plan for Mothers of Children Under 5

Key Practices (On What)	Sender (By Whom)	Channel (With What)	Venues (Where)
Hand Washing			
Wash the hands of your children under 5 with soap and air dry before every meal.	Hygiene promoters	6 pocket flyers	At community fair
	NGO workers	2 danglers	On one radio program
Wash your hands with soap and air dry every time after you go to the toilet.	Mother-to-mother	1 banner	In the community
	Mother-to-daughter	2 TV spots (urban)	At home
Wash the hands of your children under 5 with soap and air dry every time after they go to the toilet.	Daughter-to-mother	4 radio spots	Radio (TV) stations
	Husbands	1 demonstration for teaching your children hand washing with soap	
Teach your children under 5 to wash their hands with soap and air dry.	Media		
	Private sector	Home visits	
Wash your hands with soap before feeding your children.		1 drama	
Wash your hands with soap before preparing meals.			
Purifying Water			
Test your drinking water for contamination.	Hygiene promoters	5 pocket flyers	At community fair
	NGO workers	1 dangler	On one radio station
Boil your drinking water.	Mother-to-mother	1 radio spot	In the community
	Husbands	Home visits	At home
Solar heat your drinking water.	Media	1 drama	Radio stations
	Private sector		

Table 8-7. Behavior Change Communication Plan for Mothers of Children Under 5 *(Continued)*

Key Practices (On What)	Sender (By Whom)	Channel (With What)	Venues (Where)
Safely Storing Water			
Continue to cover your drinking water.	Hygiene promoters	5 pocket flyers	At community fair
Continue to place your water out of reach of your children.	NGO workers	1 dangler	On one radio program
Use a long-handled scoop to give out your drinking water.	Mother-to-mother	1 radio spot	In the community
Pour your drinking water directly from the container.	Media	Home visits	At home
Use a drinking water container with a tap.	Children	1 drama	Radio stations
Safely Disposing of Feces			
Continue to use your latrine every time you go to the bathroom. Where open defecation occurs, feces should be buried.	Hygiene promoters	1 pocket flyer	At community fair
	NGO workers	1 dangler	On one radio program
	Mother-to-mother	1 radio spot	In the community
	Media	Home visits	At home
		1 drama	Radio stations

Source: USAID (2007).

that may arise during project implementation and what their respective effects may be, in addition to how to address undesirable issues. Finally, the project design, together with the work plan, must contribute to an overall increase in community resilience. Capacity analysis, risk analysis, and resilience analysis are discussed in Chapters 9, 10, and 11, respectively. Chapter 12 discusses the execution of the final work plan when delivering services to communities.

References

Bakewell, O., and Garbutt, A. (2005). *The use and abuse of the logical framework approach*, Swedish International Development Cooperation Agency, Stockholm, Sweden. <http://www.intrac.org/data/files/resources/518/The-Use-and-Abuse-of-the-Logical-Framework-Approach.pdf> (Jan. 10, 2013).

Barton, T. (1997). "Guidelines for monitoring and evaluation: How are we doing?" CARE International in Uganda, Monitoring and Evaluation Task Force, Kampala, Uganda. <http://portals.wdi.wur.nl/files/docs/ppme/Guidelines_to_monitoring_and_evaluation%5B1%5D.pdf> (Jan. 12, 2012).

Barton, T., et al. (1997). "Our people, our resources: Supporting rural communities in participatory action research on population dynamics and the local environment." International Union for Conservation of Nature (IUCN) Publications Services, Gland, Switzerland, and Cambridge, U.K.

Booth, B. (2013). "Fostering sustainable behavior with behavior change communication (BCC)/social marketing." Lecture notes in CVEN 5929: Sustainable Community Development 2. University of Colorado at Boulder.

Cairncross, S., et al. (2003). "Health, environment and the burden of disease: A guidance note." Department for International Development, London. <http://www.lboro.ac.uk/well//resources/Publications/DFID%20Health.pdf> (May 10, 2012).

Caldwell, R. (2002). *Project design handbook*, Cooperative for Assistance and Relief Everywhere (CARE), Atlanta.

Delp, P., et al. (1977). *Systems tools for project planning*, International Development Institute, Bloomington, IN.

Department for International Development (DFID). (2002). *Tools for development*, version 15, Department for International Development, London.

EuropeAid. (2002). *Project cycle management handbook*, version 2.0, PARTICIP GmbH, Fribourg, Germany. <http://www.sle-berlin.de/files/sletraining/PCM_Train_Handbook_EN-March2002.pdf> (March 10, 2012).

Fogg, B. J. (2009). "A behavior model of persuasive design." *Proceedings of the 4th International Conference on Persuasive Technology*, Association for Computing Machinery, New York, Article No. 40.

Gawler, M. (2005). *WWF introductory course: Project design in the context of project cycle management*, Artemis Services, Prévessin-Moëns, France <http://www.artemis-services.com/downloads/sourcebook_0502.pdf> (March 15, 2012).

Glover, C., Goodrum, M., Jordan, E., Senesis, C., and Wiggins, J. (2011). "Mabu village term project." CVEN 5929: Sustainable Community Development 2, University of Colorado at Boulder.

Ho, M. (2005). "Managing project quality: Cost, control and justification." *Information Management*, October 1. <http://www.information-management.com> (Nov. 2, 2012).

Innovation Network. (2013). <http://www.innonet.org/> (Jan. 2, 2013).

Jensen, G. (2010). *The logical framework approach*, Bond for International Development, London. <http://www.dochas.ie/Shared/Files/4/BOND_logframe_Guide.pdf> (Feb, 15, 2012).

Kar, K., and Chambers, R. (2008). *Handbook on community-led sanitation*, Institute of Development Studies, Brighton, U.K.

Lewis, J. P. (2007). *Fundamentals of project management*, 3rd Ed., Amacom, New York.

McKenzie-Mohr, D. (2011). *Fostering sustainable behavior: An introduction to community based marketing*, New Society Publishers, Gabriola Islands, BC, Canada.

Oxfam. (2008). *Handbook of development and relief*, Vols. 1 and 2, Oxfam Publ., Cowley, Oxford, U.K.

Project Management Institute (PMI). (2008). *A guide to the project management body of knowledge*, PMBOK Guide, 4th Ed., Project Management Institute, Newtown Square, PA.

Rose, K. H. (2005). *Project quality management*, J. Ross Publishing, Fort Lauderdale, FL.

United Nations Development Programme (UNDP). (2009). *Handbook on planning, monitoring and evaluating for development results*, United Nations Development Programme, New York. <http://web.undp.org/evaluation/handbook/documents/english/pme-handbook.pdf> (May 22, 2012).

U.S. Agency for International Development (USAID). (2007). *Pakistan safe drinking water and hygiene promotion project: Behavior change strategy (BCS) and behavior change communication plan (BCCP)*, USAID, Washington, DC. <http://www.eawag.ch/forschung/siam/lehre/alltagsverhaltenII/pdf/PBehaviorChangeStrategy.pdf> (July 23, 2012).

U.S. Agency for International Development (USAID). (2013). "WASHPlus: Supportive environment for healthy communities." <http://www.washplus.org/> (Oct. 4, 2013).

Water and Sanitation Program (WSP). (2013). "Sanitation marketing toolkit." <http://www.wsp.org/toolkit/toolkit-home> (Oct. 4, 2013).

World Bank. (2005). *The logframe handbook: A logical framework approach to project cycle management*, World Bank, Washington, DC. <http://gametlibrary.worldbank.org/FILES/440_Logical%20Framework%20Handbook%20-%20World%20Bank.pdf> (Aug. 1, 2012).

Zahnd, A. (2012). "The role of renewable energy technology in holistic community development." Doctoral dissertation, Murdoch University, Australia.

9

Capacity Analysis and Capacity Development

9.1 From Development Aid to Capacity

Capacity and capacity development have been topics of intense discussion in the international development community over the past 50 years. More recently, both concepts received renewed attention after the 2001 New Partnership for Africa's Development initiative launched in Lusaka, Zambia, on the role of capacity in sustainable development; the 2005 Paris Declaration on Aid Effectiveness; the 2008 Accra Agenda for Action; and the 2011 High Level Forum on Aid Effectiveness in Busan, South Korea. All four meetings closely linked capacity development and aid effectiveness and focused on defining the meaning of capacity at the country level. It has been determined that capacity development must also be looked at on smaller scales: individual, organizational, and the enabling environment (OECD 2006).

Several definitions of what capacity is have been proposed in the development literature. For instance, the World Health Organization (Milèn 2001) defines capacity as "the ability of individuals, organizations or systems to perform appropriate functions effectively, efficiently and sustainably."

The Canadian International Development Agency sees capacity in its various expressions and at different scales ranging from the individual to social systems as

the abilities, skills, understandings, attitudes, values, relationships, behaviors, motivations, resources and conditions that enable individuals, organizations, networks/sectors and broader social systems to carry out functions and achieve their development objectives over time. (Bolger 2000)

The Deutsche Gesellschaft für Technische Zusammenarbeit (GTZ 2007) looks at the link between capacity and sustainable development as "the ability of people, organizations and societies to manage their own sustainable development processes."

A common element that emerges in all three definitions is that capacity is synonymous with *ability*, i.e., ability of stakeholders to achieve certain development goals and satisfy their needs. Further distinction can be made between the ability of a community to cope with various situations (*inherent capacity*) and that to adapt to new needs, challenges, changes, and opportunities (*adaptive capacity*). It is generally agreed that

- Capacity is critical to the success of human development.
- All communities have various forms of capacity that can and should be built upon.
- Capacity is acquired and built over time.
- Capacity is a strong attribute of resilient communities.
- Capacity can be assessed (qualitatively or quantitatively) using performance indicators where the performance can take multiple forms, such as "decision making, leadership, service delivery, financial management, ability to learn and adapt, pride and innovation, organizational integrity and many others" (Morgan 1998).

The other related concept that has received much interest in the field of human development is how to acquire capacity through *capacity development*. Since the 1990s, development agencies have emphasized that their main focus is no longer on development aid, technical assistance, or technical cooperation, but rather on community capacity building and capacity development. That evolution is best illustrated in Table 2-4, which shows the evolution of the UNDP approach to capacity development (UNDP 2009). In that approach, the focus of capacity development is described as "empowering and strengthening endogenous capabilities" through transformation rather than "lending and granting money to developing countries." Since the 1990s, a lot of discussion has taken place in development agencies on how to conduct capacity development so that it results in a "sustainable and authentic" process of change and transformation leading to increased individual, community, organizational, and societal capabilities that last. In this chapter, capacity building and capacity development are used together. This is not always the case in the literature, however. For instance, the UNDP makes a clear distinction between capacity building and development (UNDP 2008). This chapter looks at both in an interchangeable way.

Capacity building and capacity development can mean different things to different people and development agencies. There are still discussions about how to define capacity development, what modes of delivery exist, and how to demonstrate and verify the results of capacity development. Of all the definitions proposed in the literature, four of them have been retained. The first definition is one proposed by the World Federation of Engineering Organizations, where capacity building is referred to as *the building of human, institutional and infrastructure capacity to help societies develop secure, stable and sustainable economies, governments, and other institutions through mentoring, training, education, physical projects, the infusion of financial and other resources, and most importantly, the motivation and inspiration of people to improve their lives.* (WFEO 2010)

The UNDP (2009) defines capacity development as follows: "a process through which individuals, organizations and societies obtain, strengthen, and maintain the capabilities to set and achieve their own development objectives over time."

Another definition, used by the Canadian International Development Agency (CIDA), sees capacity development as "the approaches, strategies and methodologies used by developing countries, and/or external stakeholders, to improve performance at the individual, organizational, network/sector or boarder system level" (Bolger 2000).

Finally, a definition proposed by GTZ (2007) directly links capacity development to sustainable development as "a holistic process through which people, organizations, and societies mobilize, maintain, adapt and expand their ability to manage their own sustainable development."

Even though development agencies have somewhat different definitions of capacity building and capacity development, they seem to agree on key underlying principles, which when combined define a process that

- Does not happen by itself and is not random;
- Builds on local ownership, self-reliance, and existing local capacities;
- Promotes genuine partnership and broad-based participation;
- Accounts for the context in which it takes place;
- Understands capacity within a system and strategic management context;
- Allows for ongoing learning and adaptation and integration of complex issues;
- Ensures long-term commitments and partnership and is built to last;
- Creates a potential to act over time; and
- Is scale (physical and temporal) dependent, meaning that what works at one scale does not necessarily work at another scale.

These definitions and principles indicate that there cannot be a single approach to capacity building or capacity development that would work at all the scales of interest: individuals, communities, organizations, and society. Because the focus of this book is about small-scale community development projects, the rest of this chapter focuses on capacity development at the scale of developing communities and their components, i.e., households and individuals. Within that context, it is fair to say that given the range of issues that developing communities face (see Chapter 2), capacity development is likely to take a considerable amount of time (measured in years) and require creative design and planning tools to produce tangible results. Furthermore, within that context, capacity development is likely to start from a low-capacity and high-vulnerability baseline.

Acquiring (building) capacity within the context of small-scale projects can be seen as a participatory locally generated process at the end of which communities are expected to possess the necessary resources and knowledge to (1) address their own problems, (2) be self-motivated and self-sustaining, (3) cope and adapt to various forms of stressors and shocks, (4) satisfy their own basic needs, and (5) demonstrate livelihood security. In other words, capacity development is seen as a strategic means to an end that is about sustainable communities. In that process, capacity builds on what exists, however small that may be. From that baseline, it can be created, strengthened, and adapted to new challenges faced by the community.

Even though the emphasis is at the community level, it is important to remember that capacity development is multidimensional because there are many forms of capacity that can be addressed in a community, such as financial, technical, social, intellectual, leadership, environmental, and institutional. Often, these categories of capacity are themselves linked to each other because of the systemic nature and complexity of communities, as discussed in Chapter 6. Capacity development at the community level is likely to depend on what takes place at other scales within the community, across communities, and at the regional or national level. As a result, capacity development needs to be considered "from a systems perspective, with an appreciation of the dynamics and inter-relationships among various issues and actors in different dimensions" (Bolger 2000).

Capacity has been mentioned many times throughout this book. It was first encountered in Chapter 1 in the overall definition of risk because a community is at less risk when its capacity is higher. It was also described as an essential attribute of sustainable communities in Chapter 2. Capacity was also discussed in Chapter 5 in relation to the appraisal phase of the ADIME-E framework where capacity appraisal was presented as a methodology to define what the community baseline is in terms of assets, needs, knowledge, skills, resources, structures, and strengths. From the results of the community appraisal, the participatory action research team can determine whether or not capacity building can happen and what may be preventing it. In Chapter 11, capacity is described as an acquired attribute necessary for a community to (1) cope with or adapt to unusual conditions and transient dysfunctions associated with hazard events and (2) return to a functional balance and new normal. This process is discussed within the context of community resilience.

In this chapter, capacity analysis is presented as a tool to further understand the dynamic that exists between the current capacity of a community (its enabling environment) and its ability to support the comprehensive work plan outlined in Chapter 8. The proposed solutions outlined in that plan must match the current level of community development. More specifically, a need exists to assess whether the community has the strength, knowledge, resources, and capability to (1) accept the proposed solutions and recommendations outlined in the focused strategy and planning stages of the project, (2) implement those solutions, and (3) carry out the corresponding action plan in a sustainable way with long-term benefits. An outcome of that analysis is the identification of local weaknesses and/or potential challenges that could prevent the total or partial implementation of the recommended solutions. Finally, this step is followed by the formulation of a capacity development program to overcome the limitations that are part of the constraining environment.

Of particular interest in capacity building is how the community progresses in its development. As the community livelihood improves through capacity building or development, more sophisticated solutions can be implemented. Therefore, questions arise about (1) what level of capacity development the community aspires to (against its existing capacity), (2) over what time frame, and (3) how it addresses existing gaps between current and desired capacities. Answers to those three questions help identify, rank, plan, prioritize, and implement community development interventions in capacity building over time. These answers also help in selecting the most appropriate technologies for the community (Chapter 13).

In summary, capacity analysis helps in refining and improving the solutions and project action plan discussed in Chapter 8. At the same time, capacity analysis may also reveal missing information and issues that were ignored or overlooked in the community appraisal phase, and additional community appraisal may be needed. At the end of the process, a stronger understanding of the community emerges about what it can do, what it cannot do, and how it needs to be strengthened. In general, the results of capacity analysis can be expressed in quantitative (capacity factors) or qualitative terms (high, medium, or low).

This chapter looks at two steps involved in capacity building and development within the context of small community projects: (1) assessment of capacity assets and needs and (2) formulation of a capacity development response. Both steps are part of a larger iterative cyclical process for capacity development (Figure 9-1) proposed by the UNDP (2009), which can be used at various scales from large-sector country development programs to local projects. This chapter emphasizes a methodology originally proposed by researchers at the University of Virginia (Ahmad 2004; Bouabib 2004; Louis and Bouabib 2004) to assess the capacity of developing communities to carry out the delivery of local community municipal sanitation services (MSS) projects. As demonstrated in this chapter, the methodology is generic enough to be used to address other types of

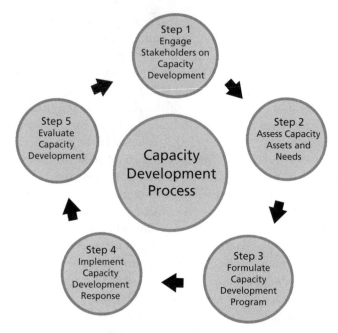

Figure 9-1. Five Steps in the Capacity Development Cycle

community services. Two additional capacity analysis frameworks proposed by the World Federation of Engineering Organizations (WFEO) and the UNDP are briefly addressed as well.

9.2 Capacity Assessment

The ability of a community to identify, evaluate, and address its own problems and needs; develop solutions; and implement an action plan in partnership with outsiders depends largely on its enabling environment. That environment can be seen as the foundation or baseline on which capacity is built over time. Clear indicators for measuring progress from the baseline along with targets and benchmarks need to be in place and integrated into the project logic discussed in Chapter 8.

As discussed in Chapter 5, the enabling environment is mapped in the community appraisal phase of the ADIME-E framework. The main goal of that appraisal is to learn as much as possible about the community through the collection of data, the transformation of data into useful information, and the analysis of that information. It provides a context of the community's enabling environment in terms of culture, leadership, level of development, and human condition. Finally, it gives some indication about the level of capacity the community is interested in achieving.

In general, the resources, knowledge, skills, assets, and strengths that contribute to the enabling environment of a community can be broken down into different but equally important categories of capacity. As remarked by Lavergne and Saxby (2001), these categories consist of *tangible* components (e.g., infrastructure, education, natural resources, health, and institutions) and *less tangible* ones (e.g., skills, social fabric, values and motivations, habits, attitudes, traditions, and culture). They may also include *core capabilities*, which "refer to the creativity, [leadership], resourcefulness and capacity to learn and adapt of individuals and social entities."

It is noteworthy that no universally accepted terminology exists among development agencies to categorize the different forms and expressions of community capacity. For instance, the Tearfund (2011) suggests identifying five forms of capacity: individual, social, natural, physical, and economic. In this chapter, we use seven groups of capacity, following a terminology proposed by Louis and Bouabib (2004): institutional, human resources, technical, economic and financial, energy, environmental, and social and cultural. An eighth group of capacity called service capacity is used to measure the level of a given service (e.g., energy; water, sanitation, and hygiene (WASH); and shelter) compared with accepted international standards. In general, the capacity components selected in the capacity analysis must be appropriate to the type of project being addressed.

UVC Framework

Professor Garrick Louis and coworkers at the University of Virginia, Charlottesville (Ahmad 2004; Bouabib 2004; Louis and Bouabib 2004) developed a detailed methodology to determine the capacity of a developing community to conduct municipal sanitation services (MSS) projects. As remarked by Bouabib (2004), these services may include

- Drinking water supply (DWS), which includes "the construction, operation and maintenance of public water systems, including production, acquisition and distribution of water to the general public for residential, commercial and industrial use";
- Wastewater and sewage services (WSS), which is defined as "the provision, operation and maintenance of sanitary and storm sewer systems, sewage disposal and treatment facilities"; and
- Management of solid waste (MSW), which is "defined as the collection, removal and disposal of garbage, refuse, hazardous, and other solid wastes."

In general, these three types of services need to be considered together at the community level. As outlined many times in the development literature and further discussed in Chapter 15, the quality and quantity of these services determine to a great extent public health and economic development in any society (SIWI 2013). The approach, referred to as UVC here, can be generalized to other types of projects and services besides MSS, such as education, health, energy, or food, as discussed by Faeh et al. (2004).

The UVC approach focuses on one group of services at a time: DWS, WSS, or MSW. For each group, eight categories of community capacity are considered, as shown in Figure 9-2, which is a more detailed version of Figure 5-5. These categories were selected because they are likely to have an influence on the type of services (e.g., MSS) being investigated. In Figure 9-2, the categories of capacity can be estimated qualitatively (high, medium, or low), or quantitatively. Furthermore, capacity cannot fall below minimum human standards, such as the Sphere standards (Sphere Project 2011).

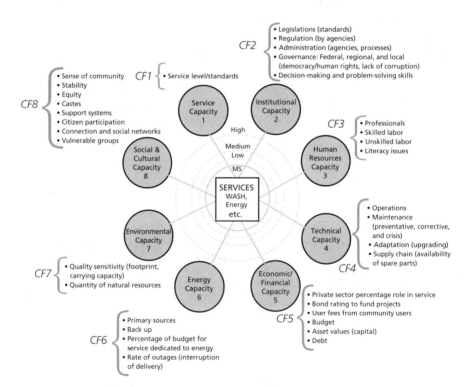

Figure 9-2. Categories of Capacity for a Community to Conduct Service Projects

Source: Adapted from Louis and Bouabib (2004), with permission from University of Virginia.

In each category of capacity, several requirements are listed that contribute to that category. The categories of capacity are described below.

- *Service capacity* measures, for a given type of service, the gap between the actual levels of service provided in the community and published standards that guarantee an "acceptable level of health concern for the community." Bouabib (2004) gives the example of water service where the water supply of the community would be compared with a 50 L/person/day standard, guaranteeing that "basic personal and food hygiene are assured as well as laundry and bathing ... and a low level of health concern." Water would be available within 100 meters of consumers. Similar standards exist for wastewater and sewage and solid waste. According to Bouabib (2004), wastewater treatment systems must be able to accommodate 80–90% of water supplied (e.g., 32–45 L/person/day). For solid waste, a capacity to process 0.5–1.5 kg/person/day is suggested. This level is discussed further in Chapter 15.

- *Institutional capacity* defines the components of the institutional framework that need to be in place to provide the services. The requirements include a body of legislation; associated regulations, regulatory standards, and codes; administrative authority; administrative process; and stable and good governance.

- *Human resources capacity* relates to the labor that is available to provide the services and its level of training. The requirements include professional, skilled labor, unskilled labor, and level of illiteracy.

- *Technical resources capacity* relates to the logistics and tactics necessary to address the components of technology that enters into the implementation of the solutions. The requirements include operations, maintenance, upgrading or adaptation, and supply chain (spare parts).

- *Economic and financial capacity* represents the financing of the services, the availability of loans, and the financial assets in the community. More specifically the requirements include percentage of the private sector providing services and the existence of bonds, user fees, budget, asset values, and debt.

- *Energy capacity* deals with the available energy, its availability, its costs, and reliability necessary to provide the services. The requirements include primary source, backup sources, percentage of budget associated with energy, and rate of outage.

- *Environmental capacity* looks at the availability of natural resources (e.g., water and forest) needed to implement the solutions, the carrying capacity of the environment, the level of stress it can sustain, and making sure that the services do not substantially affect or deplete natural resources.

- *Social and cultural capacity* deals with the community structure and components, its social networks and cohesion, its capacity of organization, the households and their interactions, and gender and equity issues.

Once the capacity categories have been identified, a capacity factor (CF) is calculated for each capacity category as the weighted sum of its requirements. Using the example proposed by Louis and Bouabib (2004), let's consider the DWS technical capacity. As shown in Table 9-1, it consists of four requirements: operations; maintenance (preventive, corrective, and crisis); adaptation (to constraints); and supply chain (spare parts). Each requirement is rated on a scale ranging between 0 and 100, broken down into five rating groups with 20 units each. Descriptors of each rating group for the four requirements are listed in Table 9-1.

The capacity factor CF_4 (4 is the fourth category of capacity in Figure 9-2) is determined as the weighted average of four requirement ratings C_{4j}, as follows:

$$CF_4 = \Sigma C_{4j} w_j \ (j = 1, 4)$$

where w_j is a weighting factor associated with requirement rating C_{4j}. Table 9-2 shows an example of technical capacity calculation for the Guatemala project discussed in Chapter 5.

In general, each one of the eight capacity factors CF_i (i = 1–8) shown in Figure 9-2 can be determined as follows:

$$CF_i = \Sigma C_{ij} w_j \ (i = 1\text{--}8 \text{ and } j = 1, n_i)$$

where n_i is the number of requirements in each capacity factor CF_i. Once calculated, the capacity factors can be plotted in the form of a radial vector diagram, such as the one shown in Figure 5-6. This diagram provides a visual quantitative map of the community capacity baseline for the selected service: DWS, WSS, or MSW. It also helps identify the strong and weak components of community capacity for a given service, where interventions are needed, and where such interventions are most likely to have a positive effect on capacity building. In Figure 5-6, for instance, technical capacity and financial/economic capacity are both low. Capacity factors and detailed requirements for DWS, WSS, and MSW can be found in Bouabib (2004).

According to the UVC framework, in the inventory of capacity categories for a given type of service, the one with the lowest capacity factor determines the so-called technology management level (TML) of the community, i.e., the stage (or level) of community development (or readiness) for that service. This conservative approach uses a weakest link criterion (or a pessimistic rule criterion). Other criteria could be used to determine the TML, as suggested by Bouabib (2004).

Table 9-1. Breakdown of Technical Capacity Factor into Four Components for Drinking Water Supply

Score	1–20	21–40	41–60	61–80	81–100
4	Technical Capacity				
C_{41} Operations	Manual collection and untreated water use	Pumping water	Pumping water Control water quality	Monitor water systems Control water quality Control pipes	Monitor water systems Control water quality Monitor pipes network Monitor treatment
C_{42} Maintenance	None	Disinfection Minor repair	Check water systems Major repair	Check/maintain water systems Major repair Maintain pipes	Check/maintain water systems Check/maintain network Check/maintain water meter
C_{43} Adaptation	None	Rarely	Occasionally	Usually	Frequently
C_{44} Supply chain	None	National supplier	Regional supplier	National manufacturer Regional supplier	National manufacturer Local supplier

Source: Louis and Bouabib (2004), reproduced with permission from University of Virginia.

Table 9-2. Example of Scoring for Technical Capacity Factor

Capacity Factor	Subfactor	Very Poor 0–20	Poor 21–40	Medium 41–60	High 61–80	Excellent 81–100	Score
Technical	Operations	No operations running	Few operations are running poorly	Few operations running well	All operations running well	All operations running well with backups in place	45
	Maintenance	No maintenance	Infrequent and/or poor maintenance	Low-level maintenance fairly regularly	Maintenance for most operations	Maintenance for all operations on regular basis	25
	Adaptation	No adaptation	Adaptation after many years or to very small degree	Adaptation as needed	Good adaptation regularly accessed	Excellent adaptation with many members involved throughout process	25
	Supply chain	No supply chain	Sporadic supplies from outside	Fairly consistent but inadequate supplies from outside	Supplies in community and adequate supplies from outside community	Supplies needed within community and easy access to other needed supplies outside community	55
						Subtotal score	37.5

Source: Guatemala project 2011.

In the UVC framework, communities are divided into five level-of-development groups based on the value of the TML, as shown in Table 9-3. As an example, the capacity diagram of Figure 5-6 shows that technical capacity has the lowest capacity factor of 37 (TML = 37; actual value is 37.5, as shown in Table 9-2). In Table 9-3, the community development level is equal to 2. Another example of capacity analysis by Ahmad (2004) is shown in Table 9-4 and is associated with a village in the Philippines where the minimum capacity factors are related to environmental and sociocultural factors. For that case study, a minimum value of 10 related to the environmental and sociocultural capacity (TML = 10) brings the community level of development down to 1, according to Table 9-3.

As we will see below, the TML is an outcome indicator that limits, for a given type of service, the range of technical solutions that could be used to provide that service. If the technology is not appropriate for the community, i.e., it does not match the ability of the community to use the technology, either the technology is inappropriate or the community's ability in its level of development has to be changed. As capacity development takes place over time, the TML and the range of appropriate solutions are expected to increase.

Other categories of capacity and requirement types besides those shown in Figure 9-2 can be introduced into the UVC model. Furthermore, not all capacity types and associated requirements are equally important on a given project and need to be included in the capacity assessment. The need depends greatly on the type of project and its scale and the community context. Finally, the capacity factors do not always have to be expressed in a quantitative manner if it is not possible to quantify the various requirements, which is often the case. Qualitative measures of capacity such as low, medium, and high would also be appropriate, as long as descriptors support the ranking.

Table 9-3. Community Development Levels

Minimum Capacity Factor Score	Level	Explanation
1–20	1	No local capacity to manage the service
21–40	2	Capacity to manage systems for small collections of residential units
41–60	3	Capacity to manage community-based systems
61–80	4	Capacity to serve multiple communities from a centralized system
81–100	5	Capacity to manage a centralized system, along with individual service to more remote units

Source: Adapted from Louis and Bouabib (2004), with permission from University of Virginia.

Table 9-4. Capacity Analysis (Capacity Scores, CS) for a Village in the Philippines

Capacity Factor	Score	Level 1 0–20	Level 2 21–40	Level 3 41–60	Level 4 61–80	Level 5 81–100	Capacity Score
Institutional	Body of legislation				70		70.0
	Assoc regulated			41			41.0
	Admin agencies	20					20.0
	Admin processes	20					20.0
	Governance	20					20.0
	Institutional CS						34.2
Human Resource	Professionals		21				21.0
	Skilled labor		21				21.0
	Unskilled labor					100	100.0
	Illiterate					100	100.0
	Human Resource CS						60.5
Technical	Operations		25				25.0
	Maintenance	5					5.0
	Adaptation and modification		21				21.0
	Supply chain— related services		21				21.0
	Technical CS						18.0
Economic	Private sector %		40				40.0
	Bonds	0					0.0
	User fees		21				21.0
	Budget			41			41.0
	Asset values			41			41.0
	Debt	20					20.0
	Economic CS						27.2
Energy	Primary source				61		61.0
	Backup	20					20.0
	% of budget				61		61.0
	Outage rate	20					20.0
	Energy CS						40.5

Community Assessment (Sample), Bacoor, Philippines

Table 9-4. Capacity Analysis (Capacity Scores, CS) for a Village in the Philippines (*Continued*)

Capacity Factor	Score	Level 1 0–20	Level 2 21–40	Level 3 41–60	Level 4 61–80	Level 5 81–100	Capacity Score
Community Assessment (Sample), Bacoor, Philippines							
Environmental	Quality & sensitivity	10					10.0
	Quantity	10					10.0
	Environmental CS						10.0
Social & Cultural	Communities	10					10.0
	Stability	10					10.0
	Equity	10					10.0
	Castes	10					10.0
	Social & Cultural CS						10.0
Service	Gap		21				21.0
	Service CS						21.0

Source: Ahmad (2004), with permission from University of Virginia.

WFEO Framework

The World Federation of Engineering Organizations (WFEO 2010) introduced six essential so-called pillars of capacity, which according to them "must always be in place if a nation is to have sufficient and stable technical and decision making capacity to meet the prerequisites of sustainability." Unlike the UVC approach, which focuses on the community at the project level, the WFEO approach focuses on what it would take for a country to have a healthy engineering infrastructure development and operation. This approach is particularly important to developing countries that need to create standards and best practices in their path to development. More often than not, such standards do not exist or are rarely enforced. The six pillars of engineering capacity are the following:

- Individual capacity, expressed in terms of technical training, information, and connectivity to the outside world;
- Institutional capacity, in terms of professional organizations, statutory boards, councils, foundations, and research and development;
- Technical capacity, in terms of standards, codes of practice, codes of ethics, technical literature, software, and hardware;
- Decision-making capacity that allows decisions to be made at different levels from individuals to governments;

- Business capacity where businesses are in place to support and contribute to infrastructure development, including retail and wholesale; and
- Resource and supply capacities in terms of access to equipment, materials, resources, raw and manufactured material, and in terms of quality and quantity.

No ratings and indicators specific to each form of capacity have been proposed by the WFEO to assess and rate qualitatively or quantitatively each capacity type listed above.

The UNDP Capacity Assessment Framework

The UNDP (2005, 2007, 2008) uses a different approach to capacity assessment that is more appropriate at the country level (programmatic and planning) rather than at the project level. A special effort has been made to show how the framework coincides with others proposed in the development literature (UNDP 2008). The more global perspective of the UNDP capacity assessment framework complements the UVC approach, which is more project specific.

Using the enabling environment as a point of entry, and once local ownership of the capacity development outcome has been established, capacity is assessed along two dimensions. In the first one, instead of considering specific capacities, the UNDP approach looks at existing core issues representing "areas where capacity change happens most frequently within and across a variety of sectors and themes" at the country level. These issues include (1) institutional arrangements, (2) leadership, (3) knowledge, and (4) accountability. The second dimension in the UNDP approach is to assess the functional capacities that are necessary "for creating and managing policies, legislations, strategies and programmes." They include stakeholders' engagement; situation assessment and vision and mandate definition; formulation of policies and strategies; budgeting, management, and implementation; and evaluation.

9.3 Capacity Development Response

According to UNDP (2009), effective capacity development response starts with addressing three basic questions: "(i) to what end do we need to develop the capacity, what will be the purpose; (ii) whose capacities need to be developed, which group or individual needs to be empowered; and (iii) what kinds of capacities need to be developed to achieve the broader development objectives." Such questions are appropriate at all scales, ranging from the country to the project levels. In the ADIME-E framework, the action plan discussed in Chapter 8, including the logframe analysis, should provide answers to those three questions. As discussed in the following, further analysis may need to be carried out to refine these answers.

Two steps in the capacity development response are outlined. The first step considers whether the proposed action plan outlined in Chapter 8 matches the current stage of community development and the solutions match the enabling environment. The second step explores strategies about what needs to be implemented to improve the enabling environment over time, and correspondingly what new solutions are more appropriate in that new environment. The UVC framework is used to illustrate these two steps, using MSS projects as an example.

Matching the Solutions with the Level of Community Development

Using the UVC framework again, once the stage of community development, measured by the TML, has been determined for a given service, the next step is to check whether the proposed solutions in the action plan discussed in Chapter 8 match the level of community development. For a given type of service, DWS, WSS, or MSW (or any other type of service) a technology requirement level (TRL) is determined. It is an indicator that essentially tells decision makers, out of all the technical solutions available to address a given service, which are likely to fit better with the current stage of community development. Those solutions serve as a starting point in the overall capacity development response. This approach is summarized in Figure 9-3 and explained in detail in the following.

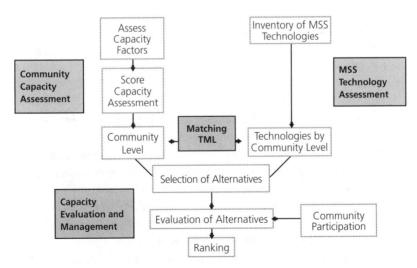

Figure 9-3. Combining Community Capacity Assessment and Technology Assessment

Source: Adapted from Ahmad (2004), with permission from University of Virginia.

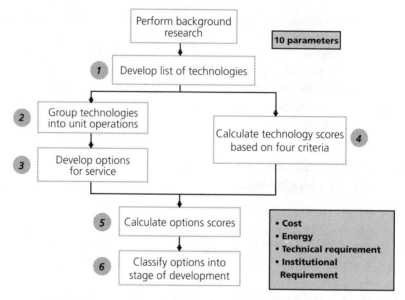

Figure 9-4. Six Steps in Technology Assessment

Source: Adapted from Ahmad (2004), with permission from University of Virginia.

Figure 9-3 shows two converging tracks. The left-hand track is used to determine the enabling environment and results in the TML, as discussed. The right-hand track, which is discussed here, starts with a review of technologies that are available for the service being addressed. A six-step process (Figure 9-4) results in classifying the service options and determining a TRL value. The entire approach is described in more detail in Ahmad (2004).

Step 1: List the Technologies
The first step is to list all of the technologies for the service of interest. In his thesis, Ahmad (2004) gives an inventory of some technologies that should be considered with DWS, WSS, and MSW services. Each technology is described in terms of various characteristics. Examples include components, blueprints, cost, operation and maintenance, energy requirements, technical knowledge requirements, institutional and societal requirements, and environmental considerations. Other characteristics may need to be added, such as performance under different past conditions or advantages and disadvantages.

As discussed in Chapter 15, the WASH literature contains many examples of technical solutions that can be used at the small-project scale. The reader may want to consult a series of booklets on innovative low-cost WASH technical solutions produced by the Netherlands Water Partnership entitled *Smart Water Solutions* (NWP 2006b); *Smart Sanitation Solutions* (NWP 2006a); and *Smart*

Hygiene Solutions (NWP 2010). Other references include Pickford (2001), Jordan (2006); Laugesen and Fryd (2009); and Mihelcic et al. (2009).

Step 2: Define Unit Operations

Technologies are simply tools involved in the process of providing a service. They are important components of the unit operations (or processes) that contribute to the provision of that service:

- In DWS, bringing water to the customer consists of five unit operations: source, procurement, storage, treatment, and distribution.
- In WSS, there are four unit operations: collection, transfer, treatment, and disposal.
- In MSW, there are four unit operations: storage, transfer, disposal, and recovery. As remarked by Ahmad (2004), there may be several disposal and recovery phases.

The term "unit operation" is used to regroup technologies toward the provision of service. Tables 9-5a, 9-5b, and 9-5c show three lines of operations for DWS, WSS, and MSW, respectively.

Step 3: Create Service Options

For any given suite of unit operations in a service, various service options can be created. A service option is defined as "a series of technologies that when used together, lead to the provision of a municipal sanitation service" (Ahmad 2004). Looking at Tables 9.5a–c, the number of service options can be quite large if all technology combinations are possible. They include 4,200 DWS service options, 1,296 WSS service options, and 525 MSW options. Not all options are possible, and the next step is to reduce them to a manageable number.

Step 4: Calculate Technology Score

The various technologies used in the unit operations are rated based on four criteria (Table 9-6): cost, energy required, technical, and institutional factors (Ahmad 2004).

- *Cost* refers to the initial cost and annual operation and maintenance costs.
- *Energy required* relates to the energy requirement of the technology.
- *Technical* is about the technical knowledge required to install, operate, and maintain the technology.
- *Institutional factors* relate to the organizational structure that needs to be in place at the community level for the technology to succeed.

Based on the rating (low, medium, high) for each criterion, a certain number of points are assigned to each technology: 1 point for low, 5 points for medium, and 10 points for high. Finally, for each technology, a normalized score (score/$[4 \times 10]$) is determined, as shown in Table 9-7 for the DWS technologies.

Table 9-5. Unit Operations for (a) DWS, (b) WSS, and (c) MSW

(a)

Source	Procurement	Storage	Treatment	Distribution
Rooftop water harvesting	Bucket	None	None	None
Ground level catchment	Handpump	Barrel	Chlorination	Household water connections
Subsurface dam	Handpump—Deep well	Tank	Slow sand filter	Piped water (gravity)
Surface water abstraction	Rope and bucket w/ windlass	Reservoir	Boiling	Piped water (pumped)
Spring water captation	Motorized pump	Cistern	Ultraviolet light	
Hand dug well	Standpump			
Drilled well				

(b)

Collection	Transfer	Treatment	Disposal
None	None	None	Burial
Bucket	Small bore/settled sewerage	Constructed wetlands	Composting
Vault/Cartage	Conventional sewerage	Soil aquifer treatment	Pit privy
Septic/Tank	Drainage field	Oxidation ditch	Ventilated improved pit latrine
		Rotating biological contractor	Double vault compost latrine
		Trickling filters	Aqua privy
		Upflow anaerobic sludge blanket	Pour flush toilet
		Activated sludge process	Cistern flush toilet
		Stabilization ponds	Drainage field

(c)

Storage	Transfer	Disposal	Recovery
None	None	None	None
Household bin	Human power	Waste discarded at source	Composting
Communal bin	Animal power	Open burning	Refuse-derived fuel
	Noncompactor trucks	Open dumps	Pyrolysis
	Compactor trucks	Controlled dumps	Recycling/reuse
		Sanitary landfilling	
		Incineration	

Source: Ahmad (2004) with permission from University of Virginia.

Table 9-6. Classification of Service Options

Criteria	Level 1 (Low)	Level 2 (Medium)	Level 3 (High)
Cost	Low cost	Moderate cost	High cost
Energy Required	No or minimal energy	Medium energy	High energy
Technical	Low level of technical knowledge	Medium level of technical knowledge	High level of technical knowledge
Institutional	No formal organization needed. Low level of organization	Moderate level of organization	High level of organization

Level	Points
Low	1
Medium	5
High	10

Source: Ahmad (2004), with permission from University of Virginia.

Step 5: Calculate Option Scores

For each of the 4,200 DWS service options, 1,296 WSS service options, and 525 MSW service options, an option score is calculated as follows:

$$\text{Option Score} = \frac{\sum x_i + w[x_1 \, x_2 \, ... \, x_N]}{N + w}$$

In this equation, x_i is the score based on the technology for each unit operation $i = 1, N$ where N is the number of unit operations for the service option. In our case and according to Table 9-5, $N = 5$ (DWS score), 4 (WSS score), and 4 (MSW score). In the equation, w is a reward factor weight that is larger than 0 when all unit operations are present. If one of them is missing, the second term in the numerator is automatically equal to zero, thus creating a built-in handicap and lower score.

Step 6: Calculate the Technology Requirement Level

The option score is then converted into a technology requirement level (TRL) value, as shown in Table 9-8. Only service options that have a TRL less than or equal to the technology management level (TML) of the community are retained as viable options. In other words, according to this methodology, only alternative service options that match the level of development of the community can be selected. This process reduces the number of potential alternative service options considerably.

Table 9-7. Examples of Technology Scores for each DWS Technology

Unit Process	Technologies	Cost	Energy	Technical	Institutional	Score	NScore
Drinking Water Supply							
Source	Rooftop water harvesting	Medium	Low	Medium	Low	12	0.30
	Ground level catchment system	High	Low	Medium	Low	17	0.43
	Subsurface dam	Medium	Low	Medium	Medium	16	0.40
	Surface water abstraction	Low	Low	Low	Low	4	0.10
	Spring water captation	Low	Low	Medium	High	17	0.43
	Hand-dug well	Medium	Low	Medium	Low	12	0.30
	Drilled well	Medium	High	High	Medium	30	0.75
Procurement	Bucket	Low	Low	Low	Low	4	0.10
	Handpump	Low	Low	Medium	Low	8	0.20
	Handpump—Deep well	Low	Low	High	Low	13	0.33
	Rope and bucket w/windlass	Low	Low	Low	Low	4	0.10
	Motorized pump	Medium	High	High	Low	26	0.65
	Standpost	High	Low	Medium	High	26	0.65
Storage	Barrel	Low	Low	Low	Low	4	0.10
	Tank	Medium	Low	Low	Low	8	0.20
	Reservoir	High	Low	Medium	Medium	21	0.53
	Cistern	Low	Low	Low	Low	4	0.10
Treatment	Chlorination	Low	Low	Medium	Medium	12	0.30
	Slow sand filter	Low	Low	Medium	Low	8	0.20
	Boiling	Low	Medium	Low	Low	8	0.20
	Ultraviolet light	Medium	High	High	High	35	0.88
Distribution	Household water connections	High	Low	High	High	31	0.78
	Piped water (gravity)	Medium	Low	High	High	26	0.65
	Piped water (pumped)	High	Medium	High	High	35	0.88

Table 9-8. Example of Service (DWS) Option Scoring Process Leading to Stage of Development

No	Source	A	Procurement	B	Storage	C	Treatment	D	Distribution	E	Score	Stage of Dvlpt
1	Rooftop water harvesting	0.3	Bucket	0.1	None	0	None	0	None	0	0.078	1
2	Rooftop water harvesting	0.3	Bucket	0.1	None	0	None	0	Household water connections	0.775	0.230	2
3	Rooftop water harvesting	0.3	Bucket	0.1	None	0	None	0	Piped water (gravity)	0.65	0.206	2
4	Rooftop water harvesting	0.3	Bucket	0.1	None	0	None	0	Piped water (pumped)	0.875	0.250	2
5	Rooftop water harvesting	0.3	Bucket	0.1	None	0	Chlorination	0.3	None	0	0.137	1
6	Rooftop water harvesting	0.3	Bucket	0.1	None	0	Chlorination	0.3	Household water connections	0.775	0.289	2
7	Rooftop water harvesting	0.3	Bucket	0.1	None	0	Chlorination	0.3	Piped water (gravity)	0.65	0.265	2
8	Rooftop water harvesting	0.3	Bucket	0.1	None	0	Chlorination	0.3	Piped water (pumped)	0.875	0.309	2
9	Rooftop water harvesting	0.3	Bucket	0.1	None	0	Slow Sand Filter	0.2	None	0	0.118	1
10	Rooftop water harvesting	0.3	Bucket	0.1	None	0	Slow Sand Filter	0.2	Household water connections	0.775	0.270	2
11	Rooftop water harvesting	0.3	Bucket	0.1	None	0	Slow Sand Filter	0.2	Piped water (gravity)	0.65	0.245	2
12	Rooftop water harvesting	0.3	Bucket	0.1	None	0	Slow Sand Filter	0.2	Piped water (pumped)	0.875	0.289	2

Table 9-8. Example of Service (DWS) Option Scoring Process Leading to Stage of Development (*Continued*)

No	Source	A	Procurement	B	Storage	C	Treatment	D	Distribution	E	Score	Stage of Dvlpt
13	Rooftop water harvesting	0.3	Bucket	0.1	None	0	Boiling	0.2	None	0	0.118	1
14	Rooftop water harvesting	0.3	Bucket	0.1	None	0	Boiling	0.2	Household water connections	0.775	0.270	2
15	Rooftop water harvesting	0.3	Bucket	0.1	None	0	Boiling	0.2	Piped water (gravity)	0.65	0.245	2
16	Rooftop water harvesting	0.3	Bucket	0.1	None	0	Boiling	0.2	Piped water (pumped)	0.875	0.289	2
17	Rooftop water harvesting	0.3	Bucket	0.1	None	0	Ultraviolet light	0.875	None	0	0.250	2
18	Rooftop water harvesting	0.3	Bucket	0.1	None	0	Ultraviolet light	0.875	Household water connections	0.775	0.402	3
19	Rooftop water harvesting	0.3	Bucket	0.1	None	0	Ultraviolet light	0.875	Piped water (gravity)	0.65	0.377	2
20	Rooftop water harvesting	0.3	Bucket	0.1	None	0	Ultraviolet light	0.875	Piped water (pumped)	0.875	0.422	3

Option Score	Stage of Development
0.0–0.2	1
0.21–0.4	2
0.41–0.6	3
0.61–0.8	4
0.81–1.0	5

Source: Ahmad (2004), with permission from University of Virginia.

Step 7: Evaluation of Alternatives

Using the aforementioned methodology, the focused strategy solutions, and the comprehensive action plan discussed in Chapter 8 for the problem at stake can be refined and improved. There is now a better understanding of the appropriateness of the solutions to the level of community development. The solutions in the final project action plan are obtained from those outlined in Step 6, combined with additional input through community participation. If necessary, a multicriteria utility matrix similar to the one discussed in Chapter 7 can be used. The criteria in the matrix may be the same as those discussed in Chapter 7, or new ones may be introduced, including new weighting factors.

Finally, a second filtering process consists of looking at the technical feasibility of the selected solutions. This step is usually done by technical personnel with expertise in the service area(s) being addressed.

Remarks

In the approach proposed by UVC, the TML defines the level of community development for a given service and is controlled by the weakest capacity category, which is a strong constraining factor. The approach also assumes that once the weakest link is resolved, the next weakest link becomes the constraining factor. In reality, this is rarely the case and a combination of capacity in different categories may contribute to the current enabling or constraining environment and may have to be addressed simultaneously. The problem is that the combination is not always well defined. This problem may require significant experience on the team and sometimes several rounds of trial and error.

Capacity Development Response Strategies

Because the enabling environment was measured using the TML, a community is limited as to the number of service options it can realistically handle in its current state of capacity. This limitation is indeed a common characteristic of small communities in developing countries because most of them are likely to rate at a development level of 1 or 2. An overall goal of capacity building and development is therefore to increase the enabling environment so that more effective solutions can be implemented over time. In the aforementioned example of the MSS project approach, stronger solutions are likely to yield better community health. According to Bouabib (2004), starting with a development level of 1 or 2, a community should seek to reach an MSS service level of 3 over a period of 2–5 years and a level of 4 or 5 over a period of 10–20 years, which is a realistic time frame in sustainable community development.

In general, the process of increasing the enabling environment of a community through capacity building takes time. As suggested by the UNDP (2009), indicators that monitor progress toward a clear, desirable outcome need to be included with verifiable means. The acronym SMART, discussed in Chapter 8, applies to these indicators as well. In developing countries, the longer the process of capacity development, the more likely it is to create challenges when dealing with the community stakeholders and external donors.

The tradeoff in small-scale community projects is likely to be between ensuring short-term and long-term solutions and balancing between quick project successes (with smaller returns) or long-term successes (with larger returns). In doing so, various strategies of capacity development may be followed (Morgan 1998). Within the context of small-scale projects in developing communities, they may consist of

• Eliminating capacity components that are more restraining than enabling,
• Making better use of and improving existing capacity,
• Building or strengthening existing capacity by adding resources, and/or
• Enabling the creation of new forms of capacity and their use through experimentation and learning.

The degree of success of the various strategies of capacity development depends on many factors and involves some components of risk. Among other things, any strategy must be owned by local stakeholders in the community who are directly involved in capacity development and are committed to its success. As remarked by Morgan (1998), no strategy works if it is imposed on "skeptical participants."

The success of capacity development strategies also depends on *what* components (tangible vs. intangible) in each category of capacity are being addressed and *how* they are being addressed. No magic formula or quick fix exists that guarantees success in capacity development. More specifically, the tangible components of capacity (e.g., infrastructure, natural resources, health, or institutions) are easier to influence, especially from the outside, which makes the results of capacity development more predictable. This notion applies to the technical MSS solutions discussed earlier in this chapter.

However, the less tangible components, such as behavior change, values, or motivation, cannot be influenced by outsiders, who can at best serve as facilitators of resourcefulness, e.g., provide resources and advise on process rather than deliver the expected outcomes (Lavergne and Saxby 2001). Such components have the potential to derail a project. As a result, special measures such as monitoring and evaluation need to be taken to prevent the creation of undesirable results that may negatively affect community livelihood.

9.4 Chapter Summary

In summary, capacity analysis helps in refining and improving the solutions and project action plans discussed in Chapter 8. More specifically, it addresses the ability of a community to handle the action plan in terms of skills, knowledge, capabilities, and resources. Capacity analysis also assesses whether the community is able to overcome any constraining environment and move forward in its development.

At the end of the capacity analysis, there is a better understanding of the community's enabling environment, what it can do, what it cannot do, and how it needs to be strengthened. In general, the results of capacity analysis can be expressed in quantitative (capacity factors) or qualitative terms (high, medium, or low). Finally, it is important to remember that the capacity development program must be created through a participatory and locally generated process.

Because of the inherent nature of developing communities, it is likely that the capacity baseline is low from the beginning. It is also likely that the rate of capacity development moves slowly as well. However, there are ways to fast-track the development if a special effort is made by outsiders, in participation with the communities, to identify using appreciative inquiry with the positive deviant individuals and groups in the community; i.e., those change makers who are successful at addressing problems because of their uncommon habits, behaviors, and attitudes (Pascale et al. 2010). The challenge then becomes understanding the reasons for success of the positive deviants and how to encourage others to adopt their behaviors and attitudes through behavior change and thinking differently.

References

Ahmad, T. T. (2004). "A classification tool for selecting sanitation service options in lower income communities." M.S. dissertation, University of Virginia, Charlottesville, VA.

Bolger, J. (2000). "Capacity development: Why, what and how." Capacity development occasional papers 1(1). Canadian International Development Agency (CIDA) policy branch, CIDA, Guatineau, Quebec.

Bouabib, A. (2004). "Requirements analysis for sustainable sanitation systems in low-income communities." M.S. dissertation, University of Virginia, Charlottesville, VA.

Faeh, C., et al. (2004). "Capacity-building for sustainable service delivery in lower-income communities." Proc. Systems and Information Systems Design Symp., M. H. Jones, S. D. Patek, and B. E. Tawney, eds., Charlottesville, VA, 55–63, IEEE, Washington, DC.

Gesellschaft für Technische Zusammenarbeit (GTZ). (2007). Understanding of capacity development: A guiding framework for corporate action, Deutsche Gesellschaft für Technische Zusammenarbeit GmbH, Bonn, Germany. <http://www.camlefa.org/documents/GTZ_understanding_of_CD_2006_Eng.pdf> (Feb. 5, 2012).

Jordan, T. D., Jr. (2006). A handbook of gravity water systems, Intermediate Technology Development Publications Ltd., London.

Laugesen, C., and Fryd, O. (2009). Sustainable wastewater management in developing countries: New paradigms and case studies from the field, ASCE Press, Reston, VA.

Lavergne, R., and Saxby, J. (2001). Capacity development: Vision and implications, Canadian International Development Agency (CIDA) policy branch. CIDA, Guatineau, Quebec. <http://www.seesac.org/sasp2/english/publications/7/capacity/3.pdf> (May 5, 2012).

Louis, G., and Bouabib, M. A. (2004). "Community assessment model for sustainable municipal sanitation services in low income communities." Presented at the 2004 International Council on Systems Engineering, Mid-Atlantic Regional Conference, Arlington, VA. ICSE, San Diego, CA.

Mihelcic, J. R., et al. (2009). *Field guide to environmental engineering for development workers: Water, sanitation, and indoor air*, ASCE Press, Reston, VA.

Milèn, A. (2001). "What do we do about capacity building? An overview of existing knowledge and good practice." World Health Organization, Geneva. <http://whqlibdoc.who.int/hq/2001/a76996.pdf> (Feb. 10, 2012).

Morgan, P. (1998). *Capacity and capacity development—Some strategies*, Canadian International Development Agency (CIDA), Policy Branch. CIDA, Guatineau, Quebec <http://portals .wi.wur.nl/files/docs/SPICAD/14.%20Capacity%20and%20capacity%20development_ some%20strategies%20(SIDA).pdf> (Aug. 1, 2012).

Netherlands Water Partnership (NWP). (2006a). *Smart sanitation solutions*, Netherlands Water Partnership, The Hague. <http://www.irc.nl/page/28448> (Sept. 10, 2012).

Netherlands Water Partnership (NWP). (2006b). *Smart water solutions*, Netherlands Water Partnership, The Hague. <http://www.irc.nl/page/28654> (Sept. 10, 2012).

Netherlands Water Partnership (NWP). (2010). *Smart hygiene solutions*, Netherlands Water Partnership, The Hague. <http://www.irc.nl/page/55200> (Sept. 10, 2012).

Organization for Economic Co-operation and Development (OECD). (2006). *The challenge of capacity development: Working towards good practice*, OECD Publishing, Paris. <http:// www.oecd.org/development/governanceanddevelopment/36326495.pdf> (Sept. 14, 2012).

Pascale, R., Sternin, J., and Sternin, M. (2010). *The power of positive deviance: How unlikely innovators solve the world's toughest problems*, Harvard University Review Press, Cambridge, MA.

Pickford, J. (2001). *Low-cost sanitation*, ITDG Publishing, London.

Sphere Project. (2011). *Humanitarian charter and minimum standards in humanitarian response*, Practical Action Publishing, Rugby, U.K.

Stockholm International Water Institute (SIWI). (2013). <http://www.siwi.org/>.

Tearfund. (2011). *Reducing risks of disasters in our community*, Tearfund, Teddington, U.K. <http://tilz.tearfund.org/Publications/ROOTS/Reducing+risk+of+disaster+in+our+ communities.htm> (Oct. 22, 2012).

United Nations Development Programme (UNDP). (2005). *A brief review of 20 tools to assess capacity*, United Nations Development Programme, New York. <http://www.unpei.org/PDF/ institutioncapacity/Brief-Review-20-Tools-to-Assess.pdf> (Dec. 1, 2012).

United Nations Development Programme (UNDP). (2007). *Capacity assessment methodology: A user's guide*, United Nations Development Programme, New York. <http://europeandcis .undp.org/uploads/public/File/Capacity_Development_Regional_Training/UNDP_ Capacity_Assessment_Users_Guide_MAY_2007.pdf> (Dec. 1, 2012).

United Nations Development Programme (UNDP). (2008). "Capacity development practice note." United Nations Development Programme, New York. <http://www.undp.org/content/dam/ aplaws/publication/en/publications/capacity-development/capacity-development-practice- note/PN_Capacity_Development.pdf> (Dec. 5, 2012).

United Nations Development Progamme (UNDP). (2009). *Capacity development: A UNDP primer*, United Nations Development Programme, New York. <http://www.undp.org/ content/dam/aplaws/publication/en/publications/environment-energy/www-ee-library/ climate-change/capacity-development-a-undp-primer/CDG_A%20UNDP%20Primer.pdf> (March 24, 2012).

World Federation of Engineering Organizations (WFEO). (2010). *Guidebook for capacity building in the engineering environment*, WFEO, Paris.

10

Risk Analysis and Management

10.1 Capacity, Vulnerability, and Risk

The previous chapter looked at how capacity building or capacity development can help communities reach a higher level of development and be able to face the consequences of adverse events through a participatory, locally generated process. This chapter looks at community vulnerability and its risk environment.

To become more resilient to adverse events, communities and their households must not only identify and increase their capacity, they must also identify and reduce their vulnerabilities. This dynamic was mentioned in Chapter 1 in relation to the household livelihood *crunch model* illustrated in Figure 1-2. It needs to be replaced by a *release model* shown in Figure 10-1, which is an adapted version of the disaster release model proposed by the Tearfund (2011). The directions of the arrows in Figure 1-2 are reversed to indicate the release process. This reversal requires deep changes in the social structures and their activities (processes) and drivers (politics, economics, culture, and environment) that affect community households. The release model also requires having appropriate measures in place to reduce poverty and mitigate the effect of hazards and adverse events.

As discussed in Section 5.8, the different forms of community vulnerability are mapped in the appraisal phase of the ADIME-E framework. In general, vulnerability appraisal is about identifying and appraising community weaknesses and security issues at the individual, household, and community levels.

As seen in Chapter 1, the relative difference between vulnerability and capacity defines the level of risk that a community is likely to face for a given adverse event or hazard and a given exposure to that event. Adverse events faced by households in poor communities are multiple and constantly changing. They can be internal or external to the community, small or large. They range between everyday events (e.g., lack of water and sanitation, poor shelter, living conditions, livelihood, illness, and the economy) and extreme events (e.g., floods, volcanoes, earthquakes, landslides, wildfires, and hurricanes). Several small-scale

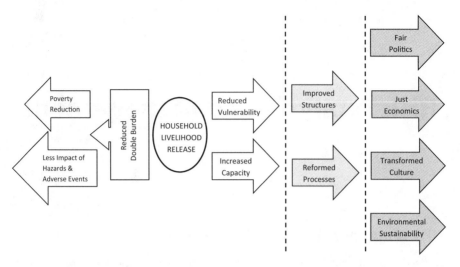

Figure 10-1. Household Livelihood Release Model

Source: Adapted from Tearfund (2011), with permission from Tearfund.

or periodic medium-scale events may arise as well, such as drought (periodic or chronic), soil degradation, deforestation, epidemics, health risks, and other hazards. Another class of adverse events deals with those associated with war or the breakdown of governments that may have disastrous consequences at the local and global levels.

In all cases, big or small, adverse events have the potential to affect the life of the community, its health, its economic well-being, its social and cultural assets, and its infrastructure. These events add to the existing burden of poverty and create a double burden for those who are disadvantaged, as shown in Figure 1-2. The consequences (impacts) of adverse events can be immediate or medium term, or in some cases affect the well-being of a community over a long period of time. In some cases, the events may create a triggering mechanism (cascading effect) for others. Determining risks associated with adverse events helps define the *risk environment* of the community, i.e., a baseline on which risk analysis can be carried out. In general, by increasing capacity and reducing vulnerability, communities are less at risk to face crises that could erode their level of development, however small they may be.

The risk environment in development projects is twofold. The first environment is defined by hazard and adverse events that exist before the project. The second environment is induced by the project itself. The former is usually mapped in the project appraisal phase. For instance, Table 5-11 shows the natural risks that exist for a community in Guatemala. Quite often, the former risk environment affects the latter to a great extent. Furthermore, there is a possibility that the project may accentuate the existing risk environment and/or create new risks in a cascading manner. As discussed at the end of this chapter, this double causality can be addressed by conducting an impact assessment pertaining to the environment, health, and social issues.

Understanding risk in development projects and including risk management in the ADIME-E framework contributes to the building of more resilient communities. This chapter is about risks and their impacts, how to integrate risk analysis (assessment, evaluation, and management) in the ADIME-E framework, and how to mitigate their effects. Like capacity analysis, risk analysis helps in refining and improving the project action plan discussed in Chapter 8 by asking the simple question: What could go wrong in the execution of the proposed action plan?

10.2 About Risks

In general, risk is the possibility that an undesired outcome (or the absence of a desired outcome) associated with an event has "adverse effects on lives, livelihoods, health, economic, social and cultural assets, services (including environmental), and infrastructure" (NRC 2012). According to Smith and Merritt (2002), at the project level, risks can be seen as "unanticipated surprises" that could jeopardize the success of a project or parts of it. Risks can stop a project at its inception, delay it, and lead to failure if not properly accounted for. Finally, risks can be real or perceived (Slovic 1987).

As discussed in Section 4.8, risk differs from uncertainty in the sense that with risks there is an idea of the odds that something will happen and a probability can be assigned for its occurrence (Knight 1921). Risks are sometimes outside the direct control of outsiders and community stakeholders. Some risks are simply uncontrollable. Sometimes, they are fully controllable and originate from negligence and cutting corners in project planning or management. Being able to understand the risk environment of a project in the early part of project management, as early as the preappraisal and feasibility phase, and monitor how it changes over time is not an option in community projects; it is an obligation.

Effective risk management requires an understanding of the risk environment, which requires (1) identifying risks and the probability of their occurrence; (2) understanding the consequences (e.g., impacts) if the risks happen and how they will happen; (3) determining the factors that drive the risks; and (4) monitoring how risks evolve over time (Smith and Merritt 2002).

Risk events cannot be predicted with certainty. If they were 100% certain, then they would become issues to be addressed explicitly. According to Smith and Merritt (2002), a risk event possesses drivers and probabilities. This dynamic is shown in Figure 10-2 as part of a standard risk model. In general, a risk event driver is a fact in the project environment that is likely to happen and create a

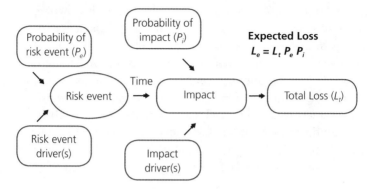

Figure 10-2. Standard Risk Model

Source: Smith and Merritt (2002), with permission from Productivity Press.

risk. The likelihood that a risk event will take place is defined by its probability (P_e). When a risk event occurs, it has an impact, which itself possesses drivers and a probability (P_i). Combining risk and impact and having an estimate of total loss (L_t), potential scenario outcomes can be explored and the expected loss (L_e) associated with various types of events can be estimated. An example of a framework to determine losses associated with various types of natural hazards was proposed in the United States by FEMA (2001). The comprehensive framework helps map hazards that may affect a community, assess the past impact of various hazard events, and calculate corresponding losses.

In the medium-high risk and complex environment of small-scale community development projects, risk is a given and can take multiple forms. Risks exist in all phases of the ADIME-E framework because of the prevailing uncertainty and complexity of the project environment. They can be internal or external to a project. For instance, in the appraisal phase of the ADIME-E framework, a risk exists that some stakeholders may create roadblocks to the execution of a project. A risk exists that the collected data are inaccurate, incomplete, poorly analyzed, or strongly biased. A risk exists that the data analysis could lead to an incomplete project hypothesis. A risk exists that the project may fail with unintended consequences, resulting in loss of life and/or resources, right after the project is completed or during the project life cycle. A risk exists that in the logical framework and project planning phase, assumptions and preconditions necessary to meet goals and objectives are not (or are partially) met and thus lead to negative results, project delays, or cost overruns, as discussed in Section 8.2. Finally, there is a risk that the project is no longer what the community needs, or, in some cases, was never needed in the first place.

Given the more subjective (intuitive) nature of decisions made by project decision makers and their adaptive approach, risks in small-scale development projects are likely to be high and have multiple drivers. The same could be said about the impact associated with those risks. Clearly, project managers undertaking projects in developing communities need to be risk averse.

10.3 Risk Management in Sustainable Community Development

Because risks are an integral part of projects in developing communities, they need to be analyzed (risk analysis) and managed (risk management). In general, risk management is a process geared toward understanding and reducing risk (once they have been identified and ranked) and recovering from risk events (PMI 2008). Whereas risk analysis is done at the planning stage of a project or at key points along the project, risk management is more of a continuous process (DFID 2002).

Within the context of community development, risk management contributes to protecting and preserving household livelihood security, well-being, and quality of life. An added value of risk management is that it helps communities become more resilient over time and creates better projects overall. As discussed by Smith and Merritt (2002), risk management consists of several steps: (1) risk identification, (2) risk analysis and prioritization (map), (3) development of risk management strategies, (4) implementation of risk strategies, and (5) monitoring and evaluation of strategies. These steps, shown in Figure 10-3, apply when assessing the existing risk environment of communities or the risks involved in planning and executing projects in these communities.

Risk Identification

One way to identify risks at different phases of the ADIME-E framework is to ask the following question: "What could go wrong at each point of the project that would prevent it from being successful?" That question should first be raised during the community appraisal phase with the various community stakeholders participating (Chapter 5). The same question arises again at other stages of the project cycle, such as in the causal analysis, logframe analysis, work plan and capacity development, and the implementation and execution phase. Even after the project is done, risk still exists in regard to guaranteeing the long-term project benefits. The mapping of risks and their respective drivers during the life cycle of a project is critical. Ignoring or choosing to overlook risks may lead to unexpected consequences (Greene 2007).

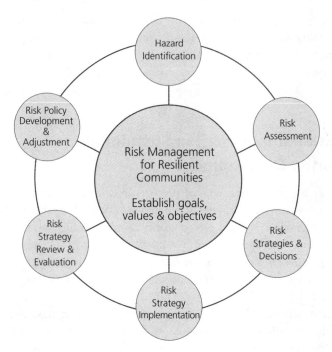

Figure 10-3. Risk Management as a Foundation for Building Resilient Communities

Source: NRC (2012), with permission from National Academies Press.

According to DFID (2002), risks originate from three sources in community development projects:
- From factors that are under the control of those involved in the project, such as poor planning, design, management, and/or execution;
- From decisions made by others, such as policy makers and institutions not necessarily directly involved in the project; and
- From uncontrollable factors, such as those associated with natural hazards or socioeconomic or political issues.

Risks can be further divided into risks that cause things to happen (source) and risks that are the consequences (effects) of other events.

Smith and Merritt (2002) proposed a strategy to conduct the inventory of project risks. It consists of establishing five basic parameters about the project itself, i.e., the scope of the project, its benefits, its purpose, the project location, and its time frame. Other questions related to the process in conducting the project include the expected outcome of the project, who will be developing the project components, where and when these components will be developed and delivered, and the decision-making process. In answering each question, potential risks are identified (type, description) along with their respective likelihood, expected level and type of impact, and a mitigation plan (methods and partners).

Table 10-1 shows an example of risk registry for a water security project in Guatemala. Table 10-2 is another example of risk identification for a project in Peru, where potential risks have been listed for major project phases along with the category of impact for each risk.

When identifying the risk environment of a community, the community-owned vulnerability and capacity assessment (COVACA) tool developed for World Vision projects (Greene 2007) can serve as an approach to identifying risk events associated with various forms of adverse events faced by a community. The approach is carried out with full participation from community members and uses many of the same participatory research tools discussed in Chapter 5. It recognizes that "communities have a vast ability to manage their own vulnerability reduction at the grassroots level" (Greene 2007). More specifically, a community should be able to (1) assess its vulnerabilities and capacities based on past and existing threats and its response to past threats, (2) identify activities that it can implement with its own capacity and coping mechanism, and (3) feel empowered in making decisions about its protection. The Tearfund (2011) follows a similar participatory approach of engagement in community risk reduction in its participatory assessment of disaster risk process by considering the enabling (capacity) and constraining (vulnerability) attributes of five types of community assets (individual, social, natural, physical, and economic) that may be affected by natural and nonnatural hazards.

Whether assessing the risk environment of a community or the risks involved in planning and executing projects in that community, at the end of the risk identification stage, project teams must have a clear indication of the following:

- *Risks* (what could go wrong?), the risk drivers (what factors make each risk real), and their likelihood (probability estimated based on risk drivers);
- *Impacts* (e.g., costs, work days, calendar days, staff months, and space), the impact drivers (what makes each impact real), and their likelihood (probability estimated based on impact drivers); and
- *Total possible loss* expressed in quantitative ways (e.g., time, money) or qualitative ways (e.g., high, medium, low, or critical).

Risk Analysis and Prioritization

Once risks, their impacts, and total possible losses have been identified and the risk environment has been acknowledged, risks must be evaluated and ranked.

Table 10-1. Inventory of Project Risk for a Water Project in Guatemala

Risk Title	Description	Type	Impact	Likelihood	Description of Impact	Mitigation Methods	Mitigation Partners
Lack of maintenance	Storage units break and are not repaired	Technical	Very high	Moderate	Short project life, waste of resources	Training, materials, contacts for technical information	Local engineers
Lack of use	Community chooses not to use the new techniques	Cultural	Very high	Low	Short project life, waste of resources	Surveys, educational sessions	Local government and community leaders
One season planting	Community continues to only plant during one season	Cultural or economic	High	Moderate	Low productivity	Education	Local farmers
Poor soil conditions	Soil runs out of nutrients because of overharvesting	Environmental	Moderate	Moderate	Lower productivity	Resources and education for farmers	Farming experts, local farmers
Landslides	Excessive flooding because of deforestation and heavy rains	Environmental	High	Moderate	Current season crops destroyed	Decrease deforestation, rainwater control	Civil and environmental engineers

Risk	Description	Category			Consequence	Mitigation	Responsible
Mosquitoes	Increase in mosquitoes because of standing water	Health	Low	Moderate	Increase in diseases	Covers for water tanks, immunizations	Ministry of Health
Inflation	Costs exceed budget because of local inflation	Economic	Moderate	Low	Higher cost/possible delay in project	Contingency	Local farmers
Permitting issues	Political conflict with installation of new infrastructure	Political	High	Very low	Delay or cancellation of project	Preplanning and approvals from government	Local and regional governments
Construction injuries	Builders or owners lack ability to construct or maintain safely	Technical or health	Moderate	Moderate	Delay in projects, injuries, and deaths	Education and training on construction and maintenance	Local and external engineers and construction workers
Change in work structure	Labor market changes because of increased farming during dry season	Socioeconomic	High	High	Lack of labor, economic disparities	Targeting appropriate farms, slow integration with monitoring and evaluation	Agricultural committee

Source: McAfee et al. (2011), reproduced with permission.

Table 10-2. Risk Identification for a Water Project in Iquitos, Peru

Detailed Risk Description	Project Phases	Category of Impact
Severe weather could flood the community and delay construction	Construction	Schedule
Cultural conflict could decrease community participation and hinder progress	Construction	Schedule
Cost of diesel could increase and affect transportation cost for materials	Procurement	Cost
Conflict with donor causes loss of funding for project	Procurement	Schedule
Conflict with NGOs could lead to loss of resources and delay the project	Procurement	Schedule
Political turmoil in region could cause withdrawal of aid organization	Construction	Schedule
The community rejects behavior changes, causing a cessation of hand washing acceptance	Startup or commissioning	Schedule

Source: Bottenberg et al. (2012), reproduced with permission.

Real and perceived risks can be plotted using an x–y diagram as suggested by Slovic (1987).

- The x-axis is a measure of risk's *impact* and how dreadful it could be. It may vary between "controllable, not dreadful, having no global catastrophic, no fatal consequences, low risk, and easily reduced" to "uncontrollable, dreadful, catastrophic, fatal, high risk, and not easily reduced."
- The y-axis is a measure of risk's *probability* (how unknown to the public the risk might be). It may vary between "observable, known to those exposed, immediate effect, old risks, known to science" to "not observable, unknown to those exposed, effect delayed, new risk, unknown to science."

The x–y diagram can be presented in the form of a *risk heat map*, also known as a *risk assessment* matrix or *risk prioritization* matrix. Various existing risk management software tools use such representation. Table 10-3 shows an example of a heat risk map for the Peru project whose risks are listed in Table 10-2. In that table, the gray-coded grid is based on two parameters: (1) one related to the level of risk impact (Table 10-4a), which can be very low (VL), low

Table 10-3. Risk Heat Map for Project in Iquitos, Peru

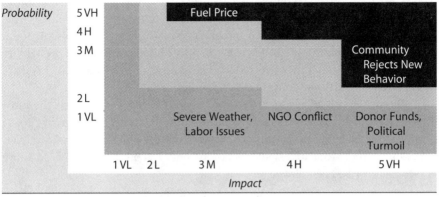

Source: Bottenberg et al. (2012), reproduced with permission.

Table 10-4. Categories of (a) Impact of Risk, (b) Probability of Risk, and (c) Level of Risk

(a) Category of Impact

	1 VL	2 L	3 M	4 H	5 VH
Schedule	Impact recoverable without affecting critical path (0%)	Impact recoverable affecting critical path (<5% delay)	Critical path affected (5–10% delay)	Restructuring of project required (10–20% delay)	Major restructuring of project required (>20% delay)
Cost	Insignificant cost increase (0%)	<5% Cost increase	5–10% Cost increase	10–20% Cost increase	>20% Cost increase

(b) Risk Probability Ranking

Ranking	Probability
5	71–100%
4	51–70%
3	31–50%
2	11–30%
1	1–10%

(c) Score

1–4	Low
5–12	Med
13–25	High

Source: Bottenberg et al. (2012), reproduced with permission.

(L), medium (M), high (H), and very high (VH); and (2) another parameter related to the probability of risk occurrence (Table 10-4b), which also varies between VL and VH.

In Table 10-3, the white, gray, and black colors correspond to low, medium, and high risks (Table 10-4c), respectively. Risks with high impact and probability of occurrence are sometimes referred to as killer risks or intolerable risks and are more likely to contribute to unsuccessful projects.

Of course, not all risks can possibly be addressed. To determine the most critical risks, priority criteria need to be introduced. Smith and Merritt (2002) introduce for each risk the concept of expected loss L_e, which is determined from the estimate of the total loss L_t, the risk probability P_e, and the impact probability P_i (see Figure 10-2) as follows:

$$L_e = L_t \times P_e \times P_i$$

For a given level of expected loss, the product $P_e \times P_i$ (e.g., the risk likelihood) is then plotted against the total loss for each identified risk (Figure 10-4). This operation creates a risk map. All risks that fall above the threshold line

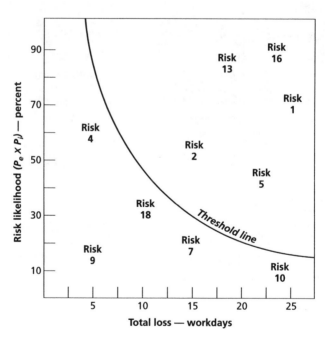

Figure 10-4. Variation of Risk Likelihood versus Total Loss for Several Risks

Note: Active risks fall above the threshold line.

Source: Smith and Merritt (2002), with permission from Productivity Press.

(defined for a given level of expected loss) need attention and are defined as active, whereas those below the line have a lesser priority and are defined as inactive. Again, the decision to address a risk is made based on an accepted level of expected loss. The higher the expected loss, the less likely the risk will fall above the threshold line. The inactive risks are not abandoned in the analysis; they are continually monitored in the event that they could become active with time.

To analyze and prioritize community risks, the community-owned vulnerability and capacity assessment (COVACA) methodology (Greene 2007) starts with creating awareness among the participants of key community needs (e.g., food, energy, shelter, and health), how those needs have been affected when various hazards and disasters have occurred in the past, how the community responded to those past events, and what the current key threats in the community are. Once those key threats are named, the next stage is to identify their causes and impact, existing community coping mechanisms, how to identify early warning signs, and how to develop and implement a mitigation plan at the grassroots level. The detailed steps recommended in this methodology can be found in the second part of Greene's report, entitled "COVACA Instruction Manual."

Risk Management Strategies

Various strategies can be developed to reduce or eliminate the impact of risks and change the risk environment. The goal is not to eliminate the risks but rather to reduce or eliminate their drivers. After all, when drivers do not exist, risks are not created. For a given risk, four strategies are possible (Smith and Merritt 2002):

- *Avoid* the risk by reversing decisions that could cause the risk (abandoning the project might be an option);
- *Transfer* the risk (or impact) to another party that may have a better potential (knowledge, resources) to tackle the problem;
- Create *redundancy*, thus reducing the effect of the risk event by providing parallel solution paths and backup options; and/or
- *Tolerate* the risk but at the same time *mitigate* the risk/impact and risk/impact drivers (to make it less severe) by developing a *prevention* plan (which works on reducing risk and risk drivers), a *contingency* plan (which works on impact and impact drivers), or a *reserve plan* (in which risk occurs and losses need to be covered).

As an example, Table 10-5 lists various strategies to handle the risks outlined in Table 10-2 for the Peru project. The four strategies mentioned require a substantial amount of advocacy work, which can take multiple forms, such as education, adoption of construction codes and best practices, and regulations.

Table 10-5. Management Strategies for Various Risks for Project in Iquitos, Peru

Risk Type	Response Strategy	Strategy Description	Responsible Individual	Pricing Strategy	Cost of Mitigation	Trigger Events
Severe weather	Avoid	Low impact and low probability do not affect project adversely	None			
Labor issues	Accept	Low impact and low probability do not affect project adversely	None	Contingency	$100	Cultural disagreements
Fuel price	Transfer	Create fuel price contract regarding price inflation between consulting firm and NGO	Planning team member	Pass through	$70	Global and local inflation rates increase
Donor funds	Mitigate	Draft contractual agreement between donor and consultant	Planning team member	Contingency	$100	Financial disagreement with donor
NGO conflict	Accept	Maintain open channels of communication or secure contracts with NGO	Planning team member	Other		Operational disagreement with NGO
Political turmoil	Accept	Medium probability, so maintain contact with local liaisons	Planning team member	Other		Political instability leads to NGO withdrawal
Community rejects new behavior	Mitigate	Use of participatory approaches to ensure community acceptance	All parties	Contingency	$500	Community support for project is lost

Source: Bottenberg et al. (2012), reproduced with permission.

According to Smith and Merritt (2002), once a risk management strategy is selected, an action plan needs to be outlined to implement the strategy. The action plan has four components: (1) an objective, (2) ways of measuring whether the objective has been achieved, (3) a time line toward completion, and (4) a designation of who is accountable. Challenges, the assumptions associated with implementing the action plan, and the potential for unintended consequences and cascading effects should also be included. In general, the action plan consists of tangible and nontangible actions. The former tends to be more objective, whereas the latter involves more policy issues.

Furthermore, to carry out the action plan, resources and knowledge are needed. The resources and knowledge available in the community, which can be related to the community capacity, dictate to great extent what levels of risks can be handled, which DFID (2002) calls the *risk appetite.*

Monitoring and Evaluating Risks

Risk management is a dynamic process because the risk drivers, risks, and associated impact are likely to change over time. Therefore, a need exists to develop indicators to monitor progress as discussed with capacity development and using the SMART acronym. As the project goes on, all activities are evaluated as to whether they are risk neutral, risk enhancing, or risk reducing.

The monitoring indicators need to (1) measure progress in the action plan leading to risk reduction, (2) identify any losses associated with the risks, (3) measure the effectiveness of the methods and action plans used to handle the risks, (4) indicate the emergence of potential new risks and/or issues (risks that are 100% certain) that may require new action plans, and (5) terminate action plans on risks that can be considered as closed. As remarked by Smith and Merritt (2002), the closing of an action plan may occur when (1) the risk event was prevented from happening, (2) the time component of the risk has passed, and (3) the risk event happened and has been managed accordingly.

10.4 Project Impact Assessment

Related to risk analysis is the issue of project impact assessment at the local, regional, and global levels. This is an issue that is regularly addressed on engineering projects in Western countries. According to the International Association for Impact Assessment (IAIA 2013), it is "the process of identifying the future (prospective) consequences of a current or proposed action. The 'impact' is the difference between what would happen with the action and what would happen without it."

That difference is often hard to measure, especially quantitatively, because it is easier to determine the contribution of a project in a qualitative anecdotal manner than it is to attribute an outcome to a specific project activity. This is especially true when trying to quantify the impact of a specific activity (e.g., transportation; pollution; water, sanitation, and hygiene; agriculture; tourism; or construction) on public health. This quantification is less of a challenge when looking, for instance, at the effect of construction activities on infrastructure development. Engineers are more familiar with the latter.

Important aspects of project impact assessment that have received a lot of attention in developed countries include the health impact assessment (HIA) and the environmental impact assessment (EIA), which deal, respectively, with the effects of projects on human health and on the biophysical environment (Birley 2011). Both HIA and EIA need to be communicated to the community in the form of public meetings before the project and as it unfolds. It is a participatory process that is unfortunately not often included in small-scale development projects. The HIA and EIA can also be initiated by the community and can force outsiders to be more transparent (Conant and Fadem 2008).

As remarked by the IAIA, the EIA is

> *a requirement in most countries in the world. It describes how a project will affect people, the environment (air, water, land, and biota) and any negative and harmful consequences of the project such as population relocation, impact on traditional cultures and livelihood, spiritual tradition and historical heritage. In some countries, there are often both national/ federal and state/regional EIA systems and regulations.* (IAIA 2013)

Health impact assessment is closely linked to health risk assessment (HRA), which looks at the impact of health-related risks and builds on the health needs assessment, which is integrated in the project appraisal phase.

In addition to the EIA and HIA, other impact assessments may need to be conducted about specific social, cultural, economic, and institutional issues. They can be conducted in parallel or integrated into one assessment study because of their overlapping nature. As remarked by Muscat (2011), being aware of various forms of project impact is critical in projects that have potential to induce conflict between different groups. According to Muscat, engineers may want to account for issues such as the following:

> *Is the project located near borders between rival groups? Will the location and design of irrigation channels impinge on divisions between different ethnic (or religious, etc.) groups? Is a project affecting areas inhabited by indigenous people? How will this affect design, cost, negotiation, and implementation? Will there be fair compensation payments/projects for people negatively affected?* Muscat (2011)

Project managers must use precaution to ensure that local and/or national regulations are also respected as projects unfold. In the absence of standards and regulations, which is often the case in the developing world, every effort should be made to minimize project impact using best practices that have had a proven record in the developed world, and which could be adapted to the situation at stake. This process also provides a unique opportunity to train local stakeholders about the importance of project impact and even influence in-country regional and national policies.

10.5 Chapter Summary

Risk analysis and management are integral parts of sustainable community development projects. Risks exist in communities before implementing projects; some are related to natural hazards and adverse events, whereas others are not natural. These risks are supplemented with additional ones related to project execution. Often the first identified risks may feed into the later ones and may enhance project-induced risks. In some cases, cascading risks can occur. In general, risk is about what could go wrong in a project. That simple issue, along with how project-related risks affect communities and their environment, should be addressed along each step of the proposed project implementation plan outlined in Chapter 8.

Methods for risk analysis and risk management are available in the risk management literature. They have been used in various industries and by development agencies. Many of these methods can be used to quantify or qualify risks in the developing community environment. In all cases, they need to be supplemented with methods to monitor and evaluate risk over time because risks do not end upon project completion.

References

Birley, M. (2011). *Health impact assessment: Principles and practice*, Earthscan Publications Ltd., London.

Bottenberg, C., et al. (2012). "Iquitos, Peru, term project report." CVEN 5929: Sustainable Community Development 2, University of Colorado at Boulder.

Conant, J., and Fadem, P. (2008). *A community guide to environmental health*, Hesperian Publ., Berkeley, CA.

Department for International Development (DFID). (2002). *Tools for international development*, Department for International Development, London.

Federal Emergency Management Agency (FEMA). (2001). "Understanding your risks: Identifying hazards and estimating losses." Publication No. 386-2, Federal Emergency Management Agency, Washington, DC.

Greene, S. (2007). "Community owned vulnerability and capacity assessment." World Vision Africa Relief Office, Nairobi, Kenya.

International Association for Impact Assessment (IAIA). (2013) <http://www.iaia.org> (March 1, 2013).

Knight, F. H. (1921). *Risk, uncertainty, and profit,* Reprinted in 2009 by Signalman Publishing, Kissimmee, FL.

McAfee, K., Sharma, A., Rowley, C., and Van Voast, C. (2011). "Paxocon, Guatemala, term project." CVEN 5929: Sustainable Community Development 2, University of Colorado at Boulder.

Muscat, R. (2011). "Peace and conflict: Engineering responsibilities and opportunities." *Global Services USA Newsletter,* September, 13(1).

National Research Council (NRC). (2012). *Disaster resilience: A national imperative,* National Academies Press, Washington, DC.

Project Management Institute (PMI). (2008). *A guide to the project management body of knowledge,* PMBOK Guide, 4th Ed., Project Management Institute, Newtown Square, PA.

Slovic, P. (1987). "Perception of risk." *Science,* 236(4799), 280–285.

Smith, P. G., and Merritt, G. M. (2002). *Proactive risk management,* Productivity Press, New York.

Tearfund. (2011). *Reducing risks of disasters in our community,* Tearfund, Teddington, U.K. <http://tilz.tearfund.org> (Oct. 22, 2012).

11

Community Resilience Analysis

11.1 About Resilience

In physics and engineering, resilience refers to the ability of a material to return to its original equilibrium after being disturbed. The concept of resilience as the ability of something to bounce back has been extended metaphorically to other nontechnical disciplines in social and psychological sciences (Norris et al. 2008; Zolli and Healy 2012), where the something could be an individual, a community, an ecological system, or an infrastructure facing disturbance (e.g., hazard, disaster, or trauma). In that context, resilience is understood by some researchers as *ability* or a *process* rather than an outcome, and *adaptability* rather than *stability* (Norris et al. 2008). Others see resilience as an outcome (Kahan et al. 2009) or as both process and outcome (Cutter et al. 2008).

This chapter explores the resilience of developing communities to risk associated with adverse events. It is an extension of the previous two chapters that dealt with capacity, vulnerability, and risk. Because of adverse events of various magnitudes and scales, communities face various risks that may affect their existence and quality of life and the well-being of their members. The events can vary from being local to crossing geopolitical borders. The consequences of those events can be immediate or long lasting. Their impact and probability of occurrence can be represented on a risk heat map, such as the one shown in Table 10-3. In some cases, there may be multiple events or the events may lead to secondary events.

Furthermore, compared with events in the developed world, adverse events in the developing world do not have to have high probabilities of occurrence and high impact to be important to communities. As seen in Chapter 2, it does not take much for people who try to step out of poverty to fall back into it because of their lack of initial capacity. Whether the consequences of events are tolerable or not depends largely on how ready communities are to face such events using their *coping resilience* and their capability to adjust to new conditions using their *adaptive resilience*. Building and maintaining both forms of resilience depend greatly on development, as long as it does not create unexpected consequences that could negatively affect the livelihood of communities. Examples include "consumption, overuse and destruction of natural resources; population growth; use of marginal lands; urbanization; pollutants; hazardous products; and misguided development projects" (UW-DMC 1997).

The concept of resilience has received a lot of attention in the developed world, especially in regard to major hazards and disasters. For instance, in the United States, various initiatives have been launched to promote a culture of resilience at the national level, a review of which can be found in a report published by the National Research Council (NRC 2012) entitled "Disaster Resilience: A National Imperative." At the international level, several initiatives on resilience are underway, such as the Hyogo Framework for Action (UNISDR 2007) launched in 2005 by the United Nations, which is designed to engage governments in developing global strategies at the national or regional level.

The main problem with such global strategies is that they do not always specifically address the challenges faced at the community level, which is of interest here. And if they do, they assume a certain baseline of inherent community capacity that is often nonexistent in the developing world. It is not that developing communities cannot be resilient. The challenge is that it takes them a tremendous amount of effort to build capacity in such a high-risk and low-capacity environment, which is often constraining rather than enabling. In general, it should be remembered that developing strategies for community resilience and putting them into practice in that context is more difficult than in the developed world.

The concept of resilience can mean different things to different people and organizations. Some see resilience as a global concept, whereas others consider different types of resilience, such as economic, social, institutional, infrastructure, or environmental. Various definitions of community resilience to hazards have been proposed in the literature (NRC 2012). They usually encompass four basic tenets: mitigation, preparedness, response, and recovery. In this book, community resilience is defined as "the ability to prepare and plan for, absorb, recover from, or more successfully adapt to [actual or potential] adverse events" (NRC 2012). Another way to look at resilience is the ability to cope with adverse events (or challenging conditions) and adapt to a *new normal*.

The literature on community resilience also calls for communities to develop *project* and *process* mitigation strategies that may be local and/or part of larger regional or national frameworks. Project mitigation involves *tangible issues* in the community that can be addressed using structural measures, such as the alteration of the built and natural environment, infrastructure (local and global) development and retrofitting, and the implementation of economic measures. However, process mitigation involves *intangible* issues related to institutional and individual behavior to reduce risk. This type of mitigation may include nonstructural measures, such as the adoption of building and construction codes, exploration of better options for land use (zoning), use of natural barriers, education of communities and individuals, reduction and elimination of inequalities, the adoption of governance and policies, climate adaptation, and meeting of immediate, medium-term, and long-term health needs. In general, both project and process strategies must be developed for a wide range of adverse events that the community is likely to experience and must be adaptable to various community needs, sizes, and types.

In general, the literature on resilience seems to recommend that civil society needs to play a more active role by engaging individuals, households, communities, and local governments (e.g., mobilizing social capital) in decision making through participation, collaboration, and collective action toward resilience building. Communities need to be more proactive by being able to map their own strengths and weaknesses. A framework based on such principles is the community-owned vulnerability and capacity assessment approach mentioned in Chapter 10 (Greene 2007).

It is important to note that abilities, skills, behaviors, and attitudes are needed at the community level to do the following:
- Convey that mitigation measures before an event determine what happens after an event,
- Mobilize all key actors along with outsiders to the community,
- Communicate community risk and uncertainties,
- Put structural and nonstructural measures in place, and
- Respond and recover from an event.

Such community attributes rarely exist, even in the developed world. In view of such complexity, more often than not, communities faced with the prospect of adverse events prefer the status quo of not changing what they are doing, or of using makeshift approaches rather than developing new comprehensive ones. This method may work for a while, until a major event strikes.

In the United States, several strategies have been proposed to make individuals and communities become more aware of (1) disasters and their effects; (2) the need for preparation; and (3) the decisions that need to be made during and after an event (e.g., strategies developed by the San Francisco Department of Emergency Management, the Aidmatrix foundation, and the Community and Regional Resilience Institute). They acknowledge that community resilience cannot rely only on government and external aid and that collaborative actions between bottom-up actions and top-down actions need to be developed, while protecting the most vulnerable. According to the Communities Advancing Resilience Tool (CART) proposed by Pfefferbaum et al. (2011), community resilience encompasses four interrelated domains:
- Connection and caring, which is about relatedness, shared values, participation, support systems, equity, justice, hope, trust, and diversity;

- Resources, which includes various forms of capital (e.g., social, economic);
- Transformative potential, which covers data collection and analysis of various forms of capital; and
- A fourth domain specific to the vulnerability that the community is facing (e.g., violence, epidemic, water, sanitation, shelter, or food shortage).

The lack of motivation of communities in the developed world to put into action a culture of resilience, despite clear evidence that such culture would pay off many times over (NRC 2012), occurs in part because there is "no simple 'blueprint' for constructing resilient communities" (Norris et al. 2011) and that people perceive risk and respond to it differently. Another reason is that building resilience takes time and requires dealing with substantial tangible and intangible issues in acquiring human and economic resources. Furthermore, the research community studying resilience does not seem to have an agreed-upon strategy. For some researchers, community resilience can be broken down into subcategories that are addressed separately. For instance, Cutter et al. (2010) consider five forms of resilience: social, economic, institutional, infrastructure, and community capital resilience. Others only consider physical resilience and social resilience. Tierney (2008) considers four resilience domains: technical, organizational, social, and economic. One common denominator in all these approaches is that they emphasize the idea that resilience is about community-acquired capacity in its many forms. But they fail to take into account the relationship among the various categories of acquired capacity and any associated synergy.

Clearly resilient communities, characterized by increased capacity and decreased vulnerability, are less at risk to see their acquired level of development erode away and are better prepared to embrace more development. As seen in Chapters 9 and 10, both capacity and vulnerability can take multiple forms: institutional, human resources, technical, economic and financial, energy, environmental, social, and cultural. According to the Tearfund (2011), they involve various issues: individual, social, natural, physical, and economic. Understanding how these various categories (and others more specific to each community) interact in a systemic way at the community level during normal and new normal conditions after an adverse event is critical to building resilient communities and developing risk strategies to cope with hazards (Sherrieb et al. 2010).

11.2 Resilience to Major Hazards and Disasters

One cannot talk about sustainable community development and the resilience of developing communities without addressing the vulnerability of poor communities to major hazards and disasters, which are extreme adverse events. There seems to be an agreement by several organizations (e.g., the Office for the Coordination of Humanitarian Affairs of the United Nations and the World Health Organization) that a disaster can be defined as "a serious disruption of the functioning of a community or a society causing widespread human, material, economic, or environmental losses which exceed the ability of the affected community or society to cope using its own resources" (WHO 2007).

Disasters can broadly be divided into natural and nonnatural events. According to Guha-Sapir et al. (2010), natural disasters can be divided into five groups: (1) biological (epidemic, insect infestation); (2) geophysical (earthquakes, volcanoes, dry mass movements); (3) hydrological (floods and wet mass movements); (4) meteorological (storms); and (5) climatological (extreme temperatures, drought, and wildfires). Nonnatural disasters may include "war and civil conflicts, displacement due to political violence and development projects such as large dams and disasters due to the collapse of existing social welfare systems as a result of wider economic and political changes" (Lloyd Jones et al. 2009).

Disasters affect communities in many different ways: environmental, health, economic, social, political, administrative, management, and psychological. They strike communities around the world irrespective of economic status, race, religion, or culture. Disasters fall into the low-to-medium probability and high-impact group of events that communities are likely to face. In general, they are difficult to mitigate, prepare for, handle, and recover from.

Since 2004, natural disasters have been on the rise with major calamities such as the 2004 tsunami in the Indian Ocean, the 2005 Hurricane Katrina in the United States, the floods in Pakistan (2010 and 2011), the earthquakes in Haiti (2010) and New Zealand (2011), the earthquake and tsunami in Japan (2011), extreme floods in Australia (2011), and the 2012 Superstorm Sandy in the United States. Global statistics about disasters worldwide and their effects can be found in a variety of reports (e.g., Guha-Sapir et al. 2010). According to the Tearfund (2011), about 240 million people each year (over the past 10 years) are affected by natural disasters worldwide. These events are characterized by their complexity, size, uncertainty, and unforeseen consequences (with cascading effects), and they affect rich and poor communities in different ways. For poor regions of the world, however, the effect of such large natural events is often compounded with other conditions of vulnerability existing beforehand, such as unplanned urbanization; nonexistent building codes and construction practices; underdevelopment; competition for limited resources; limited governance and rule of law; social, economic, and political inequality; and limited network structure and connections. As noted by the Tearfund (2011), "disasters often reverse development progress" that has been built up over several years and decades. In the 2011 UNDP/HDR report, it was noted that "a 10 percent increase in the number of people affected by an extreme event reduces a country's [human development index] HDI almost 2 percent, with larger effects on incomes and in medium HDI countries."

Furthermore, with changing population patterns and an increased level of population concentration in urban areas (more than half of the world population in 2007) and slums in those areas (1 out of 3 city dwellers lived in slums in 2005), the effect of hazards and disasters on such human concentrations has the potential to be larger in the future.

Hazards and disasters have a significant impact on the lives and livelihoods of populations of developing countries (UNDP 2004). As remarked by Peduzzi et al.,

> the least developed countries represent 11% of the population exposed to hazards but account for 53% of casualties. On the other hand, the most developed countries represent 15% of human exposure to hazards, but account for only 1.8% of all victims. (Peduzzi et al. 2009)

In general, disasters tend to amplify the inherent problems already existing in the community or society after an adverse event. When it comes to casualties caused by adverse events, there tend to be strong casualty spikes in developing communities, and "the number of people affected (those requiring immediate assistance, those who are injured, and those made homeless from the disaster) increased threefold during the first decade of the 21st century" (IFRCRC 2010). As remarked by the NRC (2012) study, the main reason for such overall differences in casualties is related to the improved awareness of population in developed countries, the existence of better warning systems, and the use and enforcement of better building codes. The implementation of such proven solutions in the developing world is still in its early stages and is not of high priority to many governments.

Even in rich countries, hazards and disasters have a disproportionate impact on the livelihood of poorer communities living side by side with well-off ones (Tierney 2009). Poor communities are more (1) vulnerable, (2) exposed to risk, and (3) affected by disasters. It takes them longer to recover (Cuny 1983; Cutter et al. 2003). Furthermore, disasters widen the gap between the poor and the rich (Mutter 2010). As remarked by the Center for Biosecurity in their December 9, 2010, policy brief on research findings on community resilience in the United States, "Hurricane Katrina's disproportionate impact on poor and working class people, and on ethnic and racial minorities, confirms a more widespread finding that broader social and economic inequalities make certain sub-groups within affected populations more vulnerable to the effects of a disaster." The same report also confirmed the same trend with the 2009 H1N1 influenza pandemic (Center for Biosecurity 2010).

11.3 Resilience as Acquired Capacity

In the literature, community resilience is often presented as acquired capacity to cope with and adapt to unusual and unscheduled conditions and transient dysfunctions and to return to a level of functional balance or a new normal. Resilience can be inherent (coping) and adaptive (Tierney 2009). *Inherent resilience* relates to the robustness of social units (households) when order is disrupted. *Adaptive resilience* relates to the effort and ingenuity of the social units and the whole community in adapting to a new normal. This type of resilience translates into ensuring continuity of operations, infrastructure, government and agencies, business, recovery plans, emergency management plans, and household livelihood by addressing needs and securing assets.

Figure 11-1 illustrates how community resilience (acquired capacity) varies with time after a major event or crisis (stressor) of some sort and at a given scale. Immediately after the crisis, the community capacity drops from its initial value to a lower level at time t_r. The magnitude of the drop depends on the severity of the stressor, its duration, and how surprising the crisis is. The drop is followed by one or several periods of recovery (new normal conditions) and adaptation, which lead, it is hoped, after a certain amount of time, to a final capacity larger than the initial one. To a great extent, the initial level of community resilience (initial acquired capacity) affects the size of the capacity drop, the recovery time, and the recovery rate. The larger the initial resilience and the better prepared a community is, the lower the drop, the faster the recovery, and the steeper the rebound. Other factors besides the initial level of resilience may control the overall shape of the curve in Figure 11-1. They include the risk of secondary disasters, the clarity of policy and direction, collective motivation, good communications, technical assistance, the availability of funds to reboot the economy, cash flow, and the reusing of salvaged material. These conclusions apply to a given community, for a given crisis, and over a given physical scale.

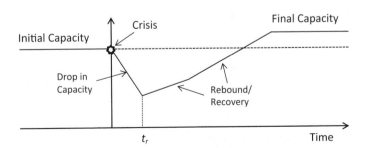

Figure 11-1. Variation of Community Capacity after a Disaster or Crisis

Source: Courtesy of Mary Ellen Hynes, adapted with permission.

Figure 11-1 is more likely to apply to communities in the developed world that are better prepared to plan for and respond to events. In the developing world, where the initial resilience (acquired capacity) of communities is low from the beginning, it is likely that the drop in capacity will be large, the time to recover will be long, and the rebound rate will be low, flat, or even negative. The aftermath of the 2010 earthquake in Haiti is the perfect illustration of this point. It is critical that under such situations, the response curve in Figure 11-1 does not go below minimum human standards, such as the Sphere standards (Sphere Project 2011), which guarantee "an acceptable level of health" for the community. It is also critical that vulnerability is not to be rebuilt in the recovery process through outdated building practices, or fixing the infrastructure that is beyond its lifetime, for instance (USAID 2012).

Even though a community may possess all of the aforementioned attributes, resilience does not prevent dysfunction or distress from happening at the community level because it may need to adapt to a series of "new normal" conditions in its trajectory of resilience and recovery. Such transient perturbations are normal in unusual events (Norris et al. 2008). Furthermore, a community as a whole may show all the signs of being resilient, yet some parts of it may be more prepared than others. The opposite is also true. Some parts of a community may be resilient, and yet the entire community is not. After the 2005 Hurricane Katrina, the city of New Orleans was a perfect example, where, even today, many neighborhoods are still lacking the knowledge and resources to return to a new normal. At a larger scale, though, all studies seem to indicate that after six years, New Orleans has achieved an average level of recovery (Brodie et al. 2006).

Prerequisites for Community Resilience

As mentioned in the introduction of this chapter, building or strengthening community resilience is not accidental, nor is it a random process; it is a process that requires sustained commitments and demands time, effort, and resources from many groups of stakeholders. According to Bruneau et al. (2003), resilient communities share four key dynamic attributes: *robustness* (ability to resist disruption with stopping), *redundancy* (presence of alternative options to provide services), *resourcefulness* (ability to mobilize resources), and *rapidity* (ability to find timely solutions to challenges). Table 11-1 shows how these attributes express themselves in four domains of resilience (or acquired capacity): technical, organizational, social, and economic (Tierney 2008).

The value added (economic and human) in having a resilient community has been demonstrated in the literature for communities in the United States. For instance, according to the Multihazard Mitigation Council (2005), $1 spent on preevent physical and process mitigation strategies translates into $4 saved in

Table 11-1. Relationship between Attributes of Resilient Communities and Domains of Acquired Capacity

Dimension or Domain	Technical	Organizational	Social	Economic
Robustness	Newer structures, built to code, older structures retrofitted	Extensiveness of emergency operations planning	Social vulnerability or resilience indicators	Economic indicators
Redundancy	Capacity of technical substitutions or "work-arounds"	Alternate sites for managing disaster operations	Availability of options for disaster victims (e.g., housing)	Ability to substitute, conserve needed inputs
Resourcefulness	Availability of materials for restoration, repair	Capacity to use information, improvise, innovate, expand	Capacity to address human needs	Economic actors' capacity to improvise and innovate
Rapidity	System downtime, restoration time	Time between impact and early recovery	Time to restore housing and jobs	Time to regain capacity and lost revenue

Source: Tierney (2008), with permission from Kathleen Tierney.

postevent damage (NRC 2012), which means that community resilience is good business. Ideally, it should be seen as a way of life for communities faced by recurring adverse events.

A community needs to work at building resilience before (in preparation), during (to absorb during), and after (to respond and recover from) adverse events. Following is a nonexhaustive list of community attributes that need to be considered when building or strengthening community resilience:

- Hazard events have been identified and made public to the entire community, including past events and their effects in terms of human, property, crops, infrastructure, and economic losses. Along this line, FEMA (2001, 2004) has developed an excellent framework that can be used to (1) create hazard event base maps using existing databases, (2) conduct an inventory of local assets that could be affected by the events, and (3) estimate associated losses.
- Governance is in place at the local level with policy measures that (1) promote leadership, participation, communication, enforcing codes

and construction practices, and zoning laws; and (2) encourage community members to be proactive and adopt a culture of resilience through incentives. Policy measures that are obsolete or have led to unintended consequences in the past need to be abandoned.

- Governance at the local level is in tune with that at the regional and national levels and is coordinated with other outside nongovernmental entities.
- The community is capacity ready to prepare for, face, and recover from an event; has invested strategically in efficient and flexible solutions; and has decision-making skills.
- A strategy is in place because the community has addressed land use, hazard mitigation, and critical infrastructure (e.g., municipal sanitation systems, transportation and energy systems, health, and communications systems).
- Community engagement and strong social networks are present for communication and response. Community members have been educated about risk and response to risk.
- An action plan and resolution of risks are in place and are clear to individual community members.
- All community sectors are involved in preparedness and response planning, and community members work together (collective resilience).
- Incentives in integrating resilience in the community are in place and are perceived as an added value and an investment rather than simply an additional cost.

In general, a community needs to be informed, have a plan, be economically strong, have an effective government in place, and be socially cohesive. In the NRC (2012) report, community resilience is seen through the lens of a healthy human body. The same way as a body is healthy not simply because of the coalescence of functional parts, a resilient community requires that all the community components (physical, social, political, economic, and environmental) are not only strong but also know how to work together in a systemic way. In theory, this makes sense. In practice, and as mentioned previously, communities in the developed world seem to have a limited view of how they should prepare for, absorb, and respond to the effects of adverse events in the future. This phenomenon can be explained by the human tendency of not investing in the future or seeing long-term benefits. Having all the aforementioned community attributes in place would be an even greater challenge for developing communities because of their limited capacity and high vulnerability.

In summary, if community resilience exists before an adverse event and project and process mitigation strategies are in place, the effects of the adverse events are likely to be less, the community will rebound faster, and the community will be more engaged in the process of recovery as long as resources are available and the community has enough capacity to do so. This conclusion applies to developed and developing communities alike.

11.4 Measuring Community Resilience

Performance metrics and indicators are needed to evaluate community resilience. If resilience were measurable (quantitatively or qualitatively), in the form of a composite index, its changes through time could be monitored from an agreed-upon baseline. Furthermore, the effectiveness of policies contributing to community resilience could be evaluated, and actions could be taken to correct unacceptable trends. The very act of having metrics and indicators could help community members and decision makers better understand the added value of resilience and could help in deciding on prioritization of actions, areas of investment, and policy changes.

A good example of performance metrics was developed for the city of San Francisco and the Bay area, in relation to how the region responds to major earthquakes. Developed by the San Francisco Planning and Urban Research Association, also known as SPUR (2009, 2011), it includes recovery objectives and a time frame for restoring the functionality of key community services and business and neighborhood activities for an "expected" earthquake and level of disruption. Three types of earthquakes were considered: (1) earthquakes that are routine events, (2) earthquakes that would occur once during the useful lifetime of the infrastructure, and (3) extreme earthquake events. SPUR introduced target states of recovery for San Francisco buildings and infrastructure after an earthquake: short-term recovery in terms of hours (3–72), medium-term recovery in terms of days (30–60), and long-term recovery in terms of months (4–36). In the SPUR model, the performance of buildings is divided into five categories: A (safe and operational), B (safe and usable during repair), C (safe and usable after repair), D (safe but not reparable), and E (unsafe). Safe and operational buildings (lifeline) associated with critical response (first responder) facilities belong to category A. Such a framework could be developed for any community, small or large. A good example would be the city of Kathmandu in Nepal, where it has been estimated that a large earthquake could result in 300,000 deaths and 60% of buildings being damaged beyond repair (A. M. Dixit, personal communication, 2014).

Even though metrics and indicators could be developed for specific purposes, such as San Francisco's built environment, it is hard to conceive of a single measure of community resilience in the form of a composite index. In Figure 11-1, for instance, a composite index of resilience could be defined as the ratio between the final overall community capacity and the initial one. The problem with such a straightforward definition is that many categories of capacity enter into community resilience. The overall initial and final capacities would have to be defined as compounded measures of several categories of capacity. Furthermore, the recovery time and the magnitude of the capacity drop in Figure 11-1 would have to be factored in as well; this factoring is particularly difficult because recovery does not have an end point per se. Finally, the index would greatly depend on the scale (spatial and temporal) being investigated.

The problem in defining a single composite index of community resilience stems from the fact that resilience is a hard-to-define and hard-to-measure concept that can be seen as a process or an outcome. Furthermore, resilience involves many systems that interact in complex and uncertain ways. If it were to exist, a single index would have to be multidimensional and combine various factors, each one carrying a different weight. The factors would need to represent what it takes for a community to reach and maintain an acceptable level of functioning structures after an event. Furthermore, a good metric would need to show several attributes, such as openness and transparency, replicability, simplicity when used by various stakeholders, and good documentation (NRC 2012).

An actual composite index of community resilience would require first agreeing upon a set of community characteristics, i.e., *baseline targets* in all categories of acquired capacity that are critical to ensure an overall level of resilience for a given type of adverse event and a well defined community size and type. A composite index would then be determined as a (weighting) function of the target values. The baseline targets would guarantee a level of community performance (quality assurance) during the response and recovery phases after an event. For each level of resilience, the actual community performance would be evaluated using a suite of established indicators for each category of capacity (Chang 2012; Jordan and Javernick-Will 2013). These aspects of resilience would include, for instance

- The speed of economic, financial, and emotional recovery;
- The performance of noncritical infrastructure and critical facilities, lifelines, and transportation systems;
- The response time of emergency resources;
- The extent of losses in terms of casualties, people displaced, and damage to public infrastructure and private property;
- How quickly critical infrastructure and community services are restored and debris is removed; and
- The wellness and security in the basic economic and social units (households) that form a community, including access to basic services of water, food, energy supply, health, and transportation.

The indicators could also be regrouped under broader categories, such as community well-being, security, robustness, redundancy, resourcefulness, and rapidity.

Once the framework is in place, the analysis of events that the community experienced in the immediate past could help frame and evaluate its current level of resilience. This evaluation is done by comparing, for each capacity category, the current acquired capacity with the baseline target associated with a given level of resilience. Once the current level of resilience is determined, decisions can be made about whether measures need to be taken to increase the current level of resilience to a higher level and close the current resilience gap. The effectiveness of the implemented measures is determined by how well a community responds and bounces back from future adverse events. In many ways, the proposed approach would resemble the approach used in Chapter 9 in determining the enabling environment (e.g., capacity) of a community to provide a service. The only difference and additional challenge is that community resilience implies that multiple interrelated services would be maintained at a certain level.

11.5 U.S. Frameworks for Community Resilience

Several metrics and indicators have been suggested in the literature to assess the degree of resilience of U.S. communities to hazards and disasters. Examples include the following:

- The Baseline Resilience Indicator for Communities (BRIC), proposed by Cutter et al. (2010), which is calculated as the arithmetic mean of five subindices related to social, economic, institutional, infrastructure, and community resilience. A similar index was proposed by Sherrieb et al. (2010) as a composite of two sets of indicators: one set of 10 indicators related to economic development capacity and another set of seven indicators related to social capacity.
- The Coastal Resilience Index proposed by Emmer et al. (2008) addresses community resilience to coastal hazards such as storm events. It is community based and participatory in the sense that communities must conduct their own self-assessment by considering two scenarios of a bad storm and a worst storm. Critical infrastructure, critical facilities, and other facilities deemed vulnerable to a storm must be assessed.
- The Argonne National Laboratory Resilience Index measures the resilience of critical infrastructure (Fisher and Norman 2010). Data are gathered on location at critical infrastructure facilities and integrated into an infrastructure survey tool covering roughly 1,500 variables. A protective measure index is calculated and ranges from 0 (lowest resilience) to 100 (highest resilience) for a given critical infrastructure or key resource sector and for a given threat.

A more detailed description of these indices and others can be found in Chapter 4 in the NRC (2012) report entitled "Disaster Resilience: A National Imperative."

11.6 International Resilience Frameworks

There has been much interest in disaster resilience at the international level, especially in relation to the needs of protecting those at the bottom of the pyramid when subject to large-scale events. The goal is to build a world that is safer from disaster and that has developed disaster risk reduction strategies that work at different scales in different regions of the world and protect the most vulnerable. Several international initiatives have been proposed since 1990, which coincided with the start of the International Decade for Natural Disaster Reduction. Presented in the following sections are the International Strategy for Disaster Reduction global strategy, two international risk indices (UNDP's Disaster Risk Index and the UN's World Risk Index), and the new Global Adaptation Index index.[1]

International Strategy for Disaster Reduction

The International Strategy for Disaster Reduction (ISDR) was introduced to reduce disaster risk and build the resilience of communities. An action plan was proposed in 2005 in Kyoto, Japan, known as the Hyogo Framework for Action (HFA) for the decade ranging between 2005 and 2015 (UNISDR 2007, 2010). Adoption of the plan by 168 nations at that time was driven by the impact of the 2004 tsunami in the Indian Ocean. A midterm review of the progress toward reaching that action plan is available in the literature (UNISDR 2011a).

The action plan recognizes that the impact of disasters is most felt at the community level. It is where risk reduction is most needed. It also recognizes that international collaboration is needed among all stakeholders interested in disaster risk reduction (i.e., states, regional organizations and institutions, international organizations, civil society, the scientific community, and the private sector).

The HFA consists of five well-defined priorities for action (UNISDR 2010) at the community level and national level:

- HFA-1: Making risk reduction a national and local priority with a strong institutional basis for implementation;
- HFA-2: Identifying, assessing, and monitoring disaster risks and enhancing early warning;
- HFA-3: Building a culture of safety and resilience using knowledge, innovation, and education at all levels;
- HFA-4: Reducing the risk in key sectors; and
- HFA-5: Strengthening disaster preparedness for effective response.

Each priority for action is itself divided into several tasks. Each task is then assigned specific measurable (local and national) indicators, a method of monitoring progress, guiding questions, and specific tools to reach the desired level of disaster risk reduction. Furthermore, implementation of each task is illustrated by one or several international case studies.

Although the Hyogo framework of action does not create specific indicators of resilience or lead to a composite index of resilience, it provides a detailed framework and disaster risk reduction strategies that can be implemented at different scales (nations, communities, individuals). It also creates a repository of best practices and information on disaster risk reduction at the international level that is currently managed by the information management unit of the United Nations ISDR secretariat (UNISDR 2011b).

UNDP Disaster Risk Index

The disaster risk index (DRI), introduced in 2004 by the United Nations Development Programme (UNDP), measures the average risk of death per country in disasters related to earthquakes, tropical cyclones, and floods. Deaths from disasters represent 39% of deaths in large- and medium-scale natural disasters at the global level. The index is based on data collected between 1908 and 2000. DRI implies that populations must be exposed to one of the three natural hazards, a sine qua non for risk to human life to exist.

DRI is a measure of vulnerability to a specific hazard. It also accounts for the role of social, technical, humanistic, and environmental issues that could be correlated to death and may point toward causal processes of disaster risks. Examples include (1) lack of economic reserves and low asset levels, (2) absence of social support mechanisms and weak social organizations, (3) poorly constructed and unsafe shelter, and (4) fragility of existing ecosystems.

The key steps in determining the DRI for a specific hazard include calculation of physical exposure in terms of the number of people exposed to a hazard event in a given year; calculation of relative vulnerability in terms of the number of people killed in relation to the number of people exposed; and calculation of vulnerability indicators using 26 variables related to the economy, quality of the environment, demography, health and sanitation, politics, early warning capacity, education, and the human development index discussed in Chapter 2. For a given specific hazard, the DRI helps rank countries based on their degree of physical exposure, relative vulnerability, and degree of risk (Peduzzi et al. 2009).

The United Nations World Risk Index

In September 2011, the United Nations University (UNU) Institute for Environment and Human Security (UNU-EHS), in collaboration with the Bündnis Entwicklung Hilft (Alliance Development Works), published the "World Risk Report 2011" and introduced the world risk index (WRI) (UNU 2011). It indicates the probability that a country or region will be affected by extreme natural events such as earthquakes, storms, flood, droughts, and sea level rise. The index consists of indicators in four areas for a population based on its

- Exposure to natural hazards (earthquakes, droughts, storms, floods, rise in sea level);
- Susceptibility to suffering harm based on its living conditions (levels of poverty, education, food security and governance, infrastructure, economic framework);
- Capacity to cope with severe and immediate disasters based on existing governance, disaster preparedness, early warning systems, medical services, social and economic security, and social networks; and
- Adaptive precautionary measures against anticipated future natural disasters.

Global Adaptation Index

The global adaptation index (GAIN) was introduced in 2011 by the Global Adaptation Institute as a measure of how ready and resilient countries are to adapt to climate change. It is described as a "navigation tool to help prioritize and measure progress in adapting to climate change and other global forces." It recognizes the idea that climate-related disasters and climate change "will lead to increased risks, and costs for business, complicate political decisions, and of most concern, threaten the quality of life for vulnerable populations around the world."

The GAIN index uses the two concepts of vulnerability and readiness (capacity) discussed in Chapters 9 and 10. It is measured as the difference between a country's readiness score and its vulnerability score and is normalized to vary between 0% and 100%. The vulnerability and readiness of each country are plotted on a readiness matrix, as shown in Figure 11-2, which is divided into four quadrants; each quadrant defines the country's challenge level to face climate change or other global forces. As of 2013, 177 countries have had their GAIN index determined. The index varies between 83.5 for Denmark and 34.3 for North Korea.

The vulnerability of a country is determined using indicators to measure three sectors that contribute to human well-being (water, food, and health) and three infrastructure sectors (coastal, energy, and transportation), as shown in Table 11-2. In the GAIN 2012 and 2013 version, two additional human well-being attributes were added (ecosystem services and urban habitat). Likewise, a country's readiness is determined using indicators divided into three categories (economic, social, and governance), as shown in Table 11-3.

Even though the GAIN index is defined at the country level, it could be formulated at the regional or community level and could be used to analyze the adaptation of individual communities to climate or to any adverse or disturbing event because of its modular approach, which is expandable and substitutable.

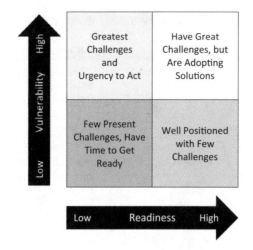

Figure 11-2. Country Readiness Matrix: Vulnerability vs. Readiness (or Capacity)

Source: Global Adaptation Institute (2011), reproduced with permission.

The GAIN index is one of the most promising indices for adaptation that is being developed at the present time because of its comprehensive nature and scientific approach.

11.7 A Systems Framework for Community Resilience

The traditional approach used in the resilience frameworks mentioned consists of enumerating a list of criteria and recommending that they are implemented in a systematic and logical way. Multicriteria decision-making tools (Keeney and Raiffa 1976) are often used in developing such metrics.

The problem with this approach is that it fails to capture the interaction, interconnectedness, and interdependence that exist among the different areas that contribute to community resilience. Second, it fails to recognize the idea that resilience is not about creating solutions to reach some form of equilibrium. Rather, it is about accommodating a constantly varying and adaptive disequilibrium, as remarked by Zolli and Healy (2012). Third, it fails to account for the possibility of multiple adverse events. Finally, it does not recognize the adaptability that communities have in preparing and responding to adverse events and how the adaptability reorganizes itself over time. These attributes are all characteristics of living systems that are complex and "operate in a state of constant dynamic disequilibrium" (Zolli and Healy 2012).

In all communities, some form of interdependence exists, whether it is in the form of local collaboration and participation of community members or in

Table 11-2. Indicators of Country Vulnerability to Climate Change

Sectors		Exposure	Sensitivity	Capacity
Water	Quantity	Projected change in precipitation	Internal and external freshwater extracted for all uses	Population with access to improved water supply
	Quality	Projected change in temperature	Mortality among 5-yr-olds caused by water-borne diseases	Population with access to improved sanitation
Food	Quantity	Projected change in agricultural (cereal) yield	Population living in rural areas	Agricultural capacity
	Quality	Coefficient of variation in cereal crop yields	Food import dependency	Children under 5 suffering from malnutrition
Health	Quantity	Estimated effect of future climate changes on deaths from diseases	Health workers per capita	Longevity
	Quality	Mortality caused by communicable diseases	Health expenditure derived from external resources	Maternal mortality
Coastal	Quantity	Land less than 10 m above sea level	Population living less than 10 m above sea level	Measured on the readiness axis
Energy	Quantity	Population with access to reliable electricity	Energy at risk	
Transport	Quantity	Frequency of floods per unit area	Roads paved	

Source: Global Adaptation Institute (2011), reproduced with permission.

the interaction of various subsystems (e.g., infrastructure, economy) that comprise the community (system) and contribute to its capacity or vulnerability. The level of interdependence that exists before any adverse event is tested during the event, and so is the community's adaptability. As mentioned in the CART framework (Pfefferbaum et al. 2011), "building a resilient community involves more than assembling a collection of resilient individuals." The relationship between the different components that contribute to community resilience and having a joint vision for global recovery are important as well.

Table 11-3. Indicators of Country Readiness (Capacity) to Climate Change

Components	Indicators
Economic	Business freedom
	Trade freedom
	Fiscal freedom
	Government spending
	Monetary freedom
	Investment freedom
	Financial freedom
Governance	Voice and accountability
	Political stability and nonviolence
	Control of corruption
Social	Mobile phones per 100 persons
	Labor freedom
	Tertiary education
	Rule of law

Source: Global Adaptation Institute (2011), reproduced with permission.

Could the system tools presented in Chapter 6 be used to simulate the interaction of the different community components that contribute to the resilience of a community? As discussed in Chapter 6, we already know that a system approach, combined with a social network analysis, represents a powerful tool to describe the existing level of community interaction. Building on that baseline, using a system approach to community resilience would allow for

- Simulating subsystems' change and adaptation to change;
- Exploring various forms of nonlinear interactions among the community subsystems;
- Identifying leverage points that may not be apparent at a first glance in the community; and
- Exploring unintended consequences to unexpected events, such as cascading events and secondary events that only occur when subsystems interact.

As discussed in Chapter 6, in systems thinking, the behavior of a system, i.e., its resilience to an adverse event, depends greatly on its structure (community), i.e., its components and the relationships among components. According to Meadows (2008), from a systems viewpoint, resilience "arises for a rich structure of many feedback loops that can work in different ways to restore a system even after a large perturbation." The existence of multiple loops creates redundancy, so that if one fails, another one kicks in. A higher level of resiliency called *meta-resilience* by Meadows (2008) can also be created to restore broken feedback loops and tap into the self-organizing and restorative nature of systems. Biological systems like the human body and natural systems possess that unique characteristic. Another question arises about how the same concept could be applied to communities that are recovering from the impact of adverse events.

Resilience of a system depends on how it interacts in a hierarchical manner with other systems and subsystems (Figure 6-1). A community is not resilient if another system does not serve its purposes and efforts, whether it is higher, lower, or at the same level in the hierarchy. Along that line, it is necessary to note that community resilience requires a bottom-up approach and a top-down approach. The first originates from the community, whereas the second is influenced by governments and other agencies. Both approaches emphasize how important middle-ground solutions are in decision making regarding how communities should prepare and plan for, absorb, recover from, or adapt to actual or potential adverse solutions. As recommended in the NRC (2012) report, "to achieve resilience, the responsibility for action resides at all levels from federal to state to communities and ultimately to individuals." This requirement may be feasible within the context of Western cultures. In the developing world context, however, this is more challenging because those at the bottom of the pyramid have limited capacity and those who govern them do not often have vested interest in developing resilience programs.

11.8 Chapter Summary

Developing communities are extremely vulnerable to adverse and hazardous events, whether they are small or cross geopolitical boundaries. Unlike communities in the developed world, developing communities are high-to-medium risk and low-capacity environments for which building resilience requires a tremendous amount of effort and time, often starting in situations that are more constraining than enabling. Developing strategies for community resilience and putting them into practice within that context is more difficult than in the developed world. Community resilience is an acquired capacity for a community to (1) cope with and adapt to unusual and unscheduled conditions and transient dysfunctions and (2) return to a level of functional balance or a new normal. The acquired capacity of a community needs to be compared with its vulnerability.

Measuring community resilience using a simple composite index is difficult because it is a hard-to-define and hard-to-measure concept. Various indices have been proposed in the literature. They focus on one or several aspects of resilience, are limited to selected adverse events, and are specific to certain community sizes. They fail to capture the complexity and uncertainty of communities, which could be captured better using a system approach rather than a cause-and-effect approach.

Sustainable development work is critical to increasing the capacity and reducing the vulnerability of communities in a sustained and long-lasting way. This change can only happen through full community engagement. A *culture of resilience* is therefore necessary in development work to ensure that developing communities are able to prepare and plan for, absorb, recover from, or adapt to actual or potential adverse events, no matter how big or small they are. Developing that culture of resilience requires changing the behavior of communities to invest in the future rather than respond in an impulsive manner to the challenges associated with adverse and hazardous events.

Note

[1]This review of the ISDR, DRI, and WRI resilience frameworks was conducted by the author in his capacity as contributor to the NRC (2012) report entitled "Disaster Resilience: A National Imperative."

References

Brodie, M., et al. (2006). "Experiences of Hurricane Katrina evacuees in Houston shelters: Implications for future planning." *American Journal of Public Health*, 96(8), 1402–1408.

Bruneau, M., et al. (2003). "A framework to quantitatively assess and enhance the seismic resilience of communities." *Earthquake Spectra*, 19(4), 733–752.

Center for Biosecurity. (2010). "Policy brief: Research findings on community resilience and their implications for federal policymaking," UMPC Center for Health Security, Baltimore, MD. <http://www.upmc-biosecurity.org/website/events/2009_resilient_american/pdf/PolicyBriefDraft.pdf> (Jan. 10, 2012).

Chang, S. E. (2012). "Urban disaster recovery: A measurement framework and its application to the 1995 Kobe earthquake." *Disasters*, 34(2), 303–327.

Cuny, F. (1983). *Disasters and development*, Oxford University Press, New York.

Cutter, S. L., Boruff, B. J., and Shirley, W. L. (2003). "Social vulnerability to environmental hazards." *Social Science Quarterly*, 84(2), 242–261.

Cutter, S. L, Burton, C. G., and Emrich, C. T. (2010). "Disaster resilience indicators for benchmarking baseline conditions." *J. Homeland Security and Emergency Management*, 7(1), Article 51.

Cutter, S. L., et al. (2008). "A place-base model for understanding community resilience to natural disasters." *Environment: Science and Policy for Sustainable Development*, 50(5), 36–47.

Emmer, R. L., et al. (2008). "Coastal resilience index: A community self-assessment. A guide to examining how prepared your community is for a disaster." NOAA Publication MAS GP-08-014, National Oceanic and Atmospheric Administration, Washington, DC. <http://research.fit.edu/sealevelriselibrary/documents/doc_mgr/434/Gulf_Coast_Coastal_Resilience_Index_-_SeaGrant.pdf> (Jan. 22, 2012).

Federal Emergency Management Agency (FEMA). (2001). "Understanding your risks: Identifying hazards and estimating losses." Publication No. 386-2, Federal Emergency Management Agency, Washington, DC.

Federal Emergency Management Agency (FEMA). (2004). "Using HAZUS-MH for risk assessment: A how-to guide." Publication No. 433, Federal Emergency Management Agency, Washington, DC.

Fisher, R. E., and Norman, M. (2010). "Developing measurement indices to enhance protection and resilience of critical infrastructure and key resources." *J. Business Continuity and Emergency Planning*, 4(3), 191–206.

Global Adaptation Institute. (2011). "Global adaptation index (GAIN): Measuring what matters." Global Adaptation Institute, Washington, DC.

Greene, S. (2007). *Community owned vulnerability and capacity assessment*, World Vision Africa Relief Office, Nairobi, Kenya.

Guha-Sapir, D., et al. (2010). *Annual disaster statistical review 2010: The numbers and trends*, Center for Research on the Epidemiology of Disasters, Louvain, Belgium. <http://reliefweb.int/sites/reliefweb.int/files/resources/fullreport_37.pdf> (April 20, 2014).

International Federation of Red Cross and Red Crescent Societies (IFRCRC). (2010). *World disasters report: Focus on urban risk*, International Federation of Red Cross and Red Crescent Societies, Geneva. <http://www.ifrc.org/en/publications-and-reports/world-disasters-report/wdr2010/> (March 1, 2012).

Jordan, E., and Javernick-Will, A. (2013). "Indicators of community recovery: A content analysis and Delphi approach." *Natural Hazards Review*, 14(1), 21–28.

Kahan, J. H., Allen, C., and George, J. K. (2009). "An operational framework for resilience." *Journal of Homeland Security and Emergency Management*, 6(1), 1–50.

Keeney, R., and Raiffa, H. (1976). *Decisions with multiple objectives: Preferences and value tradeoffs*, Wiley, New York.

Lloyd-Jones, T., et al., eds. (2009). *The built environment professions in disaster risk reduction and response: A guide for humanitarian agencies*, Max Lock Center, London. <http://www.preventionweb.net/english/professional/publications/v.php?id=10390> (Dec. 12, 2012).

Meadows, D. (2008). *Thinking in systems*, Chelsea Green Publishing, White River Junction, VT.

Multihazard Mitigation Council. (2005). "Natural hazard mitigation saves: An independent study to assess the future savings from mitigation activities." National Institute of Building Sciences, Washington, DC.

Mutter, J. (2010). "Disasters widen the rich–poor gap." *Nature*, 466(1042). <http://www.nature.com/nature/journal/v466/n7310/abs/4661042a.html> (Jan, 14, 2011).

National Research Council (NRC). (2012). "Disaster resilience: A national imperative." National Academies Press, Washington, DC.

Norris, F., et al. (2008). "Community resilience as a metaphor, theory, set of capacities, and strategy for disaster readiness." *American J. Community Psychology*, 41, 27–150.

Norris, F., Pfefferbaum, B., and Smarick, K. (2011). "An integrated conception of community resilience." National Consortium for the Study of Terrorism and Responses to Terrorism, College Park, MD. <http://www.start.umd.edu/research-projects/integrated-conception-community-resilience> (April 20, 2014).

Peduzzi, P., et al. (2009). "Assessing global exposure and vulnerability towards natural hazards: The disaster risk index." *Nat. Hazards Earth Syst. Sci.*, 9, 1149–1159.

Pfefferbaum, R. L., Pfefferbaum, B., and Van Horn, R. L. (2011). "Community advancing resilience toolkit (CART): The CART integrated system." Terrorism and Disaster Center at the University of Oklahoma Sciences Center, Oklahoma City. <http://www.oumedicine.com/docs/default-source/ad-psychiatry-workfiles/here.pdf?sfvrsn=0> (Feb. 5, 2012).

San Francisco Planning and Urban Research Association (SPUR). (2009). "The resilient city: Defining what San Francisco needs from its seismic mitigation policies." San Francisco Planning and Urban Research Association, San Francisco. <http://www.spur.org/sites/default/files/publications_pdfs/SPUR_Seismic_Mitigation_Policies.pdf> (Feb. 5, 2012).

San Francisco Planning and Urban Research Association (SPUR). (2011). "The resilient city: Ideas and actions for a better city." San Francisco Planning and Urban Research Association, San Francisco. <http://www.spur.org/policy/the-resilient-city> (Feb. 5, 2012).

Sherrieb, K., Norris, F. H., and Galea, S. (2010). "Measuring capacities for community resilience." *Social Indicators Research*, 99(2), 227–247.

Sphere Project. (2011). *Humanitarian charter and minimum standards in humanitarian response*, Practical Action Publishing, Rugby, U.K.

Tearfund. (2011). *Reducing risks of disasters in our community*, Tearfund, Teddington, U.K. <http://tilz.tearfund.org/~/media/Files/TILZ/Publications/ROOTS/English/Disaster/ROOTS%209%20Reducing%20risk%20of%20disaster.pdf> (Oct. 22, 2012).

Tierney, K. (2008). "Overview of resiliency: A working goal." Presentation at the National Earthquake Conference, Seattle, WA, April 22–26, FEMA, Washington, DC.

Tierney, K. (2009). "Disaster response: Research findings and their implications for resilience measures." Research Report 6, Community and Regional Resilience Institute (CARRI), Oak Ridge, TN. <http://resilientus.wp.in10sity.net/publications/research-reports/#> (March 5, 2012).

United Nations Development Programme (UNDP). (2004). "Reducing disaster risk: A challenge for development." United Nations Development Programme, New York. <http://www.un.org/special-rep/ohrlls/ldc/Global-Reports/UNDP%20Reducing%20Disaster%20Risk.pdf> (March 7, 2012).

United Nations Development Programme Human Development Report (UNDP/HDR). (2011). "Sustainability and equity, a better future for all." United Nations Development Programme, New York. <http://hdr.undp.org/en/media/HDR_2011_EN_Complete.pdf> (April 22, 2012).

United Nations Office for Disaster Risk Reduction (UNISDR). (2007). "Hyogo framework for action 2005–2015: Building the resilience of nations and communities to disasters." United Nations Office for Disaster Risk Reduction, Geneva. <http://www.unisdr.org/files/1037_hyogoframeworkforactionenglish.pdf> (April 22, 2012).

United Nations Office for Disaster Risk Reduction (UNISDR). (2010). "A guide for implementing the Hyogo framework for action by local stakeholders." United Nations Office for Disaster Risk Reduction, Geneva. <http://www.unisdr.org/files/13101_ImplementingtheHFA.pdf> (April 23, 2012).

United Nations Office for Disaster Risk Reduction (UNISDR). (2011a). "Hyogo framework for action 2005–2015: Building the resilience of nations and communities to disasters—Mid-term review." United Nations Office for Disaster Risk Reduction, Geneva. <http://www.unisdr.org/files/18197_midterm.pdf> (April 23, 2012).

United Nations Office for Disaster Risk Reduction (UNISDR). (2011b). "Third session of the global platform for disaster risk reduction." United Nations Office for Disaster Risk Reduction, Geneva. <http://www.unisdr.org/files/18931_gp11programme.pdf> (April 24, 2012).

United Nations University (UNU). (2011). "World risk report 2011." Bündnis Entwicklung Hilft (Alliance Development Works), Berlin. <http://www.ehs.unu.edu/file/get/9018> (April 27, 2012).

University of Wisconsin Disaster Management Center (UW-DMC). (1997). "Disasters and development: Study guide and course text for C280-DD002." Disaster Management Center, University of Wisconsin-Madison.

U.S. Agency for International Development (USAID). (2012). "Building back housing in post-disaster situations—Basic engineering principles for development professionals: A primer." USAID, Washington, DC. <http://buildchange.org/pdfs/USAID_Building_Back_Housing_in_Post-Disaster_Situations-A_Primer.pdf> (Feb. 5, 2013).

World Health Organization (WHO). (2007). "Risk reduction and emergency preparedness. WHO six-year strategy for the health sector and community capacity development." Geneva.

Zolli, A., and Healy, A. M. (2012). *Resilience—Why things bounce back*, Free Press, New York.

12

Project Execution, Assessment, and Sustainability

12.1 From Work Plan to Project Execution

Now that the project work plan discussed in Chapter 8 is complete, execution of the project is the next step. To recap, a logical framework has been established with clearly outlined project goals and objectives to respond to problems identified by outsiders in participation with the community. The project work plan includes the list of activities, indicators, means of verification, and assumptions. It contains a detailed schedule of activities and the resources or inputs (human, physical, and financial) necessary to conduct those activities. The work plan may extend over one year or several years.

In Chapters 9, 10, and 11, additional analysis tools were presented that supplement, confirm, change, and refine the decisions outlined in the work plan developed in Chapter 8. They include capacity analysis, risk analysis, and resilience analysis. Capacity analysis helps (1) ensure that the community has the capacity to move forward with the proposed action plan, (2) ensure that the solutions in the work plan match the level of community development and therefore refine what technologies and solutions are most appropriate, (3) identify the weakest links in the community, and (4) determine the necessary steps in eliminating those weak links through community capacity building so that the community can achieve a higher level of development over time. In addition, risk analysis explores the possibility of undesired outcome (or absence of desired outcome) in the different phases of the implementation plan that may result from the complex and uncertain nature of development projects. Project-related risks may supplement risks that existed before the project and in some cases may exacerbate them. The analysis of risks and their effects contributes to the identification, prioritization, resolution, and monitoring of risks, four steps that enter into risk management. Finally, resilience analysis explores how communities are resilient to various forms of adverse events by comparing their acquired capacity or readiness with their vulnerability.

Throughout the entire process described in the ADIME-E framework, one should not lose sight of community engagement and participation, especially when deciding about what will demonstrate success and the project's impact. As remarked by the Mercy Corps (2005), "only the participants themselves can define what success would mean for them and they can suggest ways that information can be collected and measured."

Success is not defined by outsiders only. Insiders must be involved in the different phases of the work plan, including project execution, closing, and ensuring its sustainability. At this particular phase in the ADIME-E framework, a convergence must occur among decisions and solutions made from the bottom up (grassroots community level), those from the top down (local and national government), and those from the outside in (consultancy from outsiders). One cannot emphasize enough the need for synergy among those three groups of stakeholders.

It is important to note that because of the uncertainty associated with development projects, even though the work plan may be thorough, it still remains a dynamic process that is likely to change during project execution. As the project proceeds and more data become available, questions arise and decisions may have to be made about changing goals and objectives. The project management team, in collaboration with community members, must be fully aware of that process and act accordingly. Furthermore, the following must be in place before project execution: (1) a monitoring and evaluation strategy to monitor and measure progress and (2) a multistakeholder decision process that enables the making of adjustments. In addition, a plan must be in place to decide when to close a project, how to ensure its long-term benefits (sustainability), and when and how to scale up the project.

12.2 Project Assessment

Throughout the entire life cycle of a project, a need exists to track or assess how implementation is taking place and review progress toward achieving the goals and objectives outlined in the logical framework and the comprehensive work plan discussed in Chapter 8. If necessary, corrective actions may need to be taken and the logical framework may need to be refined and updated, as shown in the iterative process of Figure 4-2. Being able to manage change is critical to the success of development projects.

As remarked in Chapter 4, projects in developing communities often fail for multiple reasons, some related to poor planning and management, and others because of unexpected consequences despite good planning and management. As remarked by Valadez and Bamberger (1994), "Little is known about how well projects are able to sustain the delivery of services over time, and even less about the extent to which projects are able to produce their intended impacts."

Although progress has been made to integrate assessment into the project and program frameworks of development agencies since that remark was made in 1994, assessment is still often conducted out of necessity and in an ad hoc way at the project level. As remarked by Mansouri and Rao (2012) for World Bank projects, project assessment through monitoring and evaluation remains a low priority for project and program managers, with not much incentive from the World Bank itself. In general, assessment can vary in scope and depth depending on allocated time and resources in the project life cycle. There is a need to enforce assessment standards and procedures in all development projects, no matter how big or small. After all, there is enough evidence in the planning and management of projects in the developed world to demonstrate that assessment contributes in part to project quality assurance and quality control.

In many ways, project assessment helps project stakeholders answer the question, "How well are the project and its components doing?" That question and many others need to be asked continually during the project life cycle. They may apply to the entire project and/or separate activities and programs. As the project unfolds, a need also exists to continuously review and modify (if necessary) the strategy being used and how the project is managed. Table 12-1 from Nolan (2002) shows how the nature of assessment changes as the project unfolds and lists several key questions that may arise in each step. A more detailed description of assessment in the different phases of project planning and management can be found in Valadez and Bamberger (1994).

Table 12-1. Assessment Needs in the Project Development Cycle

Point in the Cycle	Questions to Answer
Project startup	Is the design workable?
	Are resources adequate?
	Is the project team performing as expected?
	Were our assumptions and estimates correct?
Mid-project	Are things happening on schedule?
	Are there foreseen problems?
	Are people's reactions positive?
	Do any modifications need to be made now?
Project end	Are beneficiaries benefiting?
	Are objectives being achieved?
	Are modifications or design departures working?
	Have other changes occurred?
	Have any negative outcomes appeared?
Postproject	Have changes stabilized?
	Have benefits remained?
	Have new possibilities been created?

Source: Nolan (2002), with permission from Westview Press.

As noted by Nolan (2002), in the short term, assessment helps track the project and make corrections. It also provides an opportunity to work with the community if it needs assistance, decide whether assistance can be provided, and determine what kind of assistance best matches the issues faced by the community (Howard-Grabman and Snetro 2003). This time can be seen as a trouble-shooting phase where the issues that arise are addressed as they develop. According to Howard-Grabman and Snetro (2003), some of those issues can be technical (equipment, resources, and supplies) or nontechnical (blocked action from some stakeholders, loss of interest from critical groups, competing interests, and funding issues).

In the medium term, assessment helps make decisions about continuing the project, determine whether the project strategies are working, and identify whether the community is taking more responsibility and ownership for the project. Finally, in the long term, it helps to (1) learn lessons about what has worked and not worked (reflective practice), (2) design corrective measures for future projects, and (3) ensure long-term benefits (sustainability). In all cases, project assessment requires a right balance between making observations, asking questions, conducting self-evaluation, making decisions, and taking appropriate actions. Furthermore, it is an ongoing process that needs to be designed, integrated into the work plan, and used right from the start of any project. Project assessment continues well after the project has been executed to ensure that the project benefits are long-lasting.

Project assessment requires that a community baseline be established beforehand, as discussed in Chapter 5, which is an outcome of the community appraisal phase in the ADIME-E framework. It also requires that starting from that baseline,

- Indicators, benchmarks, and targets are laid out before (and not after) project execution;
- A protocol is established to collect, analyze, disseminate, and use the data generated as the project unfolds over time;
- An agreement has been reached among project stakeholders about what constitutes short-, medium-, and long-term project impact and success; and
- An action plan for making corrective decisions (feedback loop) is in place.

As for all phases of the ADIME-E framework, participatory input from the community and other project stakeholders should be included in project assessment as much as possible. It is obtained using the same participatory action research methods of data collection and analysis discussed in Chapter 5 for community appraisal. For the UNDP (2009), participation is institutionalized in project assessment. Furthermore, integrated into project assessment is an inherent reflective and adaptive practice that occurs as the project unfolds (Caldwell

2002). As discussed in Chapter 4 in relation to the work by Schön (1983), such a practice contributes to making adaptive and sound management decisions as the project (or an activity) is happening (reflection in action) or after it has been completed (reflection on action). Finally, it helps in deciding on a project exit strategy.

12.3　Project Monitoring and Evaluation

Project assessment consists of two distinct components called monitoring and evaluation, the "ME" part of the ADIME-E framework's acronym. They are often confused with each other in the development literature. Even though they have marked differences, as summarized in Table 12-2, they are also closely linked because evaluation requires that monitoring be conducted first. They are usually combined under a so-called *monitoring and evaluation work plan*, which according to Barton (1997), is defined as "the collection and management of data to be analyzed and used for regular and periodic assessment of a project's relevance, performance, efficiency, and impact in the context of its stated objectives."

Essentially, all the frameworks discussed in Chapter 4 require a mandatory monitoring and evaluation work plan, which can vary in breadth and depth across development agencies. For some agencies, it is usually designed and implemented by qualified external consultants (with knowledge, proven expertise, and no conflict of interest) who are not involved in the project implementation, thus providing more objectivity and impartiality in the project assessment. For other agencies, it relies on in-house staff with or without external evaluators. In general, the budget allocated to the monitoring and evaluation of projects is about 5–10% of the total budget (Frankel and Gage 2007).

Several guidebooks are available in the literature on the topics of monitoring and/or evaluation (Valadez and Bamberger 1994; Barton 1997; Bamberger

Table 12-2. Differences between Monitoring and Evaluation

Monitoring	Evaluation
Tracks daily events	Takes a long-range view
Accepts policies and rules	Questions policies and rules
Accepts plans and targets	Asks if plans and targets are accurate and appropriate
Checks works against targets	Checks targets against reality
Stresses input/output relationships	Stresses project purpose and goal
Looks mainly at how project delivery occurs	Looks also at unplanned things, causes, and assumptions
Reports in terms of progress	Reports in terms of lessons learned

Source: Nolan (1998), with permission from Riall W. Nolan.

et al. 2012). In addition, development agencies have their own specific guidelines on how to carry out monitoring and evaluation plans and how to interpret, report, disseminate, and integrate monitoring and evaluation results into their respective project design frameworks. These guidelines vary greatly in what is expected of monitoring and evaluation reviewers in terms of the scope (depth and breadth) of reporting, which in turn depends on the resources (financial, human) that are allocated for that phase of the ADIME-E framework. In writing this chapter, the following monitoring and evaluation frameworks were reviewed: CARE (Barton 1997, 1999; Caldwell 2002); DFID (2002); EuropeAid (2002); Health Communication Partnership (Howard-Grabman and Snetro 2003); Mercy Corps (2005); Millennium Challenge Corporation (MCC 2012); UNDP (2009); USAID (2012); World Bank (2002); and WWF (Gawler 2005).

Monitoring

According to the Mercy Corps (2005), monitoring is a "cycle of regularly collecting, reviewing, reporting, and acting on information about project implementation. Generally used to check our performance against 'targets' as well as ensure compliance with donor regulations." It is mostly a process that can be done at different scales: individuals, households (domestic), groups of interest, and community.

Monitoring helps address whether the project (or part of it) is proceeding according to plan during its implementation and whether performance criteria are met. It can be seen as a continuous process that provides real-time information and data not only about how the action plan is being developed and implemented, but more importantly, about the progress toward achieving the project goals and objectives. It can be understood as a process or formative assessment and calls for a practice of reflection in action from the project's stakeholders (Caldwell 2002). No value and judgment are placed on the collected information.

According to Barton (1997), monitoring can take multiple forms:

- *Institutional monitoring* keeps track of the financial, physical, and organizational issues of a project.
- *Context monitoring* keeps track of the context in which the project is unfolding. Changes in the context may affect the overall work plan, strategy, logistics, and tactics.
- *Results and objectives monitoring* is about keeping track of how the community is responding to the project output, meeting the objectives and goals, and noticing unanticipated side effects and unexpected consequences regarding the project itself and perceptions and behavior of the beneficiaries.

In addition, the Mercy Corps (2005) distinguishes between compliance monitoring and performance monitoring:

- *Compliance monitoring* refers to keeping track of how various project stakeholders and partners are conforming to the development agency guidelines, regulations and requirements, and terms of contracts.
- *Performance monitoring* refers to collecting data and keeping track of progress against targets and whether results are progressing against expectations.

Finally, EuropeAid (2002) recommends for their projects the monitoring of activities, means, and resources; results; assumptions; and impacts.

Evaluation

According to the Mercy Corps (2005), evaluation is an "in depth retrospective analysis of an aspect (or aspects) of a project that occurs at a single point of time. It is generally intended to measure our effects and impacts and examine how we achieve them. This process also captures our experience so that future projects can learn from it."

Evaluation is a discrete event that provides summative assessment at the end of a specific project phase or activity, but more often during midterm and end-of-project review. It is an assessment of completed and ongoing activities and of the extent to which (how well) they are achieving goals and objectives. Value and judgment about project outcomes and impacts are made. Evaluation helps decision makers understand how and to what extent a program is responsible for particular measured results, whether intended or not.

Various methods of evaluation have been proposed in the literature. According to Gawler (2005), they may consist of

- Internal evaluation done within the intervening agency,
- External evaluation using objective outside assistance not connected to the project,
- Self-evaluation by those who manage the project, and/or
- Participatory evaluation with various project stakeholders.

Evaluation calls for a practice of reflection on action from the project's stakeholders (Caldwell 2002). In addition to the *why* of the results, a need exists to determine *how* the results were acquired. Other questions addressed in the evaluation phase of the project (Howard-Grabman and Snetro 2003) may include the following:

- Is the project achieving its goals and objectives?
- Who has benefited from the project or key activities in the project?
- What elements of the project worked and did not work?
- What were noticeable successes and failures?

- What still remains to be executed?
- How has the community's capacity improved and its vulnerability decreased because of the project?
- Is the project sustainable and replicable?
- How can the community build in the future on its successes?

According to Gawler (2005), the evaluation of World Wildlife Fund's (WWF) projects must address five criteria: (1) relevance to the stakeholders; (2) effectiveness in achieving the planned results; (3) efficiency in creating results at a reasonable cost; (4) impact of the project on stakeholders, the communities, and the environment; and (5) sustainability, i.e., the long-term project benefits. Key questions and subquestions are associated with each criterion. Indicators and data to support each criterion are left to the discretion of the users.

Many components of a project need to be evaluated, including activities, tools, strategies, policies, project impact (environmental, social, cultural, economic, institutional), quality, and response of beneficiaries to the project. Evaluation can be qualitative and quantitative and can be conducted at various scales, ranging from individuals to households, community, or even larger. It is obviously limited to the scale at which the appraisal has been conducted and the baseline survey established.

Evaluation can be about performance (process) or about impact. *Performance* evaluation is about putting value and judgment on ongoing activities as the project is being implemented, for instance, halfway into a project. *Impact* evaluation is more about the value of the results (positive and negative) obtained by conducting specific activities and those obtained at the end of a project. Both forms of evaluation are related, because, as noted by Valadez and Bamberger (1994), performance or process evaluation can help "provide an early indication of whether the intended impacts are being produced and can track the trends of these impacts while the project is still under way." In other words, performance evaluation could serve as an ongoing diagnosis tool and help predict whether specific impacts can be expected. Hence, issues (economic, financial, social, and technical) could be detected and corrected well before they become too critical to the project and its stakeholders.

As noted by the World Bank (2002), impact evaluation "provides answers to some of the most central development questions—to what extent are we making a difference? What are the results on the ground? How can we do it better?" Furthermore, impact evaluation provides valuable lessons about how future projects should be conducted. Needless to say, it is of special interest to donors.

Impact evaluation can be presented in a qualitative manner or quantitatively. As an example, Table 12-3 shows the results of an impact evaluation of energy projects conducted in the village of Tulin in western Nepal (Zahnd 2012).

Table 12-3. Anecdotal Impact Evaluation of Renewable Energy Projects in Tulin, Nepal

We are happy to have indoor lighting and life has become much easier.

We can see each other inside the home and feel much more comfortable.

We can see the "dirt" now and clean the house and thus live much cleaner.

We socialize much more and discuss more around the warm stove. Neighbors drop in much more frequently and feel comfortable.

During the cold winter months, the lights are 2–3 hours per day longer in use.

No "jharro" to burn and thus less asthma, coughs, tears, chest pain, heart pain, and less "jharro" collection, resulting in fewer trees being killed.

School children can study 2–3 hours more a day.

Some women have taken up basket weaving and other small income-generating tasks, thus increasing our overall economic income.

Minimal maintenance has been carried out because RIDS-Nepal visits Tulin village periodically and checks the solar photovoltaic system. The local people need to take more interest in raising funds to maintain the system in the future.

The Tulin people cannot imagine going back to the old, traditional way of "jharro" indoor lighting, with the black, thick smoke and minimal brightness.

The house owner on whose roof top the photovoltaic system stands started to demand a monthly rent for the roof space from the local people.

Source: Zahnd (2012), with permission from Alex Zahnd.

The evaluation is presented in an *anecdotal* manner and represents the opinion of the villagers and observations after seven years of use of renewable energy. This evaluation was supplemented with a *quantitative* evaluation of renewable energy projects, i.e., whether the photovoltaic systems met the loads, the batteries discharged too quickly and why, and the effectiveness of local trainers.

In general, impact evaluation may help separate the results that can be directly attributed to the project from those that the project contributed to. This difference is particularly important when conducting the social and environmental impact assessments of a project: what and how much the project is contributing to desirable or undesirable social and environmental changes and what and how much of those changes can be attributed to the project. Being able to decide on results that can be attributed to a project and/or specific activities within that project requires the existence of a control or comparison group side by side with the implementation group. This differentiation is obviously more difficult to do than looking at the overall contribution of a project (Frankel and Gage 2007).

Finally, it is important to note that multiple parameters enter into a project that may limit the systematic evaluation of factors that affect specific outcomes, including other projects that may interact with the one being evaluated. The systems tools discussed in Chapter 6 may help, at least partially, in exploring possible links between project input, resources, and activities on one side and project impacts on the other, by running "what if" simulations. This process may help identify factors that have more impact from those that may have limited influence and can eventually be eliminated. The simulations can be run as the project unfolds. Needless to say, simulations are important when exploring various alternatives in project design, execution, and management. Such "project models" as defined by Valadez and Bamberger (1994) also have their own limitations because they require that project components and their relationships be identified and that a model boundary be selected. As emphasized in Chapter 6, the structure of a model defines its performance and limitations.

Monitoring and Evaluation Plan

Despite being distinct, as shown in Table 12-2, monitoring and evaluation have their share of complementarity: monitoring is about breadth of information, whereas evaluation is about depth. When combined under a plan, monitoring and evaluation

- Help make informed decisions on project course correction or activity adjustment;
- Create accountability and transparency to stakeholders;
- Ensure a more effective and efficient use of resources by addressing whether the cost of a project (or activities) is justified by its outcome and impact through cost–benefit and cost-effectiveness analysis (World Bank 2002; Oxfam 2008);
- Help assess impact and success and, more specifically, determine what can be attributed to the project versus what is contributed by the project;
- Enforce systematic project reporting;
- Require the collection and use of information in a timely and systemic manner;
- Create "forced learning" and knowledge generation about what works and does not work in projects in general (Narayan 1993; Barton 1999; DFID 2002; UNDP 2009); and
- Help assess the replicability and scalability of projects and solutions.

As noted by Caldwell (2002) in the CARE *Project Design Handbook*, a monitoring and evaluation plan needs to be in place at the beginning of any project (e.g., prospective assessment). Furthermore, it is highly recommended that the monitoring and evaluation plan be consistent and in line with the overall strategy and project logic expressed in the logical framework approach (LFA) discussed in Chapter 8. The LFA provides clear definitions of what represents vision, mission, goals, and objectives in a project and how, when combined, these key components yield a clear implementation road map to address the identified problems. The monitoring and evaluation plan goes one step further in looking at how much change is occurring in the logical framework during project implementation (horizontal logic) and what to do about change, especially if unintended consequences arise. As remarked by the Swedish International Development Agency (Sida 2006), "If the project has not been planned properly, with clear objectives and indicators, it will be very difficult to succeed in achieving the objectives in the implementation phase and finally, it will be difficult to evaluate the results."

As a result, the monitoring and evaluation plan uses the same verifiable performance indicators (measurements) and means of verification and relies on the same assumptions as those in the logical framework (see Section 8.2). Because an indicator is a measurement from an agreed-upon baseline, a monitoring and evaluation plan requires that reasonable and appropriate targets, benchmarks, and performance criteria be established along the life cycle of the project, i.e., the startup, implementation, midterm, and end-of-project phases. For each phase of the life cycle of the project, reporting is also required. Table 12-4 shows an example of a monitoring and evaluation plan for a project in Mabu, Nepal. The corresponding logframe matrix was shown in Table 8-1 in Chapter 8.

In addition, when planning monitoring and evaluation, the following questions must be addressed (Barton 1997):

- Why are monitoring and evaluation conducted?
- When do monitoring and evaluation start and end?
- Who is responsible for the monitoring and evaluation tasks and their reporting?
- What resources (personnel, equipment) are required to carry out monitoring and evaluation?
- How and where should the monitoring and evaluation be implemented?
- What information, data, and activities need to be monitored and evaluated to support the indicators?
- How should the data be collected and analyzed and with what methods?
- What is the time line or frequency of data collection, analysis, management, reporting, and dissemination (yearly, monthly, end of project, postproject)?
- What is the monitoring and evaluation reporting format?

In general, the monitoring and evaluation plan is a living and dynamic document that needs to be revised whenever the project action plan and/or some of its components are modified or new data and information become available (Frankel and Gage 2007). Monitoring and evaluation templates have been proposed by Barton (1999) for CARE projects, the Mercy Corps (2005, Attachments D and G), the UNDP (2009, chapters 4–6 and annexes), and Frankel and Gage (2007).

As remarked by Barton (1997) and Caldwell (2002), the execution of the monitoring and evaluation work plan creates an information management

Table 12-4. Monitoring and Evaluation Matrix for the Mabu, Nepal, Project

Hierarchy of Objectives	Type of Information	Sources of Information	Methods of Data Gathering	Who Collects	Frequency of Reporting
2.1. Electricity available for homes served by pico-hydro system	Five years after project end: Electric lighting available in 75% of homes not previously lit	All community members	Interviews	NCDC Community Volunteers	M—yearly E—yearly
2.1.1 Pico-hydro system installed	2–3 pico-hydro plants built per canal	Site visit	Site visit	SCD team/ NCDC	M—seasonally E—yearly
2.1.2 Maintenance workers trained	12–15 workers completed training program at time of plant construction	Training program records	Attendance records, etc.	SCD team/ NCDC	M—seasonally E—yearly
2.1.3 Technology adoption training completed	All houses served received training in usage of electric appliances	All community members	Interviews	SCD team/ NCDC	M—seasonally E—every 2 years
3.1. Growth of small businesses	1 small business using energy at each pico-hydro plant	NCDC observation	Interviews	SCD team/ NCDC	M—yearly E—every 2 years
3.1.1 Business training programs	50–60 people trained in new business development	Training program records	Attendance records, etc.	SCD team/ NCDC	M—every 6 months E—yearly
3.1.2 Microloan programs	20 people have received loans for new businesses and are current with repayments	Bank records	Loan financial reports	Outside microfinance organization	M—every 6 months E—yearly

Note: For goals 2 and 3 in Table 8-1. NCDC = Namsaling Community Development Center, SCD = Sustainable Community Development Team.

Source: Glover et al. (2011), reproduced with permission.

system that can be used for planning, management, and/or reporting by a range of stakeholders, including the target community members, local organizations, the government (local, regional, national), the project staff, agency country offices, and donors (funders, external supporting organizations), as the project unfolds. The information management system needs to have its own management structure in place, which could be a challenge for some projects. Another challenge is how to reach consensus in any decision process that involves multiple project stakeholders.

In general, monitoring and evaluation is not the most exciting phase of any development project and is often conducted out of necessity, especially when funding is running out and there is a need to apply for future funding from donors. In fact, it is not uncommon for development agencies to bring external reviewers in when a project is already under way (concurrent assessment) or well under way (retrospective assessment), rather than ideally at the start of the project (prospective assessment). Under such conditions, the depth, breadth, and attributes (objectivity, reliability, dependability, validity, credibility, authenticity, and transferability) of the project assessment that is being asked of the reviewers then depend largely on three parameters:

1. The amount of funding and time allocated for that phase of the project by the agency;
2. Whether a community baseline survey already exists and the quality and quantity of data available in that survey; and
3. The existence and quality of a comprehensive work plan, scope of work, and a logical framework for the project.

As remarked by Bamberger et al. (2004), the constraints of limited funds, and/or limited funding, and/or limited or no baseline data may force external reviewers to conduct so-called "shoestring evaluations" with a limited amount of data available, if any. Political constraints and time allocated for the evaluations may also contribute to how evaluations are carried out (Bamberger et al. 2012). Table 12-5 describes several scenarios of shoestring evaluation under constraint that can occur when projects unfold. Regrettably, those scenarios are more often the norm than the exception and require innovative approaches from external reviewers, project managers, and agencies to deliver "robust and useful evaluation findings when working with real-world constraints" (Bamberger et al 2004).

Likewise, carrying out the monitoring and evaluation plan is not free of challenges and biases for essentially the same reasons that were outlined in Section 5.4 regarding community appraisal. Furthermore, it should also be conducted legally, ethically, and with regard to those involved in and affected by the project. Finally, the same precaution about what represents high-quality data and information in monitoring and evaluation needs to be used, the same as in the appraisal phase (see Section 5.6).

Table 12-5. Conducting Impact Evaluations with Time, Budget, Data, or Political Constraints

Time	Budget	Data	Political	Typical Evaluation Scenarios
\multicolumn{4}{l}{Constraints under which evaluation must be conducted}				
X				The evaluator is called in late in the project and told that the evaluation must be completed by a certain date so that it can be used in a decision-making process or contribute to a report. The budget may be adequate, but it may be difficult to collect or analyze survey data within the time frame.
	X			The evaluation is allocated only a small budget, but there is no excessive time pressure. However, it will be difficult to collect sample survey data because of the limited budget.
		X		The evaluator is not called in until the project is well advanced. Consequently, no baseline survey has been conducted either on the project population or on a comparison group. The evaluation does have an adequate scope, either to analyze existing household survey data or to collect additional data. In some cases, the intended project impacts may also concern changes in sensitive areas such as domestic violence, community conflict, women's empowerment, community leadership styles or corruption, on which it is difficult to collect reliable data even when time and budget are not constraints.
			X	The funding agency or a government regulatory body has requirements concerning acceptable evaluation methods [...] In other cases, a client or funding agency may specifically request qualitative data, tests of statistical significance regarding measured program effects, or both.
			X	There is overwhelming indication that the evaluation is being commissioned for political purposes. For example, an evaluation of the effects of conservation policy might be commissioned to stall its expansion.

Table 12-5. Conducting Impact Evaluations with Time, Budget, Data, or Political Constraints (*Continued*)

Constraints under which evaluation must be conducted				
Time	**Budget**	**Data**	**Political**	**Typical Evaluation Scenarios**
			X	There is reason to suspect that the evaluation will be used for political purposes other than or contrary to those articulated in preliminary discussions. [...]
X	X			The evaluator has to operate under time pressure and with a limited budget. Secondary survey data may be available, but there is little time or few resources to analyze them.
X		X		The evaluator has little time and no access to baseline data or a comparison group. Funds are available to collect additional data, but the survey design is constrained by the tight deadlines.
	X	X		The evaluator is called in late and has no access to baseline data or comparison groups. The budget is limited, but time is not a constraint.
X	X	X		The evaluator is called in late, is given a limited budget, and has no access to baseline survey data; and no comparison group has been identified.

Source: Bamberger et al. (2012), with permission from Sage Publications Ltd.

12.4 From Assessment to Corrective Action

Monitoring and evaluation are necessary to answer the question, "How well are the project and its components doing?" If the answer is negative or partially negative during the project life cycle, the corollary question should address corrective actions that need to be implemented, how they should be implemented, who should implement them, and when and where they should be implemented.

It is impossible to have a multitude of contingency plans on hand to address all the things that could go wrong on a project. As discussed in Chapter 4, uncertainty and complexity are integral parts of small-scale projects in developing communities and need to be integrated into the design. More importantly, what counts are not just the actions that need to be taken but also how to address the process in which to take such actions. As discussed in Chapter 4, alternatives can be laid out as the project is unfolding. The engineering profession may help in identifying such alternatives. For instance, they can be based on recognizing similarities with other projects (for which solutions have been developed), which obviously requires experience from the project management team (Valadez and Bamberger 1994). Another option is to come up with unique solutions to the particular problems by examining the impact, costs, and consequences of various alternative solutions. A third and more drastic option is to reduce the scope (impact, goals, and objectives) of the project or even stop it entirely. The challenge that arises when looking at alternatives is to know when to incorporate the feedback of the various project stakeholders when trying to reach an agreement. As noted by Caldwell (2002), a reflective and adaptive approach (Table 4-3) is better suited than a rigid approach when handling unintended consequences.

12.5 Exit Strategy, Ensuring Long-Term Benefits, and Scaling Up

Project Closure

All projects come to an end sooner or later. Regrettably, more often than not in projects in developing communities, an exit strategy is not planned beforehand and is done on an ad hoc basis as the project comes to closure. Development agencies are notorious for pulling the plug on projects whether successful or not, often to the surprise of the community, the development workers, and the project stakeholders. The decision to close a project is often made by individuals who have limited vested interest in the project and for reasons that have nothing to do with the project itself, even though the project may be successful. Closing a project can be the result of a change in policies within the agency, in-country political reasons, and changes in donor priorities, among many others. In the case of charity groups and some nongovernmental organizations, their ready–fire–aim approach to projects does not even include an entry strategy or an exit strategy. This approach is driven more by activity than results.

Whether the decision to close a project is arbitrary or is based on well defined indicators of success or failure, an exit strategy (or closure plan) needs to be in place at or shortly after project start-up. The metric used to make that decision, along with its indicators, needs to have well defined tasks with targets and benchmarks and must be based on clear performance criteria. Resources must also be available to ensure a successful project closure that benefits all stakeholders in the short and long term (CH2M HILL 1996). The closure plan can be updated as the project unfolds, but it must be in place and not seen as an afterthought.

Long-Term Benefits

The long-term benefits (sustainability) of a project must be addressed as soon as possible in the project planning and management. The rationale is that the continued operation of a project must still deliver the benefits to the community of interest well after the project is officially closed and support from the external agency has ended. More specifically, project sustainability includes addressing the long-term economic, social, environmental, and institutional impact of the project on the community and ensuring that the community has the capacity (resources, skills, funding) and external assistance (if necessary) to maintain the project and handle planned or unexpected events. More often than not, the long-term benefits of projects are only partially addressed by development agencies because it is often difficult to find funding to support such a study once the project has been executed (Barton 1997). Lack of funding, along with a multitude of other technical and nontechnical reasons, may prevent a project from having a long life cycle.

EuropeAid (2002) places a special emphasis on sustainability after project closure. They have identified several quality control and assurance criteria that are likely to contribute to the long-term benefits of a project:

- The beneficiaries own the project. This is a direct extension of the participatory nature of development projects. If there is ownership and the beneficiaries know how to own the project, they are more likely to support the project and maintain it.
- Policy support exists at the local level, regional level, or country level. Once the project is in place, a question arises about whether in-country institutions will continue to invest in the project, which greatly depends on the in-country institutional capacity.
- The implemented solutions are sustainable. For infrastructure and technical projects, a need exists to have maintenance in place and to provide local maintenance training, if necessary. Human resources and technical capacity must be in place.
- The project is respectful of cultures and attitudes and ensures proper access to project services.
- Rights-based issues are being addressed by making sure that the project is not negatively affecting human dignity and gender equality.
- The project is not likely to negatively affect the environment and deplete natural resources.
- The community is able to follow up with the project in terms of management and institutions.
- The project represents an added value and a long-term investment to the community.

Table 12-6 lists several questions associated with the aforementioned criteria (EuropeAid 2002). Similar criteria have been suggested by the World Wildlife Fund (Gawler 2005) to ensure sustainability of its projects. In general, long-term project benefits require postproject monitoring and evaluation over several years, if resources (financial, human) are available.

Table 12-6. Basic Questions to Address Long-Term Project Benefits

Ownership by beneficiaries	What evidence is there that all target groups, including both women and men, support the project? How actively are and will they be involved/consulted in project preparation and implementation? How far do they agree and commit themselves to achieve the objectives of the project?
Policy support	Is there a comprehensive, appropriate sector policy by the government? Is there evidence of sufficient support by the responsible authorities to put in place the necessary supporting policies and resource allocations (human, financial, material) during and after implementation?
Appropriate technology	Is there sufficient evidence that the chosen technologies can be used at affordable cost and within the local conditions and capabilities of all types of users, during and after implementation?
Environmental protection	Have harmful environmental effects that may result from use of project infrastructure or services been adequately identified? Have measures been taken to ensure that any harmful effects are mitigated during and after project implementation?
Sociocultural issues	Does the project take into account local sociocultural norms and attitudes, including those of indigenous people? Will the project promote a more equitable distribution of access and benefits?
Gender equality	Have sufficient measures been taken to ensure that the project will meet the needs and interests of both women and men and will lead to sustained and equitable access by women and men to the services and infrastructures, and contribute to reduced gender inequalities in the longer term?
Institutional and management capacity	Is there sufficient evidence that the implementing authorities will have the capacity and resources (human and financial) to manage the project effectively and to continue service delivery in the longer term? If capacity is lacking, what measures have been incorporated to build capacity during implementation?
Economic and financial viability	Is there sufficient evidence that the benefits of the project will justify the cost involved and that the project represents the most viable way to address the needs of women and men in the target groups?

Source: EuropeAid (2002), with permission from Particip GmbH.

Scaling Up

Situations may occur when the success and continuing success of a project and its tangible benefits may warrant a scaling up, i.e., an expansion of the project at a larger scale, using the same project scope or using a new scope, within the same community level or by involving other communities at the local or regional level. As remarked by Howard-Grabman and Snetro (2003), the challenge is to expand the project without negatively affecting the initial project outcome. Scaling up can be done in incremental steps (recommended) or by rapidly jumping to a larger scale. It can take place within a community (vertical scaling) or across communities (horizontal scaling). In general, the former has a better chance of succeeding, because the solutions at the community level are more in line with the community itself. They don't always transfer to other communities.

Scaling up can be seen as reentering the general project life cycle of Figure 4-6 with considerably more information about the community than the first cycle. The various forms of analysis (e.g., capacity, vulnerability, and stakeholder) have to be reexamined any time a project scales up and new data are being considered. According to Howard-Grabman and Snetro (2003), several questions need to be raised before deciding to scale up:

- Is scaling warranted? Are there real needs the community must address or are they just perceived?
- Is there capacity to embark in scaling up while maintaining project quality?
- Is there enough political and funding support for scaling up?
- Are resources, skills, and knowledge available for scaling up to be successful?
- Will scaling up encounter policy roadblocks at the local, regional, or national level, and what would it take to overcome those constraints?

12.6 Chapter Summary

This chapter emphasizes the importance of assessment (monitoring and evaluation or M&E) in the life cycle of development projects. Assessment helps answer the questions about how well a project and its components are doing. Based on the answers to these questions, appropriate corrective actions can be taken. In the assessment plan, monitoring is a continuous process that provides real-time information and data, not only about how the action plan is being developed and implemented, but more importantly about the progress toward achieving the project goals and objectives. Evaluation is a discrete event that provides summative assessment at the end of a specific project phase or activity, but more often during midterm and end-of-project review. In general, monitoring and evaluation are complementary.

Even though monitoring and evaluation are integral components of project frameworks adopted by major development agencies, they are often seen as obligations rather than as added values in project design. In general, assessment can vary in scope and depth, depending on allocated time and resources in the project life cycle. Finally, project managers often lack incentives to conduct project assessment. Unfortunately, this dearth often leads to projects in the developing world that are short lived and of limited effectiveness. This observation is contrary to project management in the developed world, where clear evidence shows that assessment contributes in a large part to project quality assurance and quality control.

This chapter also emphasizes the importance of developing a strategy to decide on project closure and scaling up. As discussed in the last chapter, projects need to be designed for long-term benefits (sustainability) as early as possible in the project life cycle. The same can be said about deciding on when a project reaches the end of its cycle. Failing to do so as early in the project design as possible may contribute to ill-defined projects with unrealistic or disappointing outcomes. Project management in the developing world often lacks criteria for project closure, long-term benefits, and scaling up.

References

Bamberger, M., Rugh, J., and Fort, L. (2004). "Shoestring evaluation: Designing impact evaluations under budget, time, and data constraints." *American J. Evaluation*, 25, 5. <http://aje.sagepub.com/content/25/1/5.full.pdf+html> (Jan. 2, 2012).

Bamberger, M., Rugh, J., and Mabry, L. (2012). *RealWorld evaluation: Working under budget, time, data, and political constraints.* Sage Publ., Thousand Oaks, CA.

Barton, T. (1997). "Guidelines for monitoring and evaluation: How are we doing?" CARE International in Uganda, Monitoring and Evaluation Task Force, Kampala, Uganda. <http://portals.wdi.wur.nl/files/docs/ppme/Guidelines_to_monitoring_and_evaluation%5B1%5D .pdf> (Jan. 10, 2013).

Barton, T. (1999). "Project monitoring and evaluation plans." Creative Research and Evaluation Center, Kampala, Uganda. <http://gametlibrary.worldbank.org/FILES/49_Project%20M&E %20plan%20Template.pdf> (Jan. 10. 2013).

Caldwell, R. (2002). *Project design handbook*, Cooperative for Assistance and Relief Everywhere (CARE), Atlanta.

CH2M HILL (1996). "Project delivery system: A system and process for benchmark performances." CH2M HILL and Work Systems Associates, Inc., Denver.

Department for International Development (DFID). (2002). "Tools for international development." Department for International Development, London.

EuropeAid. (2002). *Project cycle management handbook*, version 2.0, PARTICIP GmbH, Fribourg, Germany. <http://www.sle-berlin.de/files/sletraining/PCM_Train_Handbook_EN -March2002.pdf> (March 10, 2012).

Frankel, N., and Gage, A. (2007). "M&E fundamentals: A self-guided minicourse." U.S. Agency for International Development (USAID), Washington, DC. <http://pdf.usaid.gov/pdf_docs/ PNADJ235.pdf> (April 22, 2012).

Gawler, M. (2005). "WWF introductory course: Project design in the context of project cycle management." Artemis Services, Prévessin-Moëns, France <http://www.artemis-services.com/downloads/sourcebook_0502.pdf> (March 15, 2012).

Glover, C., Goodrum, M., Jordan, E., Senesis, C., and Wiggins, J. (2011). "Mabu village term project." CVEN 5929: Sustainable Community Development 2, University of Colorado at Boulder.

Howard-Grabman, L., and Snetro, G. (2003). "How to mobilize communities for health and social change." Center for Communication Programs, Baltimore. <http://www.jhuccp.org/resource_center/publications/field_guides_tools/how-mobilize-communities-health-and-social-change-20> (March 22, 2012).

Mansouri, G., and Rao, V. (2012). *Localizing development: Does participation work?* World Bank Publications, Washington, DC.

Mercy Corps. (2005). *Design, monitoring and evaluation guidebook*, Mercy Corps, Portland, OR. <http://www.mercycorps.org/sites/default/files/file1157150018.pdf> (March 15, 2012).

Millennium Challenge Corporation (MCC). (2012). "Policy for monitoring and evaluation of compacts and thresholds programs." Millennium Challenge Corporation, Washington, DC. <http://www.mcc.gov/pages/results/evaluations#668> (June 24, 2012).

Narayan, D. (1993). *Participatory evaluation: Tools for managing change in water and sanitation*, World Bank, Washington, DC. <http://www.chs.ubc.ca/archives/files/Participatory%20Evaluation%20Tools%20for%20Managing%20Change%20in%20Water%20and%20Sanitation.pdf> (Dec. 10, 2012).

Nolan, R. (1998). "Projects that work: Context-based planning for community change." Unpublished report. Department of Anthropology, Purdue University, West Lafayette, IN.

Nolan, R. (2002). *Development anthropology: Encounters in the real world*, Westview Press, Boulder, CO.

Oxfam. (2008). *Handbook of development and relief*, Vols. 1 and 2, Oxfam Publ., Cowley, Oxford, U.K.

Schön, D. A. (1983). *The reflective practitioner: How professionals think in action*, Basic Books, New York.

Swedish International Development Agency (Sida). (2006). "Logical framework approach—With an appreciative approach." Swedish International Development Agency, Stockholm, Sweden. <http://www.sida.se/Publications/Import/pdf/sv/Logical-Framework-Approach—with-an-appreciative-approach_1691.pdf> (Nov. 2, 2012).

United Nations Development Programme (UNDP). (2009). *Handbook on planning, monitoring and evaluating for development results*, United Nations Development Programme, New York. <http://web.undp.org/evaluation/handbook/documents/english/pme-handbook.pdf> (May 22, 2012).

U.S. Agency for International Development (USAID). (2012). Automated Directives System (ADS) Chapter 203: "Assessing and learning." U.S. Agency for International Development, Washington, DC. <http://transition.usaid.gov/policy/ads/200/203.pdf> (Oct. 3, 2012).

Valadez, J., and Bamberger, M. (1994). *Monitoring and evaluating social programs in developing countries: A handbook for policymakers, managers, and researchers*, World Bank Publications, Washington, DC.

World Bank. (2002). "Monitoring and evaluation: Some tools, methods and approaches." World Bank, Washington, DC. <http://www.worldbank.org/oed/ecd/tools/> (April 4, 2012).

Zahnd, A. (2012). "The role of renewable energy technology in holistic community development." Doctoral dissertation, Murdoch University, Australia.

13

Service Delivery in Development Projects

13.1 Delivering Services Rather than Technology

A confusion that is often made by engineers and technologists involved in small-scale community projects in developing countries is that once the community has been appraised, problems have been identified and ranked, and solutions outlined, the next step is to deliver technology. Such an *activity-driven* approach often results in short-lived projects with limited impact. A more comprehensive and impactful approach, discussed in this chapter, is to look at development projects as delivering meaningful (*results-driven*) services to address the needs of communities, their households, and the individuals in those households.

A case in point is the difference between providing "energy" and "energy services" to a community. As remarked by Reddy (2000), "what human beings want is not oil or coal, or even gasoline or electricity per se, but the services that those energy sources provide." Possessing "energy" is not an end in itself. Integrating the energy into the development of the community and using technology to (1) provide services to its members such as cooking, heating, lighting, water pumping, refrigeration, communication, and transportation and (2) create income-generating activities is more in line with what the community is expecting. The same discussion applies to roads as well. People want good roads and reliable transportation for access to markets, employment, education, health centers, goods, and multiple services (Fox and Porca 2001; Africa Union 2005; Bryceson et al. 2008).

Of special interest in community needs are the physiological and safety needs listed in the two bottom layers of Maslow's pyramid (Figure 2-2). Addressing these needs, however, does not mean ignoring or undervaluing the higher forms of needs, such as love and belonging, self-esteem, and self-actualization. As noted by Birley (2011) in regard to the WHO definition of health, well-being is defined by many components, such as physical, social, mental, *and* spiritual. In fact, development is not development without addressing spiritual needs along with the other needs, as long as it does not require adopting belief systems that are oppressive and harmful to any segment of the population. Furthermore, as discussed in Chapter 2, higher forms of needs do not necessarily have to be addressed in a hierarchical manner *after* those at the bottom of Maslow's pyramid are met. They can be addressed at the same time, as successfully demonstrated by the Sarvodaya Shramadana Movement in Sri Lanka or the approach envisioned by the country of Bhutan in its vision for 2020 (Bhutan 1999). But it is important to acknowledge other cases, where belief systems have slowed down development and created entrapment for the poor, mostly women. The caste system and a culture of fatalism in Nepal and India are good examples of roadblocks to equitable development (Bista 2001; Zahnd 2012).

In general, the various forms of basic service delivery that enter into sustainable community development projects are related to energy; water, sanitation, and hygiene (WASH); shelter; and health. Some of those services are discussed in Chapters 14 and 15. Others include food and nutrition, transportation, clothing, and communication. All those services interact in a systemic way and are all inter- and intradependent. For instance, to provide health services, a health clinic requires electricity for light, sterilization, and water heating. It also needs access to water and ways to dispose of solid, liquid, and gaseous waste. The clinic also needs access to means of transportation, roads, and modes of communication.

Likewise, a strong link exists between water and energy. Energy is needed to pump, store, distribute, treat, and recycle water. Similarly, water is necessary and often in large quantities to produce energy from hydropower, conventional fuels (which still provide 80% of energy needs worldwide), and biofuel from crops. Water is also needed in food production and agriculture. Therefore, it is clear that water stresses can have dramatic consequences on a country's economy at the global, regional, or local level because water is related to so many issues.

Meeting the eight Millennium Development Goals (MDGs) and associated targets requires providing many complementary services. For instance, the MDGs could not be met if water and energy supply were not existent or transportation and communication infrastructure were missing. Various reports that specifically address the role of critical services necessary to meet the MDGs have been published (Africa Union 2005; Modi et al. 2005; Sustainable Sanitation Alliance 2011).

Clearly community development is not about providing, for example, just light, just water, a phone, or just food in a compartmentalized way. It requires adopting an *integrated approach* to capture the dynamics of communities and their various interrelated needs. For instance, in their projects in western Nepal, Zahnd and McKay (2008) address simultaneously the needs of clean water, sanitation, heating, and light in a concept called *the family of four*. The needs are met by introducing into the community smokeless metal cooking stoves, solar lighting, pit latrines, and access to safe drinking water systems. In the subsequent and expanded *family of four plus* approach, cooking and nutrition, access to clean water, and WASH facilities are met through the use of solar dryers, greenhouses, solar cookers, slow sand water filters, and solar water heaters. In addition, nonformal education (literacy) and scholarships are included for the beneficiaries to acquire the skills necessary to operate and maintain the technical solutions.

Within community development, health requires an integrated approach. It is mentioned several times in this book because the ultimate goal of sustainable community development projects is to build (or strengthen) communities that have the potential to become, over time, more stable, equitable, secure, prosperous, and above all more resilient and healthy. As remarked by Dodd and Munck (2002), health is at "the center of human development." Many types of services (e.g., energy, WASH, transportation, shelter, and communication) and issues (tangible and nontangible) contribute to creating healthy communities, including the service of health delivery itself. Health can be seen as a process or as an outcome, or as both. It is further discussed in Chapter 15 within the broader context of water, sanitation, and hygiene.

Another type of service that should be considered (and is often ignored in the development literature) is the *mobility* of people with physical, visual, and cognitive disabilities. In the world today, there are about 650 million people with disabilities; an estimated 520 million of them live in the developing world (UN 2006). The role of assistive technology in providing the service of mobility is a topic that does not often enter into national development priorities, despite the observation that "people with disabilities have poorer health outcomes, lower education achievements, less economic participation and higher rates of poverty than people without disabilities" (WHO 2011). As remarked by Borg et al. (2009, 2011), "assistive technology can have a positive socioeconomic effect on the lives of people with disabilities by improving access to education and increasing achievement."

All the aforementioned services require that the community has the capacity to support the services in terms of such things as human resources, skills, and regulations. In other words, the service options must be appropriate to the current level of community development. Another requirement is that *appropriate and sustainable* technologies must be included in the various service options to address community needs.

Following a brief discussion on service capacity, this chapter discusses what represents appropriate and sustainable technology in small-scale community development projects. Whether technologies are high tech or low tech is unimportant. What counts is their appropriateness with regard to the communities using them. Another important factor about such technologies is how they engage the communities through participation, education, and empowerment in creating viable, sustainable options that lead to employment. In "Small Is Working" (UNESCO 2003), it is emphasized that appropriate and sustainable technologies must ensure respect, dignity and freedom, sustainability, equity and inclusion, and creativity and innovation. Finally, the technologies must match the needs of those who use them in terms of availability, accessibility, affordability, sustainability, and scalability. Another way of looking at appropriate and sustainable technologies is to see them as *technologies with a human face* (Schumacher 1973).

13.2 Service and Service Capacity

Chapter 9 presented a methodology to assess whether a community is able to carry out the delivery of services. It was based on the approach proposed by Prof. G. Louis and his coworkers at the University of Virginia for the delivery of municipal sanitation services (MSS) that included drinking water supply, wastewater and sewage disposal, and the management of solid waste. For the services to be appropriate to the community, they need to match the level of community development, i.e., the enabling environment that is mapped in the appraisal phase of the ADIME-E framework.

In Chapter 9, it was demonstrated that when considering MSS services, the current level of *service capacity* available in the community must be compared with standards that guarantee "an acceptable level of health for the community." Once the level of community development for the service was determined, solutions were matched with that level of development. Possible technologies that enter into drinking water supply, wastewater and sewage disposal, and the management of solid waste were outlined; rated in terms of selected characteristics (costs, energy required, technical knowledge required, institutional factors); and finally assembled in what was called *unit operations*, i.e., processes and steps necessary to provide each type of service. For instance, for drinking water supply, the process included the water source, procurement, storage, treatment, and distribution, i.e., all the steps that are necessary from source to consumption. By combining the different technologies, several *service options* can be created, with each option carrying a score. Only those options with scores that match the level of community development are selected as potential candidates. The number of options is reduced by using tangible and nontangible criteria; some are community specific, whereas others are based on engineering common sense and principles and past experience on similar projects.

The last step in the strategy mentioned in Chapter 9 for MSS services was to explore how the capacity of the community could be improved toward increasing its level of development over time, i.e., 5, 10, or 20 years down the road. The rationale is that, with an increased capacity, the community will be able to adopt more advanced service options (and technologies), which will increase its capacity over time.

The summarized approach has several key features:

- It is based on options for basic service delivery and not technology.
- It starts with comparing the existing level of service at the community level with established standards, thus identifying service and capacity gaps.
- It matches the services with the current level of community development and capacity.
- It investigates ways to improve the level of community development over time through capacity building and community participation.

Furthermore, the approach requires a good knowledge of the technologies that contribute to the service options in terms of cost, range of application, limitations, past performance, rate of failure, level of expertise required, availability of spare parts, level of operation and maintenance (O&M) required, and environmental impact. All these characteristics contribute to defining the "right type and level of technology," e.g., the "appropriateness" of the technologies for a community.

In deciding what options to select for the basic delivery of services (e.g., energy, WASH, or shelter), it is important that standards for normal services be defined. The standards vary from service to service and depend on the context in which services are provided, i.e., whether the community is in an early stage of development or is well under way in its capacity development. In developed countries, standards of service are common and have been advocated in relation to such elements as water, sanitation, hygiene, energy, shelter, and health. Regulations are often in place and are enforced with well defined indicators and methods of measurement. In the area of water and wastewater, for instance, national bodies (e.g., U.S. Environmental Protection Agency) and international institutions (e.g., the World Health Organization) set those standards in partnership with the water industry and other stakeholders.

In the developing world, regulations and standards for most basic services are of limited impact, even though they may have been formulated by international agencies such as the World Health Organization. If they do exist, though, a lack of governance and rule of law make them often nonenforceable. The problem becomes even more complex in crisis situations after adverse events or hazards (e.g., in refugee situations), where standards for normal services do not exist because the "normal" does not exist anymore. Instead, minimum standards have been suggested by various humanitarian organizations (e.g., Sphere Project 2011a, 2011b) to encourage effectiveness and accountability.

Finally, the approach provides necessary (but not sufficient) conditions to meet two objectives, which according to Narayan (1993) must be included "as a rule" in all development projects and in the delivery of basic services to meet people's needs. One objective is to ensure functioning facilities. The other one concerns the sustainability, effective use, and replicability of the proposed solutions. Meeting both objectives clearly depends on the capacity of the community and of the development agencies involved in the projects. Both objectives seem obvious in the developed world. In projects in developing countries, they are rarely met, as noted, for instance, by the Water Supply and Sanitation Collaborative Council (2004) for water, sanitation, and hygiene projects. As also remarked by Schweitzer and Mihelcic (2012), "an alarmingly high percentage of drinking water systems in the developing world do not provide design service, or may even fail." This idea is discussed further in Chapter 15.

13.3 Appropriate and Sustainable Technology

A Brief History

Over the past 40 years, there has been a slow but growing interest in so-called *appropriate technology*, following the early recommendation by Schumacher (1973) in his book *Small Is Beautiful* that a need exists to develop practical technologies for the poor that stand between the sophisticated high technologies in the developed world and indigenous technologies. Schumacher introduced the term "intermediate technologies" as early as 1962 (Akubue 2000). That concept was strongly influenced by the Sarvodaya Shramadana Movement community development model promoted by Gandhi in India starting in the 1930s (Jequier and Blanc 1983). It referred to technologies characterized as being generally small in scale but appropriate in terms of the needs being met and the capacity of the local community to maintain and operate it.

The concept of intermediate technology in the 1970s quickly created an international movement that was promoted by various groups, such as the Intermediate Technology Development Group (ITDG, now Practical Action) in the United Kingdom and Volunteers in Technical Assistance (VITA) in the United States. It was also endorsed by agencies and organizations such as the U.S. Agency for International Development (USAID) in 1977 and other bilateral organizations around the world. Around 1968 (Akubue 2000), the term "intermediate" was replaced by "appropriate." An excellent review of the history behind the intermediate and appropriate technology movement can be found in the book titled *Small Is Possible* by McRobie (1981) and in a report titled "Development from the Bottom of the Pyramid" by Leland (2011).

More recently, appropriate technology for the developing world has regained interest with U.S. academia and abroad. The availability of global communication tools and social networks has contributed to making several platforms that deal with appropriate technology and its role in poverty reduction. Among many existing platforms, the reader may want to explore the Engineering for Change and Appropedia websites. A web search on "appropriate technology" conducted by the author on March 1, 2013, yielded 40.3 million results! Needless to say, appropriate technology is of interest to many people and may appear to be the new buzzword, along with capacity building and sustainability in the development literature.

Characteristics of Appropriate Technology

In reality, appropriate technology is a well thought-out concept and is certainly more than a mere buzzword. The books *Appropriate Technology* by Hazeltine and Bull (1999) and *Field Guide of Appropriate Technology* (Hazeltine and Bull 2003) provide two excellent references on defining appropriate technology, its attributes, what it can do, and its limitations. According to these authors, appropriate technology is designed to benefit a broad range of people rather than a select few through goods, services, and jobs. According to Hazeltine and Bull (1999), appropriate technology has the following unique characteristics:

- It is small-scale, energy efficient, environmentally sound, labor intensive, and controlled by the community.
- It must be simple enough to be maintained by the people using it.
- It must match both the user and the need in complexity and scale.
- It provides goods, services, and jobs that are not be provided any other way.
- It fosters self-reliance, responsibility, cooperation, and frugality.

According to the nongovernmental organization (NGO) Practical Action (2013), other characteristics of appropriate technology include the following:

- It uses local skills and materials, i.e., native capabilities.
- It paves the way for a better future (for all).
- It is affordable.
- It helps people earn a living (men and women alike).
- It meets people's needs.

Many of these attributes appear in what Zahnd (2012) calls *contextualized technology*. According to Zahnd and based on his 18 years of experience with renewable energy projects in western Nepal, a technology is contextualized "when the design has emerged based on the end-users' energy service demands, their living conditions, economic power and ability to operate and maintain the new technologies with their acquired technical skills." Zahnd (2012) provides in detail the characteristics of various technologies (e.g., photovoltaic, cookstoves, pit latrines, wind turbines, and slow sand water filtration) that make them contextualized.

In general, and as remarked by Schumacher (1973), appropriate technology should be seen as technology with a human face. It is more than a catalog of mere technical gadgets. Rather, it incorporates a symbiotic relationship among technology, those who use it and benefit from it, and the environment in which the human–technology interaction takes place. It has both *hardware* (tools, materials, equipment) and *software* (knowledge, skills, attitude, and behavior) components. Finally, and often not recognized by those in the Western world, appropriate technology is applicable to many situations, both in the United States and in the so-called Third World. Because of its afore-mentioned attributes, appropriate technology is an integral component of sus-tainable community development, whether it is used in developed or developing world projects.

From Technology to Social Entrepreneurship

Another aspect of appropriate technology that results from the aforementioned characteristics is that it is a perfect entry point into what is now called *social entrepreneurship*, another term that can mean different things to different people and has several definitions in the literature. *Civic entrepreneurship* is also used instead of social entrepreneurship (Banuri and Najam 2002). A simple definition of social entrepreneurship, proposed by the Institute of Social Entrepreneurs, is that "social entrepreneurship is the art of simultaneously pursuing both a finan-cial and a social return on investment (the double bottom line)."

Social entrepreneurship is the new mindset behind value-driven marketing, or Marketing 3.0. As a result, this double bottom line approach is a much needed alternative to aid models based on charity (Kotler et al. 2010; Thompson and MacMillan 2010). It can best be summarized as *doing well by doing good*, a state-ment that is sometimes attributed to Benjamin Franklin based on his book, *The Way to Wealth*, initially published in 1758 (Franklin 2011).

Social entrepreneurship often starts with an identified problem (e.g., water, energy, shelter, telecommunication, health, or food and nutrition) that can be solved by some existing or new technology, which under certain conditions results in entrepreneurial business success (in the form of a social enterprise) that empowers those who created the business and others who benefit from it. Social entrepreneurship is a concept that brings together appropriate technol-ogy and business models, creates a symbiotic link between profit and the enhancement of social capital, and creates a third *citizen sector* besides the two traditional public and private sectors (Drayton and Budinich 2010). Among the various groups that have promoted social entrepreneurship, the Ashoka organization, under the leadership of its founder, Bill Drayton, has been the most successful one at the international level, along with Jeff Skoll, who created the Skoll Foundation in 1999. As of 2013 and since its inception in 1980, the Ashoka network has supported more than 3,000 innovators (also known as *changemakers*) in about 60 countries around the world. The literature and the web contain countless examples of how social entrepreneurs are changing the world at the grassroots level (Banuri and Najam 2002; Bornstein 2007). A search on "social entrepreneurship" on the web conducted on March 1, 2013, yielded 6 million results!

Social entrepreneurship can be seen as the latest step in the evolution of technology for development over the past 50 years. During that time period, as noted by Hazeltine and Bull (1999), the steps have been (1) delivering technology; (2) paying more attention to users' needs and providing training to accompany technology; (3) offering business training, addressing such issues as whether a market (local and global) exists and a profit can be made and predicted; and (4) creating entrepreneurs. Overall, this evolution can be seen as changing from delivering technology to teaching people how to do it themselves, a softer approach with strong social implications. Furthermore, social entrepreneurship can be seen as a much-needed approach to the so-called *bottom of the pyramid market* (Prahalad 2006), an economically active sector that has the potential to positively affect 4 billion customers.

Social entrepreneurship is often linked to such concepts as microfinance (*Economist* 2005; Yunus 2009), social design, social innovation, social enterprise, or design for (or with) the other 90% (Polak 2008). As remarked by Thompson and MacMillan (2010), creating social enterprises is not without risk because of the uncertainty inherent in communities, imperfect markets, imperfect governance, and imperfect infrastructure. The two authors suggest six guiding principles in creating new markets in highly uncertain conditions:

- Define the ballpark of the market, identifying minimum accountable performance outcomes and rules of engagement;
- Conduct a sociopolitical analysis, identifying beneficiaries, potential allies, unconcerned, and opponents;
- Design a low-cost pilot and make a hypothesis on the path to scalability;
- Preplan disengagement, i.e., have an exit strategy;
- Anticipate unintended consequences; and
- Use an adaptive and reflective process to "learn by effectuation" by reducing uncertainty into risk.

Similar recommendations were suggested by Tata (2011) for the creation of social enterprises. They include (1) involve the customer; (2) contextualize the approach chosen; (3) think big, act small; (4) keep it simple and do not overbuild; (5) know the market; (6) know how to reach the customers; (7) take care of the "last mile" in the supply chain; (8) build and buy local, thus keeping supply lines short; (9) establish creative partnerships; (10) evaluate often and make corrections; and (11) keep it real.

Frugal and Disruptive Innovation

Over the past 10 years, there has been a growing interest in the role of social innovation and design in helping lift millions of people out of poverty and in addressing the needs of hundreds of millions by providing solutions that are appropriate to their needs and purchasing power without compromising quality. That approach recognizes that

- Solutions to problems in the developed world that require a high level of skills and expensive and complex technology are not appropriate for the poor (3–4 billion people), and
- Those who have the least are more likely to embrace innovation that is appropriate to them and that they can use to improve their livelihood.

Of particular interest is the culture of innovation and the investment in technology development today in emerging markets such as China and India, where the markets are large and diverse (Kotler et al. 2010). An excellent report on this culture can be found in *The Economist* (2010). The report emphasizes the importance of *frugal innovation* or *disruptive innovation* in emerging markets and gives many examples of successful implementation of appropriate technology in China and India within the framework of social entrepreneurship. As remarked by Immelt et al.,

emerging markets are becoming centers of innovation in fields like low cost health-care devices, carbon sequestration, solar and wind power, biofuels, distributive power generation, batteries, water desalination, microfinance, electric cars, and even ultra-low cost homes. (Immelt et al. 2009)

The products are designed and built to meet specific needs and functionality and take into account the limited budget, the capacity of the customers, and their culture and value consciousness (Radjou et al. 2012). They also account for the increase in consumer demand, which itself creates opportunities for in-country entrepreneurship (Kotler et al. 2010).

The rapid growth of mobile phones in developing countries is an excellent example of how innovation in emerging markets is changing the ecosystem of development (Sullivan 2007). According to the International Telecommunication Union (ITU 2011), there were about 6 billion cellular phone subscriptions at the end of 2011; 1 billion were in China, and another billion were in India. The services provided by cell phones are not just about placing phone calls. According to Ogunlesi (2012), mobile phones provide many other services, such as (1) banking, (2) education, (3) entertainment, (4) disaster awareness and management, (5) health care and lifestyle, (6) access to information about agricultural markets and food prices, and (7) information about weather and rain. As remarked by Ogunlesi, mobile phones have become a necessity for development in places like Africa, where the growth in usage has been multifold over a short period of time. By 2016, it is estimated that "there will be a billion mobile phones in Africa" (Ogunlesi 2012).

Readers interested in success stories in social innovation leading to social enterprise development may want to consult the Ashoka website and the book *How to Change the World* by Bornstein (2007). It is remarkable that these success stories did not happen by chance but rather by changing the perspective of what innovation should be, violating all assumptions that pertain to markets in the developed world, changing completely the rules of engagement in market analysis, and more importantly changing the mindset at the corporation level by being able to see the world with a new perspective that is more humble and respectful of others (Hart and London 2005). As remarked by Bornstein (2007), those who make the social enterprise a reality at the grassroots level are bold risk takers and creative problem solvers from the "citizen sector." Another way of looking at them is that they have embraced *creative disruption*.

The interest in frugal or disruptive innovation is not limited to small-scale enterprises and not just enterprises in emerging markets. For instance, companies such as Tata in India, General Electric in the United States, and other multinational corporations are taking social innovation seriously because they are seeing the unique opportunity that 4 to 5 billion additional customers in the world represent, individuals who "have been bypassed or damaged by globalization" (Hart and London 2005). An example of disruptive innovation is the Tata Swach (Clean) Nanotech water filter, which, for a $15–25 initial investment (depending on the model), is able to provide 3,000 liters of clean water free of waterborne diseases for a year for a family of four. Silver nanotechnology integrated into a Tata Swach bulb (at a cost of $7) is used is used to purify the 3,000 liters of water.

Likewise, General Electric has realized the market value in supporting innovation in emerging markets toward the distribution of products not only in those markets, but also in the developed world, an approach that has been called *reverse innovation* (Immelt et al. 2009) or more recently *Jugaad innovation* (Radjou et al. 2012), where "Jugaad is a colloquial Hindi word that roughly translates as 'an innovative fix; an improvised solution born from ingenuity and cleverness.'" Furthermore, "it is widely practiced in other emerging economies such as China and Brazil, where entrepreneurs are also pursuing growth in difficult circumstances."

Short of embracing innovation, there is a danger for Western companies, such as General Electric, that companies in emerging markets may dominate the international arena with products that are custom-made for various situations in terms of cost and simplicity while retaining quality, thus undercutting the market for expensive technologies. As emphasized by Radjou et al. (2012), there is a definite change in how some major Western world companies are looking at global markets and the types of technologies that make that market. They cite the example of Siemens in Germany, which developed "a new product strategy called SMART which stands for Simple, Maintenance-friendly, Affordable, Reliable, and Timely-to-Market."

It is important to note that the vision shared by the Tatas, GEs, and Siemens of the world is not yet mainstream. Despite a clear need at the international level, small-scale technology has been less attractive to the engineering industry in the Western world, which is more at ease and interested in addressing the needs of the richest segments of the world's population and keeping the status quo. This attitude is prevalent in the West because of the misconception that small-scale technology only delivers small profits, whereas large-scale complex technologies command a higher level of compensation with higher profits. As a result, small-scale technology is rarely integrated into traditional engineering education. Today, it is unfortunate that most engineering students in the United States have never heard of appropriate technology or frugal, disruptive, or reverse innovation in their four-year college education. Of course, exceptions exist and are fast increasing in number, especially when they are intertwined with social innovation and business development. Examples include the D-Lab at MIT, the Entrepreneurial Design for Extreme Affordability program at Stanford University, or the Frugal Innovation Labs at Santa Clara University. Innovative research entities in the private and nonprofit sectors addressing such cultures of innovation in partnership with academia include, for instance, the NGO Design that Matters in Boston and the for-profit IDEO in California, which both emphasize human-centered design.

A Reality Check about Appropriate Technology

After having discussed the unique features of appropriate technology in the previous sections, it is important to note that appropriate technology, when viewed as small-scale and decentralized technology, is not a panacea to address all technological needs on our planet. It is important to remember that a unique feature of appropriate technology is its appropriateness. The literature is somewhat vague about what are the indicators of appropriateness. Bauer (2013) introduced an appropriate technology assessment tool based on eight indicators: "community input, affordability, autonomy, transferability, community control, scalability, local availability of materials, and adaptability." These indicators originated from a meta-analysis of the appropriate technology literature. Using the methods of data appraisal discussed in Chapter 5 and a multicriteria decision analysis such as the MCUA presented in Chapter 7, the various indicators can be weighted in their importance to the community. For each technical solution, an appropriateness index is determined using the same mathematical formulation used to calculate the multiple-attribute value (MAV) in Chapter 7.

Technology can be appropriate and relevant in certain situations and contexts but not in others. A need still exists to have large-scale technology and centralized solutions that are more appropriate to providing high-demand services to urban areas. In other situations, like in isolated and remote rural areas or peri-urban areas, small- or medium-scale technology and decentralized solutions are more efficient and effective. A discussion of the merits of using small-scale wastewater-treatment technology in rural areas in developed and developing countries and how such technology is more conducive to achieving sustainability can be found in Ho (2005). The combination of centralized and decentralized technical solutions is often seen as a soft path in the water resource development and management literature (Gleick 2003).

Another aspect of appropriate technology that must be considered is its sustainability. A technology or solution may be appropriate but not sustainable, and vice versa. Ideally, technologies for the developing world (and developed world) need to be appropriate, sustainable, and contextual. Finally, a third aspect of appropriate technology that needs to be taken into consideration is its scalability, diffusion, and profitability.

It is important to note that the concept of appropriate technology has evolved since the early 1970s. Even today, appropriate technology is often wrongly perceived as just being low tech. A photovoltaic system, a portable small-scale health-care device, information and communications technology (ICT), or a telemedicine system probably would have been considered complex technologies for rural areas in the 1970s. Today, such high-tech tools have become more mainstream in development projects. It must be acknowledged that appropriate technology, like all forms of technology, is evolving over time.

Appropriate technology is not without flaws. Two conditions seem to be critical in properly integrating appropriate technologies in community projects. First, the technologies need to have been tested and their specifications and limitations identified before being used. Comprehensive databases on appropriate technology are limited. Many databases exist, though, which often serve as storage platforms of ideas. The problem is that they often lack the quality assurance and control that are necessary to use the products in the field with confidence. The reader is encouraged to look at the Practical Action Technical Briefs on the Practical Action website, which provide excellent technical specifications on a variety of small-scale technologies often encountered in development projects. Several handbooks on appropriate technology are also available in the literature. Among many, the *Village Technology Handbook* published by Volunteers in Technical Assistance (VITA 1988) and the *Appropriate Technology Sourcebook* (Darrow and Saxenian 1986) are still good sources of reference for do-it-yourselfers involved in development projects.

A second condition for properly integrating appropriate technologies in community projects is that the outsiders who develop the technologies and implement them must have appropriate qualifications and must not create false hopes. The developing world does not have to become the "testing ground" for technologies thought of by inventors in developed countries. Today, it is hard to differentiate between actual professionals and "wannabes" in that field. As a result, the developing world is littered with failed technology that has been given and installed by well intentioned but unqualified and somewhat naïve individuals or groups who often promised quick and cheap (or free) solutions.

13.4 From Crisis to Development

A Continuum of Intervention

Throughout this chapter, the discussion has been mostly about the delivery of basic services to address community needs within a development context. This section explores briefly how that delivery would differ in a situation where a developing community is facing an emergency situation or is recovering from a major crisis associated with natural and nonnatural disasters. Referring to Figure 11-1 and its associated discussion, the extent and characteristics of the services during the two phases of rapid response (drop in capacity) and recovery or rebound are likely to dictate the extent of the recovery phase and whether the final community capacity exceeds the initial one. As discussed in Chapter 11, the final community capacity is to a great extent a measure of its acquired ability (resilience) to face the next crisis better than it did the last one. In general, disasters present a unique opportunity to build back better, safer, and smarter and to promote more community resilience rather than repeating the mistakes of the past and rebuilding vulnerability (USAID 2012).

During the rapid response and recovery phases after a crisis, the basic community needs must be met by various lifeline systems that need to be in place and functional before the crisis. They include water, sanitation, hygiene, energy, access to medical care, transportation, and security. But being able to meet those needs requires a decision process from the service providers that matches the exceptional intensity of the situation at hand. The books titled *Engineering in Emergencies* by Davis and Lambert (2002), *The Oxfam Handbook of Development and Relief* (2008), and the UNHCR *Handbook for Emergencies* (UNHCR 2007) give a detailed description of the complexity of the situation in the aftermath of a crisis or disasters. They also provide recommendations on how to address community needs. This is also emphasized in the notebook *The Camp Management Toolkit* published by the Norwegian Refugee Council (2008) in relation to refugee camps.

In general, in the rapid response phase, decisions have to be made rapidly about a variety of issues (mostly interrelated) using existing resources and under circumstances that are far from normal. In general, postcrisis situations are complex and cut across a wide range of emotional, social, physical, and psychological issues (Sphere Project 2011a, 2011b). In such situations, there is often neither room nor time to optimize the solutions. At best, good enough (i.e., satisficing) solutions (Simon 1972) obtained from a reflective practice (Schön 1983) fit the uncertainty and complexity of the situation.

In many ways, the context described compares well with what one experiences when entering the emergency part of a hospital, in contrast to a regular medical visit. The doctors in the emergency room are capable of making decisions very rapidly using triage methods that are not needed in noncrisis situations. Whether one is an emergency service provider in a postcrisis setting or has to make decisions that may decide on the life or death of a patient, there is a risk in not thinking the whole problem entirely through and making wrong decisions based on incomplete information. In the developing world context, with a constraining rather than enabling environment, the odds are more in favor of failed solutions unless a resilience framework is already in place to prepare the community for the crisis and proper awareness and training have been instilled at all levels of the communities (individuals, households, and all governmental levels) beforehand, as discussed in Chapter 11. The reader may want to recall the SPUR (2009, 2011) initiative in San Francisco discussed in Chapter 11, where resilience plans are in place to respond to earthquakes. Communities in the developing world are less likely to have such a framework in place, especially when governance and rules of law are weak. The situation in Haiti before and after the 2010 earthquake is a perfect example.

According to Smith and Wenger (2006), disaster recovery is defined as "the differential process of restoring, rebuilding, and reshaping the physical, economic, and natural environment through pre-event planning and post-event actions." Table 13-1 shows a schematic of the different phases of recovery after a disaster (crisis) according to FEMA (2011). Three steps in the recovery are indicated. The first is the short-term recovery, which is also called the rapid response or emergency phase. It is likely to last for days or weeks and focuses more on humanitarian assistance. The first recovery phase is followed by intermediate recovery, which is expressed in weeks and months, and long-term recovery, which is expressed in months or years. Recovery is associated with the restoration of livelihoods and living conditions and the restoration of the infrastructure. For each phase of recovery, Table 13-1 lists several types of services related to shelter, debris and infrastructure, business, emotional and psychological, public health and health care, and mitigation activities. Finally, recovery should change over time into sustainable development.

The appropriateness of the technical solutions that provide the services after a crisis depends on the stage of recovery. Immediately after a crisis event, the appropriateness is defined by an immediate and pressing need to maintain security and guarantee health. The solutions do not have to be sustainable and

Table 13-1. Recovery Continuum with Description of Activities by Phase

Preparedness (Ongoing)	Short Term (Days)	Intermediate (Weeks–Months)	Long Term (Months–Years)
Predisaster recovery planning	Mass care and sheltering	Housing	Housing
Mitigation planning and implementation	Provide integrated mass care and emergency services	Provide accessible interim housing solutions	Develop permanent housing solutions
Community capacity and resilience building	Debris	Debris and infrastructure	Infrastructure
Conducting disaster preparedness exercises	Clear primary transportation routes	Initiate debris removal and plan immediate infrastructure repair and restoration	Rebuild infrastructure to meet future community needs
Partnership building	Business	Business	Business
Articulating protocols in disaster plans for services to meet the emotional and health-care needs of adults and children	Establish temporary or interim structures to support business reopening and reestablish cash flow	Support reestablishment of businesses where appropriate and one-stop centers	Implement economic revitalization strategies and facilitate funding for business rebuilding
	Emotional and psychological	Emotional and psychological	Emotional and psychological
	Identify adults and children who benefit from counseling or behavioral health services and begin treatment	Engage support networks for ongoing care	Follow-up for ongoing counseling, behavioral health, and case management services
	Public health and health care	Public health and health care	Public health and health care
	Provide emergency and temporary medical care and establish appropriate surveillance protocols	Ensure continuity of care through temporary facilities	Reestablishment of disruptive health care facilities
	Mitigation activities	Mitigation activities	Mitigation activities
	Assess and understand risks and vulnerabilities	Inform community members of opportunities to build back stronger	Implement mitigation strategies

Source: Adapted from FEMA (2011).

perfect. As time goes by and normalcy returns through increased community capacity, the temporary solutions need to be replaced by more long-lasting solutions.

The four distinctive columns in Table 13-1 do not exist in practice because the situation after a crisis implies more a continuum of intervention or of transition, with interdependent service activities rather than a sequence of piecemeal solutions dictated by some experts. The challenges in operating over that continuum are manifold. First, all service providers must work together in a coordinated way to ensure that the continuum of services is operational over a long period of time. This is rarely the case in practice because outsiders are likely to be very specialized. They address issues in the first few hours, days, or weeks of emergency and leave, without coordinating with those involved in the following stage of recovery. A case in point is related to how shelter is usually handled after a disaster. Ideally, the solutions should be implemented right after the disaster and designed to change from emergency shelter, to transitional shelter, and to permanent construction. More often than not, these three forms of construction are disconnected and not integrated because they are handled by various groups. Such disconnects have the potential to create more harm than good in the medium and long term, where emergency shelter (not meant to last) is forced on populations to become transitional and permanent construction. In early 2013, the city of Port-au-Prince in Haiti still had almost 400,000 people living in the same emergency shelter as in January 2010, right after the earthquake struck the region.

Another challenge in operating in the continuum shown in Table 13-1 is related to the level of preparedness of the community (acquired capacity or resilience) before the crisis, which as discussed in Chapter 11, is closely linked to its level of development. That level of preparedness, described in the first column in Table 13-1, essentially dictates the drop in capacity shown in Figure 11-1 and the type of recovery in the immediate, intermediate, and long-term response. As remarked by Davis and Lambert (2002), the level of response of a community to disasters depends largely on its level of development; development and response are interconnected.

Sphere Minimum Standards

When addressing the basic needs of communities in postcrisis situations and emergencies and providing services, it must be understood that all efforts should be made and services provided to alleviate human suffering and provide measures that guarantee (1) a right to life with dignity, (2) a right to humanitarian assistance, and (3) a right to protection and security. These are in essence the fundamental principles and core beliefs behind the *Humanitarian Charter and Minimum Standards in Humanitarian Response* (Sphere Project 2011a), which have been adopted by various international aid agencies. These recommendations translate into four principles to protect populations from further harm and help them "achieve greater safety and security." These principles include the following:

- Avoid the exposure of population to additional harm (physical, psychological, abuse of rights);
- Provide people access to assistance in an impartial and nondiscriminatory way;
- Protect people from violence and coercion related to physical and psychological harm; and
- Provide ways for people to claim their rights by providing information and other forms of community support.

The Humanitarian Charter and the four protection principles form the basis on which standards (and indicators) have been developed to ensure meeting the urgent survival needs of people in response to disasters. The Sphere Standards are divided into Core Standards and Minimum Standards in four sectors: (1) water supply, sanitation, and hygiene promotion; (2) food security and nutrition; (3) shelter, settlement, and nonfood items; and (4) health action.

The core standards define what needs to be in place before addressing the more specific minimum standards. In many ways, they represent guidelines on how to build different forms of acquired capacity in populations affected by disasters. They include the following:

- Having in place capacities (skills, knowledge, resources) and strategies for populations to survive with dignity;
- Developing efficient and effective means of collaboration and coordination among various stakeholders (populations, civil society, and governmental and nongovernmental agencies);
- Conducting a good appraisal of the different forms of capacities and vulnerability of affected populations;
- Analyzing appraisal results and designing solutions to minimize the effects of disasters on affected populations through increased capacity and decreased vulnerability;
- Adopting a practice of reflection in action as humanitarian projects unfold and having a monitoring and evaluation framework in place to examine the performance of humanitarian agencies; and
- Providing aid workers the support, skills, resources, and knowledge to carry out their work efficiently and effectively.

Once the core standards are addressed, the Sphere Project (2011a) provides detailed recommendations and indicators on what needs to be in place to ensure that those affected by emergencies receive the minimum level of services in WASH, food, shelter, and health issues. These issues are interrelated and should be seen as basic rights: "The right to water and sanitation is inextricably related to other human rights, including the right to health, the right to housing and the right to adequate food."

Figure 13-1 shows as an example of the minimum water, sanitation, and hygiene (WASH) standards recommended in the Sphere Project. They were developed with the overall objective to "reduce the transmission of faeco-oral diseases and exposure to disease bearing-vectors" by promoting good hygiene practices, providing clean drinking water, and reducing the health risks associated with the environment.

In addition to the Sphere Standards, companion standards have been proposed in the areas of education, economic recovery, and livestock in emergency situations (Sphere Project 2011b).

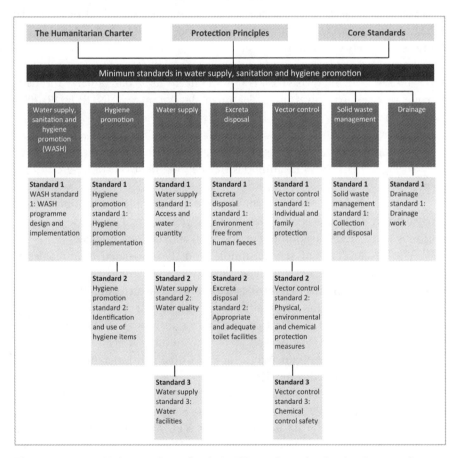

Figure 13-1. Minimum Standards in Water Supply, Sanitation, and Hygiene Promotion

Source: The Sphere Project (2011a), reproduced with permission. © The Sphere Project/www.SphereProject.org.

13.5 Chapter Summary

Service delivery in development projects can take multiple forms depending on the community level of development; its capacity (service capacity) and vulnerability; and whether the community is facing small, medium, or large adverse events. In all circumstances, the services should not fall below minimum international standards.

The technology that enters into the various service options needs to be appropriate and sustainable; it is a technology with a human face. Both attributes are important because one alone does not imply the other. Even though such technology is based on tangible parameters, it is not sufficient to guarantee successful projects. Other nontechnical (nontangible) parameters need to be accounted for as well. For instance, a new technology that does not consider behavior change communication awareness and training in maintenance and repair is not likely to last long.

Technology, as infatuating as it may be (especially for engineers and scientists who see it for its own sake), is not an end in itself but rather a means to provide services that contribute to human development (UNDP/HDR 2001; Juma and Yee-Cheong 2005). In fact, it is clear that if all problems in the world were technical, it is likely that we would have solved them by now. Having said that, it is indisputable, as discussed in Chapter 3, that science, technology, and engineering are critical in creating positive development outcomes around the planet.

In general, standards for normal service delivery in the developing world rarely exist. If they do exist, they are not usually enforced. The engineering profession is in a unique position to develop such standards of practice.

The development of service options and the associated technologies to address the needs in the developing world require a new mindset and approach, which, as discussed in this chapter, is called creative disruptive innovation. Frugal and disruptive innovation must address the appropriateness, affordability, and adaptability components of service options. As discussed in this chapter, the social entrepreneurship side of disruptive innovation is enormous. It is already well under way in emerging markets, such as China and India, but has been of less interest in developed countries.

Finally, the delivery of services to address the needs of developing communities depends on the context in which the needs are met, the nature of the needs, and how quickly the needs must be met. As seen in this chapter, in the recovery phase after a disaster associated with natural or nonnatural hazards, the needs are different and decisions (1) have to be made quickly, (2) cannot be optimized, (3) require reflective practice, and (4) are made in an environment of exceptional intensity. These decisions differ from those in a noncrisis context.

References

Africa Union. (2005). *Transport and the millennium development goals in Africa*, Africa Union/UN Economic Commission for Africa, Washington, DC. <http://www4.worldbank.org/afr/ssatp/ Resources/PapersNotes/transport_mdg.pdf> (Nov. 12, 2011).

Akubue, A. (2000). "Appropriate technology for socioeconomic development in third world countries." *Journal of Technology Studies* (26), 33–43.

Banuri, T., and Najam, A. (2002). *Civic entrepreneurship: A civil society perspective on sustainable development*, Vol. 1, Gandhara Academy Press, Islamabad, Pakistan.

Bauer, A. M. (2013). "Quantitative assessment of appropriate technology." M.S. Report. University of Colorado at Boulder.

Bhutan. (1999). "Bhutan 2020: A vision of peace, prosperity and happiness." Planning Commission, Royal Government of Bhutan, Thimphu, Bhutan.

Birley, M. (2011). *Health impact assessment: Principles and practice*, Earthscan, London.

Bista, D. B. (2001). *Fatalism and development*, Orient Longman Ltd., Hyderabad, India.

Borg, J., Lindström, A., and Larsson, S. (2009). "Assistive technology in developing countries: National and international responsibilities to implement the convention on the rights of persons with disabilities." *The Lancet*, 374, 1863–1865.

Borg, J., Lindström, A., and Larsson, S. (2011). "Assistive technology in developing countries: A review from the perspective of the convention on the rights of persons with disabilities." *Prosthetics and Orthotics International*, 35(1), 20–29.

Bornstein, D. (2007). *How to change the world*, Oxford University Press, New York.

Bryceson, D. F., Bradbury, A., and Bradbury, T. (2008). "Roads to poverty reduction? Exploring rural roads' impact on mobility in Africa and Asia." *Development Policy Review*, 26(4), 459–482.

Darrow, K., and Saxenian, M. (1986). *Appropriate technology sourcebook: A guide to practical books for village and small community technology*, Volunteers in Asia, Stanford, CA.

Davis, J., and Lambert, R. (2002). *Engineering in emergencies*, 2nd Ed., ITDG Publishing, Bourton-on-Dunsmore, Warwickshire, UK.

Dodd, R., and Munck, L. (2002). *Dying for change: Poor people's experience of health and ill-health*, World Health Organization, Geneva.

Drayton, W., and Budinich, V. (2010). "A new alliance for global change." *Harvard Business Review*, September <http://hbr.org/2010/09/a-new-alliance-for-global-change/ar/1> (March 1, 2012).

Economist. (2005). "The hidden wealth of the poor." November 5. <http://www.economist.com/ node/5079324?story_id=5079324> (Feb, 1, 2012).

Economist. (2010). "The world turned upside down: A special report on innovation in emerging markets." April 17 <http://www.economist.com/node/15879369> (May 1, 2010).

Federal Emergency Management Agency (FEMA). (2011). "National disaster recovery framework: Strengthening disaster recovery for the nation." *Federal Emergency Management Agency*, Washington, DC.

Fox, W., and Porca, S. (2001). "Investing in rural infrastructure." *International Regional Science Review*, 24(1), 103–133.

Franklin, B. (2011). *The way to wealth*, Best Success Book Publ. First published in 1758.

Gleick, P. H. (2003). "Global freshwater resources: Soft-path solutions for the 21st century." *Science*, 302, 1524–1528.

Hart, S., and London, T. (2005). "Developing native capability: What multinational corporations can learn from the base of the pyramid." *Stanford Social Innovation Review* <http:// www.ssireview.org/articles/entry/developing_native_capability/> (May 10, 2012).

Hazeltine, B., and Bull, C. (1999). *Appropriate technology: Tools, choices, and implications*, Academic Press, San Diego.

Hazeltine, B., and Bull, C., eds. (2003). *Field guide to appropriate technology*, Academic Press, San Diego.

Ho, G. (2005). "Technology for sustainability: The role of onsite, small and community scale technology." *Water Science & Technology*, 51(10), 15–20.

Immelt, J. R., Govindarajan, V., and Trimble, C. (2009). "How GE is disrupting itself." *Harvard Business Review*, October <http://hbr.org/2009/10/how-ge-is-disrupting-itself/ar/1> (May 12, 2012).

International Telecommunication Union (ITU). (2011). "ICT facts and figures." <http://www.itu.int/ITU-D/ict/facts/2011/material/ICTFactsFigures2011.pdf> (March 2, 2013).

Jequier, N., and Blanc, G. (1983). "The world of appropriate technology—A quantitative analysis." Development Centre of the Organization for Economic Co-operation and Development (OECD), Paris.

Juma, C., and Yee-Cheong, L. (2005). *Innovation: Applying knowledge in development*, Earthscan Publications Ltd., London.

Kotler, P., Kartajaya, H., and Setiawan, I. (2010). *Marketing 3.0: From products to customers to the human spirit*, John Wiley & Sons, Hoboken, NJ.

Leland, J. (2011). "Development from the bottom of the pyramid." WISE Program, University of Washington, Seattle. <http://files.asme.org/asmeorg/NewsPublicPolicy/GovRelations/Programs/29625.pdf> (May 15, 2012).

McRobie, G. (1981). *Small is possible*, Harper & Row, New York.

Modi, V., et al. (2005). "Energy services for the millennium development goals." International Bank for Reconstruction and Development/World Bank and the United Nations Development Programme, Washington, DC. <http://www.unmillenniumproject.org/documents/MP_Energy_Low_Res.pdf> (Oct. 2, 2012).

Narayan, D. (1993). *Participatory evaluation: Tools for managing change in water and sanitation*, World Bank, Washington, DC. <http://www.chs.ubc.ca/archives/files/Participatory%20Evaluation%20Tools%20for%20Managing%20Change%20in%20Water%20and%20Sanitation.pdf> (Dec. 10, 2012).

Norwegian Refugee Council. (2008). *The camp management toolkit*. Norwegian Refugee Council, Oslo, Norway. <http://www.nrc.no/camp> (Dec. 10, 2012).

Ogunlesi, T. (2012). "Seven ways mobile phones have changed lives in Africa." CNN.com., September 14 <http://www.cnn.com/2012/09/13/world/africa/mobile-phones-change-africa/index.html> (Oct. 1, 2012).

Oxfam. (2008). *The Oxfam handbook of development and relief*, Vols. 1 and 2, Oxfam Publ., Cowley, Oxford, U.K.

Polak, P. (2008). *Out of poverty: What works when traditional approaches fail?* Berrett-Koehler Publishers, San Francisco.

Practical Action. (2013). "What is appropriate technology?" <http://practicalaction.org/media/view/20259> (Oct. 15, 2013).

Prahalad, C. K. (2006). *The fortune at the bottom of the pyramid: Eradicating poverty through profits*, Wharton School Publishing, Upper Saddle River, NJ.

Radjou, N., Prabhu, J., and Ahuja, S. (2012). *Jugaad innovation*, Jossey-Bass Publishers, San Francisco.

Reddy, A. K. N. (2000). "Energy and social issues." *World energy assessment: Energy and the challenge of sustainability*, Chapter 2, United Nations Development Programme, New York. <http://www.undp.org/content/undp/en/home/librarypage/environment-energy/sustainable_energy/world_energy_assessmentenergyandthechallengeofsustainability.html> (March 1, 2012).

San Francisco Planning and Urban Research Association (SPUR). (2009). "The resilient city: Defining what San Francisco needs from its seismic mitigation policies." San Francisco Planning and Urban Research Association, San Francisco. <http://www.spur.org/sites/default/files/publications_pdfs/SPUR_Seismic_Mitigation_Policies.pdf> (Feb. 5, 2012).

San Francisco Planning and Urban Research Association (SPUR). (2011). "The resilient city—Ideas and actions for a better city." San Francisco Planning and Urban Research Association San Francisco. <http://www.spur.org/policy/the-resilient-city> (Feb. 5, 2012).

Schön, D. A. (1983). *The reflective practitioner: How professionals think in action*, Basic Books, New York.

Schumacher, E. F. (1973). *Small is beautiful*, Harper Perennial, New York.

Schweitzer, R. W., and Mihelcic, J. R. (2012). "Assessing sustainability of community engagement of rural water systems in the developing world." *J. Water, Sanitation and Hygiene for Development*, 2(1), 20–30.

Simon, H. A. (1972). "Theories of bounded rationality." *Decision and organization*, C. B. McGuire and R. Radner, eds., 161–176, North-Holland Pub., Amsterdam, Netherlands.

Smith, G. P., and Wenger, D. (2006). "Sustainable disaster recovery: Operationalizing an existing agenda." *Handbook of disaster research*, H. Rodriguez, E. L. Quarantelli, and R. Dynes, eds., 234–257, Springer, New York.

Sphere Project. (2011a). *Humanitarian charter and minimum standards in humanitarian response*, Practical Action Publishing, Rugby, U.K.

Sphere Project (2011b). "New companions standards to the Sphere Handbook." <http://www.sphereproject.org/news/new-companion-standards-to-the-sphere-handbook/> (Jan. 10, 2012).

Sullivan, N. P. (2007). *You can hear me now: How microloans and cell phones are connecting the world's poor to the global economy*, Jossey-Bass, San Francisco.

Sustainable Sanitation Alliance. (2011). "Compilation of 13 factsheets on key sustainable sanitation topics." Sustainable Sanitation Alliance (SuSanA) and GIZ, Eschborn, Germany. <http://www.susana.org/lang-en/library?view=ccbktypeitem&type=2&id=1229> (May 15, 2012).

Tata, Z. (2011). "Sustainable social enterprises." Lecture notes in CVEN 5919: Sustainable Community Development 1, University of Colorado at Boulder.

Thompson, J. D., and MacMillan, I. C. (2010). "Business models: Creating new markets and societal wealth." *Long Range Planning*, 43, 291–307.

United Nations (UN). (2006). "Convention on the rights of persons with disabilities." United Nations, New York <http://www.un.org/disabilities/convention/facts.shtml> (May 20, 2012).

United Nations Commissioner for Refugees (UNHCR). (2007). *Handbook for emergencies*, 3rd Ed., United Nations Commissioner for Refugees, Geneva.

United Nations Development Programme Human Development Report (UNDP/HDR). (2001). "Making new technologies work for human development." United Nations Development Programme, New York.

United Nations Educational, Scientific and Cultural Organization (UNESCO). (2003). "Small is working: Technology for poverty reduction." UNESCO/ITDG/TVE, Paris <http://portal.unesco.org/science/en/ev.php-url_id=3180&url_do=do_topic&url_section=201.html> (Dec. 10, 2010).

U.S. Agency for International Development (USAID). (2012). *Building back housing in post-disaster situations—Basic engineering principles for development professionals: A primer*, U.S. Agency for International Development, Washington, DC <http://buildchange.org/pdfs/USAID_Building_Back_Housing_in_Post-Disaster_Situations-A_Primer.pdf> (Feb. 5, 2013).

Volunteers in Technical Assistance (VITA). (1988). *Village technology handbook*, Volunteers in Technical Assistance, Arlington, VA.

Water Supply and Sanitation Collaborative Council (WSSCC). (2004). "Listening—To those working with communities in Africa, Asia, and Latin America to achieve the UN goals for water and sanitation." Water Supply and Sanitation Collaborative Council, Geneva <http://www.comminit.com/?q=africa/node/189342> (Jan. 2, 2013).

World Health Organization (WHO). (2011). "World report on disability." World Health Organization, Geneva <http://www.who.int/disabilities/world_report/2011/en/index.html> (May 20, 2012).

Yunus, M. (2009). *Creating a world without poverty: Social business and the future of capitalism,* PublicAffairs Books, Jackson, TN.

Zahnd, A. (2012). "The role of renewable energy technology in holistic community development." Doctoral dissertation, Murdoch University, Australia.

Zahnd, A., and McKay, K. (2008). "A mountain to climb? How pico-hydro helps rural development in the Himalayas." *Renewable Energy World,* 11(2), 118–123.

14

Energy Services for Development

14.1 Climbing the Energy Ladder

Energy and Development

Engineers know that energy is the ability of a physical system to do work. They also know that energy is constrained by the first law of thermodynamics (conservation) and the second law (entropy). The ability to do work combined with human skills and ingenuity has changed the face of the planet for better or worse since the start of the Industrial Revolution. As remarked by Reddy (2000), "history is the story of the control over energy sources for the benefit of society."

Access to reliable, safe, affordable, and sustainable energy services is "essential to meet our basic needs: cooking, boiling water, lighting, and heating. It is also a prerequisite for good health (WHO 2006). When people have access to energy services on a daily basis, they have more opportunities to go to school; communicate using rechargeable phones; pump, store, distribute, filter, and heat water; have access to better health care (especially women and children); be more productive; operate machinery; have food security; conduct business; and move and interact with each other.

A wide range of technologies can meet energy service needs in developing communities. The technologies can be service specific or combined in the form of so-called *multifunction platforms*, where a single engine (usually driven by diesel or, preferably, biodiesel) turns one or several pieces of machinery into a tool that can perform a variety of services, such as grinding corn, pumping water, or turning an electric generator (Nygaard 2010). It may also happen that different forms of energy help provide different services (Anderson et al. 1999).

Although access to energy brings many benefits to society, it is also associated with various health risks that depend on the type of energy used and how it is managed (Smith et al. 2013). Providing energy services to the poor can drastically change their lives, but energy can also have drastic consequences, especially in relation to the extraction of oil and mining. There has been clear evidence of abuses of human rights and ecological devastation associated with the practices of major Western oil and mining companies combined with corrupt institutions in the developing world. For the communities affected by the unethical practices of such organizations, energy is seen of as more of a curse than a blessing.

Fortunately, in most instances in the developing world, gaining access to energy services is a blessing. It means stepping out of powerlessness and isolation to having a voice, having hope for a better future, being able to conduct value-added economic activities in the community, and contributing to society. Figure 14-1 shows the relationship between the human development index (HDI) and per capita energy use for different countries (UNDP 2004a).

As noted by Zahnd (2012) and Yager (2012), Figure 14-1 clearly indicates an increase of the HDI with access to energy, but infinite access to energy does not necessarily create an infinite HDI. In fact, there is an area in Figure 14-1 where there is a strong increase in HDI with a small amount of energy increase, roughly for a value of HDI between 0.7 and 0.8. The reader may recall that an HDI of 0.7 is the value below which countries are somewhat defined as developing. Beyond that value are emerging market countries and Western countries where a small or large increase in energy use per capita does not seem to increase the HDI much. In fact, the added value of extra energy consumption (usually obtained from fossil fuels) is not enough of a convincing argument to create a small increase in HDI (from 0.7 to 1.0). Past examples have shown that the extra access to energy produced by burning fossil fuels too often results in environmental degradation and health issues, which have the potential to directly and negatively affect the economic growth that led to the level of human development in the first place. As remarked in the 2011 UNDP/HDR report, an increase in HDI results in a dramatic increase in CO_2 emission per capita, which, according to Yager (2012), may actually be counterproductive in the long term by reducing the HDI because of environmental degradation and quality of life.

Energy is central to human and economic development and to the efforts geared at reducing poverty. Even though it does not appear as one of the eight Millennium Development Goals per se, access to energy services is a sine qua non for achieving all of those goals (Table 14-1), as clearly summarized by Modi et al. (2005). Today, many people in rural areas in the developing world lack access to modern energy sources like oil, gas, and electricity and are considered energy poor (see Figure 2-1, the world at night). About 1.6 billion people do not have access to electricity, which represents 23% of the world's population (IEA 2011). In India alone, about 400 million people, mostly in rural areas, do not have access to electricity (Bornstein 2011). In other countries, electricity is only available sporadically and is not sufficient for sustained development.

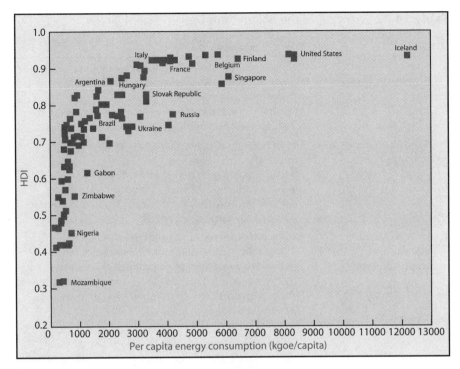

Figure 14-1. Relationship Between HDI and Per Capita Energy Consumption 1999–2000

Note: As a reference, the countries that are members of the Economic Community of West African States have an average consumption of about 88 kWh of electricity per capita per year and have HDI values ranging between 0.3 and 0.7 (ECOWAS 2005). Energy consumption is defined in kilograms of oil equivalent per capita (kgoe/capita).

Source: UNDP (2004a); with permission from United Nations Development Programme.

According to the 2011 UNDP/HDR report, about 2.6 billion people rely exclusively on biomass, such as crop residues, waste, dung, wood, and charcoal inside their homes. Such sources are used because they are more affordable and accessible than advanced fuels. Biomass is organic (as the result of photosynthesis) and carbon based and reacts with oxygen in combustion to create heat. According to Legros et al. (2009), "more than 40% of people rely on modern fuels in developed countries, but only 9% in least developed countries and 17% in Sub-Sahara Africa." In general, biomass sources of energy are less efficient than modern sources of energy and often cost a lot to acquire. In sub-Saharan Africa, 81% of households rely on wood-based biomass (AFREA 2011) and more than one-third of a household budget is set aside for this energy source (ECOWAS 2005). As a result, the development of the energy poor is held

Table 14-1. Energy and the Millennium Development Goals

MDGs	Energy Linkages
1. Eradicate extreme poverty and hunger	Energy inputs (electricity and fuels) are needed for agriculture, industrial activities, transportation, commerce, and micro-enterprises. Most staple foods must be cooked, using some kind of fuel, to meet human nutritional needs.
2. Achieve universal primary education	Teachers are reluctant to go to rural areas without electricity. After dark, lighting is needed for studying. Many children, especially girls, do not attend primary schools because they have to carry wood and water to meet family subsistence needs.
3. Promote gender equality and empower women	Adult women are responsible for the majority of household cooking and water boiling activities. This takes time away from other productive activities. Without modern fuels and stoves and mechanical power for food processing and transportation, women often remain in drudgery.
4. Reduce child mortality	Diseases caused by lack of clean (boiled) water and respiratory illnesses caused by indoor air pollution related to the use of traditional fuels and stoves directly contribute to mortality in infants and children.
5. Improve maternal health	Lack of electricity in health clinics and lack of lighting for nighttime deliveries adversely affect women's health care. Daily drudgery and the physical burdens of fuel collection and transport also contribute to poor maternal health conditions, especially in rural areas.
6. Combat HIV/AIDS, malaria, and other diseases	Electricity for radio and television can spread important public health information to combat deadly diseases. Health care facilities, doctors, and nurses need electricity for lighting, refrigeration, and sterilization to deliver effective health services.
7. Ensure environmental sustainability	Energy production, distribution, and consumption all have many adverse effects on the local, regional, and global environment, including indoor air pollution, local particulates, land degradation, acid rain, and global warming. Cleaner energy systems are needed to address all of these to contribute to environmental sustainability.
8. Develop a global partnership for development	The World Summit for Sustainable Development (WSSD) called for partnerships among public entities, development agencies, civil society, and the private sector to support sustainable development, including the delivery of affordable, reliable, and environmentally sustainable energy services.

Source: UNDP (2004b), with permission from United Nations Development Programme.

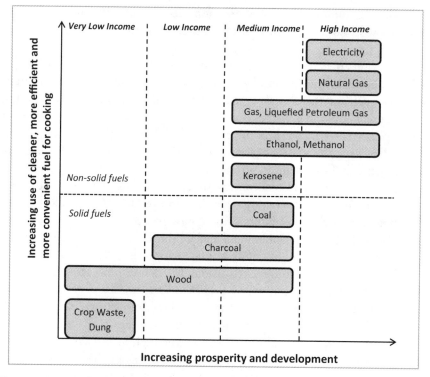

Figure 14-2. The Energy Ladder

Source: WHO (2006), with permission from World Health Organization.

back because they are trapped at the bottom of what is often referred to as the *energy ladder* (Figure 14-2). Climbing the energy ladder is synonymous with climbing out of poverty.

Alternatives to Traditional Energy Services

There are multiple consequences of burning biomass. First, it contributes to large-scale deforestation, especially in developing countries in transition. Second, using open-fire methods in the confined environment of homes (often for long periods of time) has been disastrous to people's health because of indoor air pollutants (particulate matter, CO, other gases) originating from unprocessed biomass fuels. It has been estimated that indoor air pollution resulted in 1.5 million deaths worldwide in 2002, contributing to about 2.7% of the global burden of disease (WHO 2006), mostly affecting women and children. Two recent studies conducted at the University of California Berkeley (UCB 2011) showed compelling evidence of the reduction of cases of pneumonia in homes with cookstoves connected to chimneys and a direct link between low school-aged children's IQ and prenatal maternal exposure to smoke from open fires.

A second effect of using biomass is the wasted time and energy associated with its collection, which is usually the chore of women and children. In certain situations, such as refugee camps, the collection of biomass may result in putting women and children in vulnerable and precarious conditions. In Nepal, women spend up to 40 hours a week in the collection and cutting of firewood (Zahnd 2012). Alleviating the stress of biomass collection would result in freeing time that women could use in more productive tasks, such as education, child care, and relaxation. It would also encourage women to become part of the decision process at the household level, thus encouraging gender equality and business development. Likewise, the children would have more time dedicated to education and to enjoying their childhood. Finally, a third effect of the use of biomass expresses itself in the form of deforestation, environmental degradation (MDG 7 in Table 2-3), and greenhouse gas emissions, which over time, accentuate conditions leading to increased poverty (World Bank 2013).

Alternative solutions to traditional energy service delivery, even to remote areas, do exist. They have been implemented and have shown positive results but have not necessarily been added at a pace fast enough to meet the demands of those in need. According to WHO (2006), "485,000 people need to gain access to cleaner fuels every day" to halve the number of people without access to modern cooking fuels by 2015.

In general, the alternative clean options to traditional energy services, which are discussed in greater length below, consist of three groups:

- Using traditional fuel more efficiently (modern cookstoves, charcoal use, biogas, kerosene, liquefied petroleum gas) combined with management of natural resources;
- Using renewable sources of energy, such as solar, wind, and micro- or pico-hydro, where energy is created at the point of use (distributed power generation systems); or
- Connecting to an existing grid.

These solutions vary in cost, and all require a minimum level of income per person, which according to Barnes et al. (1997) was roughly $1,000–1,500 per year ($3–5 per day) in 1997. Finally, these solutions have their own characteristics in terms of performance; capital, operation, and maintenance costs; reliability; durability; and environmental impact. The handbook entitled *Rural Energy Services* by Anderson et al. (1999) provides an excellent overview of various technical options for rural settings. As an example, Table 14-2 lists the basic characteristics of various forms of renewable energy technologies and traditional generators (USAID 2011).

Table 14-2. Characteristics of Various Forms of Energy Technology

Energy Technologies	Capital Cost	O&M Cost	Reliability	Durability	Special Considerations	Emissions	Optimal Use
Solar photovoltaic (PV) system with batteries	Very high	Low	High (if maintained properly) or low (if not)	20–30 years (PV) 5 years (batteries)	Theft (batteries or panels), vandalism (panels), availability of trained technicians	None	Small loads, areas where fuel is costly or difficult to obtain
Wind turbine with batteries	High	Low to moderate	High (if maintained properly) or low (if not)	20 years (turbine) 10 years (blades) 5 years (batteries)	Theft (batteries), lack of data on wind resources	None	Many moderate loads where resource is sufficient
Diesel generator	Moderate to high	High	High	25,000 operating hours	Fuel spills, emissions	Very high	Larger loads
Gasoline generator	Low	Very high	Moderate	1,000–2,000 operating hours	Fuel spills, emissions	Flammability high	Emergency generator
Gas generator	Moderate	High	Moderate	3,000 operating hours	Propane is of limited availability, but can use biogas	Low	Component in hybrid system or as stand-alone
Hybrid system	Very high	Low to moderate	Very high	Varies; Optimization greatly extends generator and battery life	Complexity for servicing	Low	Medium and larger loads
Grid extension	Varies	None	Varies	High	Theft, extending grid allows connection of nearby homes to grid	Not local	Where grid is reliable and not too distant

Source: USAID (2011).

Because the energy consumption of the poor is small to begin with, providing them with even a small additional amount of modern source of energy has the potential to change their livelihood. This change is particularly impactful when combined with the use of efficient appliances and light bulbs. For example, various light-emitting diode (LED) lights are now available on the market; a 4-W LED light can provide as much as 60 W of incandescent light and can last longer than incandescent light bulbs.

As remarked by Rosenthal (2010), the money saved by using home-scale energy systems such as photovoltaic (PV), biogas, and efficient cookstoves (instead of fuels such as wood, candles, and kerosene) translates into transformative opportunities at the household level in terms of economic security, better health, and education. The main limitation in acquiring such systems has to do with the high initial costs and maintenance. This hurdle could be overcome by providing small sources of credit and reducing the cost of the energy systems (Barnes et al. 1997). This phenomenon has been the case, for instance, with PV systems, which have decreased in cost over the past 20 years.

Today, many solutions exist to supply the developing world with practical and sustainable energy solutions that would allow millions of people in poverty to ascend the energy ladder. The type of project delivery, whether centralized or decentralized, must be sensitive to a suite of factors so that solutions match the enabling environment. Referring to the capacity diagram of Figure 9-2, these factors include having (1) institutions (even local) and policies in place; (2) human resources and the ability to operate and maintain the energy systems, including access to spare parts; (3) access to financial support, no matter how small; and (4) a social structure that allows the project delivery to occur while being sensitive to social issues. Finally, project delivery requires a change in behavior on the user's part, which is often the most difficult part to implement.

A wide range of initiatives in providing access to energy services to the poor has been proposed since the 2002 World Summit on Sustainable Development in Johannesburg. They have been led by international agencies such as the United Nations (UNDP, UNEP), bilateral agencies such as USAID and EuropeAid, NGOs, and the private sector. Since 2002, various global partnerships promoting local and/or international cooperation have been put in place. They consist of mobilizing various stakeholders, such as government, the public sector, businesses and the private sector, civil society organizations, and others to implement energy-related initiatives (UNDP 2004a). Examples of UNDP-related initiatives include the Global Village Energy Partnership, the LP Gas Rural Energy Challenge, and the Global Environment Facility. Other global partnerships of interest include the Global Alliance for Clean Cookstoves, the Partnership for Clean Indoor Air, and the Sustainable Ethanol for Cooking Partnership. More recently, the UN declared 2012 the International Year of Sustainable Energy for All with a goal of universal access to energy by 2030 while "controlling emissions and shifting to new and cleaner energy sources" (UNDP/HDR 2011).

Social Innovation in Energy Services

Access to energy services has also been a strong entry point into social innovation and social entrepreneurship. In general, there is no lack of case studies and technologies addressing cookstoves, biogas, renewable energy systems, and small and decentralized grid systems for the developing world. A quick Google search conducted by the author on March 1, 2013, for each technology (no distinction made between developed and developing world) revealed 347,000; 15 million; 39 million; and 65 million entries, respectively. Furthermore, there is also no lack of examples of successful enterprises that provide access to energy at the grassroots level. For instance, Bornstein (2011) describes a social enterprise by the name of Husk Power Systems that converts rice husks into electricity through a biomass digester. This grassroots enterprise, located in Bihar, India, serves the electricity needs of 30,000 households (spread into about 250 hamlets) using more than 60 power plants. Each household can afford $2 per month for energy.

Another successful social enterprise is the Solar Energy Light Company or SELCO, which since 1995 has focused on addressing the solar lighting and solar water heating needs of small communities in rural India. As remarked by Radjou et al. (2012), SELCO has provided solar energy solutions to more than 125,000 households and has demonstrated that "poor people can indeed afford and maintain renewable energy solutions."

Finally, the for-profit enterprise CleanStar in Mozambique is yet another success story of social enterprise. The company promotes the use of ethanol-based cooking fuel as a replacement for charcoal and the use of cooking stoves. It has also developed a framework that links cooking fuel to income, nutrition, slowing deforestation, and health.

14.2 Using Biomass More Efficiently

Table 14-3 summarizes three groups of solutions that can be used to reduce indoor air pollution related to the use of biomass. They include (1) changing the source of pollution, (2) improving the living environment, and (3) modifying the behavior of the users. Each one, or all three combined, can considerably affect the well-being of the poor communities that still rely on biomass. For instance, efficient, clean cookstoves that achieve high-quality combustion have been able to lower pollution by at least 90% compared with open fires and simple clay or metal stoves. These changes can be interpreted as low-hanging fruits, winnable battles with minimum investment, and investment practices that have a high

Table 14-3. Solutions to Reduce Indoor Air Pollution from Use of Biomass

Changing the Source of Pollution	Improving the Living Environment	Modifying User Behavior

Improved cooking devices
  Improved stoves without
    flues
  Improved stoves with
    flues
Alternative fuel–cooker
  combinations
  Briquettes and pellets
  Kerosene
  Liquefied petroleum gas
  Biogas
  Natural gas, producer gas
  Solar cookers
  Modern biofuels (e.g.,
    ethanol, plant oils)
Reducing need for fire
  Retained heat cooker
  Efficient housing design
    and construction
  Solar water heating
  Pressure cooker
	Improved ventilation
	  Smoke hoods
	  Eaves spaces
	  Windows
	Kitchen design and
	  placement of the stove
	  Kitchen separate from
	    house reduces
	    exposure of family
	    (less so for cook)
	  Stove at waist height
	    reduces direct
	    exposure of the cook
	    leaning over fire
		Reduced exposure by
		  changing cooking
		  practices
		  Fuel drying
		  Pot lids to conserve heat
		  Food preparation to
		    reduce cooking time
		    (e.g., soaking beans)
		  Good maintenance of
		    stoves, chimneys, and
		    other appliances
		Reduced exposure by
		  avoiding smoke
		  Keeping children away
		    from smoke (e.g., in
		    another room if
		    available and safe
		    to do so)

Source: WHO (2006), with permission from World Health Organization.

return. The aforementioned solutions can also be complemented with solar cooking solutions.

Clean Cookstoves

Social innovation in cookstoves has run wild over the past 20 years. WHO (2006) cites multiple initiatives of using modern cookstoves, such as small-scale community-led projects or even initiated by governments, such as China in the 1980s and 1990s (200 million stoves). Hundreds of initiatives have been reported in the literature about improved cookstoves to replace open fires. Cookstoves can be broadly divided into traditional cookstoves, improved cookstoves, and advanced super-clean cookstoves. Information about different types of cookstoves for the developing world and their performance in various settings can be found in Kammen (2005) and in reports published by the Global Village Energy Partnership and the Global Alliance for Clean Cookstoves, mentioned previously.

Cookstoves can run on various types of fuel, such as biomass, gas, or solar energy. When using biomass, emphasis must be placed on the quality and nature of the biomass. It has been the experience of the author that cookstoves are often designed for a specific type or range of biomass materials. When using cookstoves, special emphasis needs to be placed on the synergy between the stove itself and the biomass type and composition (e.g., crop wastes, firewood, fuel briquettes, coal, charcoal, or dung). Cookstoves rarely perform well with multiple types of biomass. As remarked in the report entitled *Igniting Change* by the Global Alliance for Clean Cookstoves (Cordes 2011), the field of cookstoves is highly fragmented, with an absence of cookstove standards. Therefore, a need for a new strategy concerning the adoption of clean cookstoves at the global level is needed. As remarked in a recent report by the World Bank (2013), cookstoves are still too expensive for poor households in developing countries. Reducing their costs would help leverage their multiple benefits, such as reduced pollution, improved health, saved lives, and a slow-down in global warming.

Liquefied Petroleum Gas

The UNDP has been successful in integrating liquefied petroleum gas (LPG) in rural development and as a replacement for biomass through its LP Gas Rural Energy Challenge, a public–private partnership between the World Liquefied Petroleum Gas Association and the UN Development Programme (UNDP 2003; Kelly and Yager 2007). LPG (butane, propane) has been found to provide a safe, clean, and efficient form of energy service for cooking, heating, refrigeration, sterilization, and lighting. It is also adaptable to local conditions and portable. It is particularly effective when other viable energy service solutions are limited, for instance, because of the remoteness of certain communities. As remarked by the UNDP (2003), "LP Gas in combination with simple, reliable appliances and proven technologies is a valuable resource that generates multiple productive services, applicable in several economic sectors" at different scales: the individual household; the community; and in the commercial, industrial, agricultural, and transportations sectors of the economy.

Biogas

Biogas has also become very popular in rural areas. It is a methane and carbon dioxide gaseous mix where the methane ranges between 40 and 70% (Zahnd 2012). The gaseous mix is generated in an airtight biomass digester by the anaerobic decay of human, livestock, food, and agricultural waste. Biodigesters can be small (compact), medium size (household level), or large industrial biogas plants. In general, biogas has been used for cooking without major side effects in many countries, especially in China, India, and Nepal. In these three countries, biogas programs have been under way since the 1930s–1950s (Fulford 1988). Biogas can also be used for generating heat, producing electricity, and refrigeration. The Biogas Sector Partnership (BSP) in Nepal has been successful at popularizing and developing the technology across the country, where more than 170,000 plants have been installed.

Among all the benefits of biogas, including clean flame, no smell, no smoke, and less particulate matter, the slurry left after full decomposition of the waste in the digester is bacteria free and has been found to be an excellent fertilizer because of its nutrients. Other benefits of biogas can be found in a technical note on the subject by Practical Action (2012). Biodigesters vary in design and include fixed-dome, rubber balloon, and floating drumdigesters (Energypedia 2013). A good review of various small-scale biogas digesters, the process of anaerobic digestion, and literature of the subject can be found on the Sustainable Sanitation and Water Management website (SSWM 2013).

Various rules of thumb have been proposed in the literature about how to design small-scale biodigesters and the potential uses of biogas. Some of them are listed below (based on personal communication with Y. Teller, 2010).

- 1 kg of manure produces 40 L of methane gas. Therefore, 1 m^3 (1,000 L) of gas is generated by 25 kg of manure, enough to support the needs (about four hours of fuel) of a family of four for one day.
- One cow generates about 9–15 kg of manure per day, which is equivalent to 0.4–0.6 m^3 of gas per day. A buffalo generates 0.45–0.48 m^3; a pig, 0.28–0.34 m^3; and 100 hens, 0.42–0.51 m^3. In comparison, one person generates about 0.02–0.03 m^3 per day.
- The energy content of biogas is about 6 kWh/m^3, which is about equivalent to that of 0.61 L of diesel fuel.
- The retention time in the biodigester varies between 40 and 100 days at a temperature ranging between 35 and 50°C.
- The slurry in the digester should be about one part manure and four parts water.
- The volume of the digester should be about 2/3 slurry and 1/3 gas.

Methods of construction of biogas digesters vary from country to country and are well documented (FUCOSOH 2012). As remarked on the Sustainable Sanitation and Water Management website (SSWM 2013), the strength of biogas digesters is that organic pollutants are removed effectively without a need to expend energy. Furthermore, both biogas and compost are generated at a reduced cost. However, biogas digesters require continuous monitoring of solid and liquid content and expert design and supervision during their construction. The book entitled *Running a Biogas Programme* by Fulford (1988) provides excellent information on how to start a biogas program and how to build, operate, and maintain various types of biogas digesters.

Other Gas Supplies

As remarked by Yager (2012), some countries have access to small amounts of gas beside LPG and biogas. Examples include biogenic gas (from swamps and marshes), dissolved methane, landfill gas, gas related to flaring reduction related to the oil and gas industry, small natural gas deposits, and ethanol cooking gas. Even though they are small, these sources can provide enough gas at the local level to sustain development.

14.3 Using Renewable Sources of Energy

In the world today, about 1.6 billion people do not have access to electricity (80% of those people live in sub-Saharan Africa and Southern Asia) because they may not be able to afford it or are isolated from any existing national or local grid (Legros et al. 2009). Instead, they rely on kerosene lamps and candles for lighting, which can cost about $40–80 each year, create indoor pollution, and cause fires (Grimshaw and Lewis 2010). Yet, the demand for electricity in the developing world is small compared with that in developed countries. As remarked by Ratterman (2007), "average home systems in the developing world are 50–75 Watts, and 'large' systems may be 120 W." In comparison, the average residential systems in the United States range between 3,000 and 5,000 W. This comparison, combined with the availability of renewal energy in the developing world, shows that small, decentralized renewable energy systems have great potential for addressing the energy needs of households in developing communities and can provide a high return on investment in changing people's livelihood with small initial investments. This improvement could further lead to "life-saving improvements in agricultural productivity, health, education, communications and access to clean water" (Grimshaw and Lewis 2010). As noted by these authors, most of Africa has around 325 days of sunlight a year, with a daily average of 6 kWh of energy per square meter.

The potential for social innovation and entrepreneurship in the field of renewable energy, whether it is in regard to designing refrigerators, water pumps, pasteurization systems, cookers, or dryers that run on renewable energy, is large. It is also a field with huge potential for vocational training and technical education capacity building. A good example is the Barefoot Solar Engineers initiative that was launched in Africa and Asia.

Three common sources of renewable energy are considered below: solar photovoltaic (PV), wind, and hydro (micro and pico). Other forms of renewable energy not addressed here (but with potential to be used in developing communities) include solar thermal, concentrated solar power systems (macro and micro), tidal and wave power, geothermal (deep and closed-loop), and the temperature differences at different depths in the oceans. These energy systems may include one source of renewable energy or multiple sources in the form of hybrid systems. The term hybrid can also signify a combination of a source of renewable energy and the use of traditional generators (diesel, gas, gasoline) as backup. As reported by the UNDP/HDR (2011), renewable sources provide a large share of energy production at the global level (45% in Sweden and 85% in Brazil) and are being adopted more regularly in developing countries.

In general, the process of delivering small-scale renewable energy systems to a community (or a facility in that community) can be broken down into four steps (USAID 2011):

1. *Determine and estimate the energy demand.* The design of all small renewable energy systems requires that electric loads be determined beforehand. In small communities, this load is determined by calculating an average daily load for each household or facility and taking into account loads associated with equipment that might be shared at the community level. Tables are available in the literature to determine the wattage (power) consumption of various appliances that may be encountered in that context (see, for instance, SEI 2004). The total energy load expressed in kilowatt-hours is determined by multiplying the wattage of each appliance by the number of hours it is used and adding all these forms of energy for a given day. Consideration must be taken into account for variations in energy needs on a weekly, monthly, or seasonal basis.

2. *Explore energy options.* Various options are available to meet the demand based on existing renewable energy sources (wind, solar, biomass, hydro) and conventional energy resources (diesel, LPG, gasoline) already operating in the community. The software HOMER, initially developed at the National Renewable Energy Laboratory in Golden, Colorado, can serve as a decision-making tool to select and rank various options to best match energy needs at the community level. The software is unique in the sense that it allows users to look at the economic feasibility of single and hybrid energy systems; conduct sensitivity analyses; and optimize capital, operation, and maintenance costs. Various technical and nontechnical criteria enter into that phase of the decision process. These criteria may include equipment-specific issues (availability of parts, technical resources, reliability, durability); geographical issues (location and elevation, forest cover, shading); and community issues (specific needs, culture, noise, emissions). A software tool similar to HOMER by the name of RETScreen was developed by Natural Resources Canada.

3. *Design, procure, and install the energy systems.*

4. *Maintain and finance the systems* to ensure their long-term reliability, operation and maintenance, and economic sustainability.

Examples of applications of the aforementioned methodology within the context of addressing the electricity needs of rural health centers can be found in the report "Powering Health: Electrification Options for Rural Health Centers" by USAID (2011). It gives an excellent description of the various steps involved in designing energy systems for health clinics with various levels of energy requirements: low (5–10 kWh/day), moderate (10–20 kWh/day), and high (20–30 kWh/day). Besides the USAID approach to addressing energy needs in developing countries, other agencies have proposed best practices in that area as well. For instance, the Clean Energy Solutions Center (2013) serves as a platform for guidelines and best practices for rural electrification.

In general, despite their *relative* simplicity, small-scale renewable energy systems require close attention to technical and nontechnical details; a lot of caution must be taken to ensure their long-term benefits (sustainability). Among other things, basic regular maintenance and servicing must be planned and executed, which may include training local personnel. The systems are very sensitive to the local geography and site conditions and must match the needs of the users.

Another important aspect of such systems is their limited capacity to store energy and the difficulty in matching supply with local demand. As remarked by Zahnd (2012), "energy is useful only if it is available when and where it is wanted." There is no way to guarantee that energy is available every time it is needed; the wind does not blow and the sun does not shine all the time. Hence, renewable energy resources are often seasonal, and the energy generated by wind, water, and the sun needs to be stored to be used during times of demand. In small community projects, this storage is accomplished using batteries, which usually have two to three days of autonomy, at best. For larger renewable-energy systems, various schemes are under consideration (DOE 2012) but are still well beyond the capacity and ability for small communities to handle.

Solar Photovoltaic Off-Grid Systems

Figure 14-3 shows an example of a simple PV off-grid system that provides direct current (DC) from sunlight for a single home. The system consists of an array of one of several solar panels, a charge controller, and a battery pack. It can be used to power DC appliances. An inverter is used to convert direct current to alternating current (AC), which can be used to run regular appliances and connect to the grid.

PV systems can also be integrated into a microgrid that serves several households and businesses clustered closely together. In microgrids, the PV systems tend to be larger than those in single homes and can be combined with other systems (wind, hydro, diesel generators) in a hybrid manner. Sizing the PV

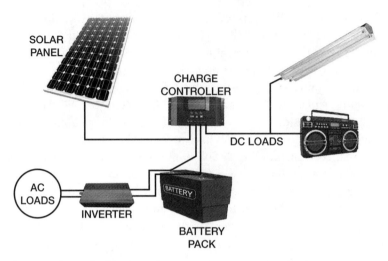

Figure 14-3. Simple Photovoltaic System

panel (its wattage) and the battery bank size (in ampere-hours, Ah) can be determined using two basic equations:

$$W_p(\text{watt}) = \frac{\text{Load (Wh)}}{\text{PSH} \times \text{PSF}}$$

$$\text{Bank Size (Ah)} = \frac{\text{Load (Wh)} \times D_{\text{aut}}}{\text{DOD} \times \text{Eff}_B \times V}$$

where W_p is the PV panel wattage necessary to provide the energy to cover the
loads.

PSH is the amount of peak sun hours (hours where energy is
1,000 Wh/m^2). This amount is determined using a pyranometer or by
consulting tables published, for instance, by NASA (2013), and data
available through SWERA (2013) or local regional solar maps, such as
those available in Colorado (DRCOG 2013).

PSF is the panel sizing factor, which can be seen as an efficiency factor that
counts for temperature (panel vs. ambient), battery efficiency, and
inverter efficiency; use 0.75 as an estimate.

D_{aut} is the number of days of autonomy expected if there is no sunshine
available; use 2 or 3 days.

DOD is the battery depth of discharge; use 0.5 or 0.6.

Eff_B is the battery efficiency; use 0.85.

V is the voltage; use 12 V or 24 V.

Figure 14-4 shows an example of a 300-W, two-axis tracking PV system
installed by Zahnd (2012) in the village of Tulin in northwest Nepal. The system
provides electricity to 180 people spread over 29 households.

Figure 14-4. Photovoltaic System in the Village of Tulin, Nepal

Source: Zahnd (2012), with permission from Alex Zahnd.

In general, the cost of PV systems is specific to the equipment used, the country, and the project location. A conservative rule of thumb for small systems with less than 10 kW (e.g., household scale) is to use $5–10 per (installed) watt in the United States and $10–15 per (installed) watt overseas. These figures include installation but not the batteries. More recently, because of the availability of more PV panels on the international market, the cost of PV systems has been reported as being below $2 per (installed) watt in the Western world.

It is important to consider adding the cost of maintenance to these figures, which can vary with local conditions and the level of expertise available in each community. Despite these reductions in installation and maintenance costs, PV systems are still expensive, especially in terms of up-front capital and installation. The good news is that they are becoming cheaper by the day and will more than likely become more affordable to a broader market in the near future. Grimshaw and Lewis (2010) noted that "more than 2.5 million homes in developing countries had access to electricity from solar systems in 2007." Furthermore, the World Bank "approved two projects in Bangladesh in 2008 to install 1.3 million solar home systems" such as the off-grid system shown in Figure 14-3.

There has also been an interesting development in very small, portable PV home systems that contain a small battery and produce 4-5 W of power using a small PV panel. These systems are capable of lighting two to three rooms using efficient LED lights and charging a cell phone. An example of such a system marketed by Duron in India requires a $130 initial investment. It provides 4 hours of light at a bright setting and 10 hours of light at a dim setting after a

Figure 14-5. The Solar Suitcase

Source: Courtesy of Hal Aronson.

one-day charge. Solar lanterns have also been found to be practical. Another example of innovative solar technology is called the Solar Suitcase from WE CARE Solar developed at Berkeley, California. It is a "portable power unit that provides health workers with highly efficient medical lighting and power for mobile communication, computer, and medical devices." The system shown in Figure 14-5 runs on a 40–80-W solar panel and a 12-Ah sealed battery.

Another aspect of PV systems that needs careful consideration has to do with batteries that are used for the limited storage of energy. In general, deep-cycle batteries are recommended for energy storage, and car batteries should be avoided. Among deep-cycle batteries, nonliquid ones are recommended because they do not require maintenance, a positive attribute despite their reduced longevity and higher cost. A comprehensive review of different types of batteries for photovoltaic systems can be found in LaForge (2008).

There is obviously more to PV systems than what is presented here. Readers interested in the practical aspect of PV systems should consult the abundant references on this topic. The book *Photovoltaics: Design and Installation Manual* published by Solar Energy International (SEI 2004) in the United States is an excellent reference, especially when used within the context of developing community projects. The U.S. Department of Commerce also published a manual that provides worksheets and specifications necessary for the design of multiple stand-alone PV systems (SNL 1995). Finally, several commercial software packages are available for the design of PV systems of different sizes.

Solar Thermal Systems

Besides providing electricity, the sun's radiation can also be concentrated on the plates of solar thermal collectors to heat water or air or to provide solar cooking. The systems are usually small and portable and have been designed to be safe and reliable and to work over a long period of time. All they need is a system of heat-absorbing pipes and a water storage tank. They are common in rural and urban areas in the Middle East, for instance.

Wind Power Systems

As remarked by Bartmann and Fink (2009), "a wind turbine extracts energy from moving air by slowing the air down and transferring this harvested energy into a spinning shaft, which then turns a generator to produce electricity." Wind power systems are ideal in situations where wind is prevalent year round or in a seasonal manner. In the latter case, wind power is used as part of hybrid systems, along with PV, generators, and other systems. Wind power systems have been installed around the world by various utility companies and the private sector. The installation of a small tower and a wind turbine can provide some energy to small-scale communities.

Figure 14-6 shows an example of a 1-kW wind turbine installed in the Bedouin community of Susya in the West Bank by the NGO Community Energy Technology in the Middle East (Hartman 2012). The turbine, combined with a

Figure 14-6. Wind Turbine Installed in a Bedouin Village in the West Bank

Source: Photograph by Tomer Appelbaum, with permission from Community Energy Technology in the Middle East (COMET-ME).

PV system, has provided the necessary energy to the Bedouin community for making butter and refrigerating dairy products that are sold in local markets. The project was initially successful despite the constant political and ethnic tension in the vicinity of the village. More recent political decisions to remove the community altogether have hampered the success of this project.

The energy produced by a single wind turbine can be determined by multiplying the power (P) generated by the wind with the time of operation expressed in hours per day. The power is expressed as follows:

$$P(W) = \frac{1}{2}C_p A \rho V^3$$

where C_p is an efficiency factor of the wind turbine ranging between 0.2 and 0.3;

A is the swept area measured in m^2;

ρ is the density of the air passing through, e.g., 1.23 kg/m³ (at sea level, 15°C); and

V is the average wind speed of the air measured in m/s, which can be measured using an anemometer or by consulting wind speed data (e.g., NREL 2013; SWERA 2013).

This equation shows that doubling the wind velocity creates an eightfold increase in power. In low wind situations, more power can only be achieved by increasing the areas swept by the rotating blades. Detailed specifications about the construction and installation of wind turbines can be found in the literature. The book entitled *Homebrew Wind Power* by Bartmann and Fink (2009) gives detailed information on the components of small wind turbines that are ideal for providing energy to small communities and households.

As for PV systems, the loads used by the community or household must be compared with the energy (power multiplied by time) produced by the wind turbine. Batteries are also used for storage and are linked to 24-V or 48-V systems rather than 12-V systems. Finally, wind systems are not usually efficient because the coefficient C_p entering into the preceding equation varies between 0.2 and 0.3. In fact, it cannot exceed 0.593, which is known as the Betz limit (Bartmann and Fink 2009).

Hydropower Systems

Hydropower systems convert the potential energy of water into mechanical and electrical energy. Depending on the amount of power generated, they can be divided into four categories:

1. Full-scale hydro systems (>10 MW) supply power to large towns and extensive grid supplies. That group can be further divided into medium-scale (10–300 MW) and large-scale (>300 MW) systems.

2. Mini and small-scale hydro systems (100 kW to 10 MW) address mid-scale power issues.
3. Micro-hydro systems (<100 kW) are ideal for remote areas away from the grid and groups of houses and small factories (minigrid).
4. Pico-hydro systems (<10 kW) can be used locally in small-scale communities.

For developing community applications, micro- (and sometimes pico-) hydro systems are of special interest because of their small sizes and their capacity to supply smaller amounts of power that respond to the demand at the local level. Micro-hydro systems provide power as long as the water is flowing. If the flow is seasonal, they can become part of hybrid systems and can be complemented with solar, wind, and/or generators.

Micro-hydro systems have been particularly attractive in mountainous regions in the world. In Nepal, many micro-hydro systems are providing local sources of energy in rural settings to about 150,000 households. Their cost varies between $2,000 and $2,500 per kW before governmental subsidies, which can be as high as 80% (H. Neopane, personal communication, Kathmandu, Nepal, 2013). As remarked by Zahnd and McKay (2008), small power generation hydro systems with a power as low as 1.1 kW combined with white light emitting diodes (WLED) can provide remote areas of Nepal sufficient lighting for schooling and everyday use at the household level. In addition, all extra energy generated by the system that is not used for electricity is diverted to heat water tanks in the community, thus contributing to improving hygienic conditions or cooking. Sanchez (2006) describes several successful models of microenterprise development for five small-scale hydroelectric plants in Peru.

In a micro-hydro system, water is diverted from a stream using an intake and is channeled under pressure to a turbine (reaction or impulsive) located in a powerhouse. The location of the intake and the powerhouse must be carefully selected to ensure good performance. The advantages of such systems include the following:

• It uses a portion of the stream flow.
• It is environmentally benign.
• AC or DC electricity can be generated.
• A flow as low as 20 L/min and a head as low as 0.6 m can be used.
• The available energy is predictable.
• It is available to meet continual demand if water is flowing.
• It has low maintenance and operating costs, in addition to being lasting and reliable.
• It can be connected to a utility grid.

The disadvantages of such systems include the following:

• Certain flow, head, and output characteristics are required.
• The design and performance are very site specific.

- A risk of damage exists in flood conditions.
- It can be sensitive to seasonal variations in flow.
- It requires knowledge and skills to sustain the technology.

The equation used to determine the power associated with a stream can be expressed as follows:

$$P(\text{kW}) = \eta \rho g h Q$$

where ρg is the unit weight of the water equal to 9.81 kN/m³;

   h is the head of water in meters, i.e., the difference in elevation between the intake and the turbine;

   Q is the flow rate in the pipe in m³/s; and

   η is the overall efficiency of the system and varies between 0 and 1; typically, 0.5 is assumed in preliminary design and site assessment.

Several excellent books have been published to provide guidelines for the design, implementation, maintenance, and operation of micro-hydro systems (Davis 2003; New 2004, 2005a, 2005b). A more comprehensive treatise on the subject is the book published by Harvey (2006) entitled *Micro-Hydro Design Manual*, which details the design of small to mid-size hydro systems.

14.4 Grid Extensions

Extending an existing grid works well if it is reliable and the overall cost is less than that associated with other solutions. The grid can be an existing national or regional grid that is maintained by a utility company. In more remote areas where electricity is needed (e.g., a local clinic, small businesses, or an agricultural processing facility), the grid can also be an autonomous local minigrid powered by a main source of energy, such as a generator combined with PV, wind, and/ or hydro (micro to small-scale) systems. Such systems are often referred to as remote area power supply (RAPS) systems or stand-alone power systems (SAPS). They provide the energy necessary to meet basic needs at the local level and in areas that are not connected to a distribution system. RAPS and SAPS are quickly becoming legitimate alternatives to distribution grids, as documented by the Lighting Africa (2013) initiative supported by the World Bank, and through the work of SELCO India.

Grid extension often has a high initial capital cost that does not vary much with the loads that need to be served. The cost can be as large as $8,000 to $10,000 per kilometer of grid extension and even higher in areas with difficult terrain. As a result, this solution is more appropriate when larger loads need to be addressed. Various loads can be served once the extension is in place, which is more relevant when serving communities rather than households. Finally, no maintenance is required from the community, assuming that the utility serving the grid provides the maintenance in a reliable and consistent way, a critical assumption in energy projects in the developing world.

14.5 Chapter Summary

Gaining access to modern sources of energy is critical to human and economic development. It is an integral part of making each of the eight Millennium Development Goals a success. Alternatives to traditional biomass for cooking and kerosene and candles for lighting are needed for the poor to climb the energy ladder and step out of poverty. Providing modern energy services for all should not be detrimental to the environment. There are indeed alternatives to traditional First World energy solutions that could provide developing communities better health, living conditions, economic development, gender empowerment, and more time in households for more productive tasks. This change can be accomplished with minimum impact on the environment and at a fraction of the cost of the systems in the developed world.

The adoption of alternative options to traditional energy services, such as those mentioned in this chapter, are less costly in the developing world than in the developed world. Furthermore, the return on investment in terms of economic development is manifold. This return occurs in part because of the fact that consumption of energy in the developing world is much less than in rich countries, where too much energy is being wasted. Another positive aspect is that the alternative forms of energy services, such as those associated with renewable energy, are becoming cheaper over time because of an increased demand. The technologies are also becoming safer and are more efficient, reliable, accessible, and affordable. Three options were presented in this chapter: (1) using traditional fuel more efficiently with management of natural resources, (2) using renewable sources of energy in the form of off-grid decentralized systems, and (3) connecting to an existing grid.

It is noteworthy that the delivery of energy services to the developing world is an entry point for considerable innovation and social enterprise at multiple levels, whether it is in the form of new technology, the services around the technology, or the technology's maintenance and long-term performance. Grassroots models of business development in the developing world have the potential to employ many and reach out to remote areas that have been neglected by governments. As remarked in the UN Human Development Report (UNDP/HDR 2011), the challenge is more about scalability, i.e., developing energy service solutions that can be expanded at a large scale and can be developed quickly enough to respond to the ever-increasing demand.

Finally, it is important to note that energy plays a critical role in development. Without it, not much would be possible. It is vital to the quality of life and security of many communities. Energy is not an end in itself but rather a component of various interrelated systems (e.g., the water–energy nexus or the health–energy nexus).

References

Africa Renewable Energy Access Program (AFREA). (2011). "Wood-based biomass energy development for sub-Saharan Africa." The International Bank for Reconstruction and Development/World Bank Group, Washington, DC. <http://siteresources.worldbank.org/extafrregtopenergy/resources/717305-1266613906108/biomassenergypaper_web_zoomed 75.pdf> (Jan. 1, 2013).

Anderson, T., et al. (1999). *Rural energy services: A handbook for sustainable energy development*, Intermediate Technology Publications Ltd., London.

Barnes, D. F., van der Plas, R., and Floor, W. (1997). "Tackling the rural energy problem in developing countries." *Finance and Development*, 34(2), 11–15.

Bartmann, D., and Fink, D. (2009). *Homebrew wind power: A hands-on guide to harnessing the wind*, Buckville Publications, Masonville, CO.

Bornstein, D. (2011). "A light in India." *New York Times*, January 10, <http://opinionator .blogs.nytimes.com/2011/01/10/a-light-in-india/> (Jan. 15, 2011).

Clean Energy Solutions Center. (2013). "Featured policy resources: Rural electrification." <https://cleanenergysolutions.org/topics/energy-access/rural-electrification> (Oct. 5, 2013).

Cordes, L. (2011). "Igniting change: A strategy for universal adoption of clean cookstoves and fuel." Global Alliance for Clean Cookstoves, Washington, DC. <http://www.cleancookstoves.org/resources/fact-sheets/igniting-change.pdf> (Jan. 30, 2011).

Davis, S. (2003). *Microhydro: Clean power from water*, New Society Publishers, Gabriola Islands, BC, Canada.

Denver Regional Council of Governments (DRCOG). (2013). "Denver regional solar map." <http://solarmap.drcog.org/> (Oct. 5, 2013).

Department of Energy (DOE). (2012). "Energy storage technology overview." <http://www .netl.doe.gov/technologies/coalpower/fuelcells/seca/tutorial/TutorialII_files/TutorialII.pdf> (Feb. 2, 2013).

Economic Community of West African States (ECOWAS). (2005). "White paper for a regional policy, geared towards increasing access to energy services for rural and periurban populations in order to achieve the Millennium Development Goals." Economic Community of West African States. <http://www.hubrural.org/IMG/pdf/cedeao_uemoa_livre_blanc_strategie_regionale_energie_rurale_eng.pdf> (March 10, 2011).

Energypedia. (2013). "Types of biogas digesters and plants." <https://energypedia.info/index.php/Types_of_Biogas_Digesters_and_Plants> (April 20, 2014).

Fulford, D. (1988). *Running a biogas programme: A handbook*, Practical Action, London.

Fundación Cosecha Sostenible Honduras (FUCOSOH). (2012). "Gas bio-digester information and construction manual for rural families", Oficina de la Coordinacion Nacional, Tegucigalpa, Honduras <http://www.wcasfmra.org/biogas_docs/6%20Biodigester%20manual.pdf > (Oct. 5, 2013).

Grimshaw, D. J., and Lewis, S. (2010). "Solar power for the poor: Facts and figures." *SciDev Net*, March 24, <http://www.scidev.net/en/features/solar-power-for-the-poor-facts-and-figures-1.html> (Dec. 12, 2011).

Hartman, S. (2012). "Like water for the thirsty … Renewable energy systems in Palestinian communities in the South Hebron Hills." Community Energy Technology in the Middle East (Comet-Me), Tel Aviv, Israel. <http://comet-me.org/wp-content/uploads/2012/09/Cmt_ShuliReport_Eng_F_spreads.pdf> (Oct. 7, 2013).

Harvey, A. (2006). *Micro-hydro design manual: A guide to small-scale water power schemes*, Practical Action, London.

International Energy Agency (IEA). (2011). *World energy outlook 2011*, OECD/IEA, Paris. <http://www.iea.org/publications/freepublications/publication/WEO2011_WEB.pdf> (May 10, 2011).

Kammen, D. M. (2005). "Cookstoves for the developing world." *Scientific American*, 273(1), 72–75.

Kelly, M., and Yager, A. (2007). "The LP gas rural energy challenge." CSD-15 IPM Partnerships Fair, United Nations, New York, <http://www.un.org/esa/sustdev/csd/csd15/PF/info/M_Kelly_LPG.pdf> (Jan. 3, 2013).

LaForge, C. (2008). "Choosing the best batteries." *Home Power*, 127, Oct/Nov, 80–88.

Legros, G., et al. (2009). "The energy access situation in developing countries." United Nations Development Programme-World Health Organization, New York. <http://www.who.int/indoorair/publications/energyaccesssituation/en/index.html> (Jan. 10, 2013).

Lighting Africa. (2013). "Catalyzing markets for modern off-grid lighting." Lighting Africa, The World Bank, Washington, DC, <http://www.lightingafrica.org/> (Oct. 5, 2013).

Modi, V., et al. (2005). "Energy services for the millennium development goals." The International Bank for Reconstruction and Development/World Bank and United Nations Development Programme, Washington, DC. <http://www.unmillenniumproject.org/documents/MP_Energy_Low_Res.pdf> (Oct. 2, 2012).

NASA Atmospheric Science Data Center. (2013). "Surface meteorology and solar energy." National Aeronautics and Space Administration, <https://eosweb.larc.nasa.gov/sse/>. (Oct. 15, 2013).

National Renewable Energy Laboratory (NREL). (2013). "Wind resource information." <http://www.nrel.gov/rredc/wind_resource.html> (Oct. 5, 2013).

New, D. (2004). "Intro to hydropower, Part 1: Systems overview." *Home Power*, 103, Oct/Nov, 14–20.

New, D. (2005a). "Intro to hydropower, Part 2: Measuring head and flow." *Home Power*, 104, January, 42–49.

New, D. (2005b). "Intro to hydropower, Part 3: Power, efficiency, transmission, and equipment selection." *Home Power*, 105, Feb/Mar, 30–35.

Nygaard, I. (2010). "Institutional options for rural energy access: Exploring the concept of the multifunction platform in West Africa." *Energy Policy*, 38, 1192–1201.

Practical Action. (2012). "Biogas: power from cow dung." <http://practicalaction.org/biogas>. (Oct. 4, 2013).

Radjou, N., Prabhu, J., and Ahuja, S. (2012). *Jugaad innovation*, Jossey-Bass Publishers, San Francisco.

Ratterman, W. (2007). "Solar electricity for the developing world." *Home Power*, 119, 96–100.

Reddy, A. K. N. (2000). "Energy and social issues." *World energy assessment: Energy and the challenge of sustainability*, Chapter 2, United Nations Development Programme, New York. <http://www.undp.org/content/undp/en/home/librarypage/environment-energy/sustainable_energy/world_energy_assessmentenergyandthechallengeofsustainability.html> (March 1, 2012).

Rosenthal, E. (2010). "African huts far from the grid glow with renewable power." *New York Times*, Dec. 5. <http://www.nytimes.com/2010/12/25/science/earth/25fossil.html?_r=0> (Jan. 10, 2011).

Sanchez, T. (2006). *Electricity services in remote rural communities*, ITDG Publishing, London.

Sandia National Laboratories (SNL). (1995). *Stand-alone photovoltaic systems: A handbook of recommended design practices*, U.S. Department of Commerce, Springfield, VA.

Smith, K. R., et al. (2013). "Energy and human health." *Annual Rev. Public Health*, 34, 25.1–25.30.

Solar and Wind Energy Resource Assessment (SWERA). (2013). Solar and Wind Energy Resource Assessment, <http://en.openei.org/apps/SWERA/> (Oct. 7, 2013).

Solar Energy International (SEI). (2004). *Photovoltaics: Design and installation manual*, New Society Publishers, Gabriola Islands, BC, Canada.

Sustainable Sanitation and Water Management (SSWM). (2013). "Sustainable sanitation and water management toolbox." Sustainable Sanitation and Water Management, <http://www.sswm.info/> (Oct. 8, 2013).

United Nations Development Programme (UNDP). (2003). "LP Gas rural energy challenge." United Nations Development Programme, New York. <http://www.undp.org/content/undp/en/home/librarypage/environment-energy/sustainable_energy/lp_gas_rural_energy challenge.html> (Oct. 10, 2013).

United Nations Development Programme (UNDP). (2004a). "World energy assessment: Overview, 2004." United Nations Development Progamme, New York. <http://www.leonardo-energy.org/world-energy-assessment-2004-update> (Oct. 11, 2013).

United Nations Development Programme (UNDP). (2004b). "Energy for sustainable development." United Nations Development Programme, New York. <http://www.undp.org/content/undp/en/home/librarypage/environment-energy/sustainable_energy/undp_and_energy _forsustainabledevelopment.html> (Oct. 11, 2013).

United Nations Development Programme Human Development Report (UNDP/HDR). (2011). *Sustainability and equity, a better future for all*, United Nations Development Programme, New York. <http://hdr.undp.org/en/media/HDR_2011_EN_Complete.pdf> (April 22, 2012).

U.S. Agency for International Development (USAID). (2011) "Powering health: Electrification options for rural health centers." U.S. Agency for International Development, Washington, DC. <http://pdf.usaid.gov/pdf_docs/PNADJ557.pdf> (April 25, 2012).

University of California, Berkeley (UCB). (2011). "Wood smoke from cooking fires linked to pneumonia, cognitive impacts." University of California, Berkeley, <http://newscenter.berkeley.edu/2011/11/10/cookstove-smoke-pneumonia-iq/> (Oct. 5, 2013).

World Bank. (2013). "On thin ice: How cutting pollution can slow warming and save lives." World Bank and the International Cryosphere Climate Initiative. Washington DC. <http://www-wds.worldbank.org/external/default/WDSContentServer/WDSP/IB/2013/11/01/000456286_20131101103946/Rendered/PDF/824090WP0v10EN00Box379869B00 PUBLIC0.pdf>(Dec. 2, 2103).

World Health Organization (WHO). (2006). "Fuel for life: Household energy and health." World Health Organization, Geneva. <http://www.who.int/indoorair/publications/fuelforlife/en/index.html> (Dec. 5, 2013).

Yager, A. (2012). Technical presentation at the 2012 Energy Justice Conference, Boulder, CO, September 17 and 18.

Zahnd, A. (2012). "The role of renewable energy technology in holistic community development." Doctoral dissertation, Murdoch University, Australia.

Zahnd, A., and McKay, K. (2008). "A mountain to climb? How pico-hydro helps rural development in the Himalayas." *Renewable Energy World*, 11(2), 118–123.

15

Water, Sanitation, and Hygiene Services for Development

15.1 The WASH Health Nexus

Access to services such as clean water supply; wastewater collection, treatment, and disposal; and solid waste management combined with good hygiene practices is critical to human and economic development. The availability and characteristics of services related to water, sanitation, and hygiene (WASH), in coordination with other services related to energy, shelter, food, and transportation, determine the health and livelihood of the members of a society and its economic well-being. Evidence exists that better living conditions; access to clean drinking water, sanitation, and hygiene; along with progress in medicine and other fields of science and technology have contributed to improving life expectancy and public health in the developed world during the twentieth century. In the United States alone, it has been estimated that water purification contributed to reducing mortality by half in the first third of the twentieth century (UNDP/HDR 2006). The challenge of this century will be to expand that success to all people in the developing world who are held back by poverty. As noted in the 2006 UNDP/HDR report, "overcoming the crisis in water and sanitation is one of the great human development challenges of the 21st century." Without adopting proper water and sanitation management and policy changes, the well-being of billions of people on our planet in the next 30 years is at stake. As remarked by the OECD (2012) in its *Environmental Outlook to 2050* report, the cost of doing nothing can be high, "not just financially, but also in terms of lost opportunities, compromised health and environmental damage."

There is a common agreement that health is at the "center of human development" (Dodd and Munck 2002) and should be considered a human right. As emphasized in the Declaration of Alma-Ata (Alma-Ata 1978), "the attainment of the highest possible level of health is a most important world-wide social goal whose realization requires the action of many other social and economic sectors in addition to the health sector." All members of the World Health Organization (WHO) and UNICEF reaffirmed, through the 1978 declaration, the need to provide primary health care (essential health care) and health for all. Since that time, many efforts have been made to promote the importance of health in poverty reduction. Yet, health is still not often perceived as a priority in the development agenda by the governments of many countries in the developing world (PEP 2008).

Health is defined by the WHO (1948) as a "state of complete physical, mental and social well-being and not merely the absence of disease or infirmity." Spiritual well-being has also been proposed as a basic component of health, even though it has not been officially included in the WHO's definition. The multiple factors that contribute to defining health and influence the state of health are referred to as *health determinants*. They are usually classified into four main types: physiological, behavioral, environmental, and institutional (Skolnik 2008; Birley 2011).

One of the many key measures of health, or rather the lack of it, is the so-called *global burden of disease*, which according to the WHO (2008) combines "years of life lost due to premature mortality and years of life lost due to time lived in states of less than full health." Both are included in a single metric of time called the *DALY*, which represents a measure of disability-adjusted life years or alternatively of lost years of healthy life caused by mortality, morbidity, and disability. DALYs associated with various diseases and risk factors have been computed for different regional groupings of countries, economies, gender, and age (WHO 2008).

In general, a range of diseases and health risk factors contributes to the overall burden of disease (Skolnik 2008). Several classifications of diseases approved by the World Health Organization have been proposed in the literature (WHO 2001, 2013), and it is not the purpose of this chapter to study them in depth. The global burden of disease is divided into three groups:

- *Group I*: Communicable diseases are self-perpetuating and multiply rapidly through various forms of contact among humans and animals. They refer to infectious diseases and parasitic diseases caused by germs (viruses, bacteria, and protozoa) and worms. As remarked by Skolnik (2008), communicable diseases spread in a variety of forms: food-borne, water-borne, sexual or blood-borne, vector-borne, inhalation, nontraumatic contact, or traumatic contact. Besides communicable diseases, this group also includes maternal, perinatal, and nutritional conditions.
- *Group II*: Noncommunicable diseases include, for instance, cardiovascular, cancer, diabetes, hypertension, stroke, cholesterol, and mental diseases.
- *Group III*: Injuries can be nonintentional (road traffic accidents, falls, burns, drowning) or intentional (self-inflicted injuries, violence, homicides, suicide).

As noted by the WHO (2008), in the world today, "among both men and women, most deaths are due to noncommunicable conditions (Group II), and they account for about 6 out of 10 deaths globally. Communicable, maternal, perinatal and nutritional conditions (Group I) are responsible for just under one third of deaths in both males and females. The largest difference between the sexes occurs for Group III, with injuries accounting for almost 1 in 8 male deaths and 1 in 14 female deaths." At the regional level, the burden of disease can be quite different from these averages. For instance, communicable diseases are the most common in sub-Saharan Africa and South Asia and in low-income countries in general. Health inequalities also contribute to a variable local burden of disease at the country or community levels, in developed and developing countries alike.

As remarked by the Third World Academy of Sciences (TWAS 2002) and the WHO (2008), communicable diseases result in millions of deaths annually, especially among children in the developing world. Acute respiratory infection (ARI), diarrheal diseases, neonatal infections, and malaria account for 50% of annual deaths of children under 5 years old (WHO 2008). In 2010 worldwide, Liu et al. (2012) reported that for that age group, ARI (mostly pneumonia) contributes to 14.1% of all 7.6 million deaths, and diarrhea contributes to 9.9% of all deaths, followed by malaria, with 7.4%. Diarrhea translates into 0.75 million deaths annually, with an uncertainty range of 0.54–1.0 million per year (about 3,000 children under 5 per day). This can be attributed to a perpetuating reinforcing cycle connecting diarrheal diseases to malnutrition and impaired immune systems while living in unsanitary conditions (WHO/UNICEF 2005).

Much evidence shows that a lack of or limited access to clean water, sanitation, and hygiene promotion, combined with a lack of health services, contributes to perpetuating communicable diseases and health risks, mostly in poor communities, in both the developed and developing worlds (Esrey 1996; WHO/UNICEF 2000; Cairncross et al. 2003; UNDP/HDR 2006). As noted by Parry-Jones and Kolsky (2005), "almost all diarrhea deaths are potentially attributable to inadequate water, sanitation and hygiene." They add to other environmental health risk factors, such as indoor and outdoor air pollution, exposure to toxicity, and factors originating from the interaction of humans and the natural and built environment (LEED-ND 2006; Birley 2011).

More specifically, WASH-related diseases can be divided into those that are water related and those associated with the fecal–oral transmission of pathogens. The first group of water-related diseases is usually divided into four subgroups based on the type of transmission route: (1) water borne (e.g., cholera, typhoid, and diarrhea); (2) water washed (e.g., cholera, dysentery, skin infections, and typhus); (3) water based (e.g., schistosomiasis and guinea worm); and (4) insect vector (e.g., malaria, yellow fever, dengue, and onchocerciasis). This differentiation is referred to as the Bradley classification (Cairncross and Feachem 1993), based on the work of David Bradley in the early 1970s.

The second group of diseases is associated with the fecal–oral transmission of pathogens via various routes, such as fingers, flies, fluids (water), food, and fields. It is often represented by the F-diagram of Figure 15-1. The diagram was initially proposed by Wagner and Lanoix (1958) and has been adapted many times by various authors. The version presented in Figure 15-1 was developed by the Water, Engineering and Development Center in the United Kingdom (WEDC 2013). It shows the complex pathways (fluids, fingers, flies, food, and floods) of movement of pathogens and various barriers to transmission related to water, sanitation, and hygiene intervention. The barriers can be primary, such as sanitation, and secondary, related to hygienic practices such as hand washing. The transmission of pathogens can be complex and can involve multiple pathways.

Besides water-related communicable diseases associated with germs and worms, some noncommunicable diseases are associated with the toxic pollution of water where the toxicity is natural, such as arsenicosis and fluorosis, caused by high concentrations of arsenic and fluoride, respectively. Other diseases can be created from water toxicity related to various industrial activities associated with mining and oil extraction.

Sustainable improvements in clean water supply, proper sanitation, and hygienic conditions contribute to a healthy environment conducive to substantially reducing the incidence of communicable diseases, especially diarrhea (Esrey et al. 1991; Cairncross et al. 2003; Fewtrell et al. 2005; WHO/WSSCC 2005; Jamison et al. 2006; Prüss-Üstün and Corvalán 2006; PEP 2008; Mihelcic et al. 2009; Günther and Fink 2010; Hunter et al. 2010; Birley 2011; Cairncross et al. 2013). Recent statistics cited by Cairncross and Valdmanis (2006) show that reductions in diarrhea attributable to WASH interventions are about 17% because of water supply from public sources, 63% if housing water connections are included, 35% because of sanitation, and 48% because of hygiene promotion. As remarked by Cairncross et al. (2013), WASH interventions (such as hand washing) seem to play an additional role in reducing the transmission of respiratory germs leading to ARI. These authors also found some evidence that diarrhea (e.g., as a consequence of inadequate WASH) could result in increasing the risk of pneumonia, thus showing that diseases can be interrelated.

There is a global consensus that as human development progresses around the world, the first group of diseases (i.e., communicable diseases) in the global burden of disease decreases worldwide (WHO 2008). This group is largely associated with poorer populations. Interestingly, diseases in this group are often replaced by diseases of the second group (i.e., noncommunicable diseases) as the populations are lifted out of poverty. In urban and peri-urban areas, however, both groups of diseases are likely to coexist among the poor.

The movement of pathogens from the **faeces** of a sick person to where they are ingested by somebody else can take many pathways, some direct and some indirect. This diagram illustrates the main pathways. They are easily memorized as they all begin with the letter 'f': **fluids** (drinking water) **food**, **flies**, **fields** (crops and soil), **floors**, **fingers** and **floods** (and surface water generally).

WATER   Barriers can stop the transmission of disease; these can be primary (preventing the initial contact with the faeces) or secondary (preventing it being ingested by a new person). They can be controlled by water, sanitation and hygiene interventions.

SANITATION

HYGIENE

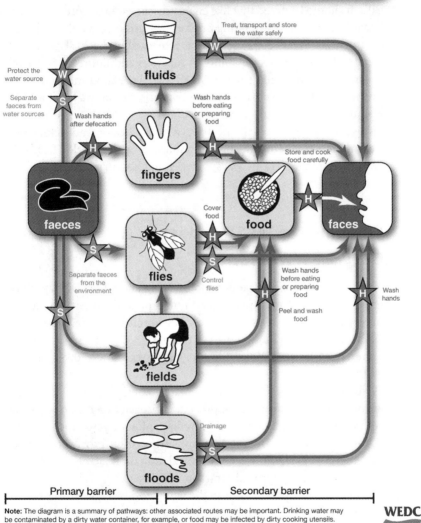

Treat, transport and store the water safely

Protect the water source

Separate faeces from water sources

Wash hands after defecation

Wash hands before eating or preparing food

Store and cook food carefully

Cover food

Separate faeces from the environment

Control flies

Wash hands before eating or preparing food

Peel and wash food

Wash hands

Drainage

| Primary barrier | Secondary barrier |

Note: The diagram is a summary of pathways: other associated routes may be important. Drinking water may be contaminated by a dirty water container, for example, or food may be infected by dirty cooking utensils.

WEDC

Figure 15-1. F-Diagram Showing Pathways for the Fecal–Oral Transmission of Pathogens

Notes: W = water, S = sanitation, and H = hygiene.

Source: Reed and Shaw (2012), with permission from WEDC, Loughborough University. <http://wedc.lboro.ac.uk/resources/factsheets/FS009_FDI_A3_Poster.pdf>

It should be noted that the topics of health, public health, environmental health, and global health in the developing world have received considerable attention in the literature, especially since 1990. The reader will find more information about those topics on the websites of the World Health Organization (WHO), UNICEF, and WEDC, among others.

Finally, it should be noted that like energy discussed in the previous chapter, access to health is critical to sustaining progress toward meeting the Millennium Development Goals (WHO 2012) because as remarked in the 2011 UNDP/HDR report, "health affects people's capability to function and flourish." Some of the MDGs directly concern health, such as MDG 4 (reduce child mortality); MDG 5 (improve maternal health); MDG 6 (combat HIV/AIDS, malaria, and other diseases); and MDG 7 (ensure environmental sustainability). The other four MDGs involve health indirectly. The link between health and achieving the MDGs is clearly illustrated in a document published by the World Health Organization in 2010 on the various ways it is helping countries reach the MDGs (WHO 2010).

15.2 Climbing the WASH Ladder

The Added Value of WASH Services

In the same way that an energy ladder and building energy security at the community level were considered in Chapter 14, it is possible to envision a WASH ladder and explore what it takes to ensure WASH security at the household and community levels. A ladder, also known as the WAT/SAN ladder, was first introduced in the WHO/UNICEF Joint Monitoring Programme (2008) report and consists of two components:

- A three-step ladder for drinking water, consisting of three levels of drinking water delivery: unimproved, improved other than piped water, and improved piped water into dwelling, plot, or yard; and
- A four-step ladder for sanitation, consisting of four forms of sanitation coverage: open defecation, unimproved sanitation facility, shared sanitation facility, and improved sanitation facility.

The distribution of populations in the world and by region for each ladder can be found in Figures 2-9 and 2-10. An agreed-upon proxy for climbing the ladder is the use of one of the improved sources of drinking water and improved sanitation listed in Table 15-1. It should be noted that there is no ladder for hygiene promotion.

At the bottom of the WASH ladder, the WASH poor are more likely to have inadequate water supplies, drink surface water, practice open defecation, and live in unhygienic conditions where trash, garbage, and rubbish littering the streets is the norm. They have access to less water in quality and quantity and pay some

Table 15-1. Improved and Unimproved Drinking Water Sources and Sanitation Facilities

	Drinking Water	Sanitation
Improved	Piped water into dwelling, yard, or plot Public tap or standpipe Tube well or borehole Protected dug well Protected spring Rainwater collection	Flush or pour-flush to piped water system septic tank pit latrine Ventilated improved pit (VIP) latrine Pit latrine with slab Composting toilet
Unimproved	Unprotected dug well Unprotected spring Cart with small tank or drum Tanker truck Surface water (river, dam lake, pond, stream, canal, irrigation channel) Bottled water	Flush or pour-flush to elsewhere (not to piped sewers, septic tank, or pit latrine) Pit latrine without slab or open pit Bucket Hanging toilet or latrine Shared facilities No facilities, bush, or field

Source: Data from WHO/UNICEF Joint Monitoring Programme website: <http://www.wssinfo.org/en/definitions-methods/watsan-categories/> (Oct. 10, 2013).

of the highest prices for it (UNDP/HDR 2006). Furthermore, they are likely to face malnutrition and hunger, have no access to modern forms of energy, and live in desperate conditions characterized by poor housing in unsanitary and unhealthy environments. They live in countries where no regulatory bodies exist, and if they do exist, they are unreliable and poorly managed. Among the WASH poor, "women and girls suffer disproportionately with lack of privacy and there are health and personal safety risks associated with not having access to household sanitation" (Hunt 2006).

Furthermore, the WASH poor have lower resistance to infection and have limited or no access to affordable health services and to basic infrastructure that allows them to access those services. They simply cannot afford the expenditure of staying healthy. As a result, the livelihood of the WASH poor is very vulnerable, and they are more susceptible to illness and death; their burden of disease consists mostly of the diseases in group I. Other health conditions related to injuries, road traffic accidents, falls, and violence (group III) add to the overall burden of disease.

The WASH poor are also energy poor, food poor, and shelter poor. They live in both developing countries and in poor communities in the developed world. They are found in rural, peri-urban, and urban areas. In urban areas, the WASH poor often reside side by side with those who have access to WASH services and other services. Among the many examples of inequities reported in the media, Ramshaw (2011) describes such conditions prevailing in 350 *colonias* on the Texas side of the U.S.–Mexico border involving more than 45,000 people. The same is true in Bedouin communities in the Negev and the West Bank, where poor communities with limited or no access to WASH services live side by side with rich Israeli settlements. The existence of inequitable conditions in peri-urban and urban settings in most big cities in the world is the norm rather than the exception and sometimes involves millions of people. It often results in unstable conditions that lead to violence and unrest.

Climbing the WASH ladder (and other ladders related to energy, food, and shelter) is synonymous with climbing out of poverty and experiencing better health (physical and psychological) at the household and community levels. In turn, good health translates into education and having access to opportunities for human and economic development (Skolnik 2008). In the 2006 UNDP/HDR report, it was mentioned that "every $1 spent in the [water and sanitation] sector creates on average $8 in productivity gain." In the "Water for Life" report (WHO/UNICEF 2005), it was estimated that meeting the MDG drinking water and sanitation target would cost $11.3 billion per year with a return on the investment of $84 billion. The socioeconomic benefits associated with access to clean water supply, sanitation, and hygiene are undeniable (Hunter et al. 2010).

Climbing the WASH ladder has nonhealth positive effects as well (Hunt 2006; Cairncross et al. 2013). At the household level, the benefits may include increased social status and prestige, convenience, safety, time savings, school attendance (especially for girls), school enrollment, learning, dignity, self-respect, privacy, and women's empowerment. At the community levels, the benefits include enhanced social development; confidence; better, safer, more stable, and healthier living conditions and environment; and reduced likelihood of the spreading of diseases.

It should be noted that the close link between economic development and health has limits, though, because growth often leads to dysfunctional forms of consumption and environmental degradation, as seen over the past 50 years in the Western world. Using the human development index (HDI) as a measure of development and considering the wealth–energy–health nexus, a wealth increase is more likely to lead to an increase in CO_2 emission because of increased energy consumption (Figure 14-1), especially in economies based on the burning of solid fuels. The increase in HDI may actually create new health problems, some psychological and some physical, the extent and nature of which depend on local attitude and cultural norms. Health problems related to the environment may rise with the HDI until a certain level of awareness is reached at the local, regional, or national level. At the more global level (planetary), issues such as global greenhouse gas emissions may create additional risk factors that are shared with everyone.

Progress in WASH Services

Data published in a report by the WHO/UNICEF (2012) Joint Monitoring Programme (JMP) show that since 1990, progress has been made in providing WASH services to address the needs of more people on the planet. In addressing more specifically MDG-7C, i.e., "halve by 2015 [from 1990 figures] the proportion of people without sustainable access to safe drinking water and sanitation," the JMP report concluded that the drinking water part of the goal has been reached but not the sanitation component. In 2010, it was estimated that 89% of the world's population was using the improved water sources listed in Table 15-1 compared with 76% in 1990. Over that period, 2 billion people gained access to improved water sources (50% of them in India and China), still leaving 780 million without it today.

Promising trends have also been observed in sanitation. During the period 1990–2010, 1.8 billion people gained access to improved sanitation facilities (40% of them in India and China); however, that still leaves 2.5 billion people (more than 50% of them in India and China) without it today. It has been estimated that 63% of the world's population use the improved sanitation facilities listed in Table 15-1, up from 49% in 1990. Finally, the WHO/UNICEF JMP report concluded that for 59 countries analyzed, more users of improved sanitation are likely to use improved water sources than the other way around. Overall, it was found that half of the 59 countries' populations use both.

Several remarks need to be made about the aforementioned statistics, if looking at the actual characteristics of service delivery. First, it is necessary to clearly define the terms "improved" water and sanitation services that enter into the technologies listed in Table 15-1. Second, the reported success stories are often region specific, some faring better than others. As seen in reference to Figures 2-9 and 2-10, care should be taken when using average numbers to describe local conditions. More often than not, major disparities exist at the national, regional, and local levels. For instance, looking at the living conditions of poor communities in all major cities in the world today, whether in the developed or developing world, demonstrates that WASH progress is not uniformly distributed. The same remark can be made when comparing WASH facilities in rural areas with those in urban areas.

A third remark about the aforementioned statistics is that there is a lack of indicators of what represents high-quality service delivery and more specifically what constitutes "safe drinking water" and "sustainable access" in the actual definition of MDG-7C. Indicators are missing in regard to the safety, reliability, functionality, and sustainability of WASH facilities. Today, there is also no clear agreement about what represents successful WASH interventions beyond the physical existence of the WASH facilities themselves. Water may be safe as it flows out of the well, but by the time it is used in the household, it may have been contaminated because of multiple exposures, such as dirty buckets, dirty hands, poor sanitation practices, and exposure to outside elements, among other things (Gundry et al. 2006; Trevett et al. 2005). As mentioned in the WHO/ UNICEF (2012) JMP report, the only proxy for safe drinking water in MDG-7C today is the physical existence of some of the improved water sources listed in Table 15-1. It is noteworthy that the same lack of indicators applies to sanitation as well.

Lastly, concerning the reported improvement of WASH services, more often than not, water, sanitation, and hygiene promotion services are not addressed together. Instead, they are addressed in a fragmented way, which often reflects the expertise, funding priorities, and belief systems of the various groups involved in providing those services.

The aforementioned remarks make the success stories reported by policy makers who wrote the WHO/UNICEF (2012) JMP report highly subjective and debatable. Not only are indicators of safety, reliability, and functionality poorly defined, there is no concrete plan in place to define what it takes to (1) ensure the medium- and long-term quality control and quality assurance of WASH interventions and (2) develop an enabling environment to ensure their sustainability and scalability. As clearly summarized by WaterAid (2011), "the neglect of the word 'sustainable' in the MDG target to 'halve by 2015 [relative to 1990] the proportion of people without sustainable access to safe drinking water and basic sanitation', needs to be corrected."

Furthermore, the aforementioned remarks make the reported JMP success stories extremely vulnerable and susceptible to be short lived, short of agreeing on implementing strategies and polices that extend and sustain the provision of WASH services (UN-Water GLAAS 2012). It is important to remember that it does not take much for those climbing the WASH or the WAT/SAN ladder to fall back down.

15.3 Sustainability of WASH Services

A History of Less than Satisfactory Projects

The short life of functioning facilities in WASH interventions in the developing world is indeed disheartening because it does not seem to have much to do with the availability of technology, even if the technology was initially selected to match the local needs at the time of installation. Technology may be appropriate but not necessarily sustainable, and vice versa. As a result, communities, mostly in the developing world, are littered with technical solutions that have not provided their intended services or have simply broken down, sometimes shortly after implementation (Baumann 2006; Harvey and Reed 2007; Schweitzer and Mihelcic 2012). According to the organization Water.org, "over 50 percent of all water projects fail and less than five percent of projects are visited, and far less than one percent have any longer-term monitoring." The Rural Water Supply Network (RWSN 2009) surveyed hand pumps in 20 selected countries in Africa and found the proportion of failed hand pumps ranging between 10% and 65%, with a vast majority around 25–30%. Detailed statistics on the level of functioning of water and sanitation systems can also be found on the websites of Improve International and the FairWater Foundation.

Despite major advances in securing the functioning of WASH facilities in the developed world, it is surprising to read that in the international development industry, a "functioning" WASH service in the developing world context "is often regarded as a system which operates at above 50% of its design capacity" (Abrams 2001). In the 2000 report on global water supply and sanitation assessment (WHO/UNICEF 2000), it was mentioned that the "median percentage of rural water supplies which are functioning at more than 50% of design capacity for more than 70% of the time on a daily basis" was 70% in Africa, 83% in Asia, 96% in Latin America and the Caribbean, compared with 97% in North America and 100% in Europe.

It must be recognized that failed projects leading to discontinuing WASH services have a negative effect on the well-being of those who have become accustomed to depending on the services by forcing them to resort to previous bad practices or to use less than safe facilities. Broken down or partially functioning WASH facilities may also create additional pathways to contamination in regard to poorly managed dry sanitation, waterborne sewage, and latrine facilities (Prüss et al. 2002). Failed projects also create confusion and mistrust among community users, service providers, and service authorities and divide communities, as illustrated in the water of Ayolé case study in Section 15.5.

Causes of Less than Satisfactory Projects

There has been ample discussion in the literature to explain why WASH services have short life spans, and a multitude of reasons have been proposed to explain the observed trends. Some can be attributed to the lack of capacity of communities themselves in handling the facilities and providing services. Others may be related to service providers or operators external to the community that fail to provide support (technical, management, or financial). Others are related to the limited engagement, commitment, and interest often shown by service authorities (local, regional, or district government) in overseeing, coordinating, planning, and supporting project execution, operation, and maintenance. Finally, external development agencies, nongovernmental organizations (NGOs), international NGOs (INGOs), and charity groups also contribute to the short life span of projects because of the way they manage, implement, monitor, and evaluate their projects.

At the community, service provider, and service authority levels, multiple reasons for short-lived WASH interventions can be identified:

- The community is not motivated and convinced about new WASH services and has not bought into ownership.
- WASH services are not affordable, practicable, and/or accessible even if available.
- The community has lost hope in any form of outside assistance because of failed promises or inadequate projects in the past.
- The community has lost interest in the projects and/or those in charge of it have left or died.
- No or limited community management of facilities and services exists because of a lack of local vision, organization, and capacity (institutional, human resources, technical, economic and financial, energy, environmental, or social and cultural).
- The service providers do not have the capacity or have limited capacity to operate and maintain the facilities.
- There is no clear understanding of how responsibilities of facilities are distributed among the community, service providers, service authorities, and outside assistance, and in particular who is responsible for what and for how long. As noted by Davis and Brikké (1995), governments in developing countries often change "their role of provider of services to that of facilitators of processes," thus placing more responsibilities on communities to manage their own projects without necessarily including any form of empowerment or training or even informing communities about the changes in service. Local or central governments may fail to provide major repair to facilities.
- There is a lack of standards of services or regulations at the national level, and if available, no enforcement of policies is in place.
- Excessive bureaucracy, at the local or global government levels (with or without corruption), may block and delay new initiatives.
- Governments fail to realize their role in safeguarding the interests of society, balancing the interests of various groups, and developing partnerships through the implementation of WASH projects (WHO/WSSCC 2005).
- Policy makers and government officials are infatuated with projects that emulate those in the Western world but are out of touch with the society they are supposed to support.
- In remote rural areas in the developing world, there is no economy of scale in the maintenance of water and sanitation systems. For water supply systems in Africa, Skinner (2009) noted that per capita management costs decrease as the number of people served in urban areas increases but not in remote rural areas.

Likewise, outside organizations that conduct projects in developing communities have contributed to incomplete projects for multiple reasons:
- No effort to match services with the current forms of community capacity and make them relevant to the beneficiaries;
- A narrow focus on technology hardware as an end in itself rather than as a means to an end;
- A lack of technical competency in service delivery;
- Failure to account for and influence human behavior in interventions;
- Failure to develop and strengthen an enabling environment to support interventions in terms of behavior change, resource allocations, financing, and shared roles and responsibilities;
- Failure to provide training in maintenance of technology and systems;
- Failure to identify and consider the full range of life-cycle costs of systems, such as initial, installation, energy, operational, maintenance and repair, downtime, environmental, and decommissioning and disposal costs;
- No interest in providing long-term benefits (sustainability) or improving the community level of development over time;
- Reinforcing past practices, even though they have proven to lead to medium- or long-term breakdown (RWSN 2009) or are not appropriate; and
- Preference for centralized WASH facilities rather than on-site systems.

Additionally, the Water Supply and Sanitation Collaborative Council published a report (WSSCC 2004) that listed deeper reasons for the chronic failure of WASH interventions. It denounced a conservative mindset in the development industry over the past 50 years characterized by:
- No willingness on the part of those who had contributed to failed projects to learn from or even be accountable for their failures;
- A commitment to "business as usual," with no incentives to change the mindset or the incapacitating bureaucracy;
- A lack of community empowerment considered in decision making; and
- A myopic expert–elite approach to solving problems using the same rational and predictable methods applied mostly on large projects in the Western world.

Finally, it is not uncommon for many of the aforementioned reasons to synergistically interact and as a result create cascading effects that are hard to predict. This problem relates to the discussion in Chapter 4 regarding how the complexity and uncertainty in any intervention (WASH in this case) combined with that of the communities often contribute to less than successful projects. This is, of course, not a reason to legitimize the short life spans of WASH projects. However, as remarked in Section 4.9, it explains how easy it is to lose track of how the complexity and uncertainty of human and environmental systems change over time and interact with the infrastructure.

The complexity of WASH projects within the context of developing communities calls for embracing an adaptive and reflective approach to design, planning, and management of such projects rather than the traditional blueprint approach often used in development agencies and governments. As discussed in Section 4.9, with an adaptive approach, uncertainty is handled through reflection, adaptation, monitoring, and evaluation and by incorporating feedback mechanisms within the project life cycle. This approach is also about finding solutions that are good enough rather than believing that the best and optimized solutions are possible, especially using sophisticated solutions from the Western world that are clearly out of context with the community's reality. Finally, the adaptive approach requires the adoption of WASH service indicators combined with performance criteria and targets that guarantee long-term services and project success.

Improving the Delivery of WASH Services

Before entering into a discussion of what it would take to ensure the sustainability of WASH services, it is important to recall the aims and objectives of WASH interventions and outline appropriate determinants of project success. By keeping the end in mind, it is easier to evaluate progress and performance. Carter et al. succinctly defined the aims and objectives of water supply and sanitation projects and programs as follows:

> The aim of such projects and programmes is to bring about health improvements, privacy in defecation, and reductions in time and effort spent in water hauling, for the whole community. Soil, surface water and groundwater are to be protected from fecal contamination. Hygiene practices are to be improved by appropriate components of such projects. These goals should be achieved at acceptable capital and recurrent costs. These goals should be realized for the foreseeable future. (Carter et al. 1999)

The jury is still out about what represents effective WASH interventions and services, and more specifically, what factors affect their sustainability over a "foreseeable future." A simple (but pragmatic) proxy used in the past has been to say that WASH services are sustainable if the facilities they depend on are functioning at the time of inspection (but not necessarily before and after the inspection). Another proxy commonly used in water supply and sanitation projects is the number of systems built and the number of people served. Obviously, there is more to sustainability than those static attributes.

According to Abrams (1998), sustainability in the context of service delivery is "whether or not something continues to work and deliver benefits over time" with the assumption that "if water flows then all of the many elements which are required for sustainability must have been in place." These elements include "technical issues, social factors, financial elements, the natural environment, durable gender equity and empowerment, and institutional arrangements." Even though Abrams' recommendation focuses on the outcome, i.e., whether it works or does not work, using a proxy for sustainability based on functionality of the system service delivery alone is limiting. It does not say anything about the components that make it work and how it could be duplicated and/or scaled up.

More sophisticated frameworks have been suggested in the development literature over the past 20 years to improve the delivery of WASH services and to ensure that those services (1) deliver the intended benefits to the recipients; (2) deliver them for an extended period (not clearly defined) after external support ends; and (3) deliver them at the appropriate cost during the service life cycle, including "investment (planning, design, construction, and equipment), operating costs, maintenance costs ... and eventually disposal and replacement costs" (Baumann 2006). Some of these frameworks are reviewed below.

Narayan (1993, 1995) proposed adopting a framework for participatory evaluation of water and sanitation programs. It is based, first, on the idea of mobilizing the community through participation and intimately involving users and stakeholders in "decision making, goal setting, design and management ... of water and sanitation facilities." A second component of Narayan's framework is that indicators must be in place to monitor and evaluate the progress of water and sanitation programs in three categories related to

- *Project sustainability* in terms of reliability of systems, human capacity development, local institutional capacity, cost sharing and unit costs, and collaboration among organizations;
- *Effective use of WASH facilities* expressed in terms of optimal use, hygienic use, and consistent use of the facilities; and
- *Replicability of solutions* in new locations in terms of ability of a community to expand services, and the transferability of agency strategies.

These three groups of indicators are further broken down into measurable subindicators and associated qualifiers.

Narayan's framework was later included in the Methodology for Participatory Assessment (MPA) proposed by the Water and Sanitation Program (WSP) at the World Bank and the IRC International Water and Sanitation Center (Mukherjee and van Wijk 2003). Special emphasis was placed on the positive role played by gender and social equity in contributing to long-term success of WASH interventions. The MPA considers the role of equity in five interrelated dimensions of sustainability:

- *Technical sustainability* related to the "reliable and correct functioning of the technology, and for water supplies, the delivery of enough water of an acceptable quality … with meeting the demands of all user groups";
- *Financial sustainability* in terms of having the resources to cover the cost of operation, maintenance, and repair, with costs shared equitably;
- *Institutional sustainability* related to the type of support institutions provide in operation, maintenance, and repair of facilities and how all stakeholders have a voice in interacting with the institutions;
- *Social sustainability* in terms of how WASH services match "sociocultural preferences and practices" and meet all people's expectations; and
- *Environmental sustainability* in regard to protection of water supply, prevention of overexploitation, and control of waste disposal, with equitable rights and responsibilities for all users.

In all the frameworks proposed to improve the sustainability of WASH services over the past 20 years, a special focus has been placed on ensuring that effective operation and maintenance (O&M) strategies are included in project design, planning, and management (Davis and Brikké 1995) where (1) *operation* refers to "the everyday running and handling of a water supply" and (2) *maintenance* is about "the activities required to sustain the water supply in a proper working condition."

Maintenance can itself be divided into preventive, corrective, and crisis maintenance, depending on the context of the project. According to the International Institute of Environment and Development (IIED), proper maintenance of water supply systems requires "the right technology, ownership by the communities involved, and local capacity to repair and maintain wells and systems" (Skinner 2009). In a broader sense, effective O&M relies on having local capacity (ability) and assets in the different sectors of capacity shown in Figure 9-2: (1) institutional (an enabling environment, regulating agencies, district and village management units); (2) human resources (skilled and unskilled workers, expertise); (3) technical resources to provide service (availability of materials and equipment, spare parts); (4) economic and financial (cost sharing); (5) energy; (6) environmental; and (7) social and cultural. The capacity appraisal and analysis discussed in Chapters 5 and 9 can help determine what forms of capacity are lacking to support O&M and what needs to occur to fill the gap between what should be in place and what is currently available. The importance of O&M was a topic of major interest in the village level operation and management of maintenance hand pumps initiative in the 1980s and 1990s (Arlosoroff et al. 1987; Reynolds 1992).

Harvey and Reed (2007) emphasized the importance of both *community participation* and *community management* in WASH interventions. Community management implies that the community is in control of the projects and shows local (1) leadership, capacity of management, and social cohesion; (2) ownership; and (3) ability to handle cost recovery and financial management in a transparent way. As discussed by Harvey and Reed, community participation alone does not necessarily lead to effective community management, and community management without participation does not often lead to sustainable and equitable solutions. Harvey and Reed concluded that both participation and management are equally important and that there is also a need for "ongoing support from an overseeing institution to provide encouragement and motivation, monitoring, participatory planning, capacity building, and specialist technical assistance." The institution can be a government (local, regional), an NGO, or the private sector. External support and oversight are indeed critical at different levels (community, local government, higher level) to avoid failed projects. As noted by Hunter et al. (2010) for water supply, a lack of support may lead to interrupted supply of water that "may be sufficient to destroy the health benefit from the provision of clean drinking water."

WaterAid developed an extensive framework for sustainable water supply, sanitation services, and hygiene behavior change in poor communities. It defines sustainability as

about whether or not WASH services and good hygiene practices continue to work and deliver benefits over time. No time limit is set on those continued services, behavior changes and outcomes. In other words, sustainability is about permanent beneficial change in WASH services and hygiene practices. (WaterAid 2011)

The WaterAid framework encompasses many of the aforementioned recommendations and in particular the synergy among the community, governmental institutions (national, regional, district, local), and external support and intervention. The framework is described in detail in Figure 15-2.

WaterAid proposed five requirements "to ensure sustainable WASH services and hygiene practices," which are listed here:
- The demand for WASH service must be real. This need is indicated by how the community uses the water and sanitation services and how its members are practicing improved forms of hygiene behavior.
- An equitable structure is in place to cover lifelong costs with "tariff structures that include the poorest and most marginalized" members of the community.
- A "functioning management and maintenance system" must be in place.
- External support is available to "community-level structures and institutions" that manage the services.

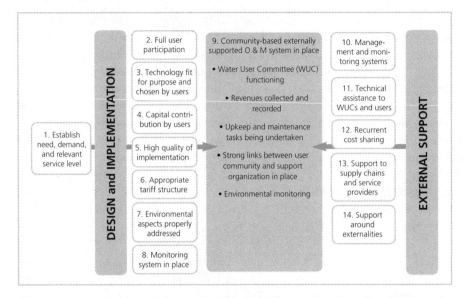

Figure 15-2. Conceptual Framework for Effective Externally Supported Community-Based Management of Rural Water Supply Services

Source: WaterAid (2011), reproduced with permission.

- Special attention is placed on "the natural resource and environmental aspects of the systems installed."

Schweitzer and Mihelcic (2012) proposed an assessment tool that goes one step further than the general recommendations mentioned previously. It is designed to score the sustainability of rural water systems in the developing world and provides a measure. In that framework, the two authors characterize sustainability as "an equitable access amongst all members of a population to continual services at acceptable levels providing sufficient benefits, and reasonable and continual contributions and collaboration from service, consumers, and external participants."

This definition calls for rural water facilities to not only "simply work" but also to be reliable, affordable, accessible, and available for an extended period of time (i.e., a foreseeable future) through community participation and engagement, community management, and outside support (government, NGO).

The sustainability assessment tool of Schweitzer and Mihelcic is based on eight indicators and 15 measures that are listed in Table 15-2. They include the activity level, participation, governance, tariff payment, accounting transparency, financial durability, repair service, and system function. Based on the values of the indicators, rural water systems are scored into three groups: sustainability likely, sustainability possible, and sustainability unlikely.

Table 15-2. Sustainability Assessment Tool

Indicators	Measures	Targets			
		Sustainability Unlikely	Sustainability Possible	Sustainability Likely	
Activity level	1. Active water committee members	1 person or fewer	2 people	3 people or more	
Participation	2. Average percent attendance at community meetings	Less than 50%	$50 \leq x < 66.6\%$	66.6% or greater	
Governance	3. Decision-making process	Minority decision No transparency	Majority decision Transparent but arbitrary process	Democratic decision Community discussion Water committee facilitates	
Tariff payment	4. Percent debtors	Greater than 80%	$80 \geq x > 10\%$	10% or less	
Accounting transparency	5. Accounting ledger 6. Report frequency	Do not use ledger AND Report less than once a year	Use ledger OR Report at least once a year	Use ledger AND Report at least once a year	
Financial durability	7. Wages 8. Costs 9. Tariff 10. Average level payment 11. Connections 12. Savings	Income $\leq$ O&M AND Less than "significant savings"	Income > O&M OR "significant savings"	Income > O&M AND "significant savings"	
Repair service	13. Downtime	More than 5 days	1 to 5 days	Less than a day	
System function	14. Average hours/day 15. Average days/week	Both Less than 8 h	Pump system $8 \leq x < 12$ Gravity systems $8 \leq x < 16$	Pump system 12 h or more Gravity systems 16 h or more	

Note: References for each measure are available in the original table.

Source: Schweitzer and Mihelcic (2012), with permission from IWA Publishing.

Another framework used to assess the effectiveness and sustainability of water supply systems was proposed by the International Water and Sanitation Center (IRC) in the United Kingdom. Referred to as the Triple-S (sustainable services at scale) framework, it focuses on what it takes to provide reliable service delivery rather than focusing on just the infrastructure (Lockwood and Smits 2011; Smits et al 2012). The framework emphasizes the importance of having service delivery performance indicators and scores at multiple intervention levels. For the case of Ghana, Adank et al. (2011) considers three levels of service:

- *At the level of consumers,* key indicators relate to "quantity, quality, accessibility and reliability [of services] over time." Based on the value of those indicators, the service level can range between no service and high service, with intermediate qualifiers such as intermediate service, basic service, and substandard service.
- *At the level of service providers* (community-based organizations, private operators, and public utilities), key indicators are related to the fulfillment of "basic technical, financial, management and organization functions necessary to deliver a sustainable service."
- *At the level of service authorities* (local, district, or national government), key indicators are related to the fulfillment of "planning, coordination, regulatory and support functions necessary to ensure the establishment of service providers."

Another determinant of WASH intervention success is that of community motivation, commitment, and the realization that it is in the best interest of the community to deliver, maintain, and operate high-quality services (Carter et al. 1999). Closely related to these community attributes is the role played by behavior change communication methods in ensuring the benefits of long-lasting WASH interventions. This role was emphasized in the more recent approaches promoted by the World Bank (the Sanitation Marketing Toolkit; WSP 2013) and the USAID through the Hygiene Improvement Project from 2006 to 2010 and the more recent WASHPlus project (USAID 2013). Of particular interest in WASHPlus is the School Promoting Learning Achievements through Water, Sanitation, and Hygiene (SPLASH) initiative in Zambia (USAID 2012). This well documented initiative emphasizes the importance of appropriate latrines, hand-washing facilities, and water sources, which when combined with behavior change leads to higher student (boys and girls) retention and attendance and a better learning and teaching environment.

Finally, before ending the discussion on what makes WASH interventions long lasting, it should be noted that no single issue contributing to their sustainability is sufficient in itself, whether it is technology, participation, management, support, or operation and maintenance. As remarked by Abrams et al. (1998), there is no "silver bullet" that guarantees success in WASH service delivery. Instead, it is more effective to adopt a service delivery approach that is holistic, systemic, and adaptive and integrates multidimensional issues cutting across technology, legislation, policy, and institutions. The approach must take into consideration the specific and interacting roles of users, service providers, service authorities, and national and international groups (Smits et al. 2012). Furthermore, it is recommended that sustainability be addressed early in the project life cycle (initiation phase) and continue to be an ongoing (continuing) objective as projects unfold. Another way of looking at it, as suggested by Abrams et al. (1998), is to ensure that water supply and sanitation systems be designed for sustainability. As remarked in Chapter 2 in relation to the discussion on the characteristics of sustainable development projects, the service delivery approach must recognize that WASH projects not only need to be *done right* but they also need to be the *right projects* for the community and the environment with which they interact (ISI 2012).

15.4 Basic Water and Sanitation Requirements

When considering people's basic needs in water, sanitation, and hygiene, what is required to meet such needs should be addressed. As noted by Gleick (2002), "people do not [necessarily] want to 'use' water. They want to drink and bathe, swim, produce goods and services, grow food, and otherwise meet human needs and desires." Likewise, "farmers do not want to use water per se; they want to grow crops for profitability." In other words, people would use less water if they could get the same benefits with less consumption using more efficient technologies.

Meeting basic requirements for drinking, cooking, and hygiene dictates the provision of WASH services. The issue of how well those services are provided, now and in the future, depends largely on the current level of community development and any potential to increase the capacity of the community toward a higher level of development in the foreseeable future. This dynamic of capacity building was explored in Chapter 9.

Water Requirements

Since the 1960s, there has been ample discussion on the basic (or minimum) water requirements to meet human needs at the household level, such as drinking water and water for hygiene, sanitation services, and food preparation. Six characteristics of water supply must be considered in the preservation of human health: quality, quantity, access, reliability, cost to the user, and ease of management (Hunter et al. 2010). These characteristics are combined in Table 15-3, which summarizes health concerns for four levels of water service.

Figure 15-3 shows a suggested hierarchy of water requirements for different domestic activities and is expressed in liters of water per person

Table 15-3. Levels of Health Concerns for Four Types of Water Service

Service Level	Access Measure	Needs Met	Level of Health Concern
No access (quantity collected often below 5 L/person/day)	More than 1,000 m away, or 30 min total collection time	Consumption—cannot be assured Hygiene—not possible (unless practiced at sources)	Very high
Basic access (average quantity unlikely to exceed 20 L/person/day)	Between 100 and 1,000 m away, or 5–30 min total collection time	Consumption—should be assured Hygiene—hand washing and basic food hygiene possible; laundry/bathing difficult to assure unless carried out at source	High
Intermediate access (average quantity about 50 L/person/day)	Water delivered through one tap on site (or within 100 m or 5 min total collection time)	Consumption—assured Hygiene—all basic personal and food hygiene ensured; laundry and bathing should also be assured	Low
Optimal access (average quantity 100 L/person/day and above)	Water supplied through multiple taps continually	Consumption—all needs met Hygiene—all needs should be met	Very low

Source: Howard and Bartram (2003), with permission from the World Health Organization.

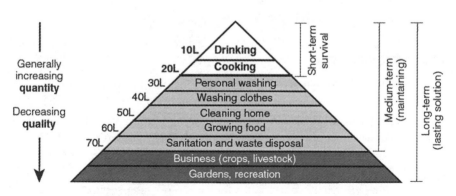

Figure 15-3. Hierarchy of Water Requirements

Source: WHO (2011b), with permission from World Health Organization.

(or capita) per day (L/person/day). Besides securing enough water to address the basic needs of households, it must be considered that water is also required in agriculture, manufacturing, energy production, and other fields that contribute to the economic development of a region or country. The relationship between a country's HDI and the amount of water required to reach a certain level of development has not been thoroughly researched. An attempt to explore that relationship was conducted by Chenoweth (2008), who concluded that "a country could meet its domestic water requirements together with its water requirements for maintaining a water efficient non-agricultural economy capable of sustaining a high level of human development with as little … as 135 l/pp/d [L/person/day]." That figure was based on data collected in Europe and assumed a 10% water loss. It cannot be extrapolated to developing countries.

Many factors determine water needs at the household level. As remarked by Gleick (1996), they include "climatic conditions, lifestyle, culture, tradition, diet, technology, and wealth." Water needs also depend on whether the community is in a development phase or in one of the many phases of recovery following a crisis situation (Table 13-1). Water needs are likely to vary for each country and region. Even within a given region, household needs are likely to be higher in an urban environment than in a rural environment. Furthermore, there are multiple predictable and unpredictable reasons for using water at the household level, some of a higher priority and level of urgency than others, e.g., water for drinking and cooking versus water to wash clothes. The amount of water also depends on the type of sanitation used in the household; for instance, flush toilets use much more water than pit latrines (WHO 2011b). In general, all these variables make it difficult to determine an exact and unique basic water requirement.

Guidelines for water quantity have been proposed by various agencies since the 1970s and have changed somewhat over time. Chenoweth (2008) gives an excellent review of various studies reported in the literature on determining per capita water requirements. For instance, a figure of 20 L/person/day was suggested in the early 1980s by the International Drinking Water and Sanitation Decade (IDWSD; UN 1985) in its goal "to provide all people with water of safe quality and adequate quantity and basic sanitary facilities by 1990." At the subsequent Dublin conference on Water and the Environment (UN 1992), that figure was raised to 40 L/person/day to include drinking, cooking, cleaning, and sanitation.

On a side note, the Dublin conference built on the outcome of the IDWSD and the 1990 New Delhi statement on Global Consultation on Safe Water and Sanitation (IELRC 1990). The Dublin conference was instrumental in establishing four basic principles that have been critical in defining international water policies over the past 20 years (WEDC 2011).

- Freshwater is a finite and renewable resource, essential to sustain life, development, and the environment.
- Water development and management should be based on a participatory approach, involving users, planners, and policy makers at all levels.
- Women play a central part in the provision, management, and safeguarding of water.
- Water has an economic value in all its competing uses and should be recognized as an economic good.

These four principles were included in Agenda 21 (UNCED 1992) at the 1992 Rio Summit. Subsequently, other conferences and initiatives emphasizing the importance of water in development have been held, all emphasizing access to water as a human right, water scarcity, and the need to use water more efficiently and effectively (Abrams 2001). Among these initiatives, the Joint Monitoring Programme (JMP) for Water Supply and Sanitation between the WHO and UNICEF has played an active role in monitoring "the use of various types of drinking-water sources and sanitation facilities at the national, regional and global levels." The reader will find up-to-date information about world water facts and figures in the United Nations *World Water Development Report* (UNESCO 2012) through the World Water Assessment Program. The latter recommends a figure of 20–50 L/person/day to meet the basic needs for drinking, cooking, and cleaning.

A basic water requirement often referred to in the literature was proposed by Gleick (1996). The author suggested adopting a standard for human domestic needs of 50 L/person/day to include all four basic human needs for water: survival, hygiene, sanitation services, and food preparation. According to that author, the 50 L/person/day breaks down into several components: (1) 5 L/person/day for basic fluid replacement, (2) 20 L/person/day for sanitation, (3) 15 L/person/day for bathing, and (4) 10–20 L/person/day for food preparation. It must be kept in mind that additional water is needed to grow the food necessary for human survival, as shown in Figure 15-3.

Most world organizations seem to have adopted the 50 L/person/day recommendation in noncrisis situations, even though that figure may not even be a reality for most people in the developing world today. An amount of 20 L/person/day is often considered a lower bound or minimum for domestic use. In emergency situations, that lower bound can decrease to 7.5–15 L/person/day (2.5–3 for drinking and food, 2–6 for basic hygiene practices, and 3–6 for basic cooking needs), as recommended in the Sphere standards (Sphere Project 2011). The lower figure corresponds to short-term *domestic* survival conditions two weeks to one month after an emergency. It increases to reach the 15–20 L/person/day within three to six months after an emergency has passed (WHO 2011b). As remarked by Roberts (1998), in postcrisis situations, it is important to have enough water to prevent diseases such as diarrhea.

The minimum water requirements in emergency situations increase well above the 15 L/person/day if nondomestic uses of water are included, such as water for health centers, schools, and livestock. For instance, USAID (2005) suggests 30 L/inpatient/day for feeding centers and 40–60 L/inpatient/day for health centers and hospitals. They also recommend that "a large quantity of reasonably safe water is preferable than a small amount of pure water."

The basic water requirements for human activities are not just about quantity. Issues such as water quality, the distance of the water supply to households (or the collection time necessary for a round trip from home to water source), the waiting time at the place of water supply, how many people use water facilities, and their level of functioning are equally important. Regarding water quality, not all forms of water supply need to be of drinking quality, as commonly used in Western countries. Water to wash clothes or clean the floor does not have to be of the same quality as drinking water and can originate from sources other than a spring or a tap. For potable water, strict guidelines have been proposed by the WHO (2011a) to ensure drinking water quality. In Western countries, regulatory institutions, such as the Environmental Protection Agency in the United States, have developed strict guidelines regarding public drinking water systems. In developing countries, water quality standards may or may not exist and be regulated. Table 15-4 shows specific uses of water originating from different sources.

The distance of water supply from households and the waiting time at a water point are critical issues that require special consideration. It has been observed that longer distances and times for collection result in less water being carried, less consumed water, fewer hygiene practices, and increased health issues. It has been found that when it takes more than 30 minutes round-trip for people to collect water for their household (a distance of about 1 km from the household, assuming a speed of 4–5 km/h), the amount of water collected is substantially less. Interestingly, for shorter times between 5 and 30 minutes, the water consumption is constant: around 15 L/person/day. But, for even shorter collection times, it increases rapidly, thus supporting the need for WASH interventions on site as much as possible, but not to exceed 100 m from homes. As noted by Carter et al. (1999), "people seem to prefer to save time than use more water" despite the observation that more water leads to better hygiene and health.

In nonemergency situations, the WHO/UNICEF (2008) JMP report recommends access from source to user's dwelling not to exceed 1,000 m. In emergency situations, the Sphere standards suggest that the maximum distance from any household to a water point is not to exceed 500 m (1,000 m in early recovery),

Table 15-4. Appropriate Water Sources for Specific Uses

Water Use	Water Source					
	Surface Water (Untreated)	Surface Water (Treated)	Groundwater[a]	Rainwater (First Flush)	Rainwater (After First Flush)	Greywater
Drinking	X	A	A	X	A	X
Cooking	X	A	A	X	A	X
Bathing, hand washing, anal ablution	A[b]	U	A	A	A	X
Irrigation	A	U	A	A	A	A
Latrine flushing	A	U	A	A	A	A

Note: A = Appropriate; U = Unnecessary expense; and X = Inappropriate.

[a]Care must be taken in areas where arsenic, fluoride, or microbial contamination of groundwater is a concern. Sanitary surveys should be used to determine the risk of microbial contamination.

[b]In areas where schistosomiasis is endemic, people's skin should not come into contact with untreated surface water. Infected snails release a larval form of the parasite that can burrow through skin.

Source: Mihelcic et al. (2009); reproduced with permission.

with a waiting time to collect water of no more than 15–30 minutes (Sphere Project 2011). In its *Field Operations Guide for Disaster Assessment and Response,* the USAID (2005) recommends one water tap stand for 250 people, with access not exceeding 100 m from the users.

Regarding the number of users, it has been suggested that a water point served by hand pumps and a water tower is good for no more than 250–500 people in noncrisis situations and 500 people in crisis situations (Sphere Project 2011). It is also important to know the characteristics of the water supply systems in terms of hours per day the service is provided. As remarked by Schweitzer and Mihelcic (2012), water supply systems should be able to operate for a minimum number of eight hours per day with a downtime not exceeding five days.

Finally, meeting minimum water requirements depends on many tangible and intangible factors, as discussed in the previous section. Geopolitical factors that are well beyond the community's control may also dictate the quality of the water, its amount, and how much can be extracted. Extracting too much water may indeed affect another community, induce environmental and health issues elsewhere, and create geopolitical tensions. Furthermore, from an environmental point of view, a need exists to consider the protection of natural ecosystems because a certain amount of runoff needs to be maintained.

Sanitation Requirements

The safe disposal of waste generated by communities must be addressed at the same time as water supply. Both issues of water and sanitation are intimately related, because a lack of the latter could have severe repercussions on water quality, the environment, and community health. Clean drinking water can easily be contaminated by the improper disposal of waste. Hygienic behavior, such as washing hands, is also required because without it, much of the drinking water and sanitation improvements would be futile.

Sanitation encompasses a lot of activities. At its most basic definition, sanitation is the safe collection, storage, treatment, disposal (including reuse and recycling) of human excreta (feces and urine) and associated sewage or black water. It can be done on site (house or immediate vicinity) or off site (removed from house or immediate vicinity). Using Table 15-1 as a guide, improved sanitation is understood by development agencies as facilities with (1) flush or pour-flush to a piped sewer system, septic tank, or pit latrine; (2) ventilated improved pit latrines; (3) pit latrines with slabs; or (4) composting toilets.

In a broader sense, sanitation is also about the management of dirty water (sullage or greywater) and storm water; drainage of water; treatment and disposal (including reuse and recycling) of greywater and sludge; and the collection and management (including recycling) of solid waste, such as garbage, refuse, and rubbish.

In general, when considering sanitation requirements, it should be realized that the characteristics of sanitation services and their management depend on many parameters (Hunt 2006):

- The users of the sanitation services (who and how);
- The nature, composition, quantity, and characteristics of the wastewater generated;
- The final whereabouts of the wastewater generated;
- The planned level of sanitation service, whether it is private sanitation or shared;
- The distance from households to sanitation facilities;
- Seasonability (floods); and
- Whether users are in a postcrisis or development context.

As long as water is being used by community members, greywater is generated on site at the household level from the domestic use associated with bathing, laundry, and dishwashing. In general, assuming that it does not contain feces and pollutants, greywater is "relatively harmless from an environmental and hygienic point of view," and if problems arise, they "are often small and local" (EcoSanRes 2004). In that case, greywater can be seen as a valuable resource for agricultural purposes, especially in regions with water scarcity or limited water supply services. Furthermore, greywater may contain nutrients that make it a partial substitute for fertilizers. These ideal conditions for greywater imply that rules are being followed by community members (Morel and Diener 2006). More often than not, such rules are not considered or, if they exist, are not enforced. Therefore, there is a risk that greywater may contain components such as salt, pathogens, oil, grease, and chemical compounds that may affect health and the quality of the soil and groundwater if the greywater is not treated before disposal. Drainage and treatment (primary, secondary) then become necessary before being discharged into the environment in surface water, subsurface infiltration, groundwater recharge, or in irrigation systems.

The cost associated with treatment of greywater depends largely on its pollution load (type and amount). The load can be reduced, as suggested by Morel and Diener (2006), by using a first line of action before treatment through adoption of control measures at the source of greywater generation, i.e., the household. Point of use measures may include (1) reducing water usage; (2) optimizing the usage of cleaning products; (3) avoiding the discharge of adverse substances (solvents, bleach, oil, fat); and (4) replacing products that are hazardous with environmentally friendly products.

If greywater is deemed safe, it can indeed be used for landscape and farming irrigation. It can be allowed to flow into drainage channels and bodies of surface water or seep into the ground and replenish the aquifer with limited or no effect. However, stagnant water on site should be avoided because it has the potential to bring additional diseases, such as those associated with breeding insects.

In regard to the amount of greywater generated, it makes sense to assume that it is close (80–100%) to the amount of water used. The organization Eco-SanRes (2004) suggests using an outflow estimate of 15–20 L/person/day in poor communities and 10 times as much in richer countries. Morel and Diener (2006) proposed a higher number of 20–30 L/person/day, with the caveat that the amount generated depends largely on the existing amount of water available, the climate and soil conditions, and the type of facilities in place. When water is plentiful and is piped into houses, 90–120 L/person/day is more realistic, but the greywater is more diluted.

Mixed water conditions (greywater and black water) require special precautions because black water contains pathogens (human excreta) that are harmful to humans. Fecal contamination may happen "through hand washing after toilet use, washing of babies and children after defecation, diaper changes or diaper washing" (Morel and Diener 2006). Proper sanitation techniques need to be implemented to remove human contact with the excreta (e.g., the primary barrier in Figure 15-1). This separation needs to be complemented with health education and hygiene promotion (e.g., the secondary barrier in Figure 15-1).

Guidelines and standards have been proposed for the collection, treatment, and disposal (management) of waste, mostly for communities in the developed world. In the developing world, this is still a work in progress and the guidelines are often inappropriate to actual field conditions. For instance, the World Health Organization (WHO 2006) published *Guidelines for the Safe Use of Wastewater, Excreta and Greywater*. As remarked by Morel and Diener (2006), the standards outlined in the WHO's guidelines can hardly be applied in developing countries "for lack of financial and human resources." The guidelines tend to favor the use of larger sanitation systems rather than smaller (on-site decentralized) ones. They also emphasize the temporary nature of smaller systems as stopgaps until more modern systems are in place (Etnier et al. 2005). The limitation of that approach is that combined with an absence of regulatory authorities and/or enforcement of regulations, large systems are highly vulnerable. Furthermore, it fails to recognize that smaller, decentralized systems such as on-site ones work as well and are more appropriate in terms of costs of operation, maintenance, and repair, especially in remote areas. This statement is conditional to having a framework in place to ensure the reliability and functionality of such systems over a foreseeable future (Etnier et al. 2005).

Guidelines and standards for sanitation projects (and hygiene promotion) in the developing world are needed but are hard to develop, because they would have to cover a wide range of socioeconomic and environmental conditions characterized by different practices of waste generation and cultural habits. They would also have to encompass multiple scales from on-plot sanitation, to on-site sanitation, to large-scale sanitation (WELL 2014). As remarked by Morel and Diener (2006), different sources of greywater in different settings cannot be managed the same way and call for different collection, treatment, disposal, and reuse or recycle options. Morel and Diener also suggest that short of having standards of regulations, there is a need to have at least basic goals in mind when planning, developing, and managing greywater systems, such as (1) protection of public health, (2) environmental protection, (3) protection of soil fertility, (4) sociocultural and economic acceptability, (5) simplicity and user-friendliness, and (6) compliance with national and international standards. In their report, Morel and Diener (2006) review several examples of greywater systems in different regions of the world ranging in size, capacity, performance, and cost.

Solid Waste Requirements

Solid waste needs to be disposed of as well. If not properly handled, it can affect the well-being of community members in many different ways during storage, transfer, processing, disposal, and even recovery. There are many ways that human health could be affected by waste through air, water, insects, animals, and direct contact transmission paths (Mihelcic et al. 2009).

In general, the amount and type of waste varies with the community (urban vs. rural) and depends on its level of development. Urban populations produce more than just garbage and trash. Waste can also include "residential, commercial, industrial, institutional, municipal, and construction and demolition waste" (World Bank 2012). Communities with a higher standard of living also generate more waste.

The waste in developing countries (especially in rural and peri-urban settings) is likely to contain more organic material than that of richer countries, where plastic, paper, and nonorganic materials (e.g., glass and metals) are dominant (Bouabib 2004; Mihelcic et al. 2009; World Bank 2012). According to the World Bank (2012), the amount of solid waste generated in developing communities ranges between 0.5 and 1.5 kg/person/day, with a wide variation depending on the community's income and region in the world: 0.65 in sub-Saharan Africa, 0.95 in East Asia and the Pacific Region, 1.1 in Central Asia, 1.1 in Latin America and the Caribbean, and 1.1 in the Middle East and North Africa (World Bank 2012). This amount should be compared with an average of 2.2 for OECD countries and an average of 1.2 worldwide, a number that is expected to increase to 1.42 in the next 15 years because of economic development.

Guidelines for the management of solid waste in developing countries are still in their infancy, compared with those in Western countries. To the author's knowledge, there are no specific guidelines for the safe management of solid waste similar to those for the safe use of wastewater proposed by the WHO

(2006). In general, collection systems do not exist and waste is disposed of in landfills and open dumps in an uncontrolled manner. This often results in contamination of surface and groundwater and exposure of people to unhealthy conditions associated with toxicity, insects, and animals. More often than not, the typical image of poor populations scavenging on trash heaps and dump sites in unhealthy and toxic conditions in the outskirts of big cities is true because of lack of regulation, laissez-faire attitudes from local and regional governments, and lack of alternative economic incentives (Thomas-Hope 1998).

15.5 The Water of Ayolé

An excellent case study that illustrates the challenges in providing drinking water supply to communities in the developing world was documented in the form of a video entitled *The Water of Ayolé*. The video was produced in 1988 and is available on the web. It focuses on the water needs (drinking and irrigation) of small village communities in Togo, Africa. It has been used in the literature to demonstrate appropriate practices (or the lack thereof) when introducing technology (e.g., a water well and pump) into a community. Furthermore, it demonstrates what could go wrong when technology is introduced for the sake of introducing technology, especially without taking into consideration the socioeconomic context for its development.

The video addresses the added value of community participation, engagement, follow-through, management, and empowerment. It also demonstrates how different stakeholders (community, government, outside aid agencies) can work together in ensuring long-lasting solutions. This gamut is captured in the video in the form of a narrative consisting of several successive stages, which are paraphrased here:

- Typical unimproved drinking water and sanitation practices in poor communities are described. People use surface water as their main source of drinking water, sometimes far from where they live. Women are in charge of collecting water and spend a considerable amount of time doing so. The water, in turn, creates health problems (guinea worms, diarrhea) that incapacitate the community.
- Government agencies take the initiative to drill water wells and install pumps in communities. People are at first satisfied with the new systems, and health improves for a while. They climb the WASH ladder. As the systems break down over time, women resort to the traditional prepump installation methods of collecting water. The health of the community deteriorates, and the community falls down the WASH ladder.
- A lack of trust develops between the community and government agencies. Both groups blame each other for the failure and do not realize that

there was never any agreement made about "who was responsible for what" before the wells were drilled and pumps installed.

- With the assistance (financial and technical) of outsiders, representatives of government agencies (extension workers) are trained in developing an action plan for operation and maintenance of the water facilities. In turn, the extension workers train local villagers. In that process, they also learn about the needs and priorities of community members. Trust is rebuilt within the community and with outside stakeholders.

- Collaboration of stakeholders (extension workers and villagers) in the project contributes to a great extent to its success. Over time, people develop a perspective of what constitutes success and why things work or don't work. They climb the WASH ladder again.

- The roles of the water and sanitation committee and of active members of the community engaged in the project contribute to more long-term success. When properly organized, people have more options to control their own destiny.

- Clean water supply leads to better health, confidence, agricultural development, profit, and investment. A new dynamic between men and women in the community is also created, with more participation and gender equality.

- A potential of scaling up the reported success story to other communities is being considered by the government.

The *Water of Ayolé* story is indeed hopeful and shows that when properly planned and managed, a water project in the developing world can be as successful as in the developed world. At the same time, one should not be naive and think that there is only one recipe for success. A success story depends on a multitude of tangible and intangible factors, such as technology, participation, engagement, trust, monitoring and evaluation, operation and maintenance, and more. Within those factors, multiple activities need to be carried out, and events take place to call a WASH project successful, as discussed in Section 15.3. Finally, the *Water of Ayolé* video does not inform the viewer about how wastewater (greywater) was handled by the community, nor does it mention the sanitation and hygiene promotion practices used in the community.

15.6 Two Paths of WASH Interventions

There are multiple ways of providing WASH-related services to communities and meeting the aforementioned requirements. At the global level, according to Gleick (2002, 2003), there are essentially two paths in the planning, development, and management of WASH programs. The first path, the *hard path* relies on "centralized infrastructure and decision making," which is more demanding in terms of capital but follows proven guidelines and regulations. The second path, the *soft path*, relies on alternative decentralized solutions at the household, neighborhood, or community scales that may complement the large infrastructure centralized solutions, when appropriate. In general, the soft path is easier to conceptualize, can more easily be designed with communities in mind, and, as a result, responds to specific needs more equitably, effectively, and efficiently, if properly addressed. The soft path is more cost-effective and can be scaled up and disseminated more rapidly, thus having a broader effect than single centralized systems.

The soft path is not a panacea because it has its own limitations. First, decentralized WASH solutions are more difficult to plan and manage (Laugesen and Fryd 2009). Second, because of the lack of regulations and/or enforcement at different levels in a country, it is more likely that those solutions are temporary and become less satisfactory over time, as discussed in Section 15.3. Finally, there are no agreed-upon practices for those decentralized and alternative solutions. As discussed in Chapter 13, the developing world literature is full of technical solutions, some of them more serious than others, tested or not, whose limitations are unknown. Nevertheless, there are serious organizations, NGOs, and research institutions that are working on developing good WASH practices for communities in the developing world.

It should be noted that the soft and hard path options in WASH are not mutually exclusive and depend greatly on the context and scale in which needs are being addressed. There is a place for large reservoirs, complex water distribution systems, and large wastewater-treatment systems for urban areas. Likewise, there are other situations where smaller or medium-sized decentralized systems are more appropriate, such as in peri-urban and rural areas. Finally, the spectrum of solutions created by implementing WASH interventions is wide. It consists of (1) *permanent* solutions in a development context; (2) *temporary* solutions in postcrisis situations; and (3) *evolving* solutions in the various phases of postcrisis recovery (in the medium and long term), as shown in Table 13-1.

15.7 Community-Based WASH Interventions

As discussed in Section 9.3 in relation to municipal sanitation service options (water supply, wastewater and sewage disposal, and management of solid waste), multiple technical alternatives enter into these options, some more realistic than others. A smaller list of reasonable alternatives can be selected by considering the community's level of development (i.e., the different forms of capacity) at the present time and the technical feasibility of each alternative. A plan can also be outlined to increase local capacity toward a higher level of development over a

10–20-year time frame. In doing so, a need exists to educate and convince community members to

- Let go of their unimproved water supply and sanitation systems and replace them with improved options (see Table 15-1);
- Consider water supply, sanitation, and hygiene promotion together rather than piecemeal; and
- Adopt healthy habits and practices that can be transmitted from one generation to the next by promoting substantial behavior change, as discussed in Section 8.7.

However, it is not the purpose of this section (or this book) to look at the details of the technologies that enter into WASH interventions. The literature on that subject is broad. There is greater awareness today, by various agencies and organizations, of the nexus between technology and the different aspects of community development. The reader will find comprehensive information on WASH technologies and practices that can be used in development work in the books written by Pickford (2001), Jordan (2006), Laugesen and Fryd (2009), Mihelcic et al. (2009), and Parten (2010), among many others.

General hubs of information specifically on WASH and health-related issues include, among others, those of the Water Sanitation Programme (WSP) at the World Bank; the USAID; the International Reference Center (IRC) for Water and Sanitation in the Netherlands; the Water, Engineering and Development Center (WEDC) in the United Kingdom; SANDEC, the Department of Water and Sanitation in Developing Countries at the Swiss Federal Institute in Switzerland; Practical Action (formerly ITDG); and the Netherlands Water Partnership. The last has produced a series of excellent booklets on innovative, low-cost WASH technical solutions entitled *Smart Sanitation Solutions* (NWP 2006a); *Smart Water Solutions* (NWP 2006b); and *Smart Hygiene Solutions* (NWP 2010). In these booklets, technology is defined as "smart" if "it can be easily manufactured and repaired in local conditions and is affordable."

Guidelines for various aspects of WASH activities and services in the developing world have been proposed by the World Health Organization (WHO 2009) and various agencies or organizations, such as USAID, Sida, IDRC, Oxfam, and UNICEF. More specifically, for humanitarian response in postcrisis situations, the Sphere Project (2011) provides guidelines and standards in various components of water supply, sanitation, and hygiene promotion, as shown in Figure 13-1. Finally, at the NGO level, various manuals have been developed that are specific to how NGOs conduct their respective WASH interventions. As an example, the NGO named Caritas in Peru has developed a detailed guide for community members that explains the basic components of potable water and sanitation systems in rural areas, how to maintain and operate those systems, and the role of hygienic practices at the household level (Caritas 2009). This guide was designed with users in mind.

The literature is also rich in books and reports that popularize do-it-yourself solutions. Even though these have been created for customers in the developed world, their practicality could be of great interest in a developing world after some adaptations. Among the many books, the reader may want to read *Cottage Water Systems* by Burns (2004), *Create an Oasis with Greywater* by Ludwig (2004), and the *Humanure Handbook* by Jenkins (1999).

A series of general recommendations follow that can be used when deciding on drinking water, sanitation, and hygiene promotion initiatives. Even though they may sound obvious to most experts in the field, these simple recommendations have the potential to improve the success of WASH interventions in the developing world. They represent low-hanging fruit that are worth considering because they are more likely to be comprehensible to the beneficiaries.

Drinking Water Recommendations

Community-based initiatives in drinking water supply cover a wide range of tasks:

- *Protection, improvement, and maintenance of the environment and infrastructure.* This includes existing (1) watersheds and water sources (catchment areas, spring boxes and source protection, lined water wells, drainage platforms, protected scoop holes); (2) water distribution systems (piped water); and (3) water storage facilities (tanks, cisterns).
- *Reduction of water loss.* As remarked by the TWAS (2002), in many countries "30% of the domestic water supply is lost to porous pipe, faulty equipment, and poorly maintained distribution systems." Water losses can be averted if basic repairs and maintenance are instituted.
- *Eliminating sources of contamination.* Possible sources include leaky septic tanks, pit toilets, waste dumps, livestock, and sources of toxicity associated with agriculture and industry. Safe distance must be kept between waste (especially excreta) and groundwater sources. A minimum distance of 30 m is often cited in the literature.
- *Increasing sources of water supply.* This increase can be done by constructing rainwater harvesting systems (reservoirs, roof catchment), drilling more wells, and tapping into additional sources of surface water (e.g., springs, rivers, and streams). Redundancy in water supply needs to be considered to reduce risk if one water source fails. An appropriate water quantity per household needs to be identified by the community that reflects the consumption over one year.
- *Ensuring availability of safe drinking water.* Various "at the source" or "point-of-use treatment" (household water treatment) processes are available after pretreatment (sedimentation, coagulation, and flocculation).

They include filtration (using cloth filters, charcoal filters, ceramic filters, or slow sand filters) and disinfection (chlorination, water boiling, solar ultraviolet radiation). Additional special treatments need to be carried out for water containing arsenic or fluoride.

- *Ensuring accessibility to safe drinking water.* The goal is to bring water closer to the community. Decisions need to be made about whether a need exists to distribute piped water to multiple households or to create a centralized place (public tap or standpipe) for water gathering.
- *Ensuring affordability of safe drinking water.* The community can decide how it plans to manage the drinking water infrastructure and how it plans to cover its costs of operation and maintenance.
- *Ensuring sustainability of water systems.* As discussed previously, many parameters contribute to the long-term performance of drinking water supply: (1) monitoring of water quality, quantity, access, consumption, and disposal; (2) promoting community participation, engagement, education, and trust; (3) ensuring operation, maintenance, and financing of systems; and (4) promoting behavior change of local populations.
- *Planning for the future.* This is where the community considers various options for expanding current systems.

Sanitation and Hygiene Initiatives

Various initiatives related to sanitation and hygiene can be carried out at the community level:

- *Building and improving on-plot and on-site solutions.* Toilets can be connected to an existing sewer system, which is rarely the case, or they can be independent in the form of pit latrines. In the latter, different low-cost options have been proposed that use little to no water: single- or double-pit compost toilets, pour-flush pit toilets, ventilated improved pit toilets, closed-pit toilets, or urine-diverting dry toilets. Choosing the right type of toilet depends on several parameters, such as depth of groundwater table and flooding risks. Of special interest in toilet design is what people want out of them: privacy, safety, comfort, cleanliness, and/or respect. The design must consider gender needs, age needs, and accessibility by vulnerable people, such as the elderly and people with disabilities. Toilets need to be placed at safe distances from springs and water wells. They can be built at ground level or elevated above the ground. A good review of the pros and cons of various types of pit latrines and design configurations can be found in Cotton et al. (1995).

- *Developing sanitation management strategies.* These strategies must match issues faced by the community and must respond to community needs. At the household level, a need exists to encourage greywater source control, develop drainage solutions, and select treatment methods that are most appropriate to the type of greywater being generated. The methods can be (1) primary (screening, sedimentation in tanks or ponds, flotation, filtration) and/or (2) secondary (anaerobic degradation and aerobic degradation by microorganisms using various forms of biofilters). A good discussion of the pros and cons of those methods can be found in Laugesen and Fryd (2009), Mihelcic et al. (2009), and Parten (2010).
- *Ensuring safe disposal of waste from point of origin to final place of discharge.* Transport and disposal of waste can take different forms, depending on the level of community development. Examples include sewers, septic tanks, and leach fields. For small household situations, disposal can be done using pipes to carry waste off site. The pipes have to be dimensioned to prevent clogging by solids. On-site disposal is another option, where greywater is used for growing trees and plants. For larger systems in urban or peri-urban environments, pretreatment of waste might be required to remove suspended solids.
- *Ensuring sustainability of sanitation systems.* As discussed previously, many parameters contribute to the long-term performance of sanitation systems: (1) monitoring of water quality, quantity, access, consumption, and disposal; (2) promoting community participation, engagement, education, and trust; (3) ensuring operation, maintenance, and financing of systems; (4) promoting behavior change of local populations; and (5) creating awareness of the links among water, sanitation, hygiene, and health.
- *Planning for the future.* This is where the community considers various options for expanding current systems.

Solid Waste Disposal Initiatives

Solid waste needs to be dealt with to reduce health risks and other outcomes, such as those associated with collection, disposal, and recycling. Waste also clogs drainage systems and waterways when not collected or if using illegal dumping practices. A possible community-based initiative in solid waste management includes reducing the amount of waste generated (waste source reduction) through behavior change and incentives. This step is followed by ensuring the safe transfer and disposal of waste. Special efforts need to be made to ensure that healthy economic alternatives to scavenging exist.

An even higher goal is to mobilize the community to create a solid waste program, i.e., an "integrated solid waste management" program, as recommended by the World Bank (2012). Such an approach considers a wide range of techniques, such as reduce, separate, compost, reuse, recycle, collect, transport and store, process, and dispose safely (landfill). Special precautions must be placed on hard-to-recycle items, such as toxic material (chemicals, asbestos, health care and infectious waste, mining and petroleum waste, and animal waste). Unfortunately, integrated solid waste management programs are still prohibitive in cost. As remarked by Ogawa (1996), many waste management initiatives in the developing world cannot sustain themselves technically, financially, and institutionally once external support has been discontinued. A need exists to develop a range of sustainable and scalable solutions that fall in between the expensive waste management techniques of Western countries and illegal dumping and landfill disposal. Such a continuum of solutions does not exist at the present time.

15.8 Chapter Summary

Access to drinking water and sanitation and adopting hygienic practices are key determinants to healthy communities. They are necessary for community development but are not sufficient because access to other services, such as shelter, food, energy, and others, are equally critical. Climbing the WASH ladder (and other ladders related to energy, food, and shelter) is synonymous with climbing out of poverty and experiencing better health (physical, psychological) at the household and community levels. In turn, good health translates into education and having access to opportunities for human and economic development (Skolnik 2008).

Recent statistics have shown that the world will meet MDG-7C by 2015 in regard to sustainable access to safe drinking water (but with regional discrepancies), but not in sanitation, despite major progress in that field. As discussed in this chapter, these promising trends seen since 1990 rely on average statistics and fail to recognize large disparities at the regional and local level and among urban, peri-urban, and rural areas. Furthermore, they are based on poorly defined indicators, i.e., functionality at the time of inspection and number of people served. Indicators are missing in regard to the safety, reliability, functionality, and sustainability of WASH facilities. Today, there is also no clear agreement about what represents successful WASH interventions beyond the physical existence of WASH facilities.

It is clear that a long way is left to go in providing high-quality WASH-related services to address the needs of everyone on the planet. A multitude of options have been proposed in the literature in regard to improving water supply (quantity and quality) and sanitation at different scales. There are still no best (or even good) practices for the developing world, no coherent databases that allow potential users to compare their actual conditions with those reported in the literature, and no agreed-upon strategy on how to approach WASH projects within the context of small communities to ensure their long-term benefits (sustainability).

This overall confusion stems from the fact that in WASH interventions, the solutions are site specific. As site conditions change, so do the most appropriate solutions to address local problems. As remarked by Abrams et al. (1998), there is no "silver bullet" that guarantees success in WASH service delivery. Instead, a more realistic approach is to adopt a service delivery that is holistic, systemic, and adaptive and integrates multidimensional issues cutting across technology, legislation, policy, institutions, users, service providers, and other stakeholders. The services of water supply and sanitation must be designed for sustainability as early as possible in the project life cycle. They must also be designed for availability, accessibility, affordability, and scalability. WASH project managers must realize that these positive project attributes need to be monitored and evaluated as projects unfold.

References

Abrams, L. (1998). "Understanding sustainability of local water services." The African Water Page, Water Web Management, Surrey, U.K. <http://www.africanwater.org/sustainability.htm> (Feb. 2, 2013).

Abrams, L. (2001). "Water for basic needs." Commissioned by the World Health Organization as Input to the 1st World Water Development Report. <http://www.africanwater.org/documents/basicneeds.pdf> (March 5, 2013).

Abrams, L., Palmer, I., and Hart, T. (1998). "Sustainability management guidelines." Prepared for Department of Water Affairs and Forestry, South Africa, Johannesburg<http://www.africanwater.org/Documents/Sustainability%20_Management_Guidelines.PDF> (March 5, 2013).

Adank, M., et al. (2011). "Service delivery indicators and monitoring to improve sustainability of rural water supplies in Ghana." Presented at the 6th Rural Water Supply Network Forum, Uganda. RWSN, St. Gallen, Switzerland. <http://www.waterservicesthatlast.org/Countries/Ghana-Triple-S-initiative/News-events/Service-delivery-indicators-and-monitoring-to-improve-sustainability-of-rural-water-supplies-in-Ghana> (April 1, 2013).

Alma-Ata. (1978). "Declaration of Alma-Ata." <http://www.euro.who.int/en/who-we-are/policy-documents/declaration-of-alma-ata,-1978> (Oct. 7, 2013).

Arlosoroff, S., et al. (1987). "Community water supply: The handpump option." World Bank, Washington, DC. <http://water.worldbank.org/publications/community-water-supply-handpump-option> (May 1, 2013).

Baumann, E. (2006). "Do operation and maintenance pay?" Waterlines, 25(1), 10–12.

Birley, M. (2011). Health impact assessment: Principles and practice, Earthscan Publications Ltd., London.

Bouabib, A. (2004). "Requirements analysis for sustainable sanitation systems in low income-communities." M.S. dissertation, University of Virginia, Charlottesville.

Burns, M. (2004). Cottage water systems, Cottage Life Books, Toronto.

Cairncross, S., and Feachem, R. (1993). *Environmental health engineering in the tropics: An introductory text*, Wiley, New York.

Cairncross, S., et al. (2003). "Health, environment and the burden of disease: A guidance note." Department for International Development, London. <http://www.lboro.ac.uk/well// resources/Publications/DFID%20Health.pdf> (April 24, 2013).

Cairncross, S., et al. (2013). "Water, sanitation and hygiene: Evidence paper." Department for International Development, London. <https://www.gov.uk/government/uploads/system/uploads/ attachment_data/file/193656/WASH-evidence-paper-april2013.pdf> (April 24, 2013).

Cairncross, S., and Valdmanis, V. (2006). "Water supply, sanitation and hygiene promotion." *Disease control priorities in developing countries*, Chapter 41 in 2nd Ed., D. T. Jamison et al., eds., World Bank, Washington, DC.

Caritas. (2009). *Agua potable en zonas rurales*, Caritas Ayaviri, Puno, Peru. <http://www .bvcooperacion.pe/biblioteca/bitstream/123456789/3502/4/BVCI0002407_1.pdf> (Feb. 9, 2013).

Carter, R. C., Tyrrel, S. F., and Howsam, P. (1999). "Impact and sustainability of community water supply and sanitation programmes in developing countries." *J. Chartered Institution of Water and Environmental Management*, 13, 292–296.

Chenoweth, J. (2008). "Minimum water requirement for social and economic development." *Desalination*, 229(1–3), 245–256.

Cotton, A., et al. (1995). "On-plot sanitation in low income urban communities: A review of literature." Water, Engineering, and Development Center (WEDC), Leicestershire, U.K. <http:// wedc.lboro.ac.uk/resources/books/On-Plot_Sanitation_-_Review_of_literature_-_ Complete.pdf> (May 5, 2013).

Davis, J., and Brikké, F. (1995). "Making your water supply work: Operation and maintenance of small water supply systems." IRC International Water and Sanitation Center, The Hague, Netherlands. <www.irc.nl/content/download/2566/26510/file/op29e.pdf> (May 5, 2013).

Dodd, R., and Munck, L. (2002). "Dying for change: Poor people's experience of health and ill-health." World Health Organization, Geneva. <http://www.who.int/hdp/publications/dying_ change.pdf> (Jan. 2, 2013).

EcoSanRes. (2004). "Introduction to greywater management." Stockholm Environment Institute, Stockholm, Sweden. <http://www.ecosanres.org/pdf_files/ESR-factsheet-08.pdf> (April 5, 2013).

Esrey, S. A. (1996). "Water, waste, and well-being: A multicountry study." *Am. J. Epidemiology*, 143(6), 608–623.

Esrey, S. A., et al. (1991). "Effects of improved water supply and sanitation on ascariasis, diarrhea, dracunculiasis, hookworm infection, schistosomiasis, and trachoma." *Bulletin of World Health Organization*, 69(5), 609–621.

Etnier, C., et al. (2005). *Decentralized wastewater system reliability analysis handbook*, Project No. WU-HT-03-57. Prepared for the National Decentralized Water Resources Capacity Development Project, Washington University, St. Louis, MO, by Stone Environmental, Inc., Montpelier, VT.

Fewtrell, L., et al. (2005). "Water, sanitation, and hygiene interventions to reduce diarrhea in less developed countries: A systematic review and meta-analysis." *The Lancet*, 5(1), 42–52.

Gleick, P. H. (1996). "Basic water requirements for human activities: Meeting basic needs." *Water International*, 21(2), 83–92.

Gleick, P. H. (2002). *The world's water 2002–2003*, Vol. 3, Island Press, Washington, DC.

Gleick, P. H. (2003). "Global freshwater resources: Soft-path solutions for the 21st century." *Science*, 302, 1524–1528.

Gundry, S. W., et al. (2006). "Contamination of drinking water between source and point-of-use in rural households of South Africa and Zimbabwe: Implications for monitoring the millennium development goal for water." *Water Practice & Technology*, 1(2). <http://shopsproject.org/sites/default/files/resources/3202_file_Gundry_Contamination_Drink Water.pdf> (June 2, 2013).

Günther, I., and Fink, G. (2010). "Water, sanitation and children's health." Policy research working paper 5275, World Bank, Washington, DC.

Harvey, P. A., and Reed, R. A. (2007). "Community-managed water supplies in Africa: Sustainable or dispensable?" *Community Development J.*, 42(3), 365–378.

Howard, G., and Bartram, J. (2003). "Domestic water quantity, service level and health." World Health Organization, Geneva. <http://www.who.int/water_sanitation_health/diseases/WSH03.02.pdf> (Jan. 28, 2013).

Hunt, C. (2006). "Sanitation and human development." Human development report, United Nations Development Programme, Office occasional paper, New York. <http://hdr.undp.org/en/reports/global/hdr2006/papers/Hunt%20Caroline.pdf> (Jan. 30, 2013).

Hunter, P. R., MacDonald, A. M., and Carter, R. C. (2010). "Water supply and health." *PLoS Medicine*, 7(11). <http://www.indiaenvironmentportal.org.in/files/Water%20supply%20and%20health.pdf> (Feb. 5, 2013).

Institute for Sustainable Infrastructure (ISI). (2012). *Envision: A rating system for sustainable infrastructure*, Version 2.0, Institute for Sustainable Infrastructure, Washington, DC. <http://www.sustainableinfrastructure.org/portal/workbook/GuidanceManual.pdf> (March 13, 2013).

International Environmental Law Research Center (IELRC). (1990). "New Delhi statement: Global consultation on safe water and sanitation." International Environmental Law Research Center, Geneva, Switzerland <http://www.ielrc.org/content/e9005.pdf> (Oct. 7, 2013).

Jamison, D. T., et al., eds. (2006). *Disease control priorities in developing countries*, World Bank and Oxford University Press, New York, <http://files.dcp2.org/pdf/DCP/DCP.pdf> (March 24, 2013).

Jenkins, J. (1999). *The humanure handbook: A guide to composting human manure*, Jenkins Publishing, Grove City, PA.

Jordan, T. D., Jr. (2006). *A handbook of gravity water systems*, Intermediate Technology Development Publications Ltd., London.

Laugesen, C., and Fryd, O. (2009). *Sustainable wastewater management in developing countries: New paradigms and case studies from the field*, ASCE Press, Reston, VA.

Leadership in Energy and Environmental Design (LEED-ND). (2006). "Understanding the relationship between public health and the built environment." Leadership in Energy and Environmental Design, Washington, DC. <http://www.usgbc.org/Docs/Archive/General/Docs1480.pdf> (Nov. 5, 2012).

Liu, L., et al. (2012). "Global, regional, and national causes of child mortality: An updated systematic analysis for 2010 with time trends since 2000." *The Lancet*, 379(9832), 2151–2161.

Lockwood, H., and Smits, S. (2011). *Supporting rural water supply: Moving toward a service delivery approach*, Practical Action Publishing Ltd., Bourton on Dusmore, UK <http://www.rural-water-supply.net/en/resources/details/419> (April 4, 2013).

Ludwig, A. (2004). *Create an oasis with greywater: Your complete guide to choosing, building, and using greywater systems*, 4th Ed., Oasis Design, Santa Barbara, CA.

Mihelcic, J. R., et al., eds. (2009). *Field guide to environmental engineering for development workers: Water, sanitation, and indoor air*, ASCE Press, Reston, VA.

Morel, A., and Diener, S. (2006). "Greywater management in low and middle-income countries: Review of different treatment systems for households or neighborhoods." Swiss Federal Institute of Aquatic Science and Technology, Dubendorf, Switzerland.<http://www.sswm.info/content/greywater-management-low-and-middle-income-countries-review-different-treatment-systems-hou-0> (June 1, 2013).

Mukherjee, N., and van Wijk, C., eds. (2003). "Sustainability planning and monitoring in community water supply and sanitation: A guide on the methodology for participatory assessment (MPA) for community-driven development programs." World Bank, Washington, DC.

Narayan, D. (1993). "Participatory evaluation: Tools for managing change in water and sanitation." World Bank, Washington, DC. <http://www.chs.ubc.ca/archives/files/Participatory%20 Evaluation%20Tools%20for%20Managing%20Change%20in%20Water%20and%20 Sanitation.pdf> (Dec. 10, 2012).

Narayan, D. (1995). "The contribution of people's participation: Evidence from 121 rural water supply projects." World Bank, Washington, DC. <http://www-wds.worldbank.org/external/ default/WDSContentServer/WDSP/IB/2006/12/29/000011823_20061229153803/Rendered/ PDF/38294.pdf> (December 10, 2012).

Netherlands Water Partnership (NWP). (2006a). *Smart sanitation solutions*, Netherlands Water Partnership, The Hague, Netherlands. <http://www.irc.nl/page/28448> (Dec. 2, 2012).

Netherlands Water Partnership (NWP). (2006b). *Smart water solutions*, Netherlands Water Partnership, The Hague, Netherlands. <http://www.irc.nl/page/28654> (Dec. 2, 2012).

Netherlands Water Partnership (NWP). (2010). *Smart hygiene solutions*, Netherlands Water Partnership, The Hague, Netherlands. <http://www.irc.nl/page/55200> (Dec. 2, 2012).

Ogawa, H. (1996). "Sustainable solid waste management in developing countries." WHO Western Pacific Regional Environmental Health Centre (EHC), Kuala Lumpur, Malaysia. <http:// www.gdrc.org/uem/waste/swm-fogawa1.htm> (March 5, 2013).

Organization for Economic Co-operation and Development (OECD). (2012). *Environmental outlook to 2050*, OECD Publishing, Paris. <http://www.oecd.org/environment/oecd environmentaloutlookto2050theconsequencesofinaction.htm> (Oct. 2, 2012).

Parry-Jones, S., and Kolsky, P. (2005). "Some global statistics for water and sanitation related disease." Water, Engineering and Development Center (WEDC), Leicestershire, U.K. <http://www.lboro.ac.uk/well/resources/fact-sheets/fact-sheets-htm/sgswsrd.htm> (March 24, 2013).

Parten, S. M. (2010). *Planning and installing sustainable onsite wastewater systems*, McGraw Hill, New York.

Pickford, J. (2001). *Low-cost sanitation*, ITDG Publishing, London. <http://www.unpei.org/PDF/ Pov-Health-Env-CRA.pdf> (June 2, 2013).

Poverty–Environment Partnership (PEP). (2008). "Poverty, health & environment: Placing environmental health on countries' development agendas." Poverty–Environment Partnership Joint Agency, Washington, DC <http://siteresources.worldbank.org/EXTENVHEA/Resources/ PovHealthEnvCRA.pdf> (April 20, 2014).

Prüss, A., et al. (2002). "Estimating the burden of disease from water, sanitation, and hygiene at a global level." *Environmental Health Perspectives*, 110(5), 537–542.

Prüss-Üstün, A., and Corvalán, C. (2006). "Preventing disease through healthy environments: Towards an estimate of the environmental burden of disease." World Health Organization, Geneva. <http://articles.healthrealizations.com/images/articles/2010/05/prevdisexecsume .pdf> (Feb. 15, 2013).

Ramshaw, E. (2011). "Major health problems linked to poverty." *New York Times*, July 9, <http:// www.nytimes.com/2011/07/10/us/10tthealth.html?_r=1> (Aug. 1, 2011).

Reed, B., and Shaw, R. (2012). "Preventing the transmission of faecal–oral diseases." Fact Sheet 9, Water, Engineering, and Development Centre, Loughborough University, Leicestershire, U.K. <http://wedc.lboro.ac.uk/knowledge/factsheets.html> (Feb. 2, 2013).

Reynolds, J. (1992). "Handpumps: Toward a sustainable technology." United Nations Development Programme–World Bank Water and Sanitation Programme, Washington, DC. <http://www .wsp.org/sites/wsp.org/files/publications/418200734405_handpumps.pdf> (Nov. 12, 2012).

Roberts, L. (1998). "Diminishing standards: How much water do people need?" International Committee of the Red Cross, Geneva. <http://www.icrc.org/eng/resources/documents/misc/57jpl6.htm> (Nov. 11, 2012).

Rural Water Supply Network (RWSN). (2009) "Myths of the rural water supply sector." Rural Water Supply Network, St. Gallen, Switzerland, <http://www.kysq.org/docs/Rural_myths.pdf> (April 12, 2013).

Schweitzer, R. W., and Mihelcic, J. R. (2012). "Assessing sustainability of community engagement of rural water systems in the developing world." *J. Water, Sanitation and Hygiene for Development*, 2(1), 20–30.

Skinner, J. (2009). "Where every drop counts: Tackling rural Africa's water crisis." International Institute of Environment and Development, London. <http://www.iied.org/pubs/display.php?o=17055IIED> (Jan. 31, 2013).

Skolnik, R. (2008). *Essentials of global health*, Jones and Bartlett Publishers, Sudbury, MA.

Smits, S., et al. (2012). "A principle-based approach to sustainable rural water services at scale: Moving from vision to action." IRC International Water and Sanitation Center, The Hague, Netherlands. <http://www.waterservicesthatlast.org/Resources/Concepts-tools/Principles-for-sustainable-services-a-conceptual-tool-for-improving-sustainability> (Feb. 5, 2013).

Sphere Project. (2011). *Humanitarian charter and minimum standards in humanitarian response*, Practical Action Publishing, Rugby, U.K.

Third World Academy of Sciences (TWAS). (2002). "Safe drinking water: The need, the problem, solutions and an action plan." Third World Academy of Sciences, Trieste, Italy. <http://twas.ictp.it/publications/twas-reports/safedrinkingwater.pdf> (Aug. 1, 2011).

Thomas-Hope, E., ed. (1998). *Solid waste management: Critical issues for developing countries*, Canoe Press, Oak Park, IL.

Trevett, A. F., Carter, R. C., and Tyrrel, S. F. (2005). "The importance of domestic water quality management in the context of fecal–oral disease transmission." *J. Water and Health*, 3, 259–270.

United Nations (UN). (1985). "General assembly: International drinking water and sanitation decade." United Nations, New York, <http://www.un.org/documents/ga/res/40/a40r171.htm> (Oct. 7, 2013).

United Nations (UN). (1992). "The Dublin statement on water and sustainable development." United Nations, New York, <http://www.un-documents.net/h2o-dub.htm> (Oct. 7, 2013).

United Nations Conference on Environment and Development (UNCED). (1992). Agenda 21. United Nations Conference on Environment and Development, New York.<http://www.un.org/esa/sustdev/documents/agenda21/english/Agenda21.pdf> (Dec. 5, 2012).

United Nations Development Programme Human Development Report (UNDP/HDR). (2006). "Beyond scarcity: Power, poverty, and the global water crisis." United Nations Development Programme, New York.

United Nations Development Programme Human Development Report (UNDP/HDR). (2011). "Sustainability and equity, a better future for all." United Nations Development Programme, New York.

United Nations Educational, Scientific and Cultural Organization (UNESCO). (2012). *World water development report: Managing water under uncertainty and risk*, Vols. 1–4, UNESCO, Paris. <http://www.unesco.org/new/en/natural-sciences/environment/water/wwap/wwdr/wwdr4-2012/> (Dec. 5, 2012).

United Nations-Water—GLAAS. (2012). "Global analysis and assessment of sanitation and drinking water: The challenge of extending and sustaining services." World Health Organization (WHO), Geneva. <http://www.un.org/waterforlifedecade/pdf/glaas_report_2012_eng.pdf> (Dec. 5, 2012).

U.S. Agency for International Development (USAID). (2005). *Field operations guide for disaster assessment and response*, Version 4.0, U.S. Agency for International Development, Washington, DC. <http://www.rmportal.net/library/content/tools/disaster-assessment-and-response-tools/da_field_guide_2005/view> (Feb. 5, 2013).

U.S. Agency for International Development (USAID). (2012). "Education projects description." U.S. Agency for International Development, Washington, DC. <http://zambia.usaid.gov/education-project-descriptions> (Jan. 2, 2013).

U.S. Agency for International Development (USAID). (2013). "WASHPlus: Supportive environment for healthy communities." U.S. Agency for International Development, Washington, DC <http://www.washplus.org/>. (Oct. 4, 2013).

Wagner, E. G., and Lanoix, J. N. (1958). "Excreta disposal for rural areas and small communities." World Health Organization, Geneva.

WaterAid. (2011). "Sustainability framework." WaterAid, London. <http://www.wateraid.org/~/media/Publications/sustainability-framework.ashx> (June 10, 2013).

Water and Sanitation Program (WSP). (2013). "Sanitation marketing toolkit." Water and Sanitation Program, The World Bank, Washington, DC <http://www.wsp.org/toolkit/toolkit-home> (Oct. 4, 2013).

Water, Engineering and Development Center (WEDC). (2011). "The WELL Guiding Principles." Fact Sheet 6, Water, Engineering and Development Center, Leicestershire, U.K. <http://wedc.lboro.ac.uk/resources/factsheets/FS006_WELL.pdf> (June 5, 2013).

Water, Engineering and Development Center (WEDC). (2013). "WEDC facts sheets and briefing notes." Water, Engineering and Development Center, Leicestershire, U.K. <http://wedc.lboro.ac.uk/knowledge/factsheets.html> (Oct. 7, 2013).

Water Supply and Sanitation Collaborative Council (WSSCC). (2004). "Listening—To those working with communities in Africa, Asia, and Latin America to achieve the UN goals for water and sanitation." Water Supply and Sanitation Collaborative Council, Geneva. <http://www.comminit.com/?q=africa/node/189342> (June 10, 2013).

WELL (2014). "On-plot sanitation in urban areas." Water and Environmental Health at London and Loughborough <http://www.lboro.ac.uk/well/resources/technical-briefs/61-on-plot-sanitation-in-urban-areas.pdf> (April 20, 2014).

World Bank. (2012). "What a waste: A global review of solid waste management." World Bank, Washington, DC. <http://web.worldbank.org/wbsite/external/topics/exturbandevelopment/0,,contentmdk:23172887~pagepk:210058~pipk:210062~thesitepk:337178,00.html> (Jan. 22, 2013).

World Health Organization (WHO). (1948). Preamble to the Constitution of the World Health Organization, as adopted by the International Health Conference, New York, June 19–July 22, 1946, signed on July 22, 1946 by the representatives of 61 States (Official Records of the World Health Organization, No. 2, p. 100) and entered into force on April 7, 1948.

World Health Organization (WHO). (2001). "Non-communicable diseases and mental health." World Health Organization, Geneva. <http://www.who.int/mip2001/files/2008/NCDDisease Burden.pdf> (Jan. 23, 2012).

World Health Organization (WHO). (2006). *Guidelines for the safe use of wastewater, excreta and greywater*, Vols. 1–4, World Health Organization, Geneva. <http://www.who.int/water_sanitation_health/wastewater/gsuww/en/index.html> (Jan. 25, 2013).

World Health Organization (WHO). (2008). "The global burden of disease: 2004 update." World Health Organization, Geneva. <http://www.who.int/healthinfo/global_burden_disease/2004_report_update/en/index.html>.

World Health Organization (WHO). (2009). *Water safety plan manual (WSP manual): Step-by-step risk management for drinking-water suppliers*, World Health Organization, Geneva. <http://www.who.int/water_sanitation_health/publication_9789241562638/en/index.html>.

World Health Organization (WHO). (2010). "Twenty ways that the World Health Organization helps countries reach the millennium development goals." World Health Organization, Geneva. <http://www.who.int/topics/millennium_development_goals/20ways_mdgs_2010 0517_en.pdf> (May 28, 2012).

World Health Organization (WHO). (2011a). "Guidelines for drinking-water quality." World Health Organization, Geneva. <http://www.who.int/water_sanitation_health/publications/2011/dwq_guidelines/en/index.html> (May 28, 2012).

World Health Organization (WHO). (2011b). "How much water is needed in emergencies? Technical notes on drinking-water, sanitation, and hygiene in emergencies." World Health Organization, Geneva. <http://www.who.int/water_sanitation_health/publications/2011/tn9_how_much_water_en.pdf> (June 5, 2012).

World Health Organization (WHO). (2012). "Millennium development goals–Fact Sheet No. 290." World Health Organization, Geneva <http://www.who.int/mediacentre/factsheets/fs290/en/> (Oct. 5, 2013).

World Health Organization (WHO). (2013). "International classification of diseases." World Health Organization, Geneva <http://www.who.int/classifications/icd/en/> (Oct. 2, 2013).

WHO/UNICEF (2000). "Global water supply and sanitation assessment." World Health Organization, Geneva. <http://www.who.int/water_sanitation_health/monitoring/globalassess/en/> (May 23, 2014).

WHO/UNICEF (2005). "Water for life: Making it happen." World Health Organization, Geneva. <http://www.who.int/water_sanitation_health/monitoring/jmp2005/en/index.html> (May 24, 2013).

WHO/UNICEF (2008). "Progress on drinking water and sanitation: Joint Monitoring Programme Update." World Health Organization, Geneva. <http://www.who.int/water_sanitation_health/monitoring/jmp2008.pdf> (May 24, 2013).

WHO/UNICEF (2012). "Progress on drinking water and sanitation: Joint Monitoring Programme Update." World Health Organization, Geneva. <http://www.who.int/water_sanitation_health/publications/2012/jmp_report/en/index.html> (March 2, 2013).

WHO/WSSCC. (2005). "Sanitation and hygiene promotion programming guidance." Water Supply and Sanitation Collaborative Council and World Health Organization, Geneva. <http://www.who.int/water_sanitation_health/hygiene/sanhygpromo.pdf> (Jan. 2, 2013).

16

Conclusions

16.1 Development Engineering

This book makes the case for the role of engineering in poverty reduction and human development in general. It introduces a framework and guidelines for conducting small-scale development projects in communities that are highly vulnerable to a wide range of adverse events and have low capacity to handle the stress associated with those events. As a result, these communities are at risk, a context that is more often a rule than an exception in the developing world. Addressing these conditions toward the betterment of mankind is an obligation for the engineering profession; it is no longer an option. But it cannot be done by the engineering profession alone and requires partnership and collaboration with a wide range of other disciplines that enter into the overall discussion regarding development.

An image that I often use to explain the role of engineering in small-scale community development projects is to imagine a conference room where development is being discussed. The room can be accessed through multiple doors. The doors are labeled things like engineering, economics, health, education, and governance. In the middle of the room there are many chairs placed around a table and each chair represents a discipline involved in development. This book was written for engineers interested in joining the conversation by entering the conference room though the engineering door. To contribute to the overall discussion, they need to not only excel in their engineering and technology skills, but they must also have a broader vision about what development is. They also need skills to communicate with individuals from other disciplines who have different opinions about development. These engineers need to be global. In this book, engineering is seen as an entry point into the development conversation, but it is not seen as an end in itself. There is much more than just engineering and technology when it comes to development work.

The framework presented in this book is unique in the sense that it combines concepts and tools that have been traditionally used by development agencies and other tools more specifically used in engineering project management.

It also emphasizes the importance of integrating systems thinking, risk analysis, capacity analysis, and resilience analysis into decision making. When combined, these tools and concepts from seemingly independent fields have the potential to better handle and model the complexity and uncertainty inherent in small-scale community development projects and in the issues faced by households in these communities. It should be noted that the proposed framework and associated guidelines are not meant to be standards. The methodology presented herein is flexible enough to be adapted by various development agencies and integrated into their specific approaches to designing, planning, and managing projects.

This book emphasizes the idea that human development calls for a new generation of engineers who can operate in unpredictable and complex environments that are different from those encountered in the developed world. It is about a new field of engineering called engineering for development, which, as suggested by Bugliarello,

> *responds to the global need for engineers who understand the problems of development and sustainability, can bring to bear on them their engineering knowledge, are motivated by a sense of the future, and are able to interact with other disciplines, with communities and with political leaders to design and implement solutions.* (Bugliarello 2008)

This new field can also be called engineering for sustainable human development (the title of this book) or simply development engineering. In this field, the fundamentals of engineering are the same, whether the community is located in a developed or developing country. However, the approach used to solve the problems in communities in low-income and lower-middle-income countries requires unique skills and approaches. In this context, engineers need to be more than value-neutral individuals capable of producing linear, blueprint, and predictable solutions delivered on time and within budget, for problems that are well defined. They need to be creative and innovative to account for uncertainty, complexity, ill-defined issues, and constraints in a cultural context to which they are not accustomed. Their designs need to be adaptable to the context in which they work.

Engineering for sustainable human development is about the delivery of projects that are technically correct and correctly done from a nontechnical point of view in regard to people and the environment. Engineering for sustainable human development is not just about technology. It is also about people, values, ethics, culture, commitment, engagement, passion, and other nontechnical issues that are not traditionally taught in engineering schools and unfortunately, not traditionally associated with the engineering profession. The practice of engineering for sustainable human development requires prospective engineers to be trained holistically by acquiring a *T*-type education (depth and breadth) rather than a traditional *I*-type education (expertise and depth only), as discussed in Chapter 3.

This chapter draws key conclusions on major themes addressed in this book. They include (1) the abnormality of poverty on our planet today, (2) understanding household livelihood and real wealth, (3) defining project success in uncertain and complex conditions, (4) improving the future practice and education of global engineers, and (5) defining sustainability and sustainable development within the context of addressing the needs of developing communities.

16.2 Poverty Is Not Normal

Engineering for sustainable human development calls for engineers to become acquainted with the field of development and its various components. The task is challenging in itself because development is still a poorly defined concept in the international community. Even though it has been agreed during the past 50 years that development is about the betterment of society and creating a better life for all, there is no clear consensus about how to reach that goal. Development may indeed be a human right, but in reality, it is still perceived by the world as "soft law," and "not legally binding" (Kirchmeier 2006). A lot of institutions are involved in the development industry, some more effectively than others. A review of the development literature shows that the dominant logic in that industry is not uniform and varies over a wide spectrum.

In the development spectrum, the humanistic approach focuses on development as people centered. It is about "creating an environment in which people can develop their full potential and lead productive, creative lives in accord with their needs and interests" (UNDP/HDR 1990). It is also about creating real wealth by "tapping into the ingenuity and creativity of the poor, and enabling them to express their own hopes for their families and their communities" (Narayan 1993). Human development is based on the core concept that "people are the real wealth of a nation" (UNDP/HDR 1990, 2006) and that it is a prerequisite for economic development and for creating stable countries. It has slowly become the dominant logic over the past 20 years, as testified by the various reports written since 1990 and current initiatives following the 1992 Rio Summit and the launch of the Millennium Development Goals in 2000.

On the other end of the development spectrum is the economic approach to development, which is capital centered rather than people centered. It is based on growth driven by markets and technology and is assumed to be quantifiable. Its dominant logic is about income and promoting consumption leading to economic growth, and it does not give much consideration to negative consequences. It is supposed to be universal and value free. Finally, it is based on the assumption that what has worked in the Western world should also work anywhere on the planet. This form of development, which emerged after World War II, is still advocated today by policy makers of major agencies.

The two aforementioned extremes of development are actually not incompatible, although they may appear that way. There is indeed room for a common ground where a new dominant logic of authentic and transformative development resides. There are many opportunities for development to return to its traditional definition of transformation of productive structure and capabilities, while at the same time paying greater attention to livelihood issues, politics, technological development, institutions, sustainability, and the environment than was done 50 years ago. Furthermore, if properly conceptualized, models of development can be flexible, contextual, and adaptive to local cultures and governments, and equally involve top-down and bottom-up approaches; both are equally important and complementary. Nolan (2002) noted that "in development work, one size does not fit all." There is room for various forms of development at different scales (households, community, regional, and national) and for different socioeconomic settings, and a need exists to see development as a two-way street where the developing world learns from the Western world and vice versa.

Since its inception after World War II, development has been a work in progress paved with many undesirable outcomes, which could have been avoided. Many reasons have been brought forth to explain the mistakes of the past, and lessons can certainly be learned from those mistakes if the various organizations involved in development are willing to be transparent and accountable. The recent decision of the World Bank to release in 2011 a database containing the evaluations of roughly 10,000 projects conducted since the 1960s is indeed commendable and represents the first step in the process of learning from past mistakes and creating better practices (IEG 2012). It is hoped that other organizations will be inspired to do the same. Sharing lessons from the past, whether good or bad, is the only way for the development industry to learn, improve, and grow. After all, the modern infrastructure in the Western world today did not emerge all at once. It is the outcome of more than 100 years of trial and error, during which various organizations and societies have been willing to learn from mistakes and adapt to change. In a nutshell, development takes time, humility, patience, and money.

Despite the various outcomes of projects in the development industry, a general consensus exists that over the past 20 years, consistent gains in human development have led to increased life expectancy, better health, access to education, and more democratic and stable governance around the world. Access to drinking water and sanitation, energy, shelter, health, education, and telecommunications is increasing rapidly. The year 2012 was marked with growing optimism and positive news in various aspects, such as the decrease of people in extreme poverty who make less than $1.25 per day, the increase in the number of people who have access to safe drinking water and sanitation, the success of emerging markets in Asia and South America, and the reduction of undernourished people worldwide. There is no reason why such progress at the global level should not continue in the next 15–30 years (Chandy et al. 2013; Ravallion 2013).

Despite these gains, clearly more work remains to be done, especially at the local level. The success stories reported by policy makers over the past 20 years are averaged at the country level and do not often reflect the reality of life at the regional or community level. There are still major discrepancies across regions, within regions, and within communities. Furthermore, development within the community is rarely equitable across social groups and gender, which can be seen in the disparity in living conditions in peri-urban areas in major cities around the word. When it comes to development and the well-being of 3–4 billion people, the world is still spiky and is certainly not flat.

Another area that needs improvement in the development industry is how to define and measure the success of development interventions (e.g., water, sanitation, and hygiene [WASH]; energy; health; and education), especially at the project level. Development solutions to community problems rarely show clear measures and indicators of affordability, accessibility, availability, sustainability, and scalability. The monitoring and evaluation of such interventions is often conducted out of necessity or as an afterthought rather than as a normal practice. More often than not, projects do not have an exit strategy, resulting in either premature closing or long-lasting and expensive commitments. In both cases, disappointing and unrealistic outcomes emerge. Integrating quality control and assurance and accountability in the different phases of development projects is critical to their successful viability.

The development industry needs to realize that it is faced with a dilemma. It is committed to making the world a better place but does not have a strategy with tangible indicators and measures to monitor and evaluate the success of its activities at different scales (household, community, regions, and countries), except at the global country-level using indices such as the gross domestic product (GDP), gross national income (GNI), or human development index (HDI). At the project level and in specific areas, such as WASH, energy, and health, the measures of success are ill defined and cannot guarantee long-lasting performance. If success cannot be measured, it cannot be managed. As a result, it is hard to define what success actually is.

Regardless of how development is defined, there is a moral (personal and institutional) obligation for people to tackle poverty head on. As discussed in Chapter 2, poverty is unnecessary pain (physical, psychological, and mental) and is a perpetual crime against humanity. Unfortunately, many of those responsible for that unnecessary pain have not been held accountable for their decisions. It should be noted that those who benefit the most from the current state of the world are often the least interested in changing the status quo, despite their rhetoric. The simple fact that people have plenty of resources (e.g., human, technical, and financial) and know-how to eliminate poverty (especially extreme poverty) in our lifetime and bring water, sanitation, energy, shelter, health, education, and other services to everyone on our planet but do not have the will and priority to do so is a true reflection of deep pathological dysfunction of the human race and its institutions. It should be made clear to our policy makers that it is not normal for

- People to live precarious livelihoods characterized by physical and psychological weaknesses;
- Children to go to bed hungry, have diarrhea, die in the cold, have no shelter, and receive no education;
- Girls and women to be treated as slaves with no rights or empowerment;
- Clinics and hospitals to lack equipment, staff, supplies, and medication;
- People to be isolated with no rights and voices in their communities;
- People to be exploited in any manner or form because of their race, gender, or age;
- Corruption to be rampant in institutions and officials to demand bribes;
- International agencies and individuals to fail to deliver their promises on time;
- Development agencies and groups to fail to be accountable for the work they are doing; and
- A great majority of projects to fail a few months after their installation.

To this list, one can easily add a multitude of other issues that are not acceptable in a so-called civilized world in the 21st century. As remarked by Forbes (2009), it should also be made clear to all policy makers that there is no added benefit to (1) not having clean drinking water, (2) living in unhealthy environments, and (3) being unhealthy. Poverty does not offer any added value to anyone who is poor! Furthermore, poor people have other things to do than climb the multiple ladders of energy, health, education, jobs, water, and sanitation only to fall back and find themselves at the bottom of those ladders when adverse events and hazards (big and small) negatively affect their lives because they do not have the capacity to handle the stress associated with difficult conditions.

It is time to think about poverty as more than the mere absence of this or that and abandon the naive idea that poverty reduction can be approached in a piecemeal way by offering quick, cheap, and nonlasting miracle "Band-Aid" solutions to separate problems. Problems are related, and so are their solutions. Poverty must be looked at from a systems point of view and seen instead as an *emergent* property of dysfunctional systems and institutions in the developed and developing world alike. As discussed in Chapter 2, a deeper root cause of that dysfunction is an *internal poverty* existing in individuals and institutions. It expresses itself in the form of destructive forms of behavior, such as greed, fear, selfishness, competitiveness, and apathy, among others. When these forms of behavior become habits, a dominant logic is created that perpetuates *external poverty* in society. External poverty is not normal, nor is internal poverty. The latter makes it almost impossible to envision a sustainable, equitable, and peaceful world any time soon, short of espousing a drastically different mindset and redefining what it is to be truly human on this planet.

It is easy to become complacent and, as a result, reinforce the dysfunctional behavior of institutions in the developed and developing world. A new mindset is indeed necessary in our everyday life, and all of us are called to become agents of change. This new mindset must acknowledge that a sustainable world for all must be a world that is equitable, peaceful, and compassionate. This vision may sound utopian to the reader, but he or she may want to deeply reflect on the day-to-day reality of the world, especially the dire reality of someone who makes less than a $1 a day (or even $2 or $4). Furthermore, poverty is not somewhere out there. It is often in our own backyards and communities, and no one is immune to it. It does not take much for anyone to fall into poverty (external or internal), even in wealthy countries. People can choose to do something about it or they can ignore it. As human beings, we can certainly do better than we are doing right now.

16.3 From Household Livelihood Crunch to Release

Seven billion people on our planet face risks on a daily basis, some more than others. As discussed in Chapter 10, risk management is about developing a list of what could go wrong. For most people in the developed world, that list may be short or long, depending on the environment in which they live (constraining versus enabling), and their respective vulnerability and capacity to handle the stress associated with unusual events and situations. Once risks are identified, they can be prioritized and managed using different strategies (avoidance, transfer, creating redundancy, tolerating risks, and mitigating them). The difference between vulnerability and capacity defines, in many ways, what each day looks like for each human being in a household or community, the risks that the person is likely to face when coping with change, and the person's ability to adapt to that change.

Without discounting and trivializing the daily risks faced by households in the developed world, the risk environment faced by those in developing communities is many times worse. Issues such as accessibility, affordability, and availability of clean water, sanitation, hygiene, energy, cooking fuels, education, health, transportation, and jobs are issues that confront the poor on a daily basis.

In some cases, a high probability exists that poor people will not make it to the end of the day. If they do, they will have likely spent the day facing malnutrition and hunger, while having no access to modern forms of energy and living in desperate conditions characterized by poor housing in unhealthy and unsanitary environments. They will have likely spent considerable time on activities to secure their basic necessities for survival. Their fate is to do it again the next day.

The poor are asked to climb a series of ladders (e.g. water, sanitation and energy, ladders) defined by the Western world without a voice and in isolation, without any support structure. As discussed within the context of the household livelihood crunch model (Figure 1-2), the poor are faced with a double burden of poverty formed by the combination of inherent isolation, powerlessness, physical and psychological weakness, and precarious livelihood, in addition to natural and nonnatural events well beyond their control. This burden creates ill health and endangered livelihoods and reinforces the downward spiral and pull of poverty.

Development is about breaking the aforementioned cycle and replacing the crunch model of Figure 1-2 with the release model of Figure 10-1. It is about simultaneously decreasing individual, household, and community vulnerability (i.e., the constraining environment) and increasing their capacity, (i.e., the enabling environment). Capacity has many expressions in terms of (1) human capital (labor power, literacy, know-how); (2) environmental capital (land, trees, forests, water, and other natural resources); (3) social capital expressed through relationships among community members as part of social networks, ties, safety nets, and time for social interaction, for family gathering and support, and for cultural identity; and (4) physical capital in the form of access to land (size and quality), ability to self-provision, household property, and valuables. As discussed in Chapter 2, these forms of capital are expressions of the *real wealth* and strength of the community, something tangible and meaningful to the community members to build on in a participatory manner, if they are empowered and encouraged to do so. Referring to the rocket-launching analogy at the beginning of Chapter 1, capacity is part of the fuel needed to create the thrust to counteract the downward drag of poverty.

This bottom-up approach to real wealth represents a strong leveraging point in capacity building when the community interacts with outside groups and in-country local, regional, and national institutions. Development is, in part, about creating a local awareness of real wealth, identifying key change makers, and getting the community to embrace change and adopt new forms of behavior that ultimately becomes habits, through behavior change communication.

The success of capacity building is only as good as the community's willingness to participate in its development and truly own the solutions to the problems it is facing. It is also only as strong as the quality and commitment that outside organizations are willing to invest. Likewise, capacity building is only as good as the role in-country institutions (national, regional, and local) are willing to support and commit. This was shown in Figure 10-1 in the form of factors external to the community. They relate to the structures and processes (e.g., government, businesses, and important individuals), which themselves are influenced by politics, economics, culture, and other factors. These external factors are probably the most difficult to comprehend for anyone involved in community development projects because of their complexity, uncertainty, volatility, and contextual nature. They cannot be ignored, but at the same time, they cannot be clearly defined. Yet, they have the potential to derail projects at any time during their execution. The systems mapping tools discussed in Chapter 6 can help outline those various factors, identify relationships, explore "what if" situations, and create an enhanced understanding of the dynamics of a community and the groups and institutions with which it interacts. Systems tools can help explore the long-term consequences of today's actions "including their environmental, cultural, and moral implications" as remarked by Sterman (2006).

However, systems tools are not a panacea and do not provide definite and predictable answers to how a community will respond to internal and external factors. System models are virtual models of reality that depend on what components are included in the model, how these components are connected, and the selected boundaries of the model. Often, all these things are decided by individuals with limited and biased perceptions of the world and its people. As discussed in Chapter 2, walking in somebody else's shoes is difficult when developing a system model of a community.

The overarching goal of sustainable community development is to build (or strengthen) communities that have the potential to become more stable, equitable, secure, prosperous, and above all, more *resilient* and *healthy* over time. In general, such communities must have the resources and knowledge to be able to acquire (or strengthen) the capacity (or ability) to (1) address their own problems, (2) be self-motivated and self-sustaining, (3) cope with and adapt to various forms of stress and shocks, (4) satisfy their own basic needs, and (5) demonstrate livelihood security for current and forthcoming generations over time.

16.4 Project Success in Complex and Uncertain Environments

A recurring remark made in this book refers to the difficulty of guaranteeing successful community development projects, especially at the small-scale level. This difficulty cannot be attributed to (1) an absence of project management tools, (2) the frameworks used by development agencies to conduct their programs and projects, or (3) access to technology. The limited success has to do with the context in which projects are carried out; the approach used by those involved in conducting these projects; and the limited predictability of human decision processes, especially when faced with complexity and uncertainty.

In the Western world, engineering projects for communities are conducted by companies that have track records of delivering projects under constraints where quality, time, and cost are taken seriously. They work in an environment where the client is educated, corruption is perceived as unethical, funding is available, and an institutional structure exists. The problems tend to be better defined and can be approached in an objective way. Such an approach does not fare well within the context of community development projects where three groups of stakeholders make decisions from (1) the bottom up (community insiders), (2) the top down (institutions), and (3) the outside in (outsiders). They form the three legs of a stool whose stability depends of the integrity of each leg. Simply put, if one leg is deficient, the stability of the stool (the community) is compromised.

There are several reasons why the Western approach to projects does not work well in the context of developing communities, especially for small-scale projects:

1. Uncertainty and complexity are inherent in community development projects. Communities are unique in their structure, cannot be known in their entirety, and are unpredictable in their behavior. No two communities are alike, and there is rarely a way to gather complete information about them. There is always a gap between what is known about the community during appraisal and how the community truly operates daily, no matter how thorough the appraisal is.

2. Communities are systems of systems that interact with other dependent systems in a hierarchical manner. In these systems, humans play a critical role and bring with them a limited potential to perceive and process information in decision making. This limitation is particularly true when decisions cut across cultural, social, and geographic boundaries.

3. Outsiders come loaded with their own challenges and biases, including their often oversimplified solutions to problems that have been crafted in a different context. They are more inclined to come up with solutions quickly that worked somewhere else, rather than spending time developing more appropriate solutions for the specific community.

4. Metrics and indicators used to define project success are poorly defined. As discussed in Chapter 15, current indicators of success in WASH projects include functionality at the time of inspection and number of people served. Indicators are missing in regard to the safety, reliability, functionality, and sustainability of projects. Likewise,

project success depends on the accessibility, availability, and afford-ability of solutions. Such attributes are rarely taken into consideration. Last but not least, different development agencies have different defini-tions of success.

5. Projects in developing communities often take place in communities that have low capacity and are subject to multiple forms of vulnerabil-ity from the beginning. It is difficult to envision successful projects when institutional capacity is nonexistent, people have limited educa-tion and resources, technical capacity is absent, energy is not available (or is available only sporadically), and the social structure prevents progress. In some cases, governments (local, regional, or national) do not have much interest in supporting, maintaining, and operating proj-ects. In some extreme cases, they may even block progress.

There are many implications to these observations:

- Because uncertainty and complexity cannot be eliminated, they need to be accounted for using appropriate decision-making tools. In particular, solutions cannot be optimized. The term "satisficing" was emphasized several times in this book to describe an approach where good-enough solutions are appropriate. The three groups of stakeholders mentioned need to understand the satisficing mindset and recognize that uncer-tainty and complexity affect (1) how community appraisals are con-ducted, (2) how the information and data obtained by the appraisal are analyzed and presented to the community, (3) how the vulnerability of the community and associated risks are perceived and interpreted, and (4) how the recommendations are received and understood by the com-munity. In turn, these implications control how projects are designed, planned, executed, monitored, and evaluated and how predictable the outcomes of those projects are.

- Scale and context are important factors to consider in development projects. Because communities are unique, different problems at differ-ent scales require different solutions. What works at the community level may not always work at the regional or national level, and vice versa. Likewise, solutions in one region may not work in another.

- Systems tools, rather than linear tools, are better suited for analyzing the "what if" of various situations encountered in development projects. They can help identify synergy, closed-loop behavior, and leverage points in community development. It makes no sense to model systems such as communities with tools that are linear and then expect predict-able answers.

- It can be said with a high level of certainty that failure is a possible option in development projects. Because nobody likes to fail and failure has a stigma attached to it, a different way of looking at failure is to see it as an opportunity to learn lessons about what has worked and what has not worked. Having said that, failure becomes a real issue when those involved in development "fail to learn from past lessons."

- Reflective practice and interactive and adaptive methods are better suited to address the changes at the community level than directive or blueprint methods. Changes are inevitable in the life cycle of development projects. As discussed in Chapter 2, sustainability is a dynamic equilibrium, and the process of sustainable development must be able to adapt to change.

- Project managers need to be more than overseers of projects with well defined deadlines and costs. Rather, they have to be managers of change and able to solve problems as they arise. Solving problems requires a need to integrate adaptive change, variability, and flexibility and introduce contingency plans and options in project planning and execution. Practitioners need to balance what is expected of them in traditional problem solving by being able to maneuver in an unpredictable environment and still deliver projects on time, within budget, of good quality, and in an environment that is constantly changing.

The unfortunate truth is that the success of small-scale community development projects cannot be guaranteed with the approach and tools used in the development industry today. Success would require adopting a long-term strategy and a dominant logic to development that emphasizes people's needs, existing community skills and capacity, participation and collaboration, integration of problems and solutions, commitment from all stakeholders over a period of 5–10 years minimum, monitoring and evaluation right from the start of any project, using tangible and intangible indicators and measures of success, guaranteeing a "good enough" level of project quality control and assurance, and defining a clear project exit strategy right from the start.

16.5 Global Engineering for a Small Planet

Over the past 15 years of my 30-year tenure in engineering education at the University of Colorado at Boulder, I have witnessed a dramatic change in what young engineers expect of their educators. My involvement with Engineers Without Borders–USA (EWB–USA) and the Mortenson Center in Engineering for Developing Communities at CU–Boulder has made me realize that today's engineering students want more meaningful and practical education and not just to be taught to solve the problems at the end of engineering book chapters. I have observed the same quest for meaning from professional engineers who volunteer as part of EWB–USA. Key words that describe this new generation of engineers include enthusiasm, passion, excellence, commitment, empathy, courage, inclusion, responsibility, purpose, impact, leadership, and a strong desire to make the world a better place through personal and professional engagement. These characteristics are those of the Millennials, also known as Gen-Y, individuals who were born between 1980 and 2000. By 2020, in the United States alone, they will comprise more than 50% of the workforce (Tarr and Ruiz 2013). It is indeed the obligation of today's elders to encourage those individuals to create new and more holistic solutions to old and new problems.

Additionally, EWB–USA involves a large proportion (40%) of women engineers. This astounding number contradicts the traditional enrollment of women in engineering programs (20%). This passion toward changing the world is not limited to U.S. students. I have seen it in many countries that I have visited where people have their own versions of Engineers Without Borders. Indeed, I am glad to report that the future of engineering is in good hands.

As is often the case, change is accompanied by resistance. It has been my observation that the enthusiasm shown by the Millennials toward engagement and volunteer work has not been matched by the institutions that are supposed to educate them. Over the past 15 years, I have witnessed students' excitement not taken seriously by academic institutions, whether in the United States or abroad.

The world is global. Problems are transboundary in nature and involve a multitude of interrelated disciplines that have traditionally been looked at independently of each other. Industry is in dire need of recruiting leaders and managers who can operate in a global context and address global issues while working together. Yet universities are still educating engineers using a nineteenth century model of engineering education that consists of four to five years of surviving courses taught independently of each other and with limited exposure to nontechnical issues and real-world problems. As remarked by a colleague of mine (Matt Jelacic, personal communication 2012), a time will come when students will not be asked about the nature of their majors, but rather about what problems they are interested in addressing. We are still a long way from this happening because it requires looking at education in a more holistic way and embracing change. It will also require a new generation of educators who will come from today's Millennials. As Max Planck said, "A scientific truth does not triumph by convincing its opponents and making them see the light, but rather because its opponents eventually die, and a new generation grows up that is familiar with it."

The same could be said about how close-minded the engineering academic establishment is to change. There is indeed a need to demystify engineering education, which is usually used to train individuals to provide value-neutral solutions to well defined problems. Such problems only exist in engineering books, university laboratories, and professors' teaching notes. Instead, there is a need to educate *global citizen engineers* who are also changemakers, peacemakers, social entrepreneurs, and facilitators of sustainable human development. Service learning, outreach, and engagement provide ways to acquire the knowledge and attitudes necessary to address global problems. Such pedagogical methods have been embraced by several pioneering universities, which are still an exception rather than the norm. Compared with the academic institutions whose mission is supposedly the education of young individuals to address the problems of the world, the engineering industry has become more open to change and has been supportive of activities such as EWB and innovative university curricula. There are three main reasons for this trend. First, innovation and creativity are driving forces in a competitive market. To stay competitive, the engineering industry needs to hire individuals with leadership skills who can address complex problems. Second, companies in the Western world are witnessing the emergence of powerful competitors in emerging markets such as India, China, and Brazil. Third, the same students who have shown an interest in social engagement during their college years want (demand) to work for companies that have espoused social values and show *genuine* corporate social responsibility. Companies are forced to change their recruiting strategies to reflect that change in an environment where more young people are interested in the engineering profession's capacity to help those in need.

16.6 Sustainability and Development for All

Since its inception in the late 1980s, the concept of sustainability has been a topic of intense discussion in various sectors and institutions. As discussed in Chapter 2, it emerged from the realization that the dominant logic behind the production–consumption systems in the Western world was not free of negative consequences and could substantially affect the environment for current and future generations. Following the 1992 Rio Summit, the concept of sustainability brought a renewed sense of purpose to the engineering profession worldwide. Since then, engineers, in partnerships with other professions, have been engaged in developing frameworks and guidelines for integrating sustainability in engineering projects, as discussed in Chapters 2 and 3.

In this book, the concept of sustainability was introduced in the discussion on sustainable community development. It was also used in the context of service delivery and how to guarantee the long-term-benefits of solutions implemented at the community level.

As discussed in Chapter 2, sustainable development is a process that leads to an end state called sustainability, which, according to Ben-Eli is

a dynamic equilibrium in the processes of interaction between a population and the carrying capacity of an environment such that the population develops to express its full potential without adversely and irreversibly affecting the carrying capacity of the environment upon which it depends. (Ben-Eli 2011)

This definition implies that sustainability is not static but changes and adapts over time. Another way of looking at sustainability is harmonizing society, the environment, and the economic and financial systems, the so-called *triple bottom line*. It has also been suggested that two additional components be added: ethics and consciousness (Clugston 2011), which are needed to keep the other three components in check.

The concepts of sustainability and sustainable development acknowledge that it is impossible to separate economic development issues from environmental and social issues because they are all interconnected. It should be noted that sustainability and sustainable development were developed within the context of the Western world. They encourage countries, communities, and organizations to adopt production–consumption models that are more intelligent while being more efficient and respectful of natural and human systems and operating within the carrying capacity of the planet. At a first glance, sustainability appears as a luxurious attribute that only rich countries can aspire to after reaching a certain level of development while making mistakes along the way. If presented in this way, there are no incentives for emerging markets to embrace sustainability at the start of their quest for rapid economic growth.

In reality, nothing prevents the developing world from leapfrogging from survival to healthy market economies with less strain on the environment by explicitly integrating sustainability in their human and economic development plans without repeating the past mistakes of the Western world. This model obviously assumes that the developing world has an opportunity to do so, is given the right to develop, and that conditions are in place for development to take root, such as equity, rule of law, and governance.

Integrating the concepts of sustainability and sustainable development into human and economic development work may represent the mindset necessary for taking that shortcut. This integration is particularly critical because vast populations in today's emerging markets have a large combined purchasing power and are experiencing rapid economic growth. The fate of our planet depends a great deal on the behavior of those new customers and the few billion people who will follow in the forthcoming decades. In emerging countries, a new approach to industrialization and consumption is needed that avoids the unintended consequences associated with the growth of the Western world. Emerging countries are challenged to make decisions that do not

- Substantially or totally deplete natural resources;
- Eliminate options for the future of natural and human systems;
- Create or reinforce inequalities and inequities among people and divide them;
- Escalate the costs of solutions to levels that all cannot afford; and
- Increase the probability of future hazards or adverse events, either natural or technological.

And such decisions have to be made at the outset of development. In other words, development needs to be *designed for sustainability* as soon as possible.

Development designed for sustainability is also key to lifting billions of people out of poverty who are not yet members of emerging markets. This lifting is where sustainable development and human development converge into sustainable human development (UNDP/HDR 2011). It is about addressing the basic needs of the people, reducing their vulnerability, and increasing their capacity and resilience. It is also about developing pathways to opportunity, industrialization, and growth starting from limited resources and using them more wisely. It is about integrating decisions into industrialization that do not deplete the carrying capacity of the environment on which people depend. In many ways, it is about a triple bottom line of people, planet, and profit. But sustainable human development is more than that. It is also about human rights, rule of law, stable governance, social well-being, peace, equity, equality, and access to a safe and secure environment.

Sustainable human development is about creating communities that

- Allow *all* their members to enjoy a quality of life and well-being where basic human needs, freedoms, rights, and meaningful work are fulfilled in a safe and secure environment;
- Have equitable access to resources and knowledge, thus becoming capable of sustaining themselves economically, socially, and environmentally;
- Are places where individuals and households have the opportunity to express their full potential without adversely and irreversibly affecting the carrying capacity of the environment upon which they depend;
- Are parts of a system where rule of law and good governance are the norm;
- Are more resilient to cope and adapt to adverse events—internal or external, small or large, routine or exceptional, natural or nonnatural, isolated or interrelated; and
- Ensure sustainable livelihood opportunities for future generations.

Sustainable human development is a multifaceted and holistic (systemic) process whose outcome is the realization of full socioeconomic and human potential for all. It is development designed not only for sustainability but also development for peace, economic growth, resilience and adaptation to hazards and adverse events, and well-being right from the start. It is a form of development that can be measured with indices other than the traditional GDP or HDI. The recent interest in using something like the gross national happiness index introduced in Bhutan more than 30 years ago (Helliwell et al. 2012) as an alternate measure of development goes well beyond the economic growth paradigm.

Finally, a second aspect of sustainability presented in this book relates to ensuring the functionality and long-term benefits of solutions (technical and nontechnical) that are implemented in projects in developing communities. As discussed in Chapter 15, less than satisfactory projects abound, and not just in the field of water, sanitation, and hygiene. They exist everywhere, including in projects related to energy, transportation, shelter, health, and food. There are multiple reasons to explain why development projects have had limited long-term benefits. No magic pill can ensure the sustainability of services and solutions. But, as discussed in the guidelines proposed in this book, steps can be taken by project and program managers in all phases of a project to reduce the odds of failure, convert inherent uncertainty to risk, and ensure a *reasonable* probability of success. Ultimately, the goal is to have projects in the developed and developing world with the same level of quality assurance and control. Sustainable human development projects need to be *done right* from a performance (technical) point of view but also need to be the *right projects* for the community and the environment with which they interact. There lies one of the great challenges for development agencies and others interested in improving the livelihood of many on our planet. This work requires a new vision and mindset that transcends the Millennium Development Goals set for 2015.

References

Ben-Eli, M. (2011). "The five core principles of sustainability." <http://www.sustainabilitylabs.org/page/sustainability-five-core-principles> (Jan. 10, 2013).

Bugliarello, G. (2008). "Engineering: Emerging and future challenges." *Engineering: Issues and challenges for development*, UNESCO, Paris.

Chandy, L., Ledlie, N., and Penciakova, V. (2013). "The final countdown: Prospects for ending extreme poverty by 2030." Brookings Institution, Washington, DC. <http://www.brookings.edu/~/media/research/files/reports/2013/04/ending%20extreme%20poverty%20chandy/the_final_countdown.pdf> (June 10, 2013).

Clugston, R. (2011). "Ethical framework for a sustainable world: Earth charter plus 10 conference and follow-up." *J. Education for Sustainable Development*, 5(2), 173–176.

Forbes, S. (2009). "Sustainable development extension plan (SUDEX)." Doctoral dissertation, University of Texas at El Paso.

Helliwell, J., Layard, R., and Sachs, J., eds. (2012). "The world happiness report." Earth Institute, New York. <http://www.earthinstitute.columbia.edu/sitefiles/file/Sachs%20Writing/2012/World%20Happiness%20Report.pdf> (Feb. 10, 2013).

Independent Evaluation Group (IEG). (2012). "World Bank project performance ratings: Projects completed in period 1981–2010." World Bank, Washington, DC. <https://databox.worldbank.org/> (Feb. 23, 2013).

Kirchmeier, F. (2006). *The right to development: Where do we stand?* Friedrich Ebert Stiftung, Geneva. <http://library.fes.de/pdf-files/iez/global/50288.pdf> (Dec. 10, 2012).

Narayan, D. (1993). "Participatory evaluation: Tools for managing change in water and sanitation." World Bank, Washington, DC. <http://www.chs.ubc.ca/archives/files/Participatory%20Evaluation%20Tools%20for%20Managing%20Change%20in%20Water%20and%20Sanitation.pdf> (May 10, 2012).

Nolan, R. (2002). *Development anthropology: Encounters in the real world*, Westview Press, Boulder, CO.

Ravallion, M. (2013). "How long will it take to lift one billion people out of poverty?" Policy research working paper 6325, World Bank Research Observer, first published online March 12. <http://www-wds.worldbank.org/external/default/WDSContentServer/IW3P/IB/2013/01/ 22/000158349_20130122091052/Rendered/PDF/wps6325.pdf> (May 10, 2013).

Sterman, J. (2006). "Learning from evidence in a complex world." *Am. J. Public Health*, 96(3), 505–514.

Tarr, C., and Ruiz, C. (2013). "The responsible generation: Leveraging corporate responsibility for employee attraction and engagement." *ICOSA Magazine*, 5(1), 46–48.

United Nations Development Programme Human Development Report (UNDP/HDR). (1990). "Concepts and measurement of human development." United Nations Development Programme, New York.

United Nations Development Programme Human Development Report (UNDP/HDR). (2006). "Beyond scarcity: Power, poverty, and the global water crisis." United Nations Development Programme, New York.

United Nations Development Programme Human Development Report (UNDP/HDR). (2011). "Sustainability and equity, a better future for all." United Nations Development Programme, New York.

Index

Page numbers followed by *e*, *f*, and *t* indicate equations, figures, and tables, respectively.

Accreditation Board for Engineering and Technology (ABET), 117, 121, 139–140
action identification matrix, for project hypothesis, 250–251, 250*t*, 256*t*
activity-driven approach, to service delivery, 381
adaptive (design-as-you-go) approach, to project design and management, 20, 22, 136, 138–139, 147, 161, 162–163*t*, 163–165, 171, 323, 375, 444
adaptive capacity, 292
adaptive resilience, 337–338, 343
ADIME-E framework, generally, 147–155, 149*f*, 150*t*, 165. *See also specific topics*
adverse events. *See* resilience, of community
aesthetic needs, 36
Africa, population growth in, 29–30
Agenda 21 (UNCED), 72, 104, 108, 454
Alma-Ata, Declaration of, 431
alternative analysis. *See* preliminary solutions
American Society of Civil Engineers (ASCE), Body of Knowledge Committee, 117–118
appraisal, of community, 169–216; appraisal differs from assessment, 169; baseline profile and, 169, 171, 178–179; challenges and biases of, 185–189; community characteristics and, 179–180; data analysis and presentation, 195–208, 197*t*, 198*f*, 199*t*, 200*f*, 202*f*, 203*f*, 205*t*, 206*f*, 206*t*, 207*t*; data collection, 189–192, 190*t*; example of, 169–170, 170*f*; forms of participation in, 180–185; goals and core information of, 172–178, 174*t*, 175*f*, 176*t*; importance of, 169; interviews for, 192–195, 193*t*,

194*t*; problem identification and ranking, 208–210, 209*t*; in proposed ADIME-E framework, 149*f*, 151; social network analysis (SNA), 210–212, 210*f*; support team for, 170*f*, 188–189, 188*t*; uncertainty and, 170–171
appropriate and sustainable technologies, 383–384, 386–394; aspects and indicators of, 392–394; attributes and characteristics of, 387–388; frugal and disruptive innovation and, 390–392; social entrepreneurship and, 388–389
archetypes, of systems, 229, 231–232
Argonne National Laboratory Resilience Index, 349
Ashoka network, 388–389, 391
assessment, of project, 360–380; corrective action resulting from, 378–379; differs from appraisal, 169; generally, 361–364, 362*t*; monitoring and evaluation and, 159, 362, 364–370, 364*t*, 368*t*, 371*t*, 372, 373–374*t*, 378–379; project closure and scaling up of, 138, 138*f*, 375–378, 377*t*; risk and, 333–335; from work plan to execution, 360–361. *See also* capacity assessment
assistive technology, service delivery and, 383
assumptions, logical framework approach (LFA) and strategy, 268–269, 269*f*
axiological needs, 36

balancing (-) loops, 228
Barcelona Declaration, 116–117, 121
baseline profile, in community appraisal, 169, 171, 178–179
Baseline Resilience Indicator for Communities (BRIC), 349

behavior change communication (BCC): in
proposed ADIME-E framework, 153;
strategy and planning and, 281–286,
284t, 287–288t
beneficiary analysis, community appraisal
and, 201
Bhutan, 36, 65, 382, 491
biocapacity debtors and creditors, 76, 77f
biogas, 412, 415–416
biomass: effects of use of, 407, 409–410;
using efficiently, 413–417, 414t
Bradley classification, of WASH-related
diseases, 433–434
Brazil, 55, 70, 391, 418, 489
Brundtland Commission, 72, 104

Canadian International Development
Agency (CIDA): definition of
capacity, 291; definition of capacity
development, 293
capacity analysis, 13–16, 13f, 207t,
291–296, 360; definitions, 291–293;
goals of building of capacity, 294; in
proposed ADIME-E framework, 149f,
153–154
capacity appraisal, 201–204, 202f, 203f,
205t, 206t, 294
capacity assessment, 296–297; forms
and expressions of, 297; UNDP
framework, 306; UVC framework, 203f,
297–300, 298f, 301t, 302t, 303, 303t,
304–305t; WFEO framework, 305–306
capacity development, in history of
economic development, 59, 60t
capacity development response, 306–307;
strategies, 315–316; University of
Virginia (UVC) framework, 60t,
292–293, 307–309, 307f, 308f, 310t,
311, 311t, 312t, 313–314t, 315
CARE. See Cooperative for Assistance and
Relief Everywhere (CARE)
causal analysis: hypothesis of project
and, 245–246, 246f, 247f, 248f, 249; in
proposed ADIME-E
framework, 151–152
causal loop diagrams, 228, 229f, 231, 234,
235–237, 236f, 249
cause-and-effect loops, 217, 228, 229f, 230f
chaotic systems, characteristics of, 224–225

charity, dependency and, 8, 67
children: lack of WASH services and death
of, 433; malnutrition in, 238f
China, 5, 12, 41, 55, 70, 78, 85, 114, 390
civic entrepreneurship, 388–389
CleanStar, 413
climate change: double causality (feedback)
and, 220, 222–223t; indicators of
vulnerability to, 354t, 355t. See also
environmental degradation
closed systems, 224
closing, of project, 138, 138f, 375–377, 377t
Coastal Resilience Index, 349
co-creation process, 18, 49–50, 180. See
also participatory research/participatory
action research
cognitive needs, 36
collaborative participation, 181, 182,
184–185
collegial participation, 181–182, 183
Colorado, University of, 119, 122–126
Communities Advancing Resilience Tool
(CART), 339–340, 354
community: community development, 66–
69; uncertainty and, 156–158; WASH
success and, 447. See also appraisal, of
community
community of life transformation, 69
community transformation, 69
community-owned vulnerability and
capacity assessment (COVACA)
tool, 325, 331
complex adaptive systems. See system
dynamics, as approach to community
development
complex systems, 224
complimentary analysis, in proposed
ADIME-E framework, 149f, 153–154
consultative participation, 181, 184
contextualized technology, 387
contractual participation, 181
cooking practices, community appraisal
and, 178
cookstoves, 412, 414–415
Cooperative for Assistance and Relief
Everywhere (CARE): benefits–harm
analysis, 155–156; project design
definition, 140; project design
framework, 144, 147–148

coping resilience, 337–338
corrective action, after monitoring and
    evaluation of project, 374–375
country income groups, 10, 11*f*
cradle-to-cradle alternative, to production-
    consumption model, 75, 76*f*
cradle-to-grave (take–make–waste)
    model, 70–71, 71*f*

data collection, in community
    appraisal, 189–192, 190*t*; analysis and
    presentation of, 195–208, 197*t*, 198*f*,
    199*t*, 200*f*, 202*f*, 203*f*, 205*t*, 206*f*, 206*t*,
    207*t*; primary and secondary data, 191–
    192; quantitative and qualitative
    methods, 189–191
Department for International Development
    Framework, in United Kingdom, 146,
    200
design-as-you go. *See* adaptive (design-as-
    you-go) approach, to project
    management
Deutsche Gesellschaft für Technische
    Zusammenarbeit (GTZ): definition of
    capacity, 291–292; definition of capacity
    development, 293
developing countries: capacity,
    vulnerability, and risk of, 13–16, 13*f*;
    small scale development projects
    and, 10, 11*f*, 12; "spiky" data and, 12–13
development aid, in history of economic
    development, 59, 60*t*
diarrhea. *See* WASH (water, sanitation, and
    hygiene) projects and services
directive (blueprint) planning
    methods, 138–139, 161, 162–163*t*, 163,
    164–165
disability-adjusted life years (DALY), 432
disasters. *See* resilience, of community
diseases, WHO classification of, 432–433
disruptive innovation, 390–392
double causality (feedback): climate
    change, poverty, engineering, and
    globalization, 220, 222–223*t*; poverty
    and, 45, 46–47*t*, 48
drinking water supply. *See* Virginia,
    University of (UVC) framework; WASH
    (water, sanitation, and hygiene) projects
    and services

Earth Charter, 72, 108
ecological footprint, 76, 77*f*, 78, 79*f*
economic concerns: economic and financial
    capacity assessment, 298*f*, 299; progress
    in human development and, 91;
    sustainability principles and economic
    domain, 74
electricity. *See energy entries*
energy capacity, in UVC capacity
    assessment framework, 298*f*, 299
energy ladder, 409*f*
energy services, 405–430; alternatives to
    traditional, 409–410, 411*t*, 412–413;
    biomass and, 407, 409–410, 413–417,
    414*t*; as central to human and economic
    development, 405–407, 407*f*, 408*t*, 409,
    409*f*; community appraisal and, 173–
    174, 175*f*; grid extension and, 423–427;
    lack of access to, 31, 32*f*; renewable
    energy sources and, 417–426; service
    delivery and, 381, 382; social innovation
    in, 413, 414
engineering, double causality (feedback)
    and, 220, 222–223*t*
Engineering for Developing Communities
    (EDC) program, 119, 122
Engineering Projects in Community
    Service (EPICS) program,
    118–119
engineers: education of, 102–103, 106,
    114–126, 115*f*, 120*t*, 477–478, 488–489;
    historic focus on large-scale projects in
    developed world, 101–103; sustainable
    human development and, 8–10,
    109–112; sustainable human
    development, design and technological
    innovation for, 112–114; uses of
    technology and effects on society, 103–
    109, 107*f*
Engineers Without Borders-USA
    (EWB-USA), 111–112, 120–121, 487,
    488
environmental concerns: capacity, in UVC
    assessment framework, 298*f*, 299;
    degradation and population
    growth, 30–31, 33; environmental
    impact assessment, 334; progress in
    human development and, 91; of WASH
    services and sustainability, 446

Envision™ Framework, 82–83
equity, environment, economics (three Es), 74, 108
esteem needs, 35, 35f, 36
EuropeAid Project Cycle Management, 145, 146f
evaluation, of project. See monitoring and evaluation
execution and assessment, of project: in project life-cycle management, 137–138, 138f; in proposed ADIME-E framework, 138f, 149f, 154; work plan and, 360–361
existential needs, 36
exit strategy: in proposed ADIME-E framework, 149f, 154–155; sustainability and scaling up of project, 375–378, 377t
external poverty, 51–52, 51f, 482
extreme poverty, defined, 41

Facility Reporting Project (FRP), 82
failure of project, uncertainty and reasons for, 138f, 158–160
family of four concepts, 382–383
F-diagram, of fecal-oral transmission of pathogens, 433–434, 435f
feasibility study. See preliminary solutions
fecal–oral transmission of pathogen diseases, 433–434, 435f
financial sustainability, of WASH services, 446
food security, community appraisal and, 178
frameworks, for development projects, 131–168; ADIME-E framework, 147–155, 149f, 150t; complex systems and, 160–161, 162–163t, 163–165; design definitions, 139–140; guiding principles, 134–136; life-cycle frameworks, 140–143, 142t, 143f; life-cycle management, 137–139, 138f; major frameworks reviewed, 143–147; rights-based approach (RBA) to, 155–156; success factors, 131–134, 133f; uncertainty and, 156–160
framing stage, of project, 159
frugal innovation, 50, 390–392

Gantt chart, for Mabu project, 271, 272–274t
gender concerns: as barrier to exit from poverty, 44; inequalities and, 33, 65; stakeholders and gender analysis, 200
genuine progress indicator (GPI), development measurement and, 62–63
Ghana, 450
global adaptation index (GAIN), 352–353, 353f, 354t, 355t
global burden of disease, 432–433
global engineers, 487–489; education and, 21–22, 114, 119–126, 120t
Global Reporting Initiative, 82
globalization: double causality (feedback) and, 220, 222–223t; economic development and, 55
grid extension, energy services and, 426–427
gross domestic product (GDP), 61–63
gross national happiness (GNH), 65
gross national income (GNI), 11f, 61–63
gross national product (GNP), 61–63
Guatemala: community appraisal in, 196, 197t, 202–204, 203f, 205t, 206–208, 206f, 206t, 207t, 320; risk identification in, 326–327t

habit loop, behavior change communication and, 285–286
Haiti, 87, 226, 226f, 344, 395, 397
hazards. See resilience, of community
Health Communication Partnership framework, 147
health conditions: basic human rights and, 431–432; health defined, 16, 381–382, 432; population growth and, 31–32; progress in human development and, 86; relationship between poverty and poor health conditions, 45, 48; service delivery and, 383; systems modeling and, 237–238, 237f; WHO definition of health, 381–382
health impact assessment (HIA), 334
hierarchy. See logical framework approach (LFA)
HOMER software, renewable energy and, 418

household livelihood crunch model, 13–16, 13*f*, 319, 320, 482–484
household livelihood release model, 16, 319, 320*f*, 482–484
household livelihood security (HLS) model, 16, 36–37
household transformation, community development and, 68–69
housing, population growth and, 33
human development concept, 55–56
human development index (HDI), 63–65, 64*t*; energy and, 406, 407*f*, 438; WASH ladder and, 438, 453
human needs spectrum, 34–37, 35*f*
human resources capacity, in UVC capacity assessment framework, 298*f*, 299
Husk Power systems, 413
hydropower systems, 424–426
hygiene. *See* WASH (water, sanitation, and hygiene) projects and services
Hyogo Framework for Action (HFA), 350–351
hypothesis (problem statement), of project, 244–257; causal analysis and, 245–246, 246*f*, 247*f*, 248*f*, 249; laying out of hypothesis, 253–255, 255*t*, 256*t*, 257; preliminary solutions and, 249–251, 250*t*, 252*t*, 253; in proposed ADIME-E framework, 149*f*, 151–152

impact evaluation, 367
income distribution, population growth and non-equitable, 31, 34
incompatibility, law of, 158
India, 5, 12, 48, 55, 66, 70, 85, 114, 183; economic development in, 55; electricity and, 406, 413; innovation and, 390
indicators, strategy and planning and, 176*t*, 267–268
inequality-adjusted human development index (IHDI), 65
inherent capacity, 292
inherent resilience, 343
initiation and identification: in project life-cycle management, 137, 138*f*; in proposed ADIME-E framework, 148–151, 149*f*, 150*f*

innovation: frugal and disruptive, 50, 390–392; social, 413, 414
institutional capacity, in UVC capacity assessment framework, 298*f*, 299
institutional sustainability, of WASH services, 446
interactive planning methods, 161, 162–163*t*, 164–165
intermediate technology concept, 386
internal poverty, 51–52, 51*f*, 482
international development, 29–100; community development and, 66–69; history of, 52–59, 57–58*t*, 60*t*; human needs spectrum and, 34–37, 35*f*; major groups of developers, 54; measuring of development, 61–66, 64*t*; need for new mindset, 59, 60*t*, 61; population growth and its consequences, 29–34, 32*f*; poverty and, 37–52; progress in human development, 84–87, 88*f*, 89*f*; sustainability and sustainable development, 69–83
International Strategy for Disaster Reduction (ISDR), 350–351
International Water and Sanitation Center (IRC), Triple-S water framework, 450
interviews, community appraisal and, 192–195, 193*t*, 194*t*
isolated systems, 224
isolation, as barrier to exit from poverty, 44, 45, 46–47*t*
iThink, modeling system dynamics with, 232–234, 232*t*, 239, 240*f*

Johannesburg summit. *See* World Summit on Sustainable Development
Jugaad innovation, 391

Liberation Theology, 66
life domain, sustainability principles and, 75
life expectancy, improvements in, 84, 86
life-cycle management: for development projects, 137–139, 138*f*; frameworks for development projects, 140–143, 142*t*, 143*f*
liquefied petroleum gas, 415
livelihood security. *See* household livelihood security (HLS) model

logframe. *See* logical framework approach (LFA)

logical framework approach (LFA): advantages and disadvantages of, 269–271; monitoring and evaluation, 369–370; to project design and management, 141–143, 142*t*, 143*f*, 145–147, 152; in proposed ADIME-E framework, 152–153; strategy and planning and, 259, 260–264*t*, 265–269

logistics, defined, 258

long-term benefits, of project, 376–377, 377*t*

love and belonging needs, 35, 35*f*, 36

Mabu Project, Nepal: action identification matrix, 250, 250*t*, 256*f*; behavior change communication and, 284, 284*t*; Gantt chart for, 272–274*t*; logical framework matrix for, 259, 260–264*t*; material cost estimate for, 275, 278*t*; monitoring and evaluation, 370, 371*t*; multi-criteria utility assessment (MCUA) matrix, 251, 252*t*, 253; problem tree, 246, 247*f*, 248*f*, 249; solution tree, 248*f*, 249; stakeholder–partner role matrix for, 275, 276–277*t*

Mali, 173–174, 175*f*

Maslow's hierarchy of human needs, 34–36, 35*f*, 381–382

material domain, sustainability principles and, 74

mental models, system dynamics and, 228, 232*t*, 233–234

Mercy Corps Framework, 144

Methodology for Participatory Assessment (MPA), of WASH services, 445–446

Millennium Development Initiative, 56–59, 72; cost of meeting goals, 5; energy and, 406, 408*t*; goals and targets of, 57–58*t*, 85–86, 104, 108, 110, 382, 436, 478

mobile phones, and development in emerging markets, 390–391

moderate poverty, defined, 41

monitoring and evaluation: differences between, 364*t*; evaluation, 366–369, 368*t*; generally, 364–365; monitoring, 365–366; plan for, 364,

369–370, 371*t*, 372, 373–374*t*; in project life-cycle management, 138, 138*f*, 159

morphological analysis. *See* preliminary solutions

Mortenson Center in Engineering for Developing Communities (MCEDC) program, at University of Colorado, 122–126, 487; core values and principle of, 124–126; courses taught at, 123–124

Mozambique, 413

multi-criteria utility assessment (MCUA) matrix, 251, 252*t*, 253

multidimensional poverty index (MPI), development measurement and, 65

multiple-attribute value (MAV), 252–253, 392

municipal sanitation services (MSS). *See* Virginia, University of (UVC) framework; WASH (water, sanitation, and hygiene) projects and services

Namsaling Community Development Center, Nepal: action research matrix, 192–193, 193*t*; community appraisal and, 169, 170*f*; geographic information system use in, 196, 197*f*; interest group matrix, 194, 194*t*; problem identification and ranking, 209, 209*t*; stakeholder analysis, 199–201, 199*t*, 200*f*

needs. *See* human needs spectrum

Nepal, 48, 200, 209, 285; Biogas Sector Partnership in, 416; family of four concept and, 382–383; feedback from, 281, 282*t*; impact evaluation, 367–368, 368*t*; photovoltaic system in, 420, 420*f*. *See also* Mabu Project, Nepal; Namsaling Community Development Center, Nepal

new Austrian tunneling method (NATM), 164

nongovernmental organizations (NGOs), approach to development, 109–110

non-participatory data collection, 191

objective tree. *See* solution tree

official development assistance (ODA) level, of "rich" countries, 6

open systems, 224
operational (implementation) plan, in
    proposed ADIME-E
    framework, 152–153
"Our Common Future" (WCED), 72,
    104

Pakistan, 286, 287t
participatory data collection, 191
participatory research/participatory action
    research (PAR), 171, 180–185
participatory rural appraisal
    (PRA), 183–184
partnership analysis, in appraisal of
    community, 201
peace, sustainable human development
    and, 7–8
people, planet, profit (three Ps), 74, 108
Performance, Cost, Time, and Scope
    (PCTS), of development project, 132–
    133, 133f
performance evaluation, 367
personal transformation, community
    development and, 68
Peru: risk identification, 328t, 329t; risk
    management strategies, 332t; stock-and-
    flow diagram of child malnutrition
    in, 238f
photovoltaic (PV) off-grid systems, 412,
    419–422, 420f, 421f, 422t
physical weakness, as barrier to exit from
    poverty, 44, 45, 46–47t
physiological needs, 34–35, 35f
planning. See strategy and planning
political rights, 155–156. See also stable
    government
politicians, common professions of, 106,
    107f
population growth: consequences
    of, 29–34, 32f; estimations of
    future, 110
poverty, 37–52; approaches to development
    and, 478–479; barriers trapping people
    in, 43–45, 46–47t, 48; cause-and-effect
    analysis of, 50–52, 51f; definitions
    of, 37–43, 40f, 42t; distribution in
    world, 40, 40f; double causality
    (feedback) and, 45, 46–47t, 48; life
    above and below $2 poverty line, 41,

42t; moral obligations to alleviate
    conditions of, 480–482; need to change
    discussion about, 48–50; poverty
    reduction defined, 42–43; progress in
    human development progress and, 86,
    87; sustainable human development
    and, 1–7
powerlessness, as barrier to exit from
    poverty, 45, 46–47t
precarious livelihoods, as barrier to exit
    from poverty, 43–44, 45, 46–47t
preconditions, logical framework approach
    (LFA) and, 268–269
preliminary solutions, hypothesis of project
    and, 249–251, 250t, 252t, 253
problem statement. See hypothesis
    (problem statement), of project
problem tree: hypothesis of project
    and, 245–246, 246f, 247f; in proposed
    ADIME-E framework, 152; poverty's
    causes and, 50–52, 51f
production–consumption model, of
    development, 70–71, 71f; cradle-to-
    cradle alternative to, 75, 76f
project development cycle, assessment of
    needs in, 362–363, 362t
project frameworks. See frameworks, for
    development projects
project sustainability management
    (PMS), 81–82
psychological weakness, as barrier to exit
    from poverty, 44, 45, 46–47t

quality planning, 279–280

rapid assessment process (RAP), 184
rapid response projects, 164
rapid rural appraisal (RRA), 184
ready–fire–aim approach, to development
    projects, 133
reflective practice, in proposed ADIME-E
    framework, 149f, 155
regulations and standards, service delivery
    and, 385–386
reinforcing (+) loops, 228
remote area power supply (RAPS)
    systems, 426–427
renewable energy sources, 417–426;
    delivery of small-scale, 418; hydropower

systems, 424–426; solar photovoltaic (PV) off-grid systems, 412, 419–422, 420f, 421f, 422t; solar thermal systems, 423; wind power systems, 423–424, 423f

resilience, of community, 337–359; as acquired capacity, 343–347, 343f, 345t; disaster risk reduction, 110–111; international frameworks for, 350–353, 353f, 354t, 355t; to major hazards and disasters, 340–342; measuring of, 343f, 347–349; in proposed ADIME-E framework, 149f, 153–154; resilience, generally, 337–340; service delivery in continuum after crises, 394–395, 396t, 397–399, 399f; in systems, 219, 225; systems framework for, 353–356; in United States, 338, 339, 349

result tree. See solution tree

results-based project management (RBM), 145

results-driven approach, to service delivery, 381

reverse innovation, 391

rights-based approach (RBA): to behavior change communication, 286; to project framework, 144, 155–156

Rio Summit (1992), 30, 72, 75, 107–108, 454, 478, 489

Rio+20 meeting (2012), 30, 65, 72, 73

risk analysis and management, 319–336, 360; analysis and prioritization and, 324f, 325, 328, 328t, 329t, 330–331, 330f; capacity and vulnerability and, 207t, 319–321; identification of risk, 323–325, 324f, 326–327t, 328t; logical framework approach and, 268–269; monitoring and evaluation of strategies, 333; project impact assessment, 333–335; in proposed ADIME-E framework, 149f, 153–154; risks, generally, 321–323, 322f; strategies for managing, 324f, 331, 332t, 333; uncertainty differs from risk, 156, 321. See also resilience, of community; vulnerability

rule of law, progress in human development and, 91

Rural Water Supply Network, 441

safety needs, 35, 35f

San Francisco Planning and Urban Research Association (SPUR), 347, 395

sanitation. See WASH (water, sanitation, and hygiene) projects

Sarvodaya Shramadana Movement (SSM), 36, 52, 382, 386

satisficing, 20, 106, 157, 395, 486

scaling up, of project, 378

security rights, 155–156

self-actualization needs, 35, 35f, 36

service capacity, 203; in UVC capacity assessment framework, 298f, 299, 384–385

service delivery, 381–404; appropriate and sustainable technology and, 383–384, 386–394; as more than technology delivery, 381–384; resilience and continuum of intervention after crises, 394–395, 396t, 397–399, 399f; service capacity and, 384–386

simple systems, 224

small scale development projects, generally, 10, 11f, 12

SMART (specific, measurable, appropriate, realistic, and timely), 392; behavior change communication and, 285; goals and, 266–267; sustainability indicators and, 81, 268, 316; risk evaluation and, 333

social and cultural capacity, in UVC capacity assessment framework, 298f, 300

social development programs, 56

social domain, sustainability principles and, 75

social entrepreneurship, 388–389

social innovation, energy services and, 413, 414

social network analysis (SNA), 210–212, 210f

social sustainability, of WASH services, 446

social well-being, progress in human development and, 91

socioeconomic rights, 155–156

Solar Energy Light Company (SELCO), 413

solar photovoltaic (PV) off-grid systems, 412, 419–422, 420f, 421f, 422t

Solar Suitcase, 422, 422*f*
solar thermal systems, 423
solid waste. *See* Virginia, University of
(UVC) framework; WASH (water,
sanitation, and hygiene) projects and
services
solution tree: hypothesis of project
and, 246, 246*f*, 248*f*, 249; in proposed
ADIME-E framework, 152
Sphere Standards, 344, 397–399, 399*f*,
454–455, 457, 464
SPICED (subjective, participatory, indirect,
cross-checked, empowering and diverse)
acronym, 268
spiritual domain, sustainability principles
and, 75
stable governance: poverty reduction
and, 44; progress in human
development and, 91
stakeholder analysis, community appraisal
and, 196, 198–201, 199*t*, 200*f*
stakeholder–partner role matrix, 271, 275,
276–277*t*
stand-alone power systems
(SAPS), 426–427
STELLA, modeling system dynamics
with, 232–234, 232*t*, 239, 240*f*
stock-and-flow diagrams, 228, 229, 230*f*,
234–235
strategy and planning, 258–290; behavior
change communication, 281–286, 284*t*,
287–288*t*; logical framework
approach, 259, 260–264*t*, 265–269;
logical framework approach, advantages
and disadvantages of, 269–271;
management activities, 279; operation-
logistics and tactics, 271, 272–274*t*, 275,
276–277*t*, 278*t*, 279; in project life-cycle
management, 137, 138*f*; in proposed
ADIME-E framework, 149*f*, 152–153;
quality planning, 279–280; strategy
defined, 258; work plan refining, 281,
282*t*
structural adjustment program (SAP), of
World Bank, 54–55
sustainability and sustainable
development, 489–490; characteristics
and attributes of sustainable
communities, 90; components of

sustainability, 73–75, 75*f*, 76*f*; ecological
footprint and, 76, 77*f*, 78, 79*f*; engineers
and, 104, 107–109, 476–478; frameworks
for, 80–83, 91; need for new mindset
and approach to industrialization,
69–70, 81*t*; process and projects of,
78–80; production–consumption
model and, 70–71, 71*f*; sustainability
movement, 71–73. *See also* sustainable
human development
sustainability quadrant, 78
sustainable human development, 490–492;
challenges to success of, 484–487;
defined, 34; engineering and, 8–10;
engineering education for, 114–119,
115*f*; engineers and, 109–114; goals
of, 482–484; peace and, 7–8; poverty
reduction and need to address with
economic development, 1–7; proposed
framework, characteristics and
caveats, 18–20; proposed framework,
goals and objectives, 16–18
Swaraj movement, 66
SWOT (SWOL, SWOC) analysis, 185, 189,
196, 197*t*, 204
system dynamics, as approach to
community development, 217–243;
cautions about, 227; modeling with
iThink and STELLA, 232–234, 232*t*, 239,
240*f*; systems archetypes, 229, 231–232;
systems components, 228–229, 229*f*,
230*f*; theoretical background, 221,
224–227, 226*f*; uses in complex adaptive
systems, 217–221, 218*t*, 219*f*, 222–223*t*,
234–239, 236*f*, 238*f*

tactics: defined, 258; in proposed
ADIME-E framework, 152
Talloires Network, 118
technology: assistance and cooperation, in
history of economic development, 59,
60*t*; community development and, 67–
68; resource capacity, in UVC
framework, 298*f*, 299; service delivery
and appropriate, 383–384, 386–394;
sustainability of WASH services, 446
threshold hypothesis concept, 62
tragedy of the commons archetype,
231–232

transcendence needs, 36
triple-bottom line, 74, 104
$2 poverty line, life above and below, 41,
42*t*

uncertainty, project frameworks and, 156–
158; community appraisal and, 170–171;
complex systems and, 160–161,
162–163*t*, 163–165; failure and, 138*f*,
158–160; risk differs from, 156, 321
undernourishment: children and, 268*f*;
progress in human development
and, 85–86
United Nations Conference on
Environment and Development
(UNCED 1992). *See* Rio Summit (1992)
United Nations Development Programme
Framework, 144–145, 292; capacity
assessment framework, 306; capacity
development approach, 60*t*; definition
of capacity development, 293; disaster
risk index of, 351
United Nations World Risk Index, 351–352
United States, resilience in, 338, 339, 349
U.S. Agency for International Development
Framework, 146–147

verification, logical framework approach
(LFA) and, 268
village self-rule (Swaraj) movement, 66
Virginia, University of, (UVC) framework:
capacity assessment, 203*f*, 297–300, 298*f*,
301*t*, 302*t*, 303, 303*t*, 304–305*t*; capacity
development response, 60*t*, 292–293,
307–309, 307*f*, 308*f*, 310*t*, 311, 311*t*, 312*t*,
313–314*t*, 315
vulnerability, 319–320; as barrier to exit
from poverty, 45, 46–47*t*; community
appraisal and, 204, 206–208, 206*f*. *See
also* risk analysis and management

WASH (water, sanitation, and hygiene)
projects and services, 59, 431–475;
community appraisal and, 177–178;
community-based interventions, 399*f*,
437*t*, 463–468; fecal–oral transmission of
pathogen diseases, 433–434, 435*f*; hard
and soft paths of, 462–463; health and
human rights issues of, 431–434, 435*f*,

436; history and causes of unsatisfactory
projects, 440–444; hubs of information
on, 464–465; progress in human
development and, 85, 87, 88*f*, 89*f*;
sanitation and hygiene initiatives, 466–
467; sanitation requirements, basic, 437*t*,
457–460; solid waste disposal
initiatives, 467–468; solid waste
requirements, basic, 460–461; Sphere
Standards and, 398–399, 399*f*;
sustainability and improved delivery
of, 444–448, 448*f*, 449*t*, 450–451; value
of and progress in WASH ladder, 88*f*,
89*f*, 436–440, 437*t*; water,
recommendations for drinking
water, 465–466; *Water of Ayolé*
video, 461–462; water requirements,
basic, 396*t*, 451, 452*f*, 452*t*, 453–455,
456*t*, 457; water-related diseases, 433–
434. *See also* Virginia, University of
(UVC) framework
Washington Consensus, 55
waste. *See* environmental concerns
wastewater and sewage services. *See*
Virginia, University of (UVC) framework
Water of Ayolé, The (video), 461–462
WaterAid, WASH framework, 447–448,
448*f*
water concerns: community appraisal
and, 174–175, 176*t*, 177; modeling
and, 238–239; population growth and
demand for, 30, 32–33; service delivery
and, 382. *See also* WASH (water,
sanitation, and hygiene) services
weak community (civil society), as barrier
to exit from poverty, 44, 45
"wealth enhancement," term preferred over
"poverty reduction," 48–50
Western values, economic development
efforts and, 52–55, 478–479
wind power systems, 423–424, 423*f*
work plan: project execution and, 360–361;
refining of, 281, 285*t*
World Federation of Engineering
Organizations (WFEO), 292–293,
305–306
World Summit on Sustainable
Development (Johannesburg, 2002), 72,
108, 412

About the Author

Bernard Amadei is Professor of Civil Engineering at the University of Colorado at Boulder, where he holds the Mortenson Endowed Chair in Global Engineering. He is also the Founding President of Engineers Without Borders–USA and the cofounder of the Engineers Without Borders–International network. Among other distinctions, Dr. Amadei is the 2007 corecipient of the Heinz Award for the Environment, the recipient of the 2008 *Engineering News-Record* Award of Excellence, an elected member of the U.S. National Academy of Engineering, and an elected Senior Knight-Ashoka Fellow. Dr. Amadei's current interests cover the topics of sustainability and development engineering.